Underground Excavations in Rock

E. Hoek

D.Sc (Eng)

Principal, Golder Associates,
224 West 8th Avenue, Vancouver V5Y 1N5, Canada

E.T. Brown

Ph.D

London University Professor of Rock Mechanics,
Imperial College of Science and Technology,
South Kensington, London SW7 2BP, England

The Institution of Mining and Metallurgy, London

1980

D
624.152
HOE

Underground Excavations in Rock

ISBN 0 900488 54 9 hard cover edition
ISBN 0 900488 55 7 soft cover edition

© Copyright

The Institution of Mining and Metallurgy
and E. Hoek and E. T. Brown

1980
Revised and reprinted 1982

Published by The Institution of Mining and Metallurgy
44 Portland Place
London W1
England

Printed by Stephen Austin and Sons Ltd., Hertford, England

Preface

Anyone setting out to write a book on the design of underground excavations in rock soon realises the enormity of the task being undertaken and the impossibility of satisfying the requirements of all possible readers. Underground excavations are constructed for a wide variety of uses in a large number of different rock types. In the mining industry, many different underground mining methods are used, depending upon the dip and the thickness of the ore body and the characteristics of the country rock. In civil engineering applications, excavations designed to store gas or oil, to house power generating equipment, for the disposal of waste materials, to conduct water or to allow the passage of vehicles must all satisfy different requirements.

The authors have attempted to identify those problems which are common to the design of all underground excavations in rock and to provide the reader with a simple and clear explanation of how these problems can be tackled. This book should be regarded as an introduction to the subject rather than a presentation of all that the engineer needs to know in order to reach the best solutions to the full range of underground excavation engineering problems. In order to develop such specialist knowledge, the reader will have to supplement what has been learned from this book with additional reading. Over 600 references have been included, and it is hoped that these will provide the reader with a useful guide to the available literature.

In general, the SI system of units has been used. However, because the Imperial system of units is still used in at least one major country, and because these units are used in many of the references cited, some illustrative examples are worked in Imperial units and some of the tables and figures are presented using both systems of units. A table of conversion factors for units is given in Appendix 7.

As in the case of *Rock Slope Engineering*, the companion volume to this book, the aim has been to keep the cost to a minimum by printing by offset lithography from a typed manuscript. Most of the typing and the preparation of all drawings and photographs was done by the senior author. Neither the authors nor the organisations for which they work receive any royalties from the sale of this book which is published on a non-profit basis by the Institution of Mining and Metallurgy.

London
September, 1980

Evert Hoek
Edwin T. Brown

Acknowledgements

The preparation of this book was commenced as a four year research project on the design of large underground excavations carried out at Imperial College of Science and Technology between 1972 and 1976. The project was sponsored by the following companies:

> Consolidated Gold Fields Ltd.,
> Gold Fields of South Africa Ltd.,
> Nchanga Consolidated Copper Mines,
> Roan Consolidated Mines,
> Selection Trust Ltd.,
>
> and the following member companies of the
> Australian Mineral Industries Research
> Association Ltd.,
> Australian Mining and Smelting Ltd.,
> Broken Hill South Ltd.,
> Consolidated Gold Fields Australia Ltd.,
> Mount Isa Mines Ltd.,
> North Broken Hill Ltd.,
> Peko-Wallsend Ltd.,
> Western Mining Corporation Ltd.

Mr. Jim May, Executive Officer of the Australian Mineral Industries Research Association, played a vital role in organising the support received from the Australian mining companies and in coordinating technical liaison with these companies. His assistance and encouragement is gratefully acknowledged.

Since 1975, the senior author has been employed by Golder Associates and the enormous amount of time required to complete the manuscript of this book has generously been donated by this company. In addition, the coverage of many of the practical aspects of underground excavation engineering has benefitted from the senior author's direct involvement in consulting assignments on Golder Associates projects. A number of staff members have made significant contributions to the text and their assistance, together with that provided by the management of the company, is hereby acknowledged.

A number of individuals have contributed critical comments, corrections, computer programs, problem solutions and discussions on specific topics. While is is not practical to specifically acknowledge all of these contributions, special thanks are due to the following:

> Dr. John Bray of Imperial College, London,
> Dr. Ross Hammett of Golder Associates, Vancouver,
> Prof. Dick Goodman of the University of California, Berkeley,
> Prof. Dick Bieniawski of Pennsylvania State University,
> Dr. Nick Barton formerly of the Norwegian Geotechnical
> Institute, Oslo,
> Dr. El Sayed Ahmed Eissa formerly of Imperial College, London,
> Dr. Grant Hocking formerly of Imperial College, London,
> Dr. Steve Priest of Imperial College, London,
> Miss Moira Knox of Imperial College, London, (who typed the
> bibliography presented in Appendix 1 and assisted with the
> final editing and compilation of the manuscript).

Finally, the encouragement and assistance provided by Mrs Theo Hoek over the many years required to complete this book is warmly acknowledged.

Contents

		Page
Chapter 1:	Planning considerations	
	Introduction	7
	Types of underground excavation	8
	Underground excavation design	9
	Bibliography on underground excavations	12
	Chapter 1 references	13
Chapter 2:	Classification of rock masses	
	Introduction	14
	Terzaghi's rock load classification	14
	Classifications by Stini and Lauffer	18
	Deere's Rock Quality Designation (RQD)	18
	Influence of clay seams and fault gouge	20
	CSIR classification of jointed rock masses	22
	NGI Tunnelling Quality Index	27
	Discussion on rock mass classification systems	34
	Chapter 2 references	36
Chapter 3:	Geological data collection	
	Introduction	38
	Study of regional geology	38
	Engineering geological maps and plans	40
	Mapping surface outcrops	40
	Geophysical exploration	43
	Diamond drilling for sub-surface exploration	45
	Index testing of core	52
	Core logging and core photography	55
	Core storage	55
	Exploratory adits and shafts	57
	Chapter 3 references	59
Chapter 4:	Graphical presentation of geological data	
	Introduction	61
	Equal area and equal angle projections	61
	Stereographic projection of a plane and its pole	63
	Definition of geological terms	63
	Construction of stereographic nets	65
	Construction of a great circle to represent a plane	72
	Determination of the line of intersection of two planes	73
	Relationship between true and apparent dip	74
	Plotting and analysis of field measurements	75
	Computer processing of structural data	79
	Sources of error in structural data collection	79
	Isometric drawings of structural planes	79
	Use of demonstration models in underground excavation design	84
	Chapter 4 references	86
Chapter 5:	Stresses around underground excavations	
	Introduction	87
	Components of stress	87
	Two dimensional state of stress	90
	In situ state of stress	93
	Stress distributions around single excavations	101
	Stresses around a circular excavation	103
	Calculation of stresses around other excavation shapes	108
	Stresses around multiple excavations	112
	Three-dimensional pillar stress problems	122

	Page
Stress shadows	124
Influence of inclination upon pillar stresses	125
Influence of gravity	125
Chapter 5 references	127

Chapter 6: Strength of rock and rock masses

Introduction	131
Brittle and ductile behaviour	133
Laboratory testing of intact rock samples	134
An empirical failure criterion for rock	137
Survey of triaxial test data on intact rock specimens	140
Simplifying assumptions	150
Anisotropic rock strength	157
Strength of rock with multiple discontinuities	163
Strength of heavily jointed rock masses	166
Use of rock mass classifications for rock strength prediction	171
Deformability of rock masses	173
Approximate equations defining the strength of intact rock and heavily jointed rock masses	175
Chapter 6 references	178

Chapter 7: Underground excavation failure mechanisms

Introduction	183
Structurally controlled instability	183
Computer analysis of structurally controlled instability	191
Optimum orientation and shape of excavations in jointed rock	194
Influence of excavation size upon structurally controlled instability	197
Influence of in situ stress on structurally controlled instability	199
Pillar failure	200
Fracture propagation in rock surrounding a circular tunnel	211
Sidewall failure in square tunnels	217
Influence of excavation shape and in situ stress ratio	221
An example of excavation shape optimisation	223
Excavation shape changes to improve stability	230
Influence of a fault on excavation stability	232
Buckling of slabs parallel to excavation boundaries	234
Excavations in horizontally bedded rock	235
Stiffness, energy and stability	236
Chapter 7 references	241

Chapter 8: Underground excavation support design

Introduction	244
Support of wedges or blocks which are free to fall	246
Support of wedges or blocks which are free to slide	247
Rock-support interaction analysis	248
Summary of rock-support interaction equations	258
Examples of rock-support interaction analysis	270
Discussion on rock-support interaction analysis	285
Use of rock mass classifications for estimating support	286
Comparison of underground excavation support predictions	298
Pre-reinforcement of rock masses	312
Suggestions for estimating support requirements	319
Additional reading	321
Chapter 8 references	325

		Page
Chapter 9:	Rockbolts, shotcrete and mesh	
	Introduction	329
	Organization of a rockbolting programme	329
	Review of typical rockbolt systems	332
	Rockbolt installation	342
	Wire mesh	351
	Shotcrete	353
	Mix design	355
	Engineering properties of shotcrete	360
	Placement of shotcrete	360
	Fibre reinforced shotcrete	363
	Chapter 9 references	365
Chapter 10:	Blasting in underground excavations	
	Introduction	367
	Basic mechanics of explosive rock breaking	367
	Creation of a free face	368
	Rock damage	370
	Smooth blasting and presplitting	372
	Design of blasting patterns	377
	Damage to adjacent underground excavations	378
	Conclusions	380
	Chapter 10 references	381
Chapter 11:	Instrumentation	
	Introduction	382
	Objectives of underground instrumentation	382
	Common inadequacies in instrumentation programmes	382
	Instrumentation for the collection of design data	384
	Monitoring of underground excavations during construction	389
	Monitoring of underground excavations after construction	393
	Monitoring of trial excavations	393
	Conclusion	394
	Chapter 11 references	395
Appendix 1:	Bibliography on large underground excavations	397
Appendix 2:	Isometric drawing charts	449
Appendix 3:	Stresses around single openings	467
Appendix 4:	Two-dimensional boundary element stress analysis	493
Appendix 5:	Determination of material constants	513
Appendix 6:	Underground wedge analysis	517
Appendix 7:	Conversion factors	523
	Index	525

Chapter 1: Planning considerations

Introduction

For many centuries miners have been excavating below the ground surface in their ceaseless search for minerals. Originally, these underground operations were simply a downward extension of the small excavations created to exploit surface outcrops. As mineral exploration methods became more sophisticated, resulting in the discovery of large ore bodies at considerable depth below surface, underground mining methods were developed to exploit these deposits. These new mining methods were evolved from hard won practical experience and one must admire the skill and courage of the mining pioneers whose only acknowledgement of the difficulties which they encountered was to admit that there were certain areas of "bad ground" in the normally "good ground" which they worked.

Most underground mining excavations were, and indeed still are, of a temporary nature. Provided that safe access can be maintained for long enough for the ore in the vicinity of the excavation to be extracted and provided that the subsequent behaviour of the excavation does not jeopardise operations elsewhere in the mine, an underground mining excavation ceases to be an asset after a relatively short space of time. Clearly, the resources allocated to investigating the stability of such an excavation and the quality and quantity of support provided must be related to the length of time for which it is required to maintain stability.

The increasing size of underground mining operations during the past few decades has led to the introduction of a concept which would have been foreign to underground miners of earlier times - the concept of *permanent* underground excavations. Major shaft systems with their surrounding complex of haulages, ore passes, pump chambers and underground crusher stations are required to remain operational for several tens of years and, from the miner's point of view, are permanent excavations. In addition to being large in size, some of these excavations can house expensive equipment and can be manned on a regular basis; consequently such excavations must be secure against rockfalls and other forms of instability.

Civil engineers are seldom concerned with temporary underground excavations since tunnels, underground power house excavations and caverns for the storage of oil or gas are all required to remain stable for periods in excess of twenty years. Because any form of instability cannot be tolerated, the resources allocated to the design and installation of support systems are normally adequate and, sometimes, even generous.

The design of underground excavations is, to a large extent, the design of underground support systems. These can range from no support in the case of a temporary mining excavation in good rock to the use of fully grouted and tensioned bolts or cables with mesh and sprayed concrete for the support of a large permanent civil engineering excavation. These two extremes may be said to represent the lower and upper bounds of underground support design and, in a book of this sort, it is necessary to consider the entire spectrum of design problems which lies between these two extremes.

Types of underground excavation

From the geotechnical engineer's point of view, the most meaningful classification of underground excavations is one which is related to the degree of stability or security which is required of the rock surrounding the excavation. This, in turn, is dependent upon the use for which the excavation is intended. Barton, Lien and Lunde[1]* suggest the following categories of underground excavations :

A Temporary mine openings.

B Vertical shafts.

C Permanent mine openings, water tunnels for hydro-electric projects (excluding high pressure penstocks), pilot tunnels, drifts and headings for large excavations.

D Storage rooms, water treatment plants, minor road and railway tunnels, surge chambers and access tunnels in hydro-electric projects.

E Underground power station caverns, major road and railway tunnels, civil defence chambers, tunnel portals and intersections.

F Underground nuclear power stations, railway stations, sports and public facilities, underground factories.

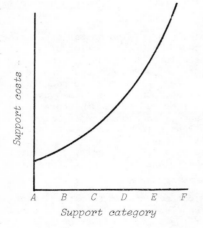

Whereas, in rock slope design the stability of the slope is expressed quantitatively in terms of the factor of safety[2], it will become clear in later chapters of this book that it is not practical to assign an equivalent quantitative stability index to underground excavations. It will be evident, however, that the stability requirements increase from category A to category F as one progresses through the list given above. For a constant set of geological conditions, the cost of support (including geological investigations, support design and installation) will be related to the excavation category in the manner illustrated diagrammatically in the sketch opposite.

In the case of large projects in the E and F categories, there is usually justification for building up a team of specialist engineers and geologists to study the wide range of geotechnical and construction problems which are likely to be encountered on site. An example of such a project is the development of the hydro-electric potential of the Snowy Mountains region of Australia. During the late 1950s and early 1960s, the team which had been set up to assist in the design of the various underground power-houses involved in this scheme made major contributions to the general advancement of underground excavation design techniques. Many of the design methods described in this book can be traced back to the activities of such teams.

The mining industry, which is normally concerned with excavations in the A and B categories, cannot justify a high level of investigation and design effort on any one particular site. On the other hand, since the number of

---

* Superscripts refer to the list of references given at the end of each chapter.

mining excavations is large and the support costs constitute a significant proportion of mining costs, the industry has tended to establish central research and development organisations which have the task of evolving general design methods which can be used throughout the industry. The appropriate divisions of the United States Bureau of Mines, the South African Chamber of Mines and the German Steinkohlenbergbauverein are typical of this type of organisation.

Underground excavation design

Given the task of designing an underground excavation or a number of such excavations, where does one start and what steps are involved in carrying the design through to completion ? A guide to the most important steps in this process is set out in the form of a chart on page 10 and the overall design philosophy is reviewed hereunder. Detailed discussion of each of the steps is given in subsequent chapters.

The basic aim of any underground excavation design should be to utilise the rock itself as the principal structural material, creating as little disturbance as possible during the excavation process and adding as little as possible in the way of concrete or steel support. In their intact state and when subjected to compressive stresses, most hard rocks are far stronger than concrete and many are of the same order of strength as steel. Consequently, it does not make economic sense to replace a material which may be perfectly adequate with one which may be no better.

The extent to which this design aim can be met depends upon the geological conditions which exist on site and the extent to which the designer is aware of these conditions and can take them into account. Hence, an accurate interpretation of the geology is an essential prerequisite to a rational design.

It is not intended, in this book, to deal with the basic geological interpretation required in this first stage of the design process. This subject matter has been covered comprehensively in text books such as that by Krynine and Judd[3] and it will be assumed that the reader is familiar with this material or that he has access to sound geological advice. The importance of this general geological background in summed up in the following quotation, taken from a paper by Wahlstrom[4] :

"Surface studies of geology, geophysical measurements, and exploratory drilling yield useful direct information, but equally important to the geologist may be a knowledge of the regional geology and the geologic history of the area, and a thorough appreciation of the manner in which rocks respond to changing geological environments. Such considerations permit him to make a very useful semi-quantitative estimate of the kinds, but not the exact locations, of the geological features which will be encountered at depth."

Although basic geological principles are not covered, it is considered necessary to discuss some of the site investigation methods which are available for the collection of geological information. The graphical presentation of these data is also an important part in the communication chain between geologists and engineers and this will be described

DESIGN OF UNDERGROUND EXCAVATIONS IN ROCK

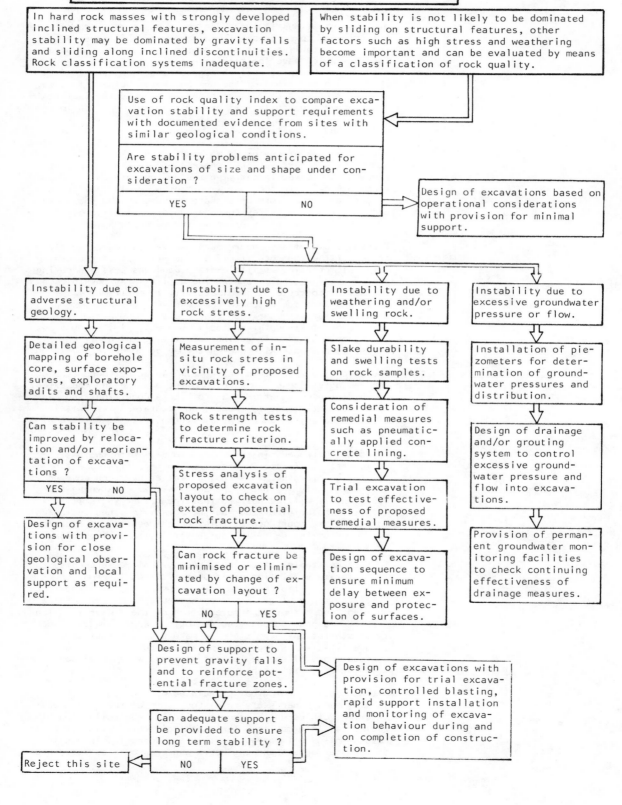

in a later chapter.

The need to make quantitative predictions on the number, inclination and orientation of geological features and of the possible mechanical properties of the rock mass containing these features has long been a requirement of geotechnical engineers. At the risk of offending the geological purists who claim that geology cannot be quantified in the way demanded by engineers, some geotechnologists have gone ahead and developed rock mass assessment systems which have proved to be very useful in the early stages of underground excavation design. When it is necessary to design a large number of A and B category excavations for a mining or underground quarrying operation, the use of some form of rock quality index to determine the support requirements may be the only design approach which is practically and economically acceptable. In recognition of this important role, rock mass classification systems have been fully reviewed and an entire chapter is devoted to their use in underground excavation engineering.

Once it has been established, on the basis of the geological interpretation, that stability problems are likely to be encountered, it becomes necessary to embark upon the more detailed steps listed in the lower half of the chart on page 10. What steps are taken will obviously depend upon the degree of risk anticipated, the category of the excavation and the practical and economic constraints within which the designer has to work.

Four principal sources of instability are identified in the chart on page 10 :
  a. Instability due to adverse structural geology tends to occur in hard rocks which are faulted and jointed and where several sets of discontinuities are steeply inclined. Stability can sometimes be improved by relocation or reorientation of the excavations but fairly extensive support is also usually required. Rockbolts, dowels and cables are particularly effective when used to support this type of rock mass, provided that the structural features are taken into account in designing the support system.

  b. Instability due to excessively high rock stress is also generally associated with hard rock and can occur when mining at great depth or when very large excavations are created at reasonably shallow depth. Unusual stress conditions such as those which may be encountered when tunnelling in steep mountain regions or unusually weak rock conditions can also give rise to stress-induced instability problems. Changes in the shape of the excavations and repositioning the excavations with respect to one another is of great assistance in overcoming these problems but support may also be required.

  c. Instability due to weathering and/or swelling is generally associated with relatively poor rock but it may also occur in isolated seams within an otherwise sound hard rock. Protection of the exposed rock surface from significant moisture changes is usually the most effective remedial measure which can be applied in this situation.

  d. Instability due to excessive groundwater pressure or

flow can occur in almost any rock mass but it would normally only reach serious proportions if associated with one of the other forms of instability already listed. Redirection of water flow by grouting and reduction of water pressure by drainage are usually the most effective remedial measures.

On a typical site, two or more of these forms of instability would occur simultaneously and it may sometimes be difficult to decide upon a rational design method. Indeed, in some cases, the optimum design to allow for one form of instability may be unsuitable in terms of another and the engineer is then faced with the task of arriving at some practical compromise.

It is appropriate, at this point, to emphasize the role of engineering judgement. A rock mass is a complex assemblage of different materials and it is very unlikely that its behaviour will approach the behaviour of the simple models which engineers and geologists have to construct in order to understand some of the processes which take place when rock is subjected to load. These models, many of which are described in this book, should only be used as an aid in the design of underground excavations and the assumptions upon which the models are based and the limitations of the models must be kept in mind at all times. A good engineering design is a *balanced* design in which all the factors which interact, even those which cannot be quantified are taken into account. To quote from a recent review paper[5] : "The responsibility of the design engineer is not to compute accurately but to judge soundly".

Bibliography on underground excavations

Faced with the formidable task of designing a large underground cavern to satisfy the requirements of a hydro-electric project or a major mining operation, the engineer or geologist will wish to draw upon the experience of others who have been through a similar process. A study of the literature on the subject, followed by visits to a few sites, where the rock conditions and excavation sizes appear to be comparable to those under consideration, would be a sound starting point for any major design project of this sort.

In order to assist the reader, who may be faced with such a design task, in finding his way through the literature, an extensive bibliography on underground excavations is presented in Appendix 1 at the end of this book. In addition to over 350 literature references, most with short abstracts, a list of major underground caverns is given. Wherever possible, details of rock types, excavation dimensions and support requirements have been included in this list.

Unfortunately, most of the literature referred to in this bibliography relates to hydro-electric projects. Relatively few papers appear to have been published on other types of underground civil engineering structures and even fewer on major underground mining excavations. In spite of this limitation, the bibliography does show the considerable success which has been achieved in underground excavation design and it should serve to encourage the reader in his own efforts.

Chapter 1 references

1. BARTON,N., LIEN,R. and LUNDE,J. Engineering classification of rock masses for the design of tunnel support. *Rock Mechanics*, Volume 6, No. 4, 1974, pages 189-236. Originally published as Analysis of rock mass quality and support practice in tunnelling. *Norwegian Geotechnical Inst.* Report No. 54206, June 1974, 74 pages.

2. HOEK,E. and BRAY,J.W. *Rock Slope Engineering*. Institution of Mining and Metallurgy, London, 2nd edition, 1977, 402 pages.

3. KRYNINE,D.P. and JUDD,W.R. *Principles of Engineering Geology and Geotechnics*. Mc Graw-Hill Book Co. Inc., New York, 1957, 730 pages.

4. WAHLSTROM,E.E. The validity of geological projection : a case history. *Economic Geology*, Volume 59, 1964, pages 465-474.

5. HOEK,E. and LONDE,P. The design of rock slopes and foundations. General Report on Theme 111. *Proceedings, Third Congress of the International Society for Rock Mechanics*, Denver, September 1974, Volume 1, part A, pages 613-752.

Chapter 2 : Classification of rock masses

Introduction

An underground excavation is an extremely complex structure and the only theoretical tools which the designer has available to assist him in his task are a number of grossly simplified models of some of the processes which interact to control the stability of the excavation. These models can generally only be used to analyse the influence of one particular process at a time, for example, the influence of structural discontinuities or of high rock stress upon the excavation. It is seldom possible theoretically to determine the interaction of these processes and the designer is faced with the need to arrive at a number of design decisions in which his engineering judgement and practical experience must play an important part.

If one is fortunate enough to have an engineer on staff who has designed and supervised the construction of underground excavations in similar rock conditions to those being considered, these design decisions can be taken with some degree of confidence. On the other hand, where no such experience is readily available, what criteria can be used to check whether one's own decisions are reasonable ? How does one judge whether the span is too large or whether too many or too few rockbolts have been specified ?

The answer lies in some form of classification system which enables one to relate one's own set of conditions to conditions encountered by others. Such a classification system acts as a vehicle which enables a designer to relate the experience on rock conditions and support requirements gained on other sites to the conditions anticipated on his own site.

The recognition of this need for rock classification systems is illustrated by the number of literature references which deal with this subject [6-29] and [1*]. Some of the most significant steps in the development of the classification systems for underground support are reviewed hereunder.

Terzaghi's rock load classification

In 1946 Terzaghi[6] proposed a simple rock classification system for use in estimating the loads to be supported by steel arches in tunnels. He described various types of ground and, based upon his experience in steel-supported railroad tunnels in the Alps, he assigned ranges of rock loads for various ground conditions. This very important paper, in which Terzaghi attempted to quantify his experience in such a way that it could be used by others, has been widely used in tunnelling in north America ever since it was published. Because of its historical importance in this discussion and also because copies of the original paper are very difficult to obtain, his classification will be treated more fully than would otherwise be justified.

In his introductory remarks in the section of his paper dealing with the estimation of rock loads, Terzaghi stresses

---

* References are numbered sequentially throughout this book and are not repeated. Hence, a reference which has been used in a previous chapter will be referred to by the number under which it first appeared.

the importance of the geological survey which should be carried out before a tunnel design is completed and, particularly the importance of obtaining information on the defects in the rock mass. To quote from his paper :

" From an engineering point of view, a knowledge of the type and intensity of the rock defects may be much more important than the type of rock which will be encountered. Therefore during the survey rock defects should receive special consideration. The geological report should contain a detailed description of the observed defects in geological terms. It should also contain a tentative classification of the defective rock in the tunnel man's terms, such as blocky and seamy, squeezing or swelling rock. "

He then went on to define these tunnelling terms as follows:

Intact rock contains neither joints nor hair cracks. Hence, if it breaks, it breaks across sound rock. On account of the injury to the rock due to blasting, spalls may drop off the roof several hours or days after blasting. This is known as a *spalling* condition. Hard, intact rock may also be encountered in the *popping* condition involving the spontaneous and violent detachment of rock slabs from the sides or roof.

Stratified rock consists of individual strata with little or no resistance against separation along the boundaries between strata. The strata may or may not be weakened by transverse joints. In such rock, the spalling condition is quite common.

Moderately jointed rock contains joints and hair cracks, but the blocks between joints are locally grown together or so intimately interlocked that vertical walls do not require lateral support. In rocks of this type, both spalling and popping conditions may be encountered.

Blocky and *seamy* rock consists of chemically intact or almost intact rock fragments which are entirely separated from each other and imperfectly interlocked. In such rock, vertical walls may require lateral support.

Crushed but chemically intact rock has the character of a crusher run. If most or all of the fragments are as small as fine sand grains and no recementation has taken place, crushed rock below the water table exhibits the properties of a water-bearing sand.

Squeezing rock slowly advances into the tunnel without perceptible volume increase. A prerequisite for squeeze is a high percentage of microscopic and sub-microscopic particles of micaceous minerals or of clay minerals with a low swelling capacity.

Swelling rock advances into the tunnel chiefly on account of expansion. The capacity to swell seems to be limited to those rocks which contain clay minerals such as montmorillonite, with a high swelling capacity.

The concept used by Terzaghi to estimate the rock load to be carried by the steel arches used to support a tunnel is illustrated in the simplified diagram presented in figure 1. During construction of the tunnel, some relaxation of the interlocking within the rock mass will occur above and on the sides of the tunnel. The loosened rock within the area a c d b will tend to move in towards the tunnel. This

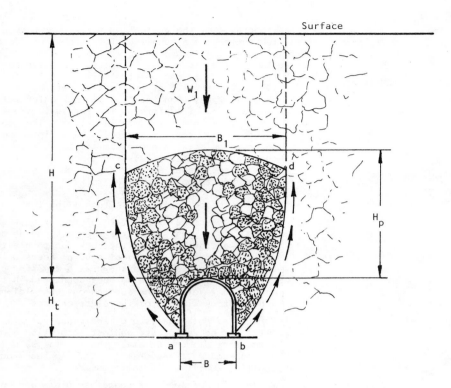

Figure 1 : Simplified diagram representing the movement of loosened rock towards a tunnel and the transfer of load onto the surrounding rock. (After Terzaghi[6]).

movement will be resisted by friction forces along the lateral boundaries a c and b d and these friction forces transfer the major portion of the overburden weight W_1 onto the material on either side of the tunnel. The roof and sides of the tunnel are required only to support the balance which is equivalent to a height H_p . The width B_1 of the zone of rock in which movement occurs will depend upon the characteristics of the rock mass and upon the tunnel dimensions H_t and B.

Terzaghi carried out numerous model tests using cohesionless sand to study the shape of what he termed the "ground arch" above the tunnel. On the basis of these tests and on his experience in steel-supported tunnels, he proposed the range of rock load values listed in Table 1. The footnotes which accompanied this table in the original paper are included for completeness.

Cording and Deere[24] report that these criteria have been widely used for the past 25 years and that they have been found to be appropriate, although slightly conservative, for steel-supported rock tunnels. However, Cecil[18] found that Terzaghi's classification was too general to permit an objective evaluation of rock quality and that it provides no quantitative information on the properties of the rock mass. He recommended that its use be limited to estimating rock loads for steel arch-supported tunnels.

TABLE 1 - TERZAGHI'S ROCK LOAD CLASSIFICATION FOR STEEL ARCH-SUPPORTED TUNNELS

Rock load H_p in feet of rock on roof of support in tunnel with width B (feet) and height H_t (feet) at a depth of more than $1.5(B + H_t)$*

Rock condition	Rock load H_p in feet	Remarks
1. Hard and intact.	zero	Light lining required only if spalling or popping occurs.
2. Hard stratified or schistose **.	0 to 0.5 B	Light support, mainly for protection against spalls.
3. Massive, moderately jointed.	0 to 0.25 B	Load may change erratically from point to point.
4. Moderately blocky and seamy.	$0.25B$ to $0.35(B + H_t)$	No side pressure.
5. Very blocky and seamy.	$(0.35$ to $1.10)(B + H_t)$	Little or no side pressure.
6. Completely crushed but chemically intact.	$1.10(B + H_t)$	Considerable side pressure. Softening effects of seepage towards bottom of tunnel requires either continuous support for lower ends of ribs or circular ribs.
7. Squeezing rock, moderate depth.	$(1.10$ to $2.10)(B + H_t)$	Heavy side pressure, invert struts required. Circular ribs are recommended.
8. Squeezing rock, great depth.	$(2.10$ to $4.50)(B + H_t)$	
9. Swelling rock.	Up to 250 feet, irrespective of the value of $(B + H_t)$	Circular ribs are required. In extreme cases use yielding support.

The roof of the tunnel is assumed to be located below the water table. If it is located permanently above the water table, the values given for types 4 to 6 can be reduced by fifty percent.

**Some of the most common rock formations contain layers of shale. In an unweathered state, real shales are no worse than other stratified rocks. However, the term shale is often applied to firmly compacted clay sediments which have not yet acquired the properties of rock. Such so-called shale may behave in a tunnel like squeezing or even swelling rock.*

If a rock formation consists of a sequence of horizontal layers of sandstone or limestone and of immature shale, the excavation of the tunnel is commonly associated with a gradual compression of the rock on both sides of the tunnel, involving a downward movement of the roof. Furthermore, the relatively low resistance against slippage at the boundaries between the so-called shale and the rock is likely to reduce very considerably the capacity of the rock located above the roof to bridge. Hence, in such formations, the roof pressure may be as heavy as in very blocky and seamy rock.

Classifications of Stini and Lauffer

Stini, in his textbook on tunnel geology[7]*, proposed a rock mass classification and discussed many of the adverse conditions which can be encountered in tunnelling. He emphasised the importance of structural defects in the rock mass and stressed the need to avoid tunnelling parallel to the strike of steeply dipping discontinuities.

While both Terzaghi and Stini had discussed time-dependent instability in tunnels, it was Lauffer[8] who emphasised the importance of the *stand-up time* of the *active span* in a tunnel. The stand-up time is the length of time which an underground opening will stand unsupported after excavation and barring down while the active span is the largest unsupported span in the tunnel section between the face and the supports, as illustrated in figure 2.

Lauffer suggested that the stand-up time for any given active span is related to the rock mass characteristics in the manner illustrated in figure 3. In this figure, the letters refer to the rock class. A is very good rock, corresponding to Terzaghi's hard and intact rock, while G is very poor rock which corresponds roughly to Terzaghi's squeezing or swelling rock.

The work of Stini and Lauffer, having been published in German, has attracted relatively little attention in the English speaking world. However, it has had a significant influence upon the development of more recent rock mass classification systems such as those proposed by Brekke and Howard[22] and Bieniawski[25] which will be discussed later in this chapter.

Deere's Rock Quality Designation (RQD)

In 1964 Deere[9] proposed a quantitative index of rock mass quality based upon core recovery by diamond drilling. This Rock Quality Designation (RQD) has come to be very widely used and has been shown to be particularly useful in classifying rock masses for the selection of tunnel support systems [18,20,21].

The RQD is defined as the percentage of core recovered in intact pieces of 100mm or more in length in the total length of a borehole. Hence :

$$\text{RQD } (\%) = 100 \times \frac{\text{Length of core in pieces} > 100\text{mm}}{\text{Length of borehole}}$$

It is normally accepted that the RQD should be determined on a core of at least 50mm diameter which should have been drilled with double barrel diamond drilling equipment. An RQD value would usually be established for each core run of say 2 metres. This determination is simple and quick and, if carried out in conjunction with the normal geological logging of a core, it adds very little to the cost of the site investigation.

---

* An English translation of the chapter entitled "The importance of rock mass structure in tunnel construction" has been prepared by the Austrian Society for Geomechanics, Translation No. 18, July 1974, 102 pages.

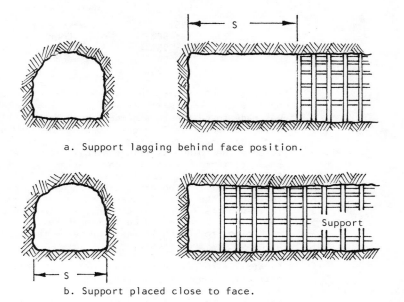

a. Support lagging behind face position.

b. Support placed close to face.

Figure 2. Lauffer's definition of active span S.

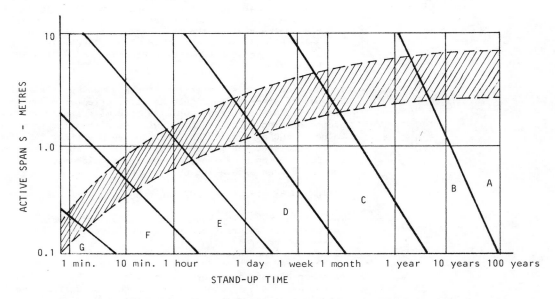

Figure 3. Relationship between active span and stand-up time for different classes of rock mass. A - very good rock, G - very poor rock. (After Lauffer[8]).

Deere proposed the following relationship between the numerical value of RQD and the engineering quality of the rock:

RQD	Rock Quality
< 25%	Very poor
25 - 50%	Poor
50 - 75%	Fair
75 - 90%	Good
90 - 100%	Very good

Since the RQD offers a means for assigning a number to the quality of a rock mass, it is not surprising that an attempt was made to relate this number to the qualitative classification proposed by Terzaghi. Cording, Hendron and Deere[27] modified Terzaghi's rock load factor and related this modified value to RQD as illustrated in figure 4. This diagram suggests that a reasonable correlation may exist between RQD and Terzaghi's rock load factor for steel-supported openings but that no correlation appears to exist between the two in the case of caverns supported by rockbolts. This supports an earlier comment that the use of Terzaghi's rock load factor should be limited very strictly to the conditions for which it was proposed - the support of tunnels by means of steel arches.

An attempt to extend the range of applicability of RQD for estimating tunnel support requirements was made by Merritt[23] and his proposals are summarised in figure 5. Although he felt that the RQD could be of great value in estimating support requirements, Merritt pointed out a serious limitation of his proposals :

" The RQD support criteria system has limitations in areas where the joints contain thin clay fillings or weathered material. Such a case might occur in near surface rock where weathering or seepage has produced clay which reduces the frictional resistance along joint boundaries. This would result in unstable rock although the joints may be widely spaced and the RQD high. "

In addition to this limitation, the RQD does not take direct account of other factors such as joint orientation which must influence the behaviour of a rock mass around an underground opening. Consequently, without detracting from the value of RQD as a quick and inexpensive practical index, it is suggested that it does not provide an adequate indication of the range of behaviour patterns which may be encountered when excavating underground.

Influence of clay seams and fault gouge

The inadequacy of the RQD index in situations where clays and weathered material occur has been discussed above. Brekke and Howard[22] point out that it is just as important - often more important - to classify discontinuities according to character as it is to note their scale parameters. They go on to discuss seven groups of discontinuity infillings which have a significant influence upon the engineering behaviour of the rock mass containing these discontinuities. Although their list does not constitute a rock mass classification, it is included in this discussion because of the important engineering consequences which can result from neglecting these facts when designing an excavation.

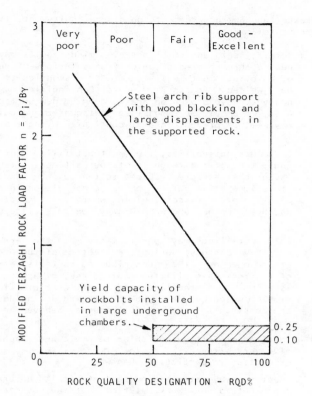
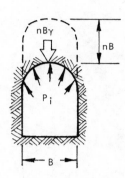

Figure 4. Approximate relationship between Terzaghi's Rock Load Factor (modified) and RQD. (After Cording, Hendron and Deere[27])

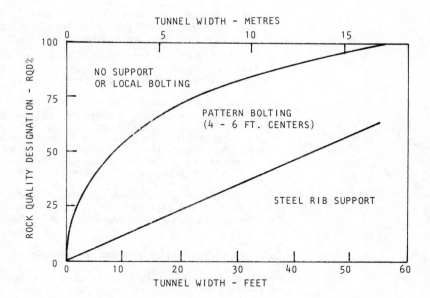

Figure 5. Proposed use of RQD for choice of rock support system. (After Merritt[23])

Brekke and Howard's comments on discontinuity infillings are as follows :

1. Joints, seams and sometimes even minor faults may be healed through precipitation from solutions of quartz or calcite. In this instance, the discontinuity may be " welded " together. Such discontinuities may, however, have broken up again, forming new surfaces. Also, it should be emphasised that quartz and calcite may well be present in a discontinuity without healing it.

2. Clean discontinuities, i.e., without fillings or coatings. Many of the rough joints or partings will have favourable character. Close to the surface, however, it is imperative not to confuse clean discontinuities with " empty " discontinuities where filling material has been leached and washed away due to surface weathering.

3. Calcite fillings may, particularly when they are porous or flaky, dissolve during the lifetime of the underground opening. Their contribution to the strength of the rock mass will then, of course, disappear. This is a long term stability (and sometimes fluid flow) problem which can easily be overlooked during design and construction. Gypsum fillings may behave the same way.

4. Coatings or fillings of chlorite, talc and graphite give very slippery (i.e. low strength) joints, seams or faults, in particular when wet.

5. Inactive clay material in seams and faults naturally represents a very weak material that may squeeze or be washed out.

6. Swelling clay may cause serious problems through free swell and consequent loss of strength, or through considerable swelling pressure when confined.

7. Material that has been altered to a more cohesionless material (sand-like) may run or flow into the tunnel immediately following excavation.

In contrast to the comment by Merritt[23] that joints containing clay fillings may occur near the surface, Brekke and Selmer-Olsen[28] report that clay fillings with a very low degree of consolidation have been encountered at considerable depth. Hence, the underground excavation designer can never afford to ignore the danger which can arise as a result of the presence of these features.

Brekke and Howard have summarised the consequences of encountering filled discontinuities during tunnel excavation in a table which has been reproduced as Table 2 on page 23.

CSIR classification of jointed rock masses

From the preceding discussion it will have become clear that no single simple index is adequate as an indicator of the complex behaviour of the rock mass surrounding an underground excavation. Consequently, some combination of factors such as RQD and the influence of clay filling and weathering appears to be necessary. One such classification system has been proposed by Bieniawski[25,26] of the South African Council

TABLE 2 - INFLUENCE OF DISCONTINUITY INFILLING UPON THE BEHAVIOUR OF TUNNELS

(After Brekke and Howard[22])

Dominant material in gouge	Potential behaviour of gouge material	
	At face	Later
Swelling clay	Free swelling, sloughing. Swelling pressure and squeeze on shield.	Swelling pressure and squeeze against support or lining, free swell with down-fall or wash-in if lining inadequate.
Inactive clay	Slaking and sloughing caused by squeeze. Heavy squeeze under extreme conditions.	Squeeze on supports of lining where unprotected, slaking and sloughing due to environmental changes.
Chlorite, talc, graphite or serpentine.	Ravelling.	Heavy loads may develop due to low strength, in particular when wet.
Crushed rock fragments of sand-like gouge.	Ravelling or running. Stand-up time may be extremely short.	Loosening loads on lining, running and ravelling if unconfined.
Porous or flaky calcite, gypsum.	Favourable conditions.	May dissolve, leading to instability of rock mass.

for Scientific and Industrial Research (CSIR). This classification will be discussed in detail since it is one of the two classifications that the authors would recommend for general use in the preliminary design of underground excavations.

Bieniawski[26] suggested that a classification for jointed rock masses should:

" 1. divide the rock mass into groups of similar behaviour;
2. provide a good basis for understanding the characteristics of the rock mass;
3. facilitate the planning and the design of structures in rock by yielding quantitative data required for the solution of real engineering problems; and
4. provide a common basis for effective communication among all persons concerned with a geomechanics problem.

These aims should be fulfilled by ensuring that the adopted classification is

1. simple and meaningful in terms; and
2. based on measurable parameters which can be determined quickly and cheaply in the field."

In order to satisfy these requirements, Bieniawski originally proposed that his "Geomechanics Classification" should incorporate the following parameters:

1. Rock Quality Designation (RQD),
2. State of weathering,
3. Uniaxial compressive strength of intact rock,
4. Spacing of joints and bedding,
5. Strike and dip orientations,

6. Separation of joints,
7. Continuity of joints, and
8. Ground water inflow.

After some experience had been gained in the practical application of the original CSIR Geomechanics Classification Bieniawski[26] modified his classification system by eliminating the state of weathering as a separate parameter since its effect is accounted for by the uniaxial compressive strength, and by including the separation and continuity of joints in a new parameter, the condition of joints. In addition, the strike and dip orientations of joints were removed from the list of basic classification parameters and their effects allowed for by a rating adjustment made after the basic parameters had been considered.

The five basic classification parameters then became:

1. *Strength of intact rock material*
   Bieniawski uses the classification of the uniaxial compressive strength of intact rock proposed by Deere and Miller[10] and reproduced in Table 3. Alternatively, for all but very low strength rocks the *point load index* (determined as described on page 52 of this book) may be used as a measure of intact rock material strength.

2. *Rock Quality Designation*
   Deere's RQD is used as a measure of drill core quality.

3. *Spacing of joints*
   In this context, the term *joint* is used to mean all discontinuities which may be joints, faults, bedding planes and other surfaces of weakness. Once again, Bieniawski uses a classification proposed by Deere[13] and reproduced here in Table 4.

4. *Condition of joints*
   This parameter accounts for the separation or aperture of joints, their continuity, the surface roughness, the wall condition (hard or soft), and the presence of in-filling materials in the joints.

5. *Ground water conditions*
   An attempt is made to account for the influence of ground water flow on the stability of underground excavations in terms of the observed rate of flow into the excavation, the ratio of joint water pressure to major principal stress or by some general qualitative observation of groundwater conditions.

The way in which these parameters have been incorporated into the CSIR Geomechanics Classification for jointed rock masses is shown in Part A of Table 5 on page 26. Bieniawski recognised that each parameter does not necessarily contribute equally to the behaviour of the rock mass. For example, an RQD of 90 and a uniaxial compressive strength of intact rock material of 200 MPa would suggest that the rock mass is of excellent quality, but heavy inflow of water into the same rock mass could change this assessment dramatically. Bieniawski therefore applied a series of *importance ratings* to his parameters following the concept used by Wickham, Tiedemann and Skinner[21]. A number of points or a rating is

TABLE 3 - DEERE AND MILLER'S CLASSIFICATION OF INTACT ROCK STRENGTH

Description	Uniaxial Compressive Strength			Examples of rock types
	lbf/in^2	Kgf/cm^2	MPa	
Very low strength	150-3500	10-250	1-25	Chalk, rocksalt.
Low strength	3500-7500	250-500	25-50	Coal, siltstone, schist.
Medium strength	7500-15000	500-1000	50-100	Sandstone, slate, shale.
High strength	15000-30000	1000-2000	100-200	Marble, granite, gneiss.
Very high strength	>30000	>2000	>200	Quartzite, dolerite, gabbro, basalt.

TABLE 4 - DEERE'S CLASSIFICATION FOR JOINT SPACING

Description	Spacing of joints		Rock mass grading
Very wide	> 3m	> 10ft	Solid
Wide	1m to 3m	3ft to 10ft	Massive
Moderately close	0.3m to 1m	1ft to 3ft	Blocky/seamy
Close	50mm to 300mm	2in to 1ft	Fractured
Very close	< 50mm	< 2in	Crushed and shattered

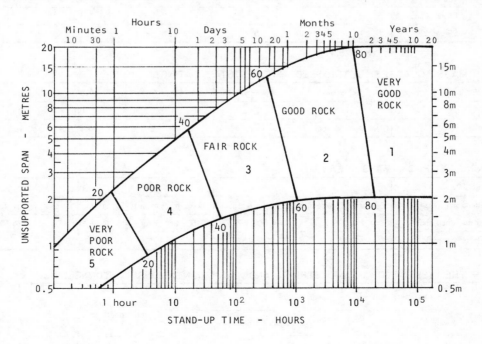

Figure 6. Relationship between the stand-up time of an unsupported underground excavation span and the CSIR Geomechanics Classification proposed by Bieniawski[26].

TABLE 5 – CSIR GEOMECHANICS CLASSIFICATION OF JOINTED ROCK MASSES

A. CLASSIFICATION PARAMETERS AND THEIR RATINGS

	PARAMETER		RANGES OF VALUES					
1	Strength of intact rock material	Point load strength index	> 8 MPa	4 – 8 MPa	2 – 4 MPa	1 – 2 MPa	For this low range – uniaxial compressive test is preferred		
		Uniaxial compressive strength	> 200 MPa	100 – 200 MPa	50 – 100 MPa	25 – 50 MPa	10-25 MPa	3-10 MPa	1-3 MPa
	Rating		15	12	7	4	2	1	0
2	Drill core quality RQD		90% – 100%	75% – 90%	50% – 75%	25% – 50%	< 25%		
	Rating		20	17	13	8	3		
3	Spacing of joints		> 3 m	1 – 3 m	0.3 – 1 m	50 – 300 mm	< 50 mm		
	Rating		30	25	20	10	5		
4	Condition of joints		Very rough surfaces Not continuous No separation Hard joint wall rock	Slightly rough surfaces Separation <1 mm Hard joint wall rock	Slightly rough surfaces Separation <1 mm Soft joint wall rock	Slickensided surfaces OR Gouge <5 mm thick OR Joints open 1–5 mm Continuous joints	Soft gouge >5mm thick OR Joints open >5mm Continuous joints		
	Rating		25	20	12	6	0		
5	Ground water	Inflow per 10m tunnel length	None	< 25 litres/min.	25 – 125 litres/min.	> 125 litres/min.			
		Ratio $\frac{\text{joint water pressure}}{\text{major principal stress}}$	0	0.0 – 0.2	0.2 – 0.5	> 0.5			
		General conditions	Completely dry	Moist only (interstitial water)	Water under moderate pressure	Severe water problems			
	Rating		10	7	4	0			

B. RATING ADJUSTMENT FOR JOINT ORIENTATIONS

Strike and dip orientations of joints		Very favourable	Favourable	Fair	Unfavourable	Very unfavourable
Ratings	Tunnels	0	-2	-5	-10	-12
	Foundations	0	-2	-7	-15	-25
	Slopes	0	-5	-25	-50	-60

C. ROCK MASS CLASSES DETERMINED FROM TOTAL RATINGS

Rating	100 — 81	80 — 61	60 — 41	40 — 21	< 20
Class No.	I	II	III	IV	V
Description	Very good rock	Good rock	Fair rock	Poor rock	Very poor rock

D. MEANING OF ROCK MASS CLASSES

Class No.	I	II	III	IV	V
Average stand-up time	10 years for 5m span	6 months for 4 m span	1 week for 3 m span	5 hours for 1.5 m span	10 min. for 0.5m span
Cohesion of the rock mass	> 300 kPa	200 – 300 kPa	150 – 200 kPa	100 – 150 kPa	< 100 kPa
Friction angle of the rock mass	> 45°	40° – 45°	35° – 40°	30° – 35°	< 30°

TABLE 6 – THE EFFECT OF JOINT STRIKE AND DIP ORIENTATIONS IN TUNNELLING

Strike perpendicular to tunnel axis				Strike parallel to tunnel axis		Dip 0° – 20° irrespective of strike
Drive with dip		Drive against dip				
Dip 45° – 90°	Dip 20° – 45°	Dip 45° – 90°	Dip 20° – 45°	Dip 45° – 90°	Dip 20° – 45°	
Very favourable	Favourable	Fair	Unfavourable	Very unfavourable	Fair	Unfavourable

allocated to each range of values for each parameter and an overall rating for the rock mass is arrived at by adding the ratings for each of the parameters. This overall rating must be adjusted for joint orientation by applying the corrections given in Part B of Table 5. An explanation of the descriptive terms used for this purpose is given in Table 6. Part C of Table 5 shows the class and description given to rock masses with various total ratings. The interpretation of these ratings in terms of stand-up times for underground excavations and rock mass strength parameters is given in Part D of Table 5.

Bieniawski has related his *rock mass rating* (or total rating score for the rock mass) to the stand-up time of an active unsupported span as originally proposed by Lauffer[8]. The proposed relationship is shown in figure 6 on page 25 and a practical example involving the use of this figure is discussed below. The application of the CSIR Geomechanics Classification to the choice of underground support systems will not be dealt with here but will be discussed in a later chapter dealing with rock support.

Practical example using CSIR Geomechanics Classification

Consider the example of a granitic rock mass in which a tunnel is to be driven. The classification has been carried out as follows:

Classification Parameter	Value or Description	Rating
1. Strength of intact material	150 MPa	12
2. RQD	70%	13
3. Joint spacing	0.5m.	20
4. Condition of joints	Slightly rough surfaces Separation <1mm. Hard joint wall rock	20
5. Ground water	Water under moderate pressure	4
	Total score	69

The tunnel has been oriented such that the dominant joint set strikes perpendicular to the tunnel axis with a dip of 30° against the drive direction. From Table 6, this situation is described as unfavourable for which a rating adjustment of -10 is obtained from Table 5B. Thus the final rock mass rating becomes 59 which places the rock mass at the upper end of Class III with a description of fair.

Figure 6 gives the stand-up time of an unsupported 3 metre tunnel in this rock mass as approximately 1 month.

NGI Tunnelling Quality Index

On the basis of an evaluation of a large number of case histories of underground excavation stability, Barton, Lien and Lunde[1] of the Norwegian Geotechnical Institute (NGI) proposed an index for the determination of the tunnelling quality of a rock mass. The numerical value of this index Q is defined by :

$$Q = \left(\frac{RQD}{J_n}\right) \times \left(\frac{J_r}{J_a}\right) \times \left(\frac{J_w}{SRF}\right)$$

where
- RQD is Deere's Rock Quality Designation as defined on page 18,
- J_n is the joint set number,
- J_r is the joint roughness number,
- J_a is the joint alteration number,
- J_w is the joint water reduction factor, and
- SRF is a stress reduction factor.

The definition of these terms is largely self-explanatory, particularly when the numerical value of each is determined from Table 7.

In explaining how they arrived at the equation used to determine the index Q, Barton, Lien and Lunde offer the following comments :

" The first quotient (RQD/J_n), representing the structure of the rock mass, is a crude measure of the block or particle size, with the two extreme values (100/0.5 and 10/20) differing by a factor of 400. If the quotient is interpreted in units of centimetres, the extreme "particle sizes" of 200 to 0.5 cms are seen to be crude but fairly realistic approximations. Probably the largest blocks should be several times this size and the smallest fragments less than half the size. (Clay particles are of course excluded).

The second quotient (J_r/J_a) represents the roughness and frictional characteristics of the joint walls or filling materials. This quotient is weighted in favour of rough, unaltered joints in direct contact. It is to be expected that such surfaces will be close to peak strength, that they will tend to dilate strongly when sheared, and that they will therefore be especially favourable to tunnel stability.

When rock joints have thin clay mineral coatings and fillings, the strength is reduced significantly. Nevertheless, rock wall contact after small shear displacements have occurred may be a very important factor for preserving the excavation from ultimate failure.

Where no rock wall contact exists, the conditions are extremely unfavourable to tunnel stability. The "friction angles" given in Table 7 are a little below the residual strength values for most clays, and are possibly downgraded by the fact that these clay bands or fillings may tend to consolidate during shear, at least if normally consolidated or if softening and swelling has occurred. The swelling pressure of montmorillonite may also be a factor here.

The third quotient (J_w/SRF) consists of two stress parameters. SRF is a measure of: 1. loosening load in the case of an excavation through shear zones and clay bearing rock, 2. rock stress in competent rock and 3. squeezing loads in plastic incompetent rocks. It can be regarded as a total stress parameter. The parameter J_w is a measure of water pressure, which has an adverse effect on the shear strength of joints due to a reduction in effective normal stress. Water may, in addition, cause softening and possible outwash in the case of clay-filled joints. It has proved impossible to combine these two parameters in terms of inter-block effective normal stress, because paradoxically a high value of effective normal stress may sometimes signify less stable conditions than a low value, despite the higher shear strength. The quotient (J_w/SRF) is a complicated empirical factor describing the "active stresses".

It appears that the rock tunnelling quality Q can now be considered as a function of only three parameters which are crude measures of :

1. block size   (RQD/J_n)
2. inter-block shear strength   (J_r/J_a)
3. active stress   (J_w/SRF)

Undoubtedly, there are several other parameters which could be added to improve the accuracy of the classification system. One of these would be joint orientation. Although many case records include the necessary information on structural orientation in relation to excavation axis, it was not found to be the important general parameter that might be expected. Part of the reason for this may be that the orientations of many types of excavation can be, and normally are, adjusted to avoid the maximum effect of unfavourably oriented major joints. However, this choice is not available in the case of tunnels, and more than half the case records were in this category. The parameters J_n, J_r and J_a appear to play a more important general role than orientation, because the number of joint sets determines the degree of freedom for block movement (if any), and the frictional and dilational characteristics can vary more than the down-dip gravitational component of unfavourably orientated joints. If joint orientation had been included the classification would have been less general, and its essential simplicity lost."

The large amount of information contained in Table 7 may lead the reader to suspect that the NGI Tunnelling Quality Index is unnecessarily complex and that it would be difficult to use in the analysis of practical problems. This is far from the case and an attempt to determine the value of Q for a typical rock mass will soon convince the reluctant user that the instructions are simple and unambiguous and that, with familiarity, Table 7 becomes very easy to use. Even before the value of Q is calculated, the process of determining the various factors required for its computation concentrates the attention of the user onto a number of important practical questions which can easily be ignored during a site investigation. The qualitative "feel" for the rock mass which is acquired during this process may be almost as important as the numerical value of Q which is subsequently calculated.

In order to relate their Tunnelling Quality Index Q to the behaviour and support requirements of an underground excavation, Barton, Lien and Lunde defined an additional quantity which they call the *equivalent dimension* D_e of the excavation. This dimension is obtained by dividing the span, diameter or wall height of the excavation by a quantity called the *excavation support ratio* ESR.

Hence :
$$D_e = \frac{\text{Excavation span, diameter or height (m)}}{\text{Excavation Support Ratio}}$$

The excavation support ratio is related to the use for which the excavation is intended and the extent to which some degree of instability is acceptable. Barton[29] gives the following suggested values for ESR :

Excavation category	ESR
A. Temporary mine openings	3 - 5
B. Permanent mine openings, water tunnels for hydro power (excluding high pressure penstocks) pilot tunnels, drifts and headings for large excavations.	1.6
C. Storage rooms, water treatment plants, minor road and railway tunnels, surge chambers, access tunnels.	1.3
D. Power stations, major road and railway tunnels, civil defence chambers, portals, intersections.	1.0
E. Underground nuclear power stations, railway stations, sports and public facilities, factories.	0.8

The ESR is roughly analogous to the inverse of the *factor of safety* used in the design of rock slopes[2].

The relationship between the Tunnelling Quality Index Q and the Equivalent Dimension D_e of an excavation which will stand unsupported is illustrated in figure 7. Much more elaborate graphs from which support requirements can be estimated were presented by Barton, Lien and Lunde[1] and Barton[29]. A discussion of these graphs will be deferred to a later chapter in which excavation support will be discussed more fully.

Practical example using the NGI Tunnelling Quality Index.

An underground crusher station is to be excavated in the limestone footwall of a lead-zinc ore body and it is required to find the span which can be left unsupported. The analysis is carried out as follows :

Item	Description	Value
1. Rock Quality	Good	RQD = 80%
2. Joint sets	Two sets	J_n = 4
3. Joint roughness	Rough	J_r = 3
4. Joint alteration	Clay gouge	J_a = 4
5. Joint water	Large inflow	J_w = 0.33
6. Stress reduction	Medium stress	SRF = 1.0

Hence $$Q = \frac{80}{4} \times \frac{3}{4} \times \frac{0.33}{1} = 5$$

TABLE 7 - CLASSIFICATION OF INDIVIDUAL PARAMETERS USED IN THE NGI TUNNELLING QUALITY INDEX

Description	Value	Notes
1. ROCK QUALITY DESIGNATION	RQD	
A. Very poor	0 - 25	1. Where RQD is reported or measured as ≤ 10 (including 0), a nominal value of 10 is used to evaluate Q.
B. Poor	25 - 50	
C. Fair	50 - 75	2. RQD intervals of 5, i.e. 100, 95, 90 etc are sufficiently accurate.
D. Good	75 - 90	
E. Excellent	90 - 100	
2. JOINT SET NUMBER	J_n	
A. Massive, no or few joints	0.5 - 1.0	
B. One joint set	2	
C. One joint set plus random	3	
D. Two joint sets	4	
E. Two joint sets plus random	6	
F. Three joint sets	9	1. For intersections use $(3.0 \times J_n)$
G. Three joint sets plus random	12	2. For portals use $(2.0 \times J_n)$
H. Four or more joint sets, random, heavily jointed 'sugar cube', etc	15	
J. Crushed rock, earthlike	20	
3. JOINT ROUGHNESS NUMBER	J_r	
a. Rock wall contact and		
b. Rock wall contact before 10 cms shear.		
A. Discontinuous joints	4	
B. Rough or irregular, undulating	3	
C. Smooth, undulating	2	
D. Slickensided, undulating	1.5	1. Add 1.0 if the mean spacing of the relevant joint set is greater than 3m.
E. Rough or irregular, planar	1.5	
F. Smooth, planar	1.0	2. $J_r = 0.5$ can be used for planar, slickensided joints having lineations, provided the lineations are orientated for minimum strength.
G. Slickensided, planar	0.5	
c. No rock wall contact when sheared.		
H. Zone containing clay minerals thick enough to prevent rock wall contact.	1.0	
J. Sandy, gravelly or crushed zone thick enough to prevent rock wall contact.	1.0	
4. JOINT ALTERATION NUMBER	J_a $\quad \phi_r$(approx.)	
a. Rock wall contact.		
A. Tightly healed, hard, non-softening, impermeable filling	0.75 —	

	J_a	ϕ_r (approx.)
B. Unaltered joint walls, surface staining only	1.0	(25° - 35°)
C. Slightly altered joint walls non-softening mineral coatings, sandy particles, clay-free disintegrated rock, etc	2.0	(25° - 30°)
D. Silty-, or sandy-clay coatings, small clay-fraction (non-softening)	3.0	(20° - 25°)
E. Softening or low friction clay mineral coatings, i.e. kaolinite, mica. Also chlorite, talc, gypsum and graphite etc., and small quantities of swelling clays. (Discontinuous coatings, 1-2mm or less in thickness)	4.0	(8° - 16°)

1. Values of ϕ_r, the residual friction angle, are intended as an approximate guide to the mineralogical properties of the alteration products, if present.

b. Rock wall contact before 10 cms shear.

	J_a	ϕ_r (approx.)
F. Sandy particles, clay-free disintegrated rock etc	4.0	(25° - 30°)
G. Strongly over-consolidated, non-softening clay mineral fillings (continuous, < 5mm thick)	6.0	(16° - 24°)
H. Medium or low over-consolidation, softening, clay mineral fillings, (continuous, < 5mm thick)	8.0	(12° - 16°)
J. Swelling clay fillings, i.e. montmorillonite (continuous, < 5 mm thick). Values of J_a depend on percent of swelling clay-size particles, and access to water	8.0 - 12.0	(6° - 12°)

c. No rock wall contact when sheared.

	J_a	ϕ_r (approx.)
K. Zones or bands of disintegrated L. or crushed rock and clay (see M. G,H and J for clay conditions)	6.0 8.0 8.0 - 12.0	(6° - 24°)
N. Zones or bands of silty- or sandy clay, small clay fraction, (non-softening)	5.0	
Q. Thick, continuous zones or P. bands of clay (see G, H and R. J for clay conditions)	10.0 - 13.0 13.0 - 20.0	(6° - 24°)

5. JOINT WATER REDUCTION FACTOR	J_w	approx. water pressure (Kgf/cm²)
A. Dry excavations or minor inflow, i.e. < 5 lit/min. locally	1.0	< 1.0
B. Medium inflow or pressure, occasional outwash of joint fillings	0.66	1.0 - 2.5
C. Large inflow or high pressure in competent rock with unfilled joints	0.5	2.5 - 10.0
D. Large inflow or high pressure, considerable outwash of fillings	0.33	2.5 - 10.0
E. Exceptionally high inflow or pressure at blasting, decaying with time	0.2 - 0.1	> 10
F. Exceptionally high inflow or pressure continuing without decay	0.1 - 0.05	> 10

1. Factors C to F are crude estimates. Increase J_w if drainage measures are installed.

2. Special problems caused by ice formation are not considered.

6. STRESS REDUCTION FACTOR

a. Weakness zones intersecting excavation, which may cause loosening of rock mass when tunnel is excavated.

	SRF
A. Multiple occurrences of weakness zones containing clay or chemically disintegrated rock, very loose surrounding rock (any depth)	10.0
B. Single weakness zones containing clay, or chemically disintegrated rock (excavation depth < 50m)	5.0
C. Single weakness zones containing clay, or chemically disintegrated rock (excavation depth > 50m)	2.5
D. Multiple shear zones in competent rock (clay free), loose surrounding rock (any depth)	7.5
E. Single shear zones in competent rock (clay free), (depth of excavation < 50m)	5.0
F. Single shear zones in competent rock (clay free), (depth of excavation > 50m)	2.5
G. Loose open joints, heavily jointed or 'sugar cube' (any depth)	5.0

1. Reduce these values of SRF by 25 - 50% if the relevent shear zones only influence but do not intersect the excavation.

b. Competent rock, rock stress problems

	σ_c/σ_1	σ_t/σ_1	SRF
H. Low stress, near surface	>200	>13	2.5
J. Medium stress	200-10	13-0.66	1.0
K. High stress, very tight structure (usually favourable to stability, may be unfavourable for wall stability)	10-5	0.66-0.33	0.5-2
L. Mild rock burst (massive rock)	5-2.5	0.33-0.16	5-10
M. Heavy rock burst (massive rock)	<2.5	<0.16	10-20

2. For strongly anisotropic virgin stress field (if measured) : when $5 \leq \sigma_1/\sigma_3 \leq 10$, reduce σ_c to $0.8\sigma_c$ and σ_t to $0.8\sigma_t$. When $\sigma_1/\sigma_3 > 10$, reduce σ_c and σ_t to $0.6\sigma_c$ and $0.6\sigma_t$, where σ_c = unconfined compressive strength, and σ_t = tensile strength (point load) and σ_1 and σ_3 are the major and minor principal stresses.

3. Few case records available where depth of crown below surface is less than span width. Suggest SRF increase from 2.5 to 5 for such cases (see H).

c. Squeezing rock, plastic flow of incompetent rock under the influence of high rock pressure

	SRF
N. Mild squeezing rock pressure	5-10
O. Heavy squeezing rock pressure	10-20

d. Swelling rock, chemical swelling activity depending upon presence of water

	SRF
P. Mild swelling rock pressure	5-10
R. Heavy swelling rock pressure	10-20

ADDITIONAL NOTES ON THE USE OF THESE TABLES

When making estimates of the rock mass quality (Q) the following guidelines should be followed, in addition to the notes listed in the tables:

1. When borehole core is unavailable, RQD can be estimated from the number of joints per unit volume, in which the number of joints per metre for each joint set are added. A simple relation can be used to convert this number to RQD for the case of clay free rock masses :

    $RQD = 115 - 3.3J_v$ (approx.)   where J_v = total number of joints per m^3
    (RQD = 100 for $J_v < 4.5$)

2. The parameter J_n representing the number of joint sets will often be affected by foliation, schistosity, slaty cleavage or bedding etc. If strongly developed these parallel "joints" should obviously be counted as a complete joint set. However, if there are few "joints" visible, or only occasional breaks in the core due to these features, then it will be more appropriate to count them as "random joints" when evaluating J_n.

3. The parameters J_r and J_a (representing shear strength) should be relevant to the *weakest significant joint set or clay filled discontinuity* in the given zone. However, if the joint set or discontinuity with the minimum value of (J_r/J_a) is favourably oriented for stability, then a second, less favourably oriented joint set or discontinuity may sometimes be more significant, and its higher value of J_r/J_a should be used when evaluating Q . *The value of J_r/J_a should in fact relate to the surface most likely to allow failure to initiate.*

4. When a rock mass contains clay, the factor SRF appropriate to *loosening loads* should be evaluated. In such cases the strength of the intact rock is of little interest. However, when jointing is minimal and clay is completely absent the strength of the intact rock may

TABLE 7 CONTINUED

> become the weakest link, and the stability will then depend on the ratio rock-stress/rock-strength. A strongly anisotropic stress field is unfavourable for stability and is roughly accounted for as in note 2 in the table for stress reduction factor evaluation.
> 5. The compressive and tensile strengths (σ_c and σ_t) of the intact rock should be evaluated in the saturated condition if this is appropriate to present or future in situ conditions. A very conservative estimate of strength should be made for those rocks that deteriorate when exposed to moist or saturated conditions.

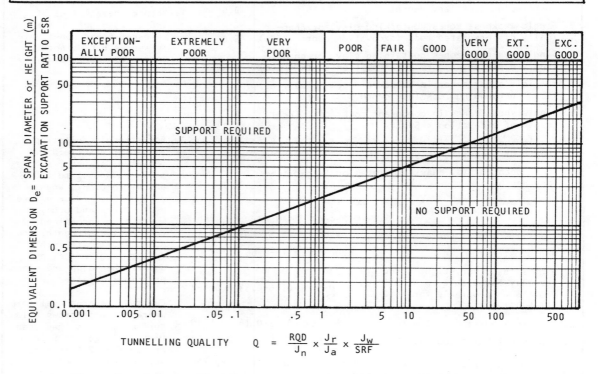

Figure 7. Relationship between the maximum equivalent dimension D_e of an unsupported underground excavation and the NGI tunnelling quality index Q. (After Barton, Lien and Lunde[1])

From figure 7, the maximum equivalent dimension D_e for an unsupported excavation in this rock mass is 4 metres. A permanent underground mine opening has an excavation support ratio ESR of 1.6 and, hence the maximum unsupported span which can be considered for this crusher station is ESR $\times D_e$ = 1.6 $\times$ 4 = 6.4 metres.

Discussion on rock mass classification systems

Of the several rock mass classification systems described in this chapter, the CSIR system proposed by Bieniawski [25,26] and the NGI system proposed by Barton, Lien and Lunde[1] are of particular interest because they include sufficient information to provide a realistic assessment of the factors which influence the stability of an underground excavation. Bieniawski's classification appears to lay slightly greater emphasis on the orientation and inclination of the structural features in the rock mass while taking no account of the rock stress. The NGI classification does not include a joint orientation term but the properties of the most unfavourable joint sets are considered in the assessment of the joint roughness and the joint alteration numbers, both of which represent the shear strength of the rock mass.

Both classification systems suggest that the influence of structural orientation and inclination is less significant than one would normally tend to assume and that a differentiation between *favourable* and *unfavourable* is adequate for most practical purposes. While this may be acceptable for the majority of situations likely to be encountered in the field, there are a few cases in materials such as slate where the structural features are so strongly developed that they will tend to dominate the behaviour of the rock mass. In other situations, large blocks may be isolated by a small number of individual discontinuities and become unstable when the excavation is created. In such cases, the classification systems discussed in this chapter may not be adequate and special consideration may have to be given to the relationship between the geometry of the rock mass and that of the excavation. This subject will be dealt with in chapter 7 of this book.

The authors have used both the CSIR and the NGI systems in the field and have found both to be simple to use and of considerable assistance in making difficult practical decisions. In most cases, both classifications are used and both the Rock Mass Rating (RMR) and the Tunnelling Quality (Q) are used in deciding upon the solution to the problem. It has been found that the equation $RMR = 9 \log_e Q + 44$ proposed by Bieniawski[26] adequately describes the relationship between the two systems.

When dealing with problems involving extremely weak ground which result in squeezing, swelling or flowing conditions (see Terzaghi's classification in Table 1 on page 17), it has been found that the CSIR classification is difficult to apply. This is hardly surprising given that the system was originally developed for shallow tunnels in hard jointed rock. Hence, when working in extremely weak ground, the authors recommend the use of the NGI system.

In discussing the CSIR and NGI classification systems, the authors have concentrated upon the basic rock mass classification and on the indication given by this classification of whether support is required or not. Bieniawski[25,26] and Barton, Lein and Lunde[1] went on to apply these classifications to the choice of specific support systems. The detailed design of support for underground excavations, including the use or rock mass classifications to assist in the choice of support systems, will be discussed in chapter 8 of this book.

Chapter 2 references

6. TERZAGHI,K. Rock defects and loads on tunnel supports. In: *Rock Tunnelling with Steel Supports*. Editors R.V.Proctor and T.White. Published by Commercial Shearing and Stamping Co., Youngstown, 1946, pages 15-99. Also Harvard University, Graduate School of Engineering, Publication 418 - Soil Mechanics Series 25.

7. STINI,I. *Tunnelbaugeologie*. Springer-Verlag, Vienna, 1950, 366 pages.

8. LAUFFER,H. Gebirgsklassifizierung fur den Stollenbau. *Geologie und Bauwesen*, Volume 24, Number 1, 1958, pages 46-51.

9. DEERE,D.U. Technical description of rock cores for engineer-purposes. *Rock Mechanics and Engineering Geology*. Volume 1, Number 1, 1964, pages 17-22.

10. DEERE,D.U. and MILLER,R.P. Engineering classification and index properties for intact rock. *Technical Report No. AFNL-TR-65-116, Air Force Weapons Laboratory*, New Mexico,1966.

11. HANSAGI,H. Numerical determination of mechanical properties of rock and rock masses. *Intnl. J. Rock Mechanics and Mining Sciences*, Volume 2, 1965, pages 219-223.

12. HAGERMAN,T.H. Different types of rock masses from rock mechanics point of view. *Rock Mechanics and Engineering Geology*, Volume 4, 1966, pages 183-198.

13. DEERE.D.U. Geological considerations. In: *Rock Mechanics in Engineering Practice*. Editors K.G.Stagg and O.C.Zienkiewcz. Published by John Wiley & Sons, London, 1968, pages 1-20.

14. COON,R.F. Correlation of engineering behaviour with the classification of in situ rock. *Ph.D Thesis, University of Illinois*, Urbana, 1968.

15. MERRITT,A.H. Engineering classification of in situ rock. *Ph.D Thesis, University of Illinois*, Urbana, 1968.

16. STAPLEDON,D.H. Classification of rock substances. *Intnl. J. Rock Mechanics and Mining Sciences*, Volume 5, 1968, pages 71-73.

17. VOIGHT,B. On the functional classification of rocks for engineering purposes. *Intnl. Symposium on Rock Mechanics*, Madrid, 1968, pages 131-135.

18. CECIL,O.S. Correlation of rockbolts - shotcrete support and rock quality parameters in Scandinavian tunnels. *Ph.D. Thesis, University of Illinois*, Urbana, 1970, 414 pages.

19. IKEDA,K.A. Classification of rock conditions for tunnelling. *Proc. 1st Intnl. Congress of Engineering Geology*, Paris, 1970, pages 1258-1265.

20. DEERE,D.U., PECK,R.B., PARKER,H.W., MONSEES,J.F. and SCHMIDT,B. Design of tunnel support systems. *Highway Research Record*, Number 339, 1970, pages 26-33.

21. WICKHAM,G.E., TIEDEMANN,H.R. and SKINNER,E.H. Support determination based on geological predictions. *Proc. First North American Rapid Excavation and Tunnelling Conference*, AIME, New York, 1972, pages 43-64.

22. BREKKE, T.L. and HOWARD, T. Stability problems caused by seams and faults. *Proc. First North American Rapid Excavation and Tunnelling Conference*, AIME, New York, 1972, pages 25-41.

23. MERRITT, A.H. Geologic prediction for underground excavations. *Proc. First North American Rapid Excavation and Tunnelling Conference*, AIME, New York, 1972, pages 115-132.

24. CORDING, E.J and DEERE, D.U. Rock tunnel supports and field measurements. *Proc. First North American Rapid Excavation and Tunnelling Conference*, AIME, New York, 1972, pages 601-622.

25. BIENIAWSKI, Z.T. Geomechanics classification of rock masses and its application in tunnelling. *Proc. Third International Congress on Rock Mechanics*, ISRM, Denver, Volume 11A, 1974, pages 27-32.

26. BIENIAWSKI, Z.T. Rock mass classification in rock engineering, *Proc. Symposium on Exploration for Rock Engineering*, Johannesburg, Volume 1, 1976, pages 97-106.

27. CORDING, E.J., HENDRON, A.J and DEERE, D.U. Rock engineering for underground caverns. *Proc. Symposium on Underground Rock Chambers*, Phoenix, Arizona, 1971, published by ASCE, 1972, pages 567-600.

28. BREKKE, T.L. and SELMER-OLSEN, R. Stability problems in underground construction caused by montmorillonite carrying joints and faults. *Engineeering Geology*, Volume 1, Number 1, 1965, pages 3-19.

29. BARTON, N. Recent experiences with the Q-system of tunnel support design. *Proc. Symposium on Exploration for Rock Engineering*, Johannesburg, Volume 1, 1976, pages 107-117.

Chapter 3 : Geological data collection

Introduction

The worst type of problem with which an underground excavation designer can be faced is the *unexpected* problem. Within the confined space of a tunnel or mine, it is both difficult and dangerous to deal with stability or water problems which are encountered unexpectedly. On the other hand, given sufficient warning of a potential problem, the engineer can usually provide a solution by changing the location or the geometry of the excavation, by supporting or reinforcing the rock mass around the opening or by draining or diverting accumulations of groundwater.

While it is impossible to anticipate all the geological conditions which can give rise to problems during the excavation of an underground opening, it is clearly necessary that every reasonable effort should be made to obtain a complete picture of the rock mass characteristics at an early stage in any project. This means that sufficient resources, both financial and manpower, and enough time must be allowed for the geological data collection and site investigation phase of an underground excavation project. Failure to do so will result in an inadequate basis for the design and could be very costly when unexpected problems are encountered at a later stage in the project.

In this chapter, techniques which can be used for geological data collection are reviewed in a general manner. Since each site will have its own peculiarities and since the site investigation equipment available locally may differ from that described in this chapter, the reader will have to adapt the information given to suit his own requirements.

Study of regional geology

The structural geological conditions which occur on any particular site are a product of the geological history of the surrounding region. Hence the rock types, folds, faults and joints in the relatively small rock volume with which the designer is concerned form part of a much larger pattern which reflects the major geological processes to which the region was subjected. A knowledge of these major geological processes can sometimes be of great benefit in building up a detailed geological picture of the site since it will tend to suggest structural trends which may not be obvious from the mass of detailed information available at a more local level.

Geological studies have been carried out in most areas of the world and these studies are generally recorded in papers submitted to scientific journals or on maps which may be available in local libraries, universities or government geological organisations. It is important that any such information covering the area under study should be located and studied as early as possible in the project.

It is also important that local knowledge should be utilised as far as possible. Local prospectors, miners, quarrymen, building contractors and amateur geologists may be able to provide useful information on previous mining or quarrying operations, unusual groundwater conditions or other factors which may be relevant. Geological departments at universities are frequently an important source of local geological information and the staff of such departments are usually very willing to help in this early but important data collection process.

Many parts of the world have been photographed from the air for military or civilian purposes and good quality air photographs can provide very useful information on structural features and on some sub-surface phenomena. Faults and other major linear features are usually fairly easy to identify but a skilled photogeologist may also be able to locate subsidence areas or caving cracks from old underground mines or the surface reflection of solution cavities which have slightly altered the local drainage, resulting in slight changes in the colour or distribution of surface vegetation.

Stereoscopic examination of adjacent pairs of air photographs is useful in areas of significant topographic relief since it may be possible to locate old landslips or other surface features which may be important in the overall design of the project. Contour maps of the area can also be prepared from air photographs and figure 8 illustrates an inexpensive stereoviewer which can be used for the approximate measurement of surface elevation differences. Accurate contour maps can usually be prepared by specialist survey organisations which have sophisticated equipment for elevation measurement and, in many cases it is worth placing a contract for air photography and contour map preparation of a particular site with such an organisation.

The reader who is interested in learning more about photogeology and the interpretation of air photographs is referred to the book on this subject by Miller[30].

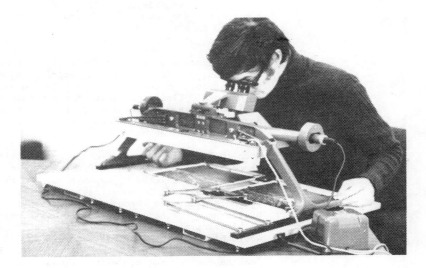

Figure 8. A stereoviewer being used to examine an adjacent pair of air photographs. This instrument can be used to measure surface elevation differences and is a model SB180 folding mirror stereoscope manufactured by Rank Precision Industries Ltd., P.O.Box 36, Leicester, England.

Engineering geological maps and plans

Maps produced as a result of the regional geological studies described in the previous section are normally to a scale of between 1:10000 and 1:100000. In order to provide the more detailed information required for the engineering design of an underground excavation, it is necessary to produce maps and plans to a scale of about 1:1000 or even 1:100. In addition, the type of information included on such plans and in the accompanying logs and notes should be such that a classification of the rock mass (as described in the previous chapter) can be produced.

A review of engineering geological mapping techniques, which are currently used in civil engineering practice (mainly for surface excavations) in the United Kingdom, has been published by Dearman and Fookes[31]. Moye[32] has given a very full description of the engineering geological investigations carried out for the Tumut 1 underground power station in the Snowy Mountains in Australia. These papers, together with the textbooks on engineering geology included in the list of references at the end of this chapter, deal with the subject of engineering geological mapping more fully than is possible in a book of this type in which the emphasis is placed on engineering design. It is recommended that the reader should study at least one of these references in order that he may become familiar with the range of techniques which are available.

Mapping surface outcrops

At an early stage in an underground excavation design project, access may not be available to the rock mass at the depth at which the excavation is to be created. Under these circumstances, the rock which outcrops on the surface must be utilised to obtain the maximum amount of information on rock types and on the structural features in the rock mass.

Stream beds are usually an important source of outcrop information, particularly where fast flowing mountain streams have cut through superficial material to expose the underlying rock. Figure 9 illustrates a stream bed exposure in horizontally bedded sedimentary rocks and many of the characteristics of the individual beds in the sequence are evident in such an exposure.

When the amount of surface exposure is limited or when it is considered that those outcrops which are available have been severely altered by weathering, the excavation of a trench or a shaft is sometimes advisable. Figure 10 shows a trench excavated through surface deposits by means of a bulldozer. Further excavation into the bed-rock by means of blasting may be required in such cases although care would have to be taken that the information being sought was not destroyed in the basting process. Sometimes, cleaning the exposed surface with a pressurised jet of water or air is sufficient to reveal the rock mass for the purpose of structural mapping.

In addition to the identification of rock types, surface outcrops should be used for the measurement of the inclination (dip) and orientation (dip direction) of structural features such as bedding planes, cleavage and joint planes.

Figure 9. Horizontally bedded sedimentary rock exposed in the bed of a mountain stream. Such exposures provide excellent facilities for surface mapping.

Figure 10. Trial trench excavated through surface materials by means of a bulldozer.

The Clar geological compass which is used for measurement of the dip and dip direction of geological planes.

A great deal of time and energy can be saved if these measurements are carried out with an instrument which is specifically designed for this purpose. Several such instruments are available but one of the most convenient is the compass illustrated in the photograph reproduced opposite. This compass was designed by Professor Clar and is manufactured by F.W.Briethaupt & Sohn, Adolfstrasse 13, Kassel 3500, West Germany.

The folding lid of the compass is placed against the plane to be measured and the target bubble is used to level the body of the instrument. The dip of the plane is indicated on the circular scale at the end of the lid hinge. With the instrument body level, the compass needle clamp is depressed (by the user's thumb in the photograph opposite) and the well-damped needle quickly establishes its magnetic north-south orientation. It is locked in this position by releasing the clamp and a friction clutch in the lid hinge holds the lid in position. The instrument can now be removed from the rock face and the dip and dip direction read off the two scales. This ability to retain the readings after the instrument has been removed from the rock face is important when the compass is being used in difficult positions.

Field measurements are usually recorded in a field note book but a portable tape recorder can sometimes provide a very effective means of making field notes. It is important that information recorded in the field should be transferred onto maps, plans or other more permanent forms at regular intervals, preferably daily. This will ensure that apparent anomalies can be checked while access to the outcrop is still readily available and erroneous information, which could be dangerously misleading at a later stage in the project, can be eliminated.

Since it is probable that the field mapping will be carried out by a geologist and that the data may be used, at some later stage, by an engineer for design purposes, it is essential that an effective means of communication be established between these two persons. It is also important that these data should be intelligible to other engineers and geologists who may be concerned with other aspects of the project but who may become involved in occasional discussions on geotechnical problems. The presentation of these data graphically or by means of models, is considered to be so important that the whole of the next chapter is devoted to this subject.

When rock exposures occur in the form of very steep faces which are difficult and dangerous to work on, photogrammetric techniques can be used to obtain information on the dip and dip direction of visible structural features. Two photographs are taken, from accurately surveyed base positions, of a rock face on which reference targets have been marked. These photographs are combined in a stereocomparator, a highly sophisticated version of the instrument illustrated in figure 8, and measurements of the three coordinates of visible points on the rock face are made. By taking a number of measurements on a plane in the photographs, the dip and dip direction of the plane can be computed by rotating an imaginary plane in space until it fits the measured coordinates. This technique has been fully described by Ross-Brown, Wickens and Markland[38] and the equipment available for photogrammetric studies has been reviewed by Moffitt[39].

Confirmation of the usefulness of this photogrammetric technique of geological mapping was obtained in one study* carried out on a very steep quarry face to which access was not available. The information gathered during this study was used in a stability analysis of the rock face as part of a feasibility study for a major project. Several years later, when access to the face was available, the information obtained in the photogrammetric study was checked by geological mapping and the differences were found to be insignificant.

Geophysical exploration

Because of the high cost of sub-surface exploration by diamond drilling or by the excavation of trial shafts or adits, the site of a proposed underground excavation is seldom investigated as fully as a design engineer would wish. Geophysical methods can be used to obtain an initial overall assessment of the site which can assist the project staff in optimising the site exploration programme.

Mossman and Heim[40] have reviewed a range of geophysical techniques which are applicable to underground excavation engineering and their summary of available methods is given in Table 8 on page 44. This table is largely self-explanatory but a few additional comments may be helpful.

Geophysical methods involving the use of gravity meters, magnetometers and electrical resistivity can be used to obtain estimates of rock properties such as porosity and density. However, these methods give relatively little indication of the structural characteristics of the rock mass and the results can sometimes be very difficult to interpret.

Seismic methods will not give satisfactory results in all geological environments and they are the most expensive of geophysical methods. On the other hand, when geological conditions are suitable, seismic methods can give valuable information on the structural attitude and configuration of rock layers and on the location of major geological discontinuities such as faults.

The interpretation of both geophysical and seismic measurements is a complex process and a great deal of practical experience is required of the operator before the results can be regarded as reliable. For this reason, do-it-yourself geophysical and seismic studies are not recommended. Where such studies are considered to be appropriate, the employment of a specialist contractor is advisable.

Once sub-surface access is available through boreholes, the usefulness of geophysical exploration techniques can be extended. Several techniques developed by the oil industry are now available for civil engineering and mining applications[41] and seismic measurements in and between boreholes can give useful information of the local characteristics of the rock mass[42].

The use of seismic techniques to locate rockbursts in underground mining situations is a special form of instrumentation which will not be dealt with in this book.

---

* By D.M.Ross-Brown in a consulting assignment with E.Hoek.

TABLE 8 - GEOPHYSICAL EXPLORATION TECHNIQUES FOR UNDERGROUND EXCAVATION ENGINEERING (After Mossman and Heim[40])

Method	Principle	Geological environment	Applications	Limitations	Cost
GRAVITY METER	Measures total density of rocks. Measurement in 10^{-8} gals. Accuracy $\pm 1 \times 10^{-7}$ gals. Coverage is spherical around point.	Any. Effective depth in excess of 3000 feet. Intensity of signal decreases as square of depth.	Measurement of lateral changes of rock types. Location of caverns.	Does not provide direct measurement of geometry of rocks.	Intermediate.
MAGNETO-METER	Measures total magnetic intensities in gammas to ± 1 gamma for total field, 2.5-10 gamma for vertical field, ± 10 gamma for horizontal field. Coverage is point, measures field intensities.	Any, but primarily igneous. Effective depth not selective but field strength decreases as square of distance from observer.	Discloses presence of local metallic bodies. Useful for mapping buried pipelines, may also indicate faulting and minor igneous intrusions.	Does not provide direct measurement of geometry of rocks.	Low to intermediate.
ELECTRICAL RESISTIVITY	Measures relative electrical conductivity of rocks in ohms from 3×10^{-3} to 10^4 ohms, generally $\pm 2 \times 10^{-1}$ sensitivity. Coverage linear over short dist.	Any, but primarily for overburden and groundwater evaluation. Effective depth to 3000 feet depending upon type of sediments and instrument used.	Exploration for ore bodies, aquifer location, gravel deposits and bedrock profiles.	Results often ambiguous	Intermediate.
ELECTROMAGNETIC	Measures amplitude and phase angle of electromagnetic field. Measurement in scale readings. Point coverage.	Any. Effective depth surficial.	Aquifer location.	Restricted application, ambiguous results.	Low to intermediate.
RADIOMETRIC (SCINTILLO-METER)	Measures gamma-ray radiation - 2.5×10^{-2} to 5 milliroentgens/hour, at up to 4000 counts/sec. Point coverage.	Any. Effective depth surficial.	Prospecting for radioactive ore. Can yield data on shale constituency.	Measures surface manifestations only. Often used in boreholes.	Low, increasing with area.
SEISMIC REFRACTION	Measures travel times of induced energy from explosives, vibrator in 10^{-3} seconds. Accuracy 2×10^{-3} sec = ± 10 to 30 feet. Coverage is linear at any desired horizontal spacing.	Sedimentary, igneous or metamorphic rocks. Effective depth 0 - 500 feet. Greater depths require large horizontal extension of operation.	Measuring depth to bedrock along extended lines. Determination of S and P-wave velocities in refracting zone for derivation of rock properties. Configuration and continuity of rock surfaces.	Velocity calibration required for depth determinations. Poor for steep dips. Uneconomical for small jobs.	High, but covers large area.
SEISMIC REFLECTION	Measures travel times of induced energy from various sources in 10^{-3} seconds. Accuracy $\pm 2 \times 10^{-3}$ sec. = 5 to 25 feet, decreasing with depth. Coverage linear at any desired horizontal spacing.	Primarily sedimentary rocks. Effective depth ± 500 feet to unlimited depth.	Measures depth and continuity of rock layers. Locates discontinuities such as faults. Provides data on stratigraphic conditions.	Velocity calibration required for determination of depth.	High.

Diamond drilling for sub-surface exploration

The recovery of core by diamond drilling is one of the most important methods of sub-surface exploration. A comprehensive discussion on drilling equipment and techniques would exceed the scope of this book and the remarks which follow are intended to provide overall guidance rather than detailed instructions.

Commercial diamond drilling services are available throughout the world and, while the quality of these services varies very widely, the reader should have no difficulty in locating a local service to meet his particular needs. Unless the project under consideration is part of a long term development programme of a large organisation with constant need for diamond drilling, for example, a major mine, it would not be worth purchasing drilling equipment for a single job - it would always be cheaper to employ a specialist drilling contractor. This is because the skill and the practical experience of the drill operator is a major factor in successful diamond drilling and the availability of drilling equipment without the supporting staff is a recipe for a series of very expensive mistakes.

Most drilling equipment manufacturers will respond to a request for information by supplying very detailed data sheets on their products. These data sheets are a better source of information on diamond drilling than most publications on the subject and the interested reader is advised to contact a number of manufacturers for up-to-date data on their products.

Diamond drilling contracts

Contracts for mineral exploration are normally negotiated on the basis of a fixed rate of payment per unit length drilled. This is because the main purpose of such drilling is to recover intact pieces of rock and there is relatively little interest in the weak seams between the intact pieces. On the other hand, the main aim of geotechnical drilling is to study the weaknesses in the rock mass and hence the core recovery should be as complete as possible. Consequently, diamond drilling contracts for structural purposes should be negotiated on the basis of payment for core recovered rather than drill hole length.

It is difficult to give detailed guidance on the contract conditions which should be specified because these will depend very much upon local conditions and upon the attitudes of drilling contractors in the country in which the project is located. In some cases, it may be possible to negotiate on the basis that payment will only be made when the percentage core recovery exceeds a certain level - say 90%. In other cases it may be preferable to settle on an hourly rate for drilling with a bonus for percentage core recovery. In all cases, the aim should be to encourage the driller to aim for a high percentage of core recovery rather than for the maximum length of drill hole per shift.

A very rough rule is that high quality diamond drilling for geotechnical purposes will cost roughly twice as much as the equivalent length of hole for mineral exploration purposes.

Figure 11. A large surface-mounted drilling machine for deep hole sub-surface exploration. Note the carefully prepared foundation and working area.

Figure 12. A hydraulically operated diamond drilling machine suitable for producing high quality 56mm core in confined underground locations.

Both photographs reproduced with permission from Atlas Copco, Sweden.

Drilling machines

In addition to providing a financial incentive to encourage the driller to aim for high core recovery, it is essential that he should also be provided with a drilling machine which has adequate capacity for the job in hand and which has been properly maintained so that it is in good working order. This may appear to be an obvious statement but it is surprising how often under-rated machines and poor equipment are used in site investigation work. An inspection of the contractor's equipment will often give a good indication of the quality of work which he is likely to produce and such an inspection is a sound precaution to take before the award of a contract.

Hydraulic feed drilling machines are essential for high quality core recovery. The independent control of thrust permits the drilling bit to adjust its penetration rate to the hardness of the rock being drilled and, in particular, to move rapidly through weathered rock and fault zones before they are eroded by the drilling water. It is also important that a wide range of drilling speeds are available to permit adjustment to the manufacturer's recommendation on rotational speeds for various bits.

Many large surface drilling rigs, such as that illustrated in figure 11, are fitted with hydraulically operated chucks which permit rapid coupling and uncoupling of rods - a job which requires the expenditure of considerable energy if it is done manually with wrenches. Hydraulic chucks are also fitted to the smaller Atlas Copco Diamec 250 machine illustrated in figure 12 and these permit very rapid rod changing - an important consideration when a large number of short drill rods have to be used in a confined underground location. Light-weight aluminium rods are also normally used on this machine and this makes it possible for one man to carry out the complete drilling operation, once the machine has been set up.

The bulky nature of hydraulic drilling machines such as that illustrated in figure 11 makes them unsuitable for use underground, except in major caverns. Consequently, until compact machines such as that illustrated in figure 12 were introduced, most underground drilling, particularly in mines, was done with screw feed machines. Because the thrust on these machines is not easy to control, they are less suitable than the hydraulic machines for high quality structural drilling. However, with careful control by the operator, good core recovery can be achieved with these machines[43].

A very large proportion of diamond drilling is carried out using water to cool the diamond bit and to flush the chippings out of the hole. In some cases, the use of air as a cooling and flushing medium is preferred, particularly when very poor rock conditions with the danger of rapid deterioration of the rock due to moisture changes are encountered. The use of air requires a special design of the ducts in the drill bit since a higher volume of air is required in order to achieve the same effect as water flush. Relatively few manufacturers offer equipment suitable for the use of air but it is anticipated that more of this equipment will become available in the future as its advantages for special applications are recognised.

Core barrels for diamond drilling

The design of a diamond impregnated drill bit for good core recovery is a highly specialised process which is undertaken by a number of manufacturers and no attempt will be made to discuss this process here. However, the results achieved by using the best available bit can be completely spoiled if the barrel in which the core is caught is poorly designed. A lamentably frequent sight is to see a driller removing a core by up-ending the barrel and either shaking it or thumping it with a hammer in order to remove the core. The "undisturbed" core, which has been recovered at great expense, is usually deposited in the core box as a meaningless jumble of pieces as a result of this type of operation.

The aim of a geotechnical drilling programme is to reconstruct the complete core sample from the rock mass in a state as close to its original condition as possible. This can only be achieved if the core passes into a non-rotating inner tube in the core barrel so that the rotation of the outer barrel, to which the drill bit is attached, does not twist the fragile core. Most manufacturers can supply a variety of double or triple tube core barrels in which the inner barrel is mounted on a bearing assembly which prevents the rotation of the outer barrel being transferred to the inner unit.

The most desirable construction of the innermost barrel is to have the tube in two matching halves which are held together by means of steel clips. When the full barrel is recovered from the drill hole, these clips are removed and the barrel is split to reveal the core which can then be tranferred into a prepared core box. One such spit double tube core barrel, manufactured by Mindrill of Australia, is illustrated in figure 13. This particular barrel has been fully described in a paper by Jeffers[44].

Wireline drilling

When drilling deep holes from surface, a great deal of time and energy can be expended in removing the drill rods from the hole at the end of each drilling run. Much of this effort can be avoided by the use of wireline equipment which allows the full core barrel only to be removed at the end of each drilling run. This barrel is lowered down the centre of the drill string by a wire and a series of clamps are used to attach the barrel to the bit. These clamps are released when the barrel has been filled and the drilling system is left in place while the core is recovered. Wireline drilling has become very common in high quality mineral exploration and site investigation work and many drilling contractors now have equipment available for this work.

Core orientation

It should already have become obvious to the reader that the orientation and inclination of structural discontinuities in the rock mass are extremely important factors to be considered in relation to the design of an underground excavation. Hence, however successful a drilling programme has been in terms of core recovery, very valuable information will have been lost if no attempt has been made to orient the core.

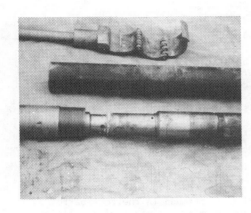

a. Core barrel with outer tube removed to reveal bearing assembly to which inner split tube is attached. Note water holes in bearing assembly.

b. Full core barrel with diamond drilling bit removed to show end of split inner barrel projecting from outer barrel. Note circular spring clip holding two halves of split barrel together.

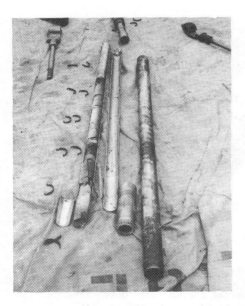

c. Split inner barrel removed from outer barrel and uncoupled from bearing assembly. Circular clips have been removed and the barrel split to reveal nearly undisturbed run of core.

Figure 13. Use of split double tube core barrel for recovery of core for geotechnical purposes.

Phillips[47] and Ragan[48] have described methods for establishing the orientation and inclination of strata from recognisable marker bands or bedding planes which are intersected by two or more non-parallel boreholes. These techniques are familiar to most geologsits and can be very useful in certain circumstances. Engineering readers should consult Phillips[47] for details.

Orientation of core from a single borehole usually depends upon the use of some form of orientation device which is used during the drilling programme. The Christensen-Hugel method utilises a scribing mechanism which marks parallel lines on the core as it is forced into the inner barrel of the drill[49]. The Atlas Copco-Craelius core orientation system uses a tool which is clamped in the core barrel as this is lowered into the hole at the start of a drilling run. A number of pins parallel to the drill axis project ahead of the drill bit and take up the profile of the core stub left by the previous drilling run, as illustrated in figure 14. The orientation of the device is determined relative to the drill rod position at the collar of the hole or, in an inclined hole, by means of a ball bearing marker which defines a vertical plane through the borehole axis. When the core has been recovered, the first piece of core is matched to the profile of the pins and the remainder of the core is pieced together to obtain the orientation of other structural features in relation to the first piece.

Figure 14. An Atlas Copco-Craelius core orienter. Clamped inside the diamond bit, the pins take up the profile of the core stub left by the previous drilling run. Pressure on the spring loaded cone locks the pins, actuates the ball bearing marker and releases the tool so that it can move up the drill barrel ahead of the core.

More elaborate core orientation systems involve drilling a small diameter hole at the end of the hole left by the previous drill run. A compass can be bonded into this hole for recovery in the next core run[45] or an oriented rod can be grouted into the hole to provide reinforcement for the core as well as orientation. This latter technique, known as the integral sampling method, has been described by Rocha[50] and can be used to produce high quality oriented core in very poor rock. However, it is both expensive and time consuming and would only be used to evaluate extremely critical areas in the rock mass.

Examination of the walls of boreholes by means of cameras[51] or television systems has been used for core orientation but the results obtained from these devices are seldom satisfactory. A great deal of time can be wasted as a result of mechanical and electrical breakdowns in equipment which was not originally designed to operate under such severe conditions. A more promising borehole inspection tool is the Televiewer which was originally developed by the oil industry[52]. This instrument, which works in a mud-filled hole, produces a television type picture as a result of the attenuation of a sonic signal by fractures in the rock surrounding the hole. The high cost of this instrument limits its application to special studies and it could not be considered for routine site investigation work.

An inexpensive tool for obtaining an impression of the inside of a diamond drilled hole has recently been developed by Hinds[53,54]. Figure 15 is a reproduction of an impression taken in a 3 inch diameter hole in sandstone and this shows the coarse grain of the rock as well as several open fissures. The impression material is a thermoplastic film called Parafilm M* which is pressed against the borehole wall by an inflatable rubber packer. Linking this device to some form of orientation system or borehole survey instrument provides information on the orientation of fractures in the rock mass which is independent of disturbance of the core.

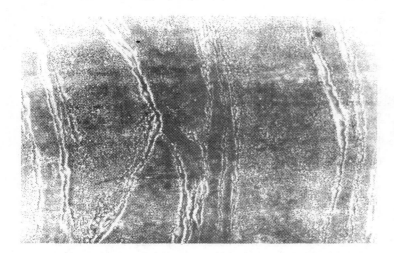

Figure 15. Photographic reproduction of the impression taken on thermoplastic film inside a 3 inch diameter diamond drill hole in sandstone.

---

* Manufactured by The American Can Company, American Lane, Greenwich, Conn. 06830, USA.

From the comments in this section, it will be obvious that core orientation is a difficult operation which frequently yields unsatisfactory results. In spite of these difficulties, a knowledge of the orientation and inclination of discontinuities in the rock mass is very important for the design of underground excavations and the reader should not give up too easily in his efforts to obtain this information.

Index testing of core

In addition to information on the discontinuities in the rock mass, it is also important to obtain estimates of the strength of the intact rock and on the weathering characteristics of this rock.

A reasonable estimate of the uniaxial compressive strength of the rock can be obtained by means of the point load test. A piece of core is loaded across its diameter between two hardened steel points as illustrated in figures 16 and 17 which show two alternative commercially available point load testing machines. The Point Load Index[55] is given by:

$$I_s = P/D^2$$

where P is the load required to break the specimen and
D is the diameter of the core.
Note that the length of the core piece should be at least 1.5 times the diameter of the core.

If the diameter D of the core is expressed in millimetres, an approximate relationship between the point load index I_s and the uniaxial compressive strength σ_c is given by[56]:

$$\sigma_c = (14 + 0.175 D) I_s$$

Because the load required to break a rock core under point load conditions is only about one tenth of that required for failure of a specimen subjected to uniaxial compressive stress, the point load equipment is light and portable and is ideal for use in the field during logging of the core.

The uniaxial compressive strength of the rock, estimated from the point load index, can be used for rock classification according to Table 3 on page 25. The value of σ_c can also be used in a more detailed analysis of rock strength which will be discussed in a later chapter.

Figure 18 shows a box of core recovered from an interbedded series of sandstones, siltstones and mudstones. The photograph was taken approximately six months after drilling and it shows that the mudstone (the dark coloured material in the central trays) has disintegrated completely in this time. This tendency to weather on exposure can have serious consequences if such a material is left unprotected in an excavation and it is important that the engineer should be made aware of this danger in good time so that he can specify appropriate protective measures. Franklin and Chandra[57] have described a Slake Durability test which is carried out in the apparatus illustrated in figure 19. This test, which can be carried out in a field laboratory, involves determining the weight loss of a number of pieces of rock which have been rotated, under water, in the sieve drums at the ends of the instrument.

Figure 16. A Point Load testing machine used for strength index testing in the field. This apparatus is manufactured by Engineering Laboratory Equipment Ltd., Hemel Hempstead, Hertfordshire, England.

Figure 17. An alternative form of Point Load tester manufactured by Robertson Research Mineral Technology Limited, Llandudno, Gwynedd, Wales.

Figure 18. Core recovered from interbedded mudstones, siltstones and sandstones. The photograph was taken approximately six months after drilling and shows the severe weathering of the siltstones and mudstones which occupy the central trays of the core box.

Figure 19. Slake Durability test apparatus in which samples of rock are rotated in the two sieve drums in water and the weight loss indicates the sensitivity of the rock to weathering. This equipment is manufactured by Engineering Laboratory Equipment Ltd., Hemel Hempstead, Hertfordshire, England.

Core logging and photography of core

The time span between the preliminary investigation of a site and the commencement of active mining or construction may be of the order of ten years and the need to refer to the original geological information may well last for several tens of years. Because the geologists involved in the site investigation are unlikely to be available throughout this period, it is essential that the geological data should be recorded in such a way that meaningful interpretation of these data can be carried out by others. It would not be appropriate to suggest the exact form which should be used for core logging since this will vary according to the nature of the project, the design approach adopted and the overall geological conditions on site. The procedure adopted in the case of an underground mining operation in massive lead-zinc ore bodies will probably differ significantly from that used for an underground powerhouse in horizontally bedded coal measure sedimentary rocks. However, it is strongly recommended that the project management and the geologists should give serious consideration to the core logging procedures to be used and to the presentation of regular carefully prepared reports. An example of a carefully planned and well presented core log is reproduced, from a paper by Moye[46], in figure 20.

Wherever possible, standard symbols should be used for the graphical presentation of geological data[58] and it is useful to include a list of such symbols in every geological report.

The preparation of a core log or a geological report requires a certain amount of judgement on the part of the geologist and subsequent users of this information may question some of these judgements. Some of the uncertainty in geological interpretation can be eliminated if the core logs and reports are accompanied by good quality colour photographs of the core. Considering the very high cost of good quality core recovery, it is invariably worth spending a little more to provide for routine core photography before the cores are placed in storage. A rigid stand for a good 35mm single lens reflex camera can be set up in a corner of the core shed and each core tray photographed as it arrives from the field. Flood lighting or the use of an electronic flash will enable the operator to set a standard exposure on the camera and will ensure consistent results. Several trial exposures may be required to test the system and to ensure that a good colour balance is achieved in the final results. In most cases it will be found convenient to use colour reversal film, balanced to the light source used, for producing a set of colour slides as the original record. Copies of these slides or colour prints from the slides are readily available commercially. It is important that each core box should be adequately identified by means of a legible label which should appear in the photograph. It is also useful to include a colour chart in the photograph so that some compensation can be made if fading or discolouration of the slide or print occurs.

Core storage

Having spent a great deal of money on diamond drilling to recover high quality core, care should be taken that this

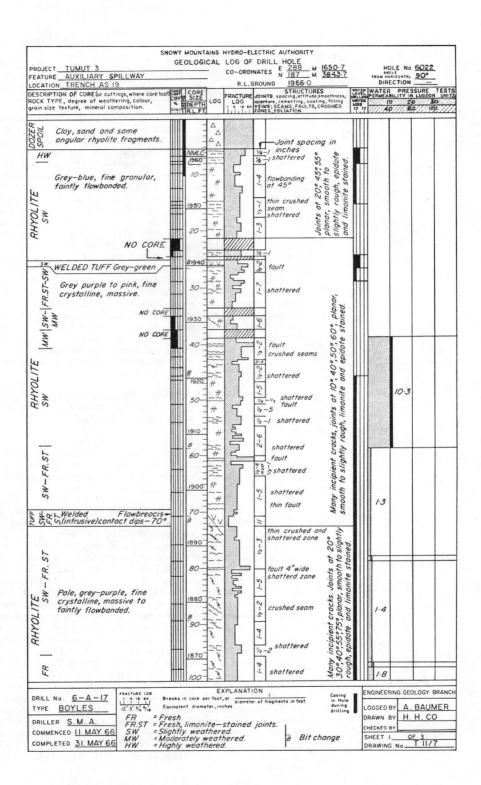

Figure 20. Example of a carefully planned and well presented core log reproduced from a paper by Moye[46] on the Snowy Mountains scheme.

Figure 21. A good example of core storage provided during the exploration of an ore body for a large underground mining operation.

core is stored in such a way that it is protected from the weather and from vandals and that it is possible to gain access to any particular core tray without major physical effort. Laying the core out for inspection is reasonable during the drilling operation, as shown in the photograph opposite, but it cannot be considered adequate for permanent core storage. The storage racks illustrated in figure 21 are a good example of the type of core storage which is considered necessary on a major project.

Exploratory adits and shafts

This chapter on geological data collection would not be complete without mention of the use of exploratory adits and shafts. While these may not be economically justified during the preliminary site investigation work, there comes a stage in the design of a major underground excavation when physical access to the rock at the excavation location becomes essential. The uncertainty of projecting geological information obtained from surface mapping and from diamond drilling is such that the choice of the optimum location and orientation and the detailed design of the support system for a large underground cavern cannot be made with confidence on the basis of this information only. Since some form of underground access will eventually be required, the excavation of an adit or a shaft is usually the best way in which to provide access to the rock mass at this stage of the project. From the discussion on rock classification systems in chapter 2, it will be clear that a small diameter shaft or tunnel can usually be driven through rock with very much less support than would be required for a large excavation and hence the cost of exploratory access can be kept low while maintaining a high rate of advance. An exception to this principle of keeping exploratory adits and shafts small occurs when a firm decision has been made to proceed with the project and when access for major construction plant and equipment is required at an early stage in the project. In such circumstances, a larger access shaft or tunnel may be excavated and used to provide geological information as well as plant access. The one disadvantage of this arrangement is that the access shaft or tunnel tends to become a very busy thoroughfare and the geologist may find it difficult to gain access to the exposed rock faces in order to carry out his mapping. The geologist's task becomes almost impossible when, in the interests of appearance and/or safety, the resident engineer insists that all exposed rock be covered as quickly as possible with pneumatically applied concrete (shotcrete or gunite).

Chapter 3 references

30. MILLER, V.C. *Photogeology*. International Series on Earth Sciences. McGraw Hill Book Company. New York, 1961, 248 p.

31. DEARMAN, W.R. and FOOKES P.G. Engineering geological mapping for civil engineering practice in the United Kingdom. *Quart. J. Engineering Geology*, Vol. 7, 1974, pages 223-256.

32. MOYE, D.G. Engineering geology for the Snowy Mountains scheme. *Journal of the Institution of Engineers, Australia*, Vol. 27, 1955, pages 287-298.

33. BLYTH, F.G.H. and DE FREITAS, M.H. *A Geology for Engineers*. (VI Edition), E. Arnold, London, 1974, 557 pages.

34. ATTEWELL, P.B. and FARMER, I.W. *Principles of Engineering Geology*. Chapman & Hall, London, 1976, 1074 pages.

35. DUNCAN, N. *Rock Mechanics and Engineering Geology*. Volumes 1 and 2, International Text Book Co, London, 1969, 560 pages.

36. LEGGET, R.F. *Geology and Engineering*. McGraw Hill Book Co., New York, 2nd Edition, 1962, 884 pages.

37. ZARUBA, Q. and MENCL, V. *Engineering Geology*. Elsevier Scientific Publishing Co., Amsterdam, 1976, 504 pages.

38. ROSS-BROWN, D.M., WICKENS, E.H and MARKLAND, J.T. Terrestial photogrammetry in open pits, part 2. An aid to geological mapping. *Trans. Inst. Mining and Metallurgy*, London, Sect. A, Vol. 83, No. 803, 1973, pages 115-130.

39. MOFFITT, F.R. *Photogrammetry*. International Textbook Co., Scranton, Pennsylvania, 1967, 540 pages.

40. MOSSMAN, R.W. and HEIM, G.E. Seismic exploration applied to underground excavation problems. *Proc. First North American Rapid Excavation and Tunnelling Conference*, AIME, New York, 1972, pages 169-192.

41. BALTOSSER, R.W. and LAWRENCE, H.W. Application of well logging techniques in metallic mineral mining. *Geophysics*, Vol. 35, 1970, pages 143-152.

42. WANTLAND, D. Geophysical measurements of rock properties in situ. *Proc. Conference on the State of Stress in the Earth's Crust*, Santa Monica, California, 1963. American Elsevier Publishing Company Inc., 1964, pages 409-448.

43. ROSENGREN, K.J. Diamond drilling for structural purposes at Mount Isa. *Industrial Diamond Review*, Vol. 30, No. 359, 1970, pages 388-395.

44. JEFFERS, J.P. Core barrels designed for maximum core recovery and drilling performance. *Proc. Australian Diamond Drilling Symposium*, Adelaide, August 1966.

45. HUGHES, M.D. Diamond drilling for rock mechanics investigations. *Rock Mechanics Symposium*. University of Sydney, Australia, 1969, pages 135-139.

46. MOYE, D.G. Diamond drilling for foundation exploration. *Civil Engineering Transactions, Institution of Engineers, Australia*, Vol. CE9, 1967, pages 95-100.

47. PHILLIPS,F.C. *The use of Stereographic Projections in Structural Geology*. Edward Arnold, London, 3rd Edition, 1971, 90 pages.

48. RAGAN,D.M. *Structural Geology - an Introduction to Geometrical Techniques*. John Wiley & Sons, New York, 2nd Edition, 1973, 220 pages.

49. KEMPE,W.F. Core orientation. *Proc. 12th Exploration Drilling Symposium*, University of Minnesota, 1967.

50. ROCHA,M. A method of integral sampling of rock masses. *Rock Mechanics*, Vol. 3, No. 1, 1967, pages 1-12.

51. BURWELL,E.B.and NESBITT,R.H. The NX borehole camera. *Trans. American Inst. Mining Engineers*, Vol.194, 1954, pages 805-808.

52. ZEMANEK, J. *et al*. The borehole Televiewer - a new logging concept for fracture location and other types of borehole inspection. *Trans. Soc. Petroleum Engineers*, Vol. 246, 1969, pages 762-774.

53. HINDS,D.V. A method of taking an impression of a borehole wall. *Imperial College Rock Mechanics Research Report* Number 28, November 1974, 10 pages.

54. BARR,M.V. and HOCKING,G. Borehole structural logging employing a pneumatically inflatable impression packer. *Proc. Symp. Exploration for Rock Engineering*, Johannesburg, 1976, Published by A.A.Balkema, Rotterdam, 1977, pages 29-34.

55. BROCH,E. and FRANKLIN,J.A. The point load strength test. *Intnl. J. Rock Mechanics and Mining Sciences*, Vol. 9, 1972, pages 669-697.

56. BIENIAWSKI,Z.T. The point load test in geotechnical practice. *Engineering Geology*, Vol. 9, 1975, pages 1-11.

57. FRANKLIN,J.A. and CHANDRA,R. The slake durability test. *Intnl. J. Rock Mechanics and Mining Sciences*, Vol. 9, 1972, pages 325-341.

58. ANON. Graphical symbols for use on detailed maps, plans and geological cross-sections. Part 1. General rules of representation. 1968. Part 11. Representation of sedimentary rocks. 1968. Part 111. Representation of magmatic rocks. 1970. *International Organisation for Standardisation*. Documents ISO/R 710/I - 1968(E), 3-4., ISO/R 710/11 - 1968(E) 5-15. and ISO/R 710/III - 1970(E), 5 - 11.

Chapter 4 : Graphical presentation of geological data

Introduction

The effective utilisation of geological data by an engineer depends upon that engineer's ability to comprehend the data, to digest them and to incorporate them into his design. The communication between geologists and engineers is particularly important when the stability of the rock mass surrounding an underground excavation is likely to be controlled by through-going structural features such as faults or well developed joints. In such cases, the three-dimensional geometrical relationship between structural features and the roof and walls of the excavation is very important since this relationship will control the freedom of blocks to fall or slide.

Most geologists are familiar with the use of spherical projections for the presentation and analysis of structural geology data but many engineering readers may not be familiar with this technique. In order to assist such readers, the principles and uses of stereographic projections are reviewed in this chapter. In addition, a method for the construction of isometric views of structural features is presented.

(Equal area and equal angle projections)

Figure 22 shows a sphere with one quarter removed and with meridional and polar nets projected onto the exposed vertical and horizontal faces. There are two types of projection which are used to generate the meridional and polar nets and these are the *equal area* and the *equal angle* projections. These projections are described below.

The equal area projection, also known as the *Lambert* or *Schmidt* projection, is generated by the method shown in the upper margin sketch. A point A on the surface of the sphere is projected to point B by swinging it in an arc centred at the point of contact of the sphere and a horizontal surface upon which it stands. If this process is repeated for a number of points, defined by the intersection of equally spaced longitude and latitude circles on the surface of the sphere, an equal area net will be generated. This net has a larger diameter than the sphere and, in order to reduce its diameter to that of the sphere, the radius of each point on the net is reduced by $1/\sqrt{2}$.

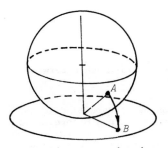

Equal area projection

The equal angle projection, also known as a *Stereographic* or *Wulff* projection, is obtained by the method illustrated in the lower margin sketch. The projection C of a point A on the surface of the sphere is defined by the point at which the horizontal plane passing through the centre of the sphere is pierced by a line from A to the zenith of the sphere. The zenith is the point at which the sphere is pierced by its vertical axis.

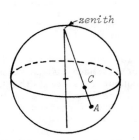

Equal angle projection

Both types of projection are used for the analysis of structural geology data. In general, the equal area projection is preferred by geologists because, as the name implies, the net is divided into units of equal area and this permits the statistical interpretation of structural data. (Engineers tend to prefer the equal angle projection because geometrical constructions required for the solution of engineering problems are simpler and more accurate on this projection than for the equal area projection.) The authors have conducted extensive trials on the speed, convenience and accuracy of

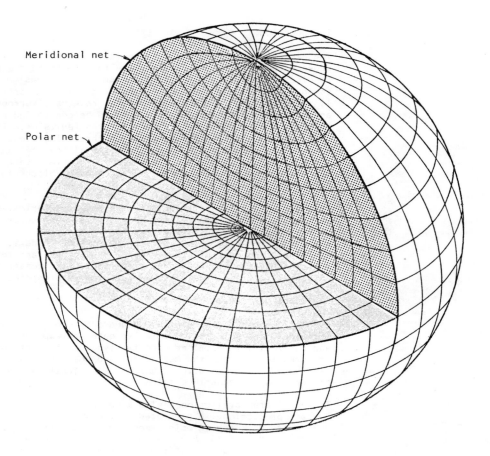

Figure 22 : Sectioned isometric view of a sphere showing the relationship between the meridional and polar nets.

both statistical interpretation of structural data and geometrical construction using both types of projection and have convinced themselves that, for the applications discussed in this book, there is no practical advantage to be gained by choosing one type of projection in preference to the other. The advantages and disadvantages associated with using either of these projections balance out when the nets are used for the total solution of a problem rather than for an analysis of part of that problem.

Since the equal area projection was used exclusively in *Rock Slope Engineering* [2], the authors have chosen to use the equal angle projection throughout this book. The techniques of using these projections are identical and the reader will have no difficulty in converting from one system to the other. The only warning which must be issued is that the same type of projection *must* be used throughout a particular analysis. A total shambles would result from an attempt to analyse data originally plotted on an equal area net by means of an equal angle net or vice versa. In order to avoid such embarrassment, it is advisable to note the type of projection used on all diagrams.

Stereographic projection of a plane and its pole

Imagine a sphere which is free to move in space so that it can be centred on an inclined plane as illustrated in figure 23. The intersection of the plane and the surface of the sphere is a *great circle* which is shaded in the figure. A line, passing through the centre of the sphere in a direction perpendicular to the plane, pierces the sphere at two diametrically opposite points which are called the *poles* of the great circle representing the plane.

Because the same information appears on both the upper and the lower parts of the sphere, only one hemisphere need be used for the presentation of structural geology information. In engineering geology, the *lower reference hemisphere* is usually used and this convention will be followed throughout this book.

Figure 24 shows the method of construction of the stereographic projection of a great circle and its pole and figure 25 shows the appearance of these projections. The inclination and orientation of an inclined plane are defined uniquely by either the great circle or the pole of that plane. As will be shown later in this chapter, poles are usually plotted when collecting geological data in the field and the corresponding great circles are normally used when analysing these data for engineering purposes.

Definition of geological terms

An inclined geological plane is defined by its inclination to the horizontal or *dip* and by its orientation with respect to north which may be defined by the *strike* or by the *dip direction* of the plane. The relationship between these terms is illustrated in the margin sketch.

The strike of a plane is the trace of the intersection of that plane and a horizontal surface and it is used by most geologists to define the orientation of a plane. In order to eliminate any possible ambiguity when using strike, it is necessary to define the direction in which a plane dips. Hence, a plane is fully defined if it is recorded as having a strike of N 30 W and dip of 20 SW. On the other hand, if it were recorded as having a dip of 20°, it would not be clear whether this dip was towards the south-west or the north-east. Several conventions are used by geologists to eliminate this problem when using dip and strike and the authors would not presume to offer an opinion upon which of these conventions is best. The geologist should use that convention with which he is most familiar but he should take care that he includes sufficient information on his records and logs to ensure that anyone else working with his data knows what convention has been used.

Geotechnical engineers, particularly those who make extensive use of computers in their analyses, have tended to use dip direction in preference to strike as a means of defining the orientation of planes. If the dip direction and dip of a plane are recorded as 240/20, there can be no confusion on the orientation and inclination of that plane and this notation is more concise than that for strike and dip - an important consideration when processing large quantities of geological data by computer.

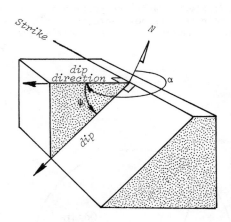

Definition of geological terms

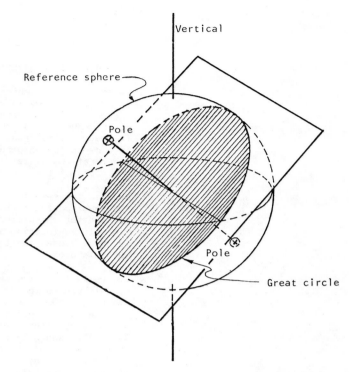

Figure 23 : Great circle and its poles which define the inclination and orientation of an inclined plane.

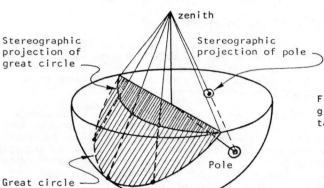

Figure 24: Stereographic projection of a great circle and its pole onto the horizontal plane of the lower reference hemisphere.

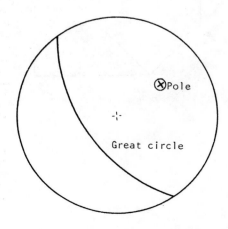

Figure 25 : Stereographic projection of a great circle and its pole.

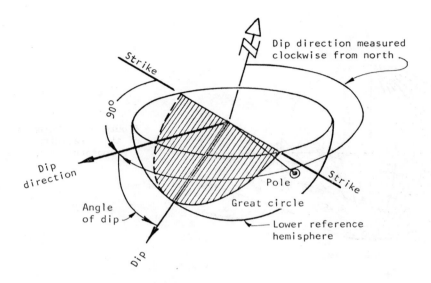

Figure 26 : Definition of terms used in conjunction with the lower reference hemisphere stereographic projection.

Figure 26 shows the dip, dip direction and strike conventions used in conjunction with the lower reference hemisphere stereographic projection. Note that dip direction is always measured clockwise from north and that the strike line is at 90° to the dip direction of a plane.

Wherever possible, in this book, both strike and dip direction will be used when presenting the basic data on sample problems.

Construction of stereographic nets

The construction of great and small circles on a meridional stereographic net is illustrated in figure 27. The centres of the great circle arcs are defined by the intersection of the east-west line through the centre of the net and the various chords shown in the figure. The centres of the small circles are defined by the intersections of the north-south line through the centre of the circle and the various tangents to this circle.

The relationship between the polar and the meridional nets can be deduced from figure 22.

High quality computer drawn meridional and polar stereographic nets are reproduced in figures 28 and 29. Photographic or machine copies of these nets can be used for the plotting and analysis of structural data, as discussed later in this chapter. Note that some photocopy machines introduce significant distortion in prints and care should be taken to ensure that any copies used for geological data analysis are free of such distortion.

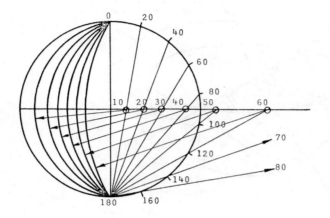

Figure 27a : Method of construction of the great circles on a meridional stereographic net.

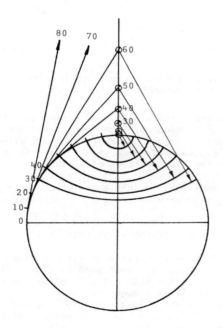

Figure 27b : Method of construction of the small circles on a meridional stereographic net.

(Both construction methods after Ragan[48])

A counting net for use in conjunction with a polar stereographic net is reproduced in figure 30. For practical applications, the most convenient form in which this counting net can be used is as a transparent overlay and the reader is advised to have a number of such overlays prepared. Once again, care should be taken to ensure that the copies are not distorted. The use of this counting net for the analysis of structural data is discussed later in this chapter.

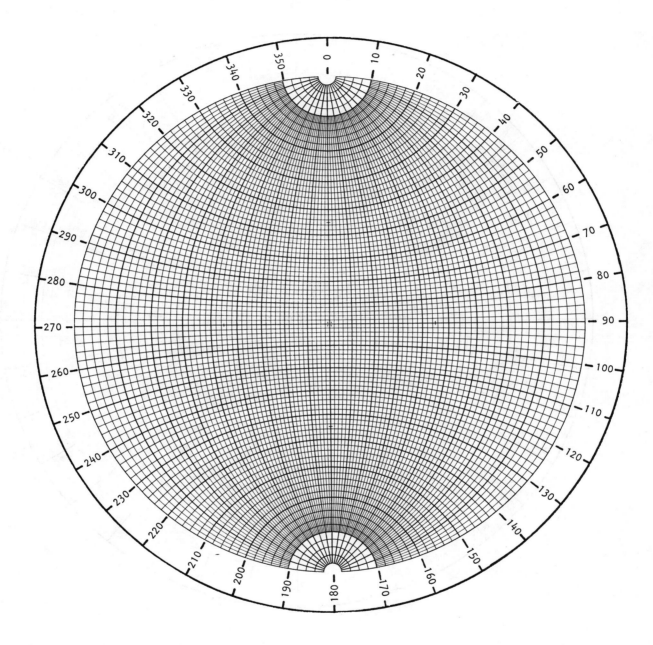

Figure 28 : Meridional stereographic net for the analysis of structural geology data.

Computer drawn by Dr. C.M.St John of the Royal School of Mines, Imperial College, London.

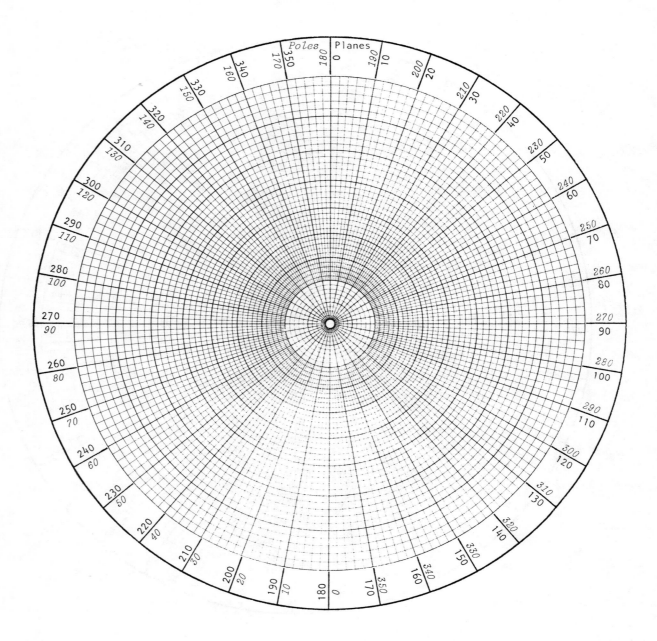

Figure 29 : Polar stereographic net on which the poles of geological planes can be plotted.

Computer drawn by Dr. C.M.St John of the Royal School of Mines, Imperial College, London.

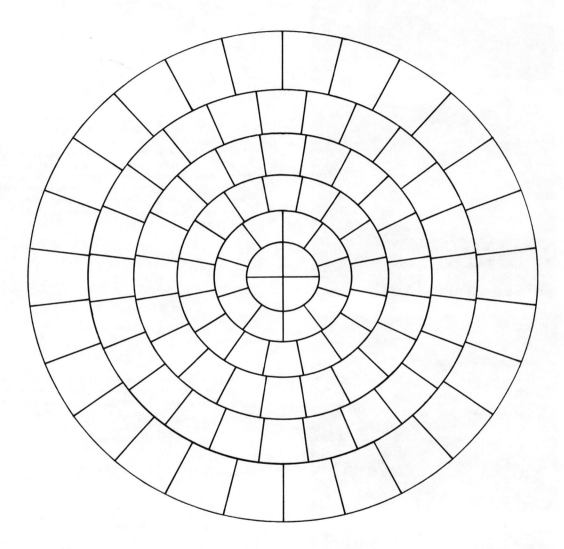

Figure 30 : Counting net for use in conjunction with the polar stereographic net given in figure 29.

Reproduced with permission of Pierre Londe, Coyne & Bellier, Paris.

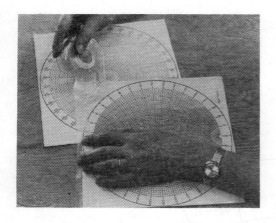

Figure 31a: Meridional and polar stereographic nets can be mounted on either side of a piece of hardboard or plywood by means of transparent adhesive tape. Transparent plastic sheet coverings will help to protect the net for field application.

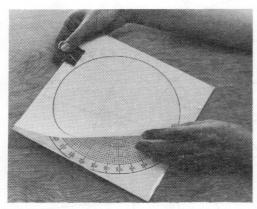

Figure 31b: A piece of tracing paper is placed over the polar net and attached to the board by means of a spring clip. A stock of tracing paper on which circles corresponding to the outer circumference of the net have been marked will be found useful for field work.

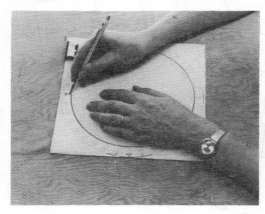

Figure 31c: When the tracing paper has been located over the polar net, the centre of the circle and the north point are marked.

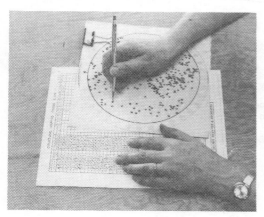

Firgure 31d: The poles of planes are plotted onto the tracing paper from field logs. Different symbols should be used to represent different types of structural features.

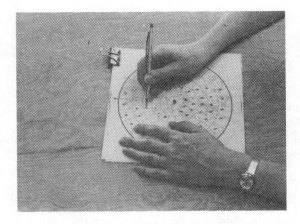

Figure 31e: A transparent copy of the counting net reproduced in figure 30 is placed over the pole plot and a piece of tracing paper placed over the counting net. The counting net is free to rotate about the centre pin but the tracing paper is fixed with respect to the pole plot. The number of poles falling in each 1% area square is counted and noted on the tracing paper. Rotation of the counting net allows maximum pole concentrations to be located more accurately than using a fixed counting net.

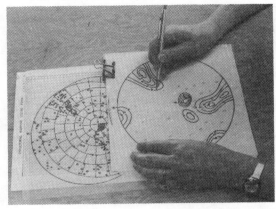

Figure 31f: On a fresh piece of tracing paper, contours of equal pole concentration are constructed by joining equal pole counts. The north point and the centre of the net must be marked on this tracing.

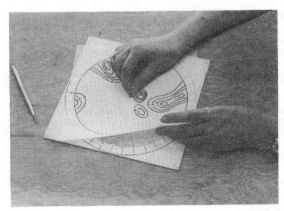

Figure 31g: The piece of tracing paper on which the pole concentration contours have been drawn is transferred onto the meridional stereographic net and it is located on the net by means of a centre pin.

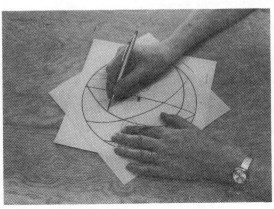

Figure 31h: Great circles representing the most important structural features are constructed by rotating the tracing around the centre pin until the pole of each plane, represented by the centre of each pole concentration, falls on the east west axis of the net. The corresponding great circle is traced as shown.

Construction of a great circle to represent a plane

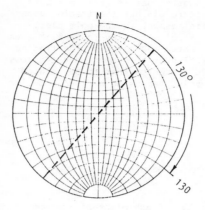

Consider a plane defined by a dip direction of 130° and a dip of 50°. This may be written 130/50 in a field note book or on a drawing. Alternatively, the plane is defined by a strike of N 40 E and a dip of 50 SE. The great circle representing this plane is constructed as follows :

Locate a piece of tracing paper over the meridional net by means of a centre pin. Mark the north point and the centre of the net on the tracing paper. When a number of stereographic analyses are to be carried out, sheets of tracing paper on which the outer circumference of the net, the centre point and the north point have already been marked will be useful.

Measure off 130° clockwise from north around the circumference of the net and mark this point on the tracing paper. Alternatively, measure off 40° and mark in the strike line, shown dashed in the upper margin sketch.

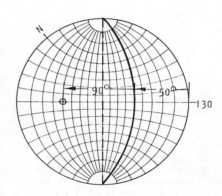

Rotate the tracing paper about the centre pin by 40° until the 130° mark lies on the east-west axis of the net, i.e. until this mark on the tracing paper coincides with the 90° mark on the net. Count 50 degree divisions along the east-west axis, starting from the outer circumference of the net, and trace the great circle at this position.

The pole representing the plane is located by counting a further 90 degree divisions along the east-west axis, while the 130° mark on the tracing paper is still aligned with this axis.

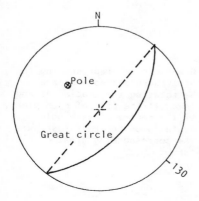

Remove the tracing paper from the net and rotate it so that the north point is again vertical. The final appearance of the stereographic projection of the great circle and its pole is illustrated in the lower margin sketch.

Determination of the line of intersection of two planes

The plane defined by a dip direction and dip of 130/50 (or strike and dip of N 40 E and 50 SE) intersects a plane which is defined by a dip direction and dip of 250/30 (strike and dip of N 20 W and 30 SW). It is required to find the *plunge* and the *trend* of the line of intersection of these two planes. For clarity of presentation, only dip direction markings will be shown on the construction.

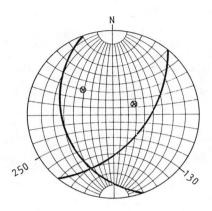

Relocate the tracing paper on the net by means of the centre pin and measure off 250° clockwise from north. Rotate the tracing paper through 20° until the 250° mark on the tracing paper coincides with the 270° mark on the net.

Count off 30 degree divisions, starting from the 270° mark on the net and counting inwards towards the centre of the net. Trace the great circle which occurs at this position. Count a further 90 degree divisions along the west-east axis and mark the pole position for the second plane.

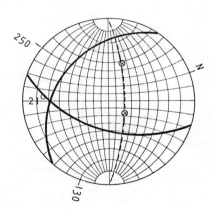

Rotate the tracing until the intersection of the two great circles, which defines the line of intersection of the two planes, lies on the west-east axis of the net. It will be found that the north point on the tracing paper now lies at the 70° position on the net.

The plunge of the line of intersection is found to be 21° by counting the number of degree divisions from the outer circumference of the net to the great circle intersection. This counting is from the 270° mark on the circumference of the net inwards along the west-east axis towards the intersection point.

Note that, with the tracing in this position, the poles of the two planes lie on the same great circle. This fact provides an alternative means of locating the line of intersection of two planes since this is given by the pole of the great circle passing through the poles of the two planes.

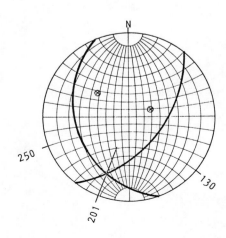

The tracing is now rotated until the north point on the tracing paper coincides with the north point on the net. The trend of the line of intersection is found to be 201°, measured clockwise from north.

Relationship between true and apparent dip

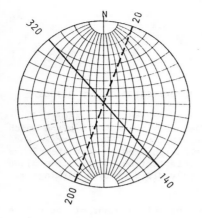

Consider a square tunnel in which the vertical sidewalls trend in a direction from 320° to 140°. The apparent dip of a joint plane which intersects the vertical sidewall, defined by the trace of the joint on the sidewall, is 40 SE. The same joint plane may be seen in the horizontal roof of the tunnel and its strike is measured as N 20 E. It is required to find the true dip and the dip direction of this joint plane.

Mark the direction of the tunnel sidewall, from 320° to 140°, on a piece of tracing paper located over the meridional net by means of a centre pin. Also mark the strike of the joint on the horizontal roof by a line running from 200° to 20°, passing through the centre of the net. Ensure that the north point is marked on the tracing.

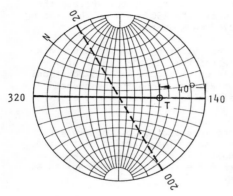

Rotate the tracing so that the 320° to 140° line lies along the west-east axis of the net, i.e. the tracing is rotated anti-clockwise through 50° until the 140° mark on the tracing coincides with the 90° mark on the net.

Measure off 40 degree divisions, starting on the outer circumference of the net and measuring inwards towards the centre of the net from the 90° mark. The point marked T defines the apparent dip of the joint plane in the direction of 140°. Note that this apparent dip is defined by the line of intersection of the joint plane with the vertical sidewall surface and that, provided the joint plane does not curve in the span of the tunnel, this apparent dip will be identical on both sidewalls.

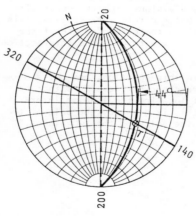

The tracing is now rotated so that the 200° to 20° line, defining the trace of the joint plane on the horizontal tunnel roof, lies along the vertical north-south axis of the net. Since the trace of a plane on a horizontal surface defines the strike of that plane (see figure 26 on page 65), a great circle can now be drawn to represent the plane. Since the point T, representing the apparent dip of the joint plane, must also lie on this great circle, the position of the great circle is defined as shown on the drawing opposite.

The true dip of the plane, which is at right angles to the strike line, is found to be 44°.

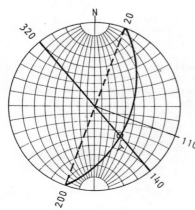

The tracing is now rotated so that the north point on the tracing coincides with the north point on the net and the dip direction of the joint plane is shown to be 110°.

Plotting and analysis of field measurements

In plotting field measurements of dip direction and dip or strike and dip, it is convenient to work with poles of planes rather than great circles since the poles can be plotted directly on a polar net such as that reproduced in figure 29. Suppose that a plane is defined by dip direction and dip values of 050/60 (strike and dip of N 40 W and 60 NE). Its pole is located on the stereonet as follows : Use the dip direction value of 050 given in *italics* on the circumferential scale around the polar net to locate the direction of the pole (dip direction of plane ± 180º). Measure the dip value of 60º *from the centre of the net* along the radial line defined by *50* and mark the pole position.

Note that no rotation of the tracing paper, centred over the polar stereonet, is required for this operation and, with a little practice, pole plotting can be carried out very quickly. There is a temptation to plot compass readings directly onto the polar net, without the intermediate step of entering the measurements into a field note book. The authors would advise against this short-cut because these measurements may be required for other purposes, such as a computer analysis, and it is a great deal easier to work from recorded numbers than from the pole plot.

Figure 32 shows a plot of 351 poles which have been plotted directly onto a polar stereographic net from a set of field data. It will be noted that different symbols have been used to represent different types of geological features. This is particularly important when these features have different characteristics and when it is important to isolate certain families or even individual features. For example, in the case of the data plotted in figure 32, the stability of an underground excavation may be controlled by the single fault and it is clearly important to differentiate between this one feature and the remaining 350 poles. In the case of the data plotted in figure 32, the characteristics of the bedding planes and the joints were very similar and their poles are treated as a single population in the analysis which follows.

The counting net presented in figure 30 was devised by Coyne and Bellier, consulting engineers in Paris, and contains 100 "squares" which represent 100 equal areas on the sphere. The derivation of this net is analogous to that used by Denness[59,60] for his curvilinear cell counting net for use with equal area projections and the interested reader is referred to these papers for further details. Obviously, because of the different types of projection, explained on page 61, the counting cells for use with the stereographic projection are significantly different from those used by Deness for equal area projections.

The most convenient method for using the counting net is to have a transparent overlay prepared and to centre this overlay over the pole plot by means of a pin through the centre of the net. A piece of tracing paper is mounted on top of the overlay, pierced by the centre pin but fixed by a piece of adhesive tape so that it cannot rotate with respect to the pole plot. Hence one has a sandwich in which the transparent counting net can rotate freely between the pole plot and the piece of tracing paper which are fixed together.

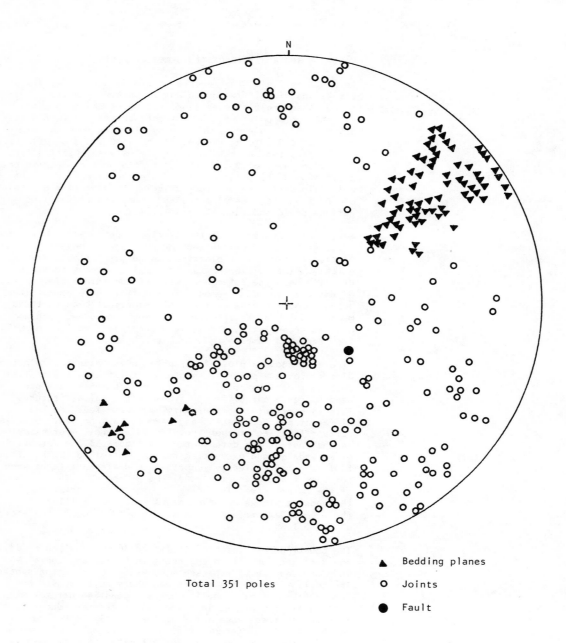

Figure 32 : Plot of 351 poles representing geological planes in a hard rock mass.

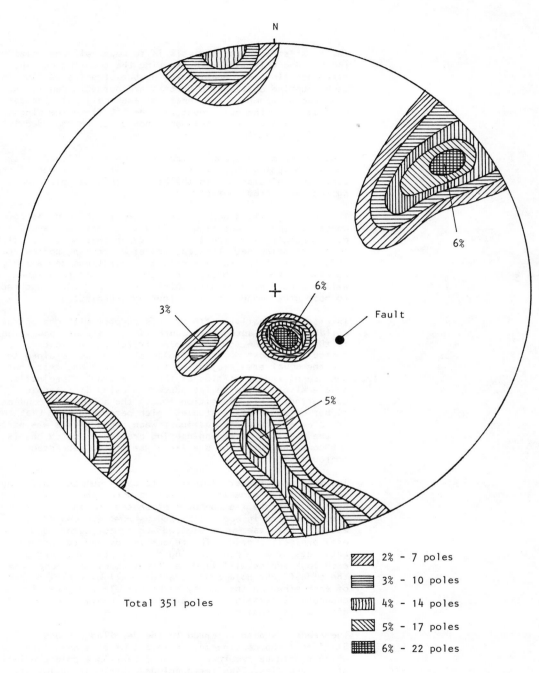

Figure 33 : Contours of pole concentrations determined from the pole plot given in figure 32.

The first step in the analysis is to count all the poles on the net and this is done by keeping the counting net in a fixed position and counting the number of poles falling in each counting cell. These numbers are noted in pencil on the tracing paper at the centre of each cell. In the case of figure 32, the pole count will be 350 since the single fault is treated separately and should not be included in the pole population.

Once the total pole population has been established, the numbers of poles which make up different percentages of the total are calculated . In the case of this example, these numbers are noted in figure 33.

The counting net is now rotated to centre the densest pole concentrations in counting cells and the maximum percentage pole concentrations are located. By further small rotations of the counting net, the contours of decreasing percentage which surround the maximum pole concentrations can quickly be established. With practice, this counting technique will produce rapid results which are of comparable accuracy to most other manual pole contouring techniques.

In discussing early drafts of this chapter with geological colleagues, the authors encountered very strong opposition to the use of the stereographic projection for pole contouring. Most of these geologists had been trained to use the equal area projection for this task and felt that, because of the distortion of the stereographic projection, severe errors would be introduced in using it for pole contouring. This opposition forced the authors to conduct a number of comparative studies which persuaded them that these fears were groundless, at least when the analysis was part of the solution of an engineering problem in which one is not concerned with the precise shape of low percentage contours.

The sceptical reader is invited to carry out the following comparison for himself. The data plotted in figure 32 and used to generate the contours presented in figure 33 were also presented in *Rock Slope Engineering* (Second edition)[2] but, in that case, were plotted and contoured on equal area projection grids. The reader is invited to contour both sets of pole plots, using the methods recommended in each book and he will find, as the authors have done, that the significant pole concentrations fall within ±1 degree of each other on the two types of net. This degree of accuracy is certainly adequate for the analysis of any engineering problem.

The reader is also referred to the detailed discussion by Stauffer[61] who considered the levels of pole concentration which should be regarded as significant in a pole population of a given size. Stauffer concluded that most geologists attempt to contour pole populations which are far too small and he offered a set of guidelines on the choice of pole population size. These guidelines are too detailed to justify inclusion in this discussion but, briefly, Stauffer suggests that contouring should not be attempted on pole populations of less than 100 and that, for very weak preferred orientations, as many as 1000 poles may be required to give a reliable result. A full discussion of Stauffer's recommendations is given in *Rock Slope Engineering*[2].

Computer processing of structural data

Consulting organisations and individuals involved in the processing of large volumes of structural geology data have turned to the computer as an aid. Details of some of these computer techniques have been published by Spencer and Clabaugh[62], Lam[63], Attewell and Woodman[64] and Mahtab et al[65]. Many of these techniques utilise the coordinates of the pole on the surface of the sphere and so eliminate the distortion which is inherent in any of the projections which reduce this spherical surface to a two dimensional plane. The reader who is likely to become involved in a large amount of structural geology processing is advised to explore the possibility of using these computer techniques in his own work.

Sources of error in structural data collection

Before leaving the subject of structural geology interpretation, two common sources of error are worthy of brief mention.

A frequent source of error in joint surveys is the inclusion on the same plot of poles from different structural domains. Hence, in mapping a tunnel, a geologist may pass from one set of geological conditions into another. Working under poor visibility conditions, this transition may be missed unless the geologist has carried out a preliminary reconnaissance to establish the limits of each structural domain. It is very important that only those poles representing geological features within one domain should be plotted on the same stereonet.

A second source of error lies in the direction of the face being mapped relative to the orientation of the structural features in the rock mass. If mapping is confined to a single adit, a major feature running parallel to the adit may never be detected until it appears unexpectedly in the face of an excavation which is larger than the adit. This problem of joint sampling was discussed by Terzaghi[66] who suggested a method of correcting for the error by weighting joint measurements in favour of those almost parallel to the direction of the exposure on which measurements are made. The authors consider that the Terzaghi correction is appropriate for joint measurements on borehole core and when the only access underground is a single straight tunnel with smooth walls. A preferred method for minimising this error is to carry out the mapping in tunnels driven in different directions or to supplement the tunnel mapping with boreholes drilled at right angles to the tunnel direction. In this way, most of the structures which exist in the rock mass will be exposed and the danger of encountering unexpected features will be minimised.

Isometric drawings of structural planes

Many engineers find it extremely difficult to visualise structural features when these are presented in the form of great circles or poles on a stereographic projection. Frequently, important points are missed by these engineers when working on the design of an underground excavation because the geologist has failed to present his data in a form which can be understood by the engineer. This problem can become acute when the data are being reviewed by a non-

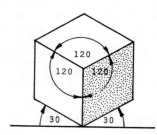

Isometric projection of a cube

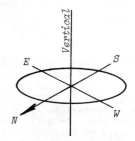

Direction convention for isometric projection

technical panel which may happen if the design becomes the subject of a legal dispute as a result of an accident or a contractual claim.

In order to overcome this communication problem, the authors have sometimes used isometric drawings of intersecting planes in an effort to present structural geology information in a relatively familiar pictorial form. A simple technique for constructing such isometric drawings is presented below.

The margin sketch shows an isometric view of a cube and it will be noted that the three visible faces are equally oblique to the line of sight. In this projection, none of the edges of the cube are true lengths but all are equally reduced by a factor of 0.8165. Graph paper with isometric projection markings is readily available commercially. The second margin sketch shows the direction convention which has been adopted for the presentation of isometric projections of structural planes.

Figure 34a shows an isometric view of a set of planes dipping towards the west at various angles and striking north-south (dip direction 270°). These planes are all identical in size, being square in a true plan view, and it will be noted that their corners generate an ellipse as they are rotated about the strike line in an isometric drawing. This fact has been used in the construction of a simple set of figures which can be used to generate isometric views of planes and which are presented in Appendix 2 at the end of this book. Figure 34b gives a superimposed view of the planes shown in figure 34a and the construction figure shown in figure 34c.

The construction figure for a particular dip direction value, in this case 270°, consists of a strike line (shown with starred ends) and a set of dip lines at 10° dip increments. These dip lines represent a line marked through the centre of the square plane, parallel to one of its edges. The outer circle in the construction figure serves no purpose other than to form a frame for the drawing. An essential feature of the construction figure is the vertical line marked on each figure.

The use of the construction figure is best illustrated by means of a practical example. It is required to construct an isometric view of a square plane defined by dip direction and dip values of 270/50 (strike N-S and dip 50 W). Using the 270° dip direction construction figure, trace the strike line and the line representing the 50° dip , as shown in figure 35a. Mark the vertical line on the tracing paper. Using a parallel rule, draw the edges of the square by drawing lines parallel to the strike and dip direction lines at the ends of these lines, as shown in figure 35b. The final appearance of the 270/50 plane in isometric projection is shown in figure 35c.

Figure 36a gives an isometric view of three planes defined by dip direction and dip values of 010/90, 270/50 and 120/70. The intersection lines in this figure are fairly obvious in the construction but, in more complex cases, it may be difficult to visualise the intersection lines. In such cases, the plunge and trend of the lines of intersection can be determined on a stereographic projection , as described on page 73, and these values used to locate a construction

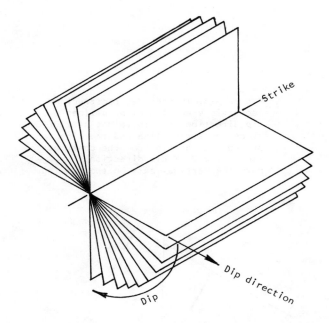

Figure 34a : Isometric view of a set of planes striking north-south and dipping at various angles between 0 and 90°.

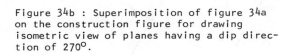

Figure 34b : Superimposition of figure 34a on the construction figure for drawing isometric view of planes having a dip direction of 270°.

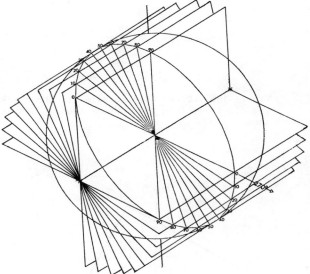

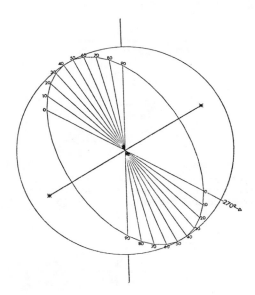

Figure 34c : Construction figure for isometric drawings of planes with a dip direction of 270°.

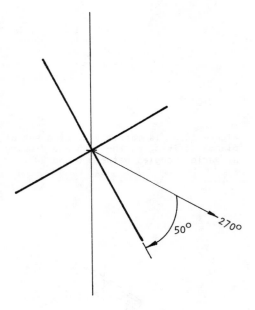

Figure 35a: In order to construct an isometric view of a plane defined by dip direction and dip values of 270/50, trace the strike line and the line representing the dip from the construction figure for 270° dip direction.
Mark the vertical axis on the tracing.

Figure 35b : Draw lines parallel to the strike and dip lines from the ends of these lines as shown.

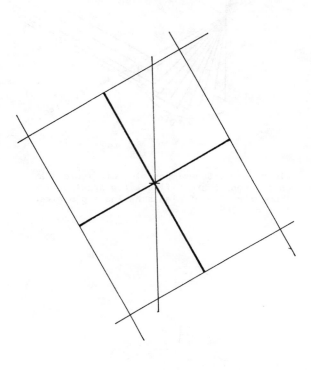

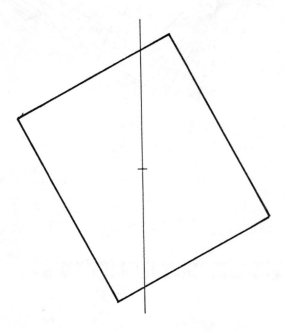

Figure 35c : Final appearance of an isometric view of a plane defined by dip direction and dip values of 270/50

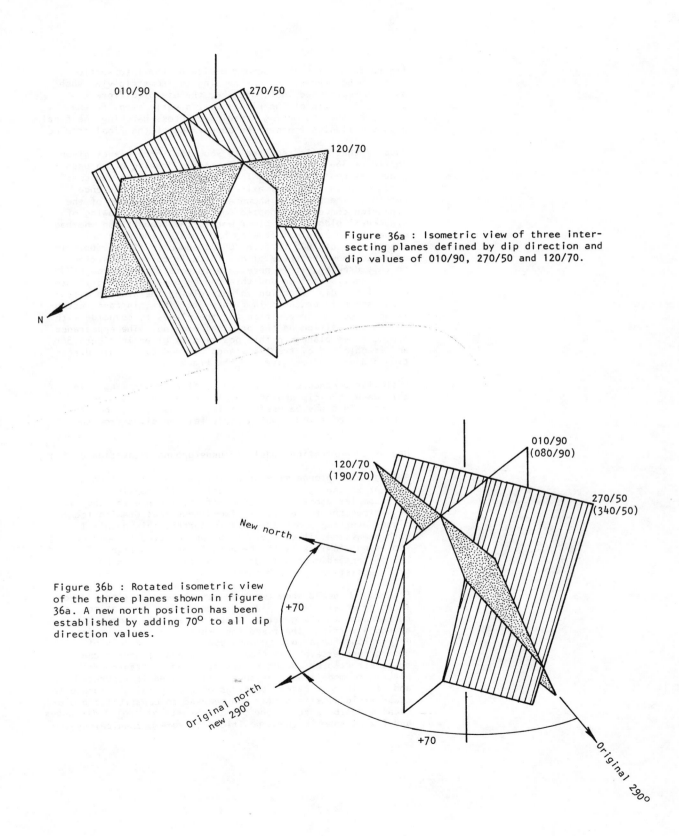

Figure 36a : Isometric view of three intersecting planes defined by dip direction and dip values of 010/90, 270/50 and 120/70.

Figure 36b : Rotated isometric view of the three planes shown in figure 36a. A new north position has been established by adding 70° to all dip direction values.

figure from which the isometric view of the intersection line can be traced. In constructing an isometric view such as that reproduced in figure 36a, the use of different colours for each of the planes will greatly simplify the construction. Colouring, shading or cross-hatching the final planes will also improve the appearance of the final drawing.

Suppose that it is required to show the three planes given in figure 36a on an existing isometric drawing of an underground excavation design. This isometric drawing was constructed using a tunnel axis trend of 290° as reference but this direction was shown as the north position of the direction convention adopted for the isometric drawing of structural planes. In other words, the draughtsman who had prepared the original drawing had rotated the original tunnel axis at 290° through 70° as shown in figure 36b. In order to show the three planes in their correct positions on this drawing, it is necessary to rotate them through 70° in the same direction and this can be done by adding 70° to each of the dip direction values, giving the new dip directions shown in brackets in figure 36b. This, in fact, establishes a new north position for the planes to coincide with the north position on the original drawing. The appearance of the three planes in this position is given in figure 36b and is obtained by tracing planes defined by the dip direction and dip values given in the brackets.

A similar procedure can be used to tilt the drawing . In this case, the dip angles are all increased by a fixed amount. This may be useful if it is required to show the appearance of a wedge which could fall or slide from the roof of an excavation.

Use of demonstration models in underground excavation design

A large underground mine or a civil engineering structure such as an underground power-house usually consists of a very complex geometrical layout of inter-connected cavities of various shapes and sizes. Two-dimensional drawing techniques are simply not adequate to portray this detail in such a way that it can quickly be appreciated by someone who is not familiar with the details of the design. The use of three-dimensional demonstration models can be of great assistance in such cases.

Figures 37 and 38 show two types of three-dimensional demonstration model used in planning underground excavations, one concerned with the general layout of service excavations around a vertical shaft and the other with the intersecting excavations for an underground power plant. Models of underground excavations showing structural features can be constructed by adding sheets of coloured transparent plastic to models such as those illustrated in figures 37 and 38. In some cases, tracings of geological sections onto clear rigid plastic sheets can be used to construct a geological model by spacing the plastic sheets at scaled distances and fixing them in these positions by some mechanical system.

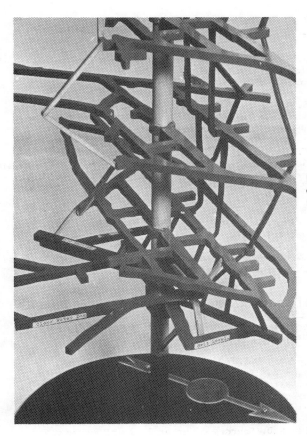

Figure 37 : Three-dimensional plastic model of the service excavations around a vertical shaft in a large underground mine.

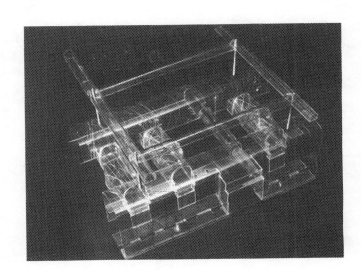

Figure 38 : Plastic model of the excavation layout for a large underground hydroelectric power project. Note that projected structural features have been drawn on the model.

(Model by M. Mac Fadyen)

Chapter 4 references

59. DENNESS, B. A method of contouring stereograms using variable curvilinear cells. *Geological Magazine*, Vol. 107, No. 1, 1970, pages 61-65.

60. DENNESS, B. A revised method of contouring stereograms using variable curvilinear cells. *Geological Magazine*, Vol. 109, No. 2, 1972, pages 157-163.

61. STAUFFER, M.R. An empirical-statistical study of three-dimensional fabric diagrams as used in structural analysis. *Canadian Journal of Earth Sciences*, Vol. 3, 1966, pages 473-498.

62. SPENCER, A.B. and CLABAUGH, P.S. Computer programs for fabric diagrams. *American Journal of Science*, Vol. 265, 1967, pages 166-172.

63. LAM, P.W.H. Computer methods for plotting beta diagrams. *American Journal of Science*, Vol. 267, 1969, pages 1114-1117.

64. ATTEWELL, P.B. and WOODMAN, J.P. Stability of discontinuous rock masses under polyaxial stress systems. *Proc. 13th Symposium on Rock Mechanics*, E.J.Cording, ed., ASCE, New York, 1972, pages 665-683.

65. MAHTAB, M.A., BOLSTAD, D.D., ALLDREDGE, J.R. and SHANLEY, R.J. Analysis of fracture orientations for input to structural models of discontinuous rock. *U.S. Bureau of Mines Report of Investigations*, No. 7669, 1972.

66. TERZAGHI, R.D. Sources of error in joint surveys. *Géotechnique*, Vol. 15, 1965, pages 287-304.

Chapter 5: Stresses around underground excavations

Introduction

The stresses which exist in an undisturbed rock mass are related to the weight of the overlying strata and the geological history of the rock mass. This stress field is disturbed by the creation of an underground excavation and, in some cases, this disturbance induces stresses which are high enough to exceed the strength of the rock. In these cases, failure of the rock adjacent to the excavation boundary can lead to instability which may take the form of gradual closure of the excavation, roof falls and slabbing of sidewalls or, in extreme cases, rockbursts. Rockbursts are explosive rock failures which can occur when strong brittle rock is subjected to high stress.

The various failure processes that can occur around underground excavations and the remedial measures that can be taken to improve the excavation stability will be discussed later in this book. In order to understand the mechanics of stress-induced instability and the measures required to control this instability, it is necessary to understand some of the basic concepts of stress and strength. This chapter deals with stresses around underground excavations while Chapter 6 is concerned with failure of rock materials and rock masses.

The subject of stress in solid bodies has been dealt with in great detail in numerous books and papers and the enthusiast could fill a modest library with publications on this subject. A few of the books which the authors have found to be useful in their own studies are listed as references 67-73 at the end of this chapter. It would be impossible to condense this material into a single chapter and, hence, the treatment presented in this book is very much simplified and is intended to be introductory only. The discussion is restricted to the use of elastic theory which means in practical terms that it is limited to use in excavation design in hard rock. The reader who is concerned with excavations in weak ground and in materials such as salt or potash which have time-dependent properties, is advised to consult publications which deal specifically with the behaviour of these types of materials.

At the outset it is important that the term elastic be clearly defined. In its most general sense, the term is used to describe materials in which the work done on the body is fully recoverable when the forces or stresses causing deformation are removed. In this book, as in many others, an elastic rock will be taken as one in which the strains are not only fully recoverable, but are also directly proportional or linearly related to the stresses causing them. The relationships between stress and strain will be pursued later in this chapter following a discussion of the fundamental concept of stress in a solid.

Components of stress*

Surface forces

Stresses are defined in terms of the forces acting at a point or on a surface. Consider the forces acting on an

---

* The form of presentation used here was suggested to the authors by Dr. J.W. Bray.

inclined surface within a rock mass. This surface may be
(i) an external surface forming part of the boundary of a
structure, (ii) an internal structural feature such as a
joint or fault, or (iii) an imaginary internal surface. In
general, the distribution of forces over this surface will
vary, and it is therefore convenient to consider the forces
applied to a small rectangular element of the surface as
shown in the margin sketch. Axes l and m are set up parallel
to the sides of the element, and axis n in the direction of
the normal to the surface.

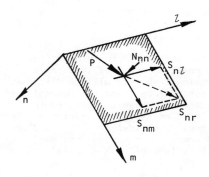

Let P be the total force applied to the surface element. In
general this force will be inclined to the surface, but it
can be resolved into (i) a normal component N_{nn}, acting
perpendicular to the surface, and (ii) a shear component S_{nr},
acting tangential to the surface. To obtain all component
forces parallel to the l, m and n axes, the shear force S_{nr}
may be resolved into two components S_{nl} and S_{nm}. Two
subscripts are attached to each force component; the first
(n) indicates the direction of the normal to the plane,
while the second gives the direction of the component. For
the normal component, the second subscript is the same as
the first, and this unnecessary duplication is avoided by
dropping the second subscript, replacing N_{nn} by N_n.

The sign convention adopted depends on the direction in which
the force component acts and the direction of the inwards
normal to the face on which it acts. The direction of the
inwards normal is that of a compressive normal force. If
the directions of the inwards normal and the force component
are both in either a positive or a negative co-ordinate
direction, then the component is considered positive. If
either the inwards normal or the component acts in a positive
co-ordinate direction and the other in a negative co-ordinate
direction, then the force component is taken to be negative.
All force components shown in the margin sketch are positive
according to this sign convention.

Surface tractions

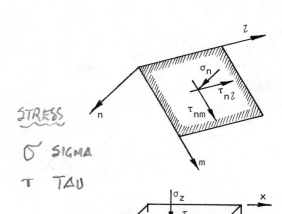

The intensity of a force applied to a surface element is
obtained by dividing the force by the area of the element, A.
In terms of the component forces defined above, we write

$$\sigma_n = \frac{N_n}{A}, \quad \tau_{nl} = \frac{S_{nl}}{A}, \quad \tau_{nm} = \frac{S_{nm}}{A} \qquad (1)$$

where σ_n, τ_{nl} and τ_{nm} are called *components of surface
traction* or *components of applied stress* acting on the
surface. σ_n is referred to as a *normal* or *direct stress*,
and τ_{nl} and τ_{nm} as *shear stresses*. The subscripts and sign
convention used are the same as those adopted for surface
forces.

Stress at a point

To fully define the state of stress at a point within a solid
body such as a rock mass, it is necessary to consider a small
element of volume enclosing the point in question. The
margin sketch shows such an element chosen with its edges
parallel to the x, y and z axes. The surface tractions shown
on the three visible faces are all positive. For the vertical face parallel to the y-z plane, $+\sigma_x$ acts in the negative

x direction and $+\tau_{xy}$ and $+\tau_{xz}$ act in the negative y and z directions repectively, because the inwards normal to this face acts in the negative x direction.

The sides of the element are taken to be vanishingly small so that the components of traction on each hidden face are the same as those on the corresponding exposed face. This means that the conditions of translational equilibrium are automatically satisfied. It can be shown that to satisfy conditions of rotational equilibrium,

$$\tau_{yx} = \tau_{xy} , \quad \tau_{zy} = \tau_{yz} , \quad \tau_{xz} = \tau_{zx} \qquad (2)$$

The pairs of shear stresses (τ_{xy}, τ_{yx}), (τ_{yz}, τ_{zy}) and (τ_{xz}, τ_{zx}) are referred to as *conjugate shear stresses*. To completely define the state of stress acting on this element, we need to know the values of the six independent quantities σ_x, σ_y, σ_z, τ_{xy}, τ_{yz} and τ_{zx}, known as the *components of stress at a point*.

The choice of a volume element with edges parallel to the x, y and z co-ordinate axes was completely arbitrary. In practice, it may be necessary to choose an element with edges parallel to a set of local axes inclined to the global axes. It may be necessary, for example, to orient the element so that one pair of faces is parallel to a structural feature such as a joint set on which the stresses are to be calculated. Alternatively, the choice may be conditioned by the orientation of the boundary surface.

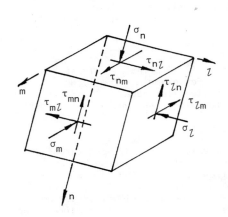

The margin sketch shows an element with edges parallel to a set of co-ordinate axes l, m and n. Referred to this set of axes, the components of stress at a point are σ_l, σ_m, σ_n, τ_{lm}, τ_{mn} and τ_{nl}. This set of components provides an alternative but equivalent definition of the state of stress at a point to the components expressed in terms of the x,y,z axes.

Transformation equations

It often happens that the engineer knows the components of stress referred to one set of axes (x,y,z) and wishes to determine another set of components (l,m,n). To effect this, a set of transformation equations is required, three for normal stresses of the type

$$\sigma_l = l_x^2 \sigma_x + l_y^2 \sigma_y + l_z^2 \sigma_z + 2(l_x l_y \tau_{xy} + l_y l_z \tau_{yz} + l_z l_x \tau_{zx}) \qquad (3)$$

and three for shear stresses of the type

$$\tau_{lm} = l_x m_x \sigma_x + l_y m_y \sigma_y + l_z m_z \sigma_z + (l_x m_y + l_y m_x) \tau_{xy} + (l_y m_z + l_z m_y) \tau_{yz} + (l_z m_x + l_x m_z) \tau_{zx} \qquad (4)$$

where l_x = cosine of the angle between the l- and x- axes,

l_y = cosine of the angle between the l- and y- axes,

etc.

To obtain the equations for σ_m and τ_{mn} from equations 3 and 4, replace l by m and m by n. A further replacement of m by n and n by l gives the expressions for σ_n and τ_{nl}. This process is called *cyclic permutation*. The order in which one subscript is replaced by another to yield the correct sequence of equations is shown by the diagram in the margin sketch.

If the l, m, n components are known and x,y,z components are required, a set of equations of similar form to equations 3 and 4 may be used. The x,y and z axes permute in accordance with the cyclic diagram shown opposite.

Principal planes

The values of the six components of stress at a point will vary with the orientation of the axes to which they are referred. Whatever the state of stress at a point, it is always possible to find a particular orientation of the co-ordinate axes for which all shear stress components vanish. These axes are called the *principal axes of stress*, and the corresponding planes parallel to the faces of the volume element are called the *principal planes*. The stresses on the faces of the element are purely normal, and are called the *principal stresses*. They are customarily denoted by the symbols σ_1, σ_2 and σ_3. By convention, σ_1 is chosen for the largest positive or *major principal stress*, σ_3 is chosen for the smallest positive or *minor principal stress*, and σ_2 for the *intermediate principal stress*. Thus

$$\sigma_1 \geqq \sigma_2 \geqq \sigma_3 \qquad (5)$$

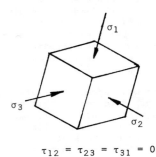

$\tau_{12} = \tau_{23} = \tau_{31} = 0$

Principal element

This scheme is followed even if all principal stresses are not positive (compressive). For example, if all the principal stresses were tensile (i.e. negative), σ_1 would be the smallest principal tensile stress and σ_3 the largest.

Two-dimensional state of stress

The burden of calculation involved in studying a three-dimensional stress problem can often be reduced by considering the two-dimensional stress distribution in one of the principal planes. Even when we are not totally justified in making this simplification, two-dimensional stress analyses can provide a useful guide to the nature of three-dimensional stress distributions.

Plane stress

A state of plane stress is defined as one in which all stress components acting on one of the three orthogonal planes at a point are zero. If with reference to the lower margin sketch on page 88, the plane on which stresses do not exist is the vertical plane perpendicular to the y-axis, then the conditions for plane stress may be defined as $\sigma_y = \tau_{yx} = \tau_{yz} = 0$. Such a state of stress exists in photoelastic or other physical models of excavations in which a perforated plate is loaded by forces applied in the plane of the plate.

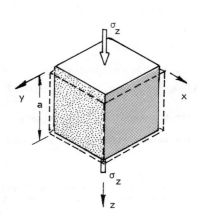

Consider a cube of rock loaded vertically by an average axial stress, σ_z, as shown in the margin sketch. Assume that the rock is free to expand laterally and that it behaves elastically as most hard rocks do at stress levels below their compressive strengths. The vertical dimension will decrease by an amount w, while the lateral dimensions will increase by an amount u = v.

The linear vertical strain in the cube is defined as the deformation per unit length and is given by $\varepsilon_z = w/a$. For a linear elastic material, this strain is related to the vertical stress by the equation

$$\varepsilon_z = \frac{\sigma_z}{E} \qquad (6)$$

where E is the *Young's modulus* of the material.

The lateral strain $\varepsilon_x = \varepsilon_y = -\frac{u}{a}$ is related to the vertical stress by the equation

$$\varepsilon_x = \varepsilon_y = -\frac{\nu \sigma_z}{E} \qquad (7)$$

where ν is the *Poisson's ratio* of the material, and compressive strains are taken as positive.

Young's modulus and Poisson's ratio are material properties generally referred to as the *elastic constants*. For typical hard rock materials, Young's modulus lies in the range 5 to 15 x 10^6 lb/in^2 (35 to 105 x 10^3 MPa), and Poisson's ratio varies from about 0.15 to 0.30.

If instead of being free to deform laterally, the cube of rock is restrained in the x direction by the application of a normal stress, σ_x, the linear strains will be

$$\varepsilon_x = \frac{1}{E}(\sigma_x - \nu\sigma_z) \qquad (8)$$

$$\varepsilon_z = \frac{1}{E}(\sigma_z - \nu\sigma_x) \qquad (9)$$

$$\text{and } \varepsilon_y = -\frac{\nu}{E}(\sigma_x + \sigma_z) \qquad (10)$$

If the general state of plane stress applied to the cube of rock is now completed by the application of the pair of conjugate shear stresses $\tau_{xz} = \tau_{zx}$, a *shear strain* in the x - z plane will result. This shear strain, γ_{xz}, may be defined as the change, measured in radians, in an angle that was originally a right angle. It is a measure of the distortion suffered by the cube and is related to the elastic constants by the equation

$$\gamma_{xz} = \frac{2(1+\nu)}{E}\tau_{xz} \qquad (11)$$

$$\text{or } \gamma_{xz} = \frac{\tau_{xz}}{G} \qquad (12)$$

where $G = \frac{E}{2(1+\nu)}$ is the *shear modulus* or *modulus of rigidity* of the material. If for a typical hard rock, $E = 100 \times 10^3$ MPa or 100 GPa, and $\nu = 0.25$, then $G = 40 \times 10^3$ MPa or 40 GPa.

Plane strain conditions

Consider a situation in which, prior to any excavation, the principal stresses at a certain depth below the ground surface are constant at p_x, p_y and p_z. Let a tunnel of arbitrary but constant cross-section be driven parallel to the y-axis. Obviously, the excavation of the tunnel will produce a redistribution of stress. However, except for regions close to the ends of the tunnel, the pattern of stress around the tunnel will be virtually the same for all cross-sections.

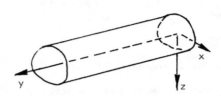

This situation closely approximates the theoretical idealisation known as plane strain. The distinguishing features of plane strain are that during the process of excavation, all displacements occur in one plane (taken as the x - z plane in the present case), and that the pattern of displacement is the same for all cross-sections. If in the situation shown in the margin sketch, u, v and w are the x, y and z components of displacement resulting from the excavation, then v = 0, and u and w are functions of x and z but not of y in the plane strain case. For linear elastic materials, this gives the following relationships between strains and the stresses *induced* by the creation of the excavation :

$$\varepsilon_x = \frac{1}{E'} (\sigma_x - \nu' \sigma_z) \qquad (13)$$

$$\varepsilon_z = \frac{1}{E'} (\sigma_z - \nu' \sigma_x) \qquad (14)$$

$$\varepsilon_y = 0 \qquad (15)$$

$$\text{and} \quad \gamma_{xz} = \frac{2(1 + \nu')}{E'} \tau_{xz} \qquad (16)$$

$$\text{where} \quad E' = \frac{E}{1 - \nu^2} \qquad (17)$$

$$\text{and} \quad \nu' = \frac{\nu}{1 - \nu} \qquad (18)$$

Comparison of equations 8-11 and 13-16 shows that the stress-strain relationships for plane stress and plane strain are of the same form but with different coefficients. It will be seen later in this chapter that the elastic stress distribution around an excavation is independent of the elastic constants. It follows that, for the same boundary conditions, a plane stress model gives the same form of stress distributions as that produced under plane strain conditions.

Two-dimensional stress transformation

It has been shown that under plane stress or plane strain conditions, we can restrict our attention to the three stress components associated with one pair of axes, e.g. σ_x, σ_z and τ_{zx} associated with the x and z axes. There are occasions, however, when it becomes necessary to introduce inclined axes l and m and the associated stress components σ_l, σ_m and τ_{lm}. If the new axes lie in the x-z plane and the angle between the l and x axes is α as shown in the margin sketch, the transformation equations given by equations 3 and 4 become

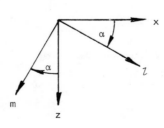

$$\sigma_l = \tfrac{1}{2}(\sigma_x + \sigma_z) + \tfrac{1}{2}(\sigma_x - \sigma_z)\cos 2\alpha + \tau_{zx} \sin 2\alpha \qquad (19)$$

$$\sigma_m = \tfrac{1}{2}(\sigma_x + \sigma_z) - \tfrac{1}{2}(\sigma_x - \sigma_z)\cos 2\alpha - \tau_{zx} \sin 2\alpha \qquad (20)$$

$$\tau_{lm} = \tau_{zx} \cos 2\alpha - \tfrac{1}{2}(\sigma_x - \sigma_z)\sin 2\alpha \qquad (21)$$

The magnitudes of the corresponding principal stresses are found by determining the value of α at which $\tau_{lm} = 0$ and σ_l and σ_m take maximum and minimum values. It is found that the principal stresses are

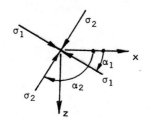

$$\sigma_1 = \tfrac{1}{2}(\sigma_x + \sigma_z) + \sqrt{\tfrac{1}{4}(\sigma_x - \sigma_z)^2 + \tau_{zx}^2} \qquad (22)$$

$$\text{and} \quad \sigma_2 = \tfrac{1}{2}(\sigma_x + \sigma_z) - \sqrt{\tfrac{1}{4}(\sigma_x - \sigma_z)^2 + \tau_{zx}^2} \qquad (23)$$

and their directions are given by

$$\alpha_{1,2} = \tfrac{1}{2} \arctan\{2\tau_{zx}/(\sigma_x - \sigma_z)\} \qquad (24)$$

$$\text{or} \quad \alpha_1 = \arctan\{(\sigma_1 - \sigma_x)/\tau_{zx}\} \qquad (25)$$

where $\alpha_2 = \alpha_1 + 90°$ \qquad (26)

Mohr's circle diagram

The relationships given by equations 19 to 26 may be represented graphically by a construction known as *Mohr's stress circle diagram* shown in figure 39. The circle is constructed on vertical and horizontal axes of τ and σ, anti-clockwise senses of τ being plotted above the horizontal axis and clockwise below. The same stress scale must be used for both axes.

The normal and shear stresses acting on any plane are plotted as $OF = \sigma_x$ and $FK = \tau_{xz}$. The centre of the circle is located by making $OC = \tfrac{1}{2}(\sigma_x + \sigma_z)$ and the circle is drawn centred on C and passing through K. The point P, obtained by the intersection of the circle and a line through K parallel to the plane on which σ_x and τ_{xz} act, is called the *origin of planes*.

If it is required to find the stresses on a set of planes associated with the l and m axes, the following construction may be used. A line is drawn through P parallel to the m axis, intersecting the circle at G. A second line in drawn through P parallel to the l axis, intersecting the circle at H. To scale, G represents the point (σ_l, τ_{lm}) and H represents the point (σ_m, τ_{ml}). When determining the directions of the stresses given by the points G and H, it is important to recognise that the lines PG and PH give the orientations of the *planes* on which the stresses (σ_l, τ_{lm}) and (σ_m, τ_{ml}) act. A correctly oriented element on which these stresses act is shown in figure 39.

Figure 39 also gives the magnitudes and directions of the principal stresses. Clearly, the principal stresses are the maximum and minimum normal stresses given by points A and B at which the shear stresses are zero. A line drawn from the point P through A gives the orientation of the plane on which σ_1 acts. A correctly oriented principal element based on this line is shown in figure 39.

In situ state of stress

As noted in the introduction to this chapter, the rock in the earth's crust is subjected to an in situ state of stress and when an excavation is made in the rock, these stresses are disturbed and re-distributed in the vicinity of the excavation. Therefore, before the distribution of stresses around any man-made excavation in rock can be calculated, the pre-existing state of stress must be measured or estimated.

Terzaghi and Richart's approach

Consider a cube of rock in the earth's crust subjected to a

DATA :

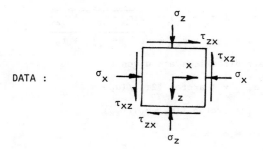

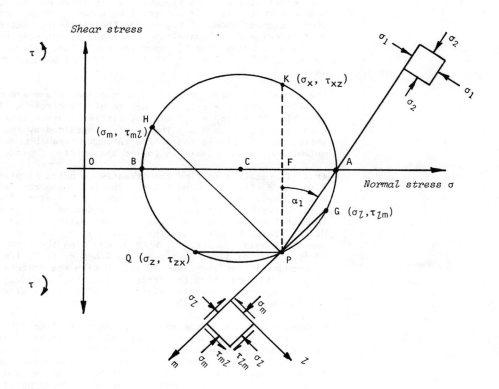

$OF = \sigma_x$, $FK = \tau_{xz}$

$OC = \frac{1}{2}(\sigma_x + \sigma_z)$

$OA = \sigma_1 = \frac{1}{2}(\sigma_x + \sigma_z) + \{\frac{1}{4}(\sigma_x - \sigma_z)^2 + \tau_{zx}^2\}^{\frac{1}{2}}$

$OB = \sigma_2 = \frac{1}{2}(\sigma_x + \sigma_z) - \{\frac{1}{4}(\sigma_x - \sigma_z)^2 + \tau_{zx}^2\}^{\frac{1}{2}}$

$\tan \alpha_1 = (\sigma_1 - \sigma_x)/\tau_{zx}$

Figure 39 : Mohr's stress circle.

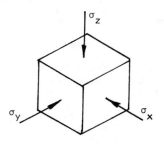

vertical stress σ_z and lateral compressive stresses σ_x and σ_y as shown in the margin sketch. The vertical and lateral direct strains associated with these stresses are given by

$$\varepsilon_z = \frac{1}{E}\{\sigma_z - \nu(\sigma_x + \sigma_y)\} \qquad (27)$$

$$\varepsilon_x = \frac{1}{E}\{\sigma_x - \nu(\sigma_y + \sigma_z)\} \qquad (28)$$

$$\text{and} \quad \varepsilon_y = \frac{1}{E}\{\sigma_y - \nu(\sigma_x + \sigma_z)\} \qquad (29)$$

In 1952, Terzaghi and Richart[74] suggested that in the case of sedimentary rocks in geologically undisturbed regions where the strata were built up in horizontal layers in such a way that the horizontal dimensions remained unchanged, the lateral stresses σ_x and σ_y are equal and are given by putting $\varepsilon_x = \varepsilon_y = 0$ in equations 28 and 29. Hence

$$\sigma_x = \sigma_y = \frac{\nu}{1-\nu}\sigma_z \qquad (30)$$

For a typical rock having a Poisson's ratio $\nu = 0.25$, equation 30 shows that the lateral stresses σ_x and σ_y are each equal to one third of the vertical stress σ_z, provided that there has been no lateral strain. As will be shown later in this chapter, measured in situ horizontal stresses are very seldom as low as those predicted by equation 30, and it must be concluded that the basic assumptions made in deriving this equation do not apply to real geological situations.

Terzaghi and Richart's suggestion has been included because of the important role it played in rock mechanics thinking for almost twenty years. The reader will find many papers, published in the late 1950s and early 1960s, which utilised equation 30 without questioning the basis for its derivation. So complete was the acceptance of this equation by workers in hard rock mechanics that, when Hast[75] (1958) reported measuring horizontal stresses of several times the vertical stress in Scandinavia, his results were treated with extreme scepticism; it is only recently that these results have been accepted as being realistic.

Heim's rule

Talobre[76], referring to an earlier suggestion by Heim[77] (1912), suggested that the inability of rock to support large stress differences together with the effects of time-dependent deformation of the rock mass can cause lateral and vertical stresses to equalise over periods of geological time. Heim's rule, as this suggestion has come to be known, is widely used by workers in weak rocks such as coal measures and evaporites, and it has been found to give a good approximation of the in situ stress field in these materials. As will be shown in the next section, Heim's rule also appears to give a reasonable estimate of the horizontal stresses which exist at depths in excess of one kilometre.

Results of in situ stress measurements

During the past twenty years, several techniques for in situ stress measurement have been developed and have been used to measure rock stresses in various parts of the world. Some of these techniques will be described in a later chapter dealing with instrumentation.
The senior author was marginally involved in the development

of stress measuring equipment in South Africa, knows many of the men who have worked in this field, and is familiar with many of the sites where rock stress measurements have been carried out. With this background, the authors have worked through the literature on this subject and have selected the measurements presented in Table 8 and figures 40 and 41. In selecting these data, measurements obtained in extremely unusual geological environments, such as areas of very recent tectonic activity, have been omitted and care has been taken to use only those results which are supported by reasonably detailed back-up data.

Figure 40 shows that the measured vertical stresses are in fair agreement with the simple prediction given by calculating the vertical stress due to the overlying weight of rock at a particular depth from the equation

$$\sigma_z = \gamma z \qquad (31)$$

where γ is the unit weight of the rock (usually in the range 20 to 30 kN/m^3) and z is the depth at which the stress is required.

At shallow depths, there is a considerable amount of scatter which may be associated with the fact that these stress values are often close to the limit of the measuring accuracy of most stress measuring tools. On the other hand, the possibility that high vertical stresses may exist cannot be discounted, particularly where some unusual geological or topographic feature may have influenced the entire stress field.

Figure 41 gives a plot of k, the ratio of the average horizontal to vertical stress, against depth below surface. It will be seen that, for most of the values plotted, the value of k lies within the limits defined by

$$\frac{100}{z} + 0.3 < k < \frac{1500}{z} + 0.5 \qquad (32)$$

The plot shows that, at depths of less than 500 metres, the horizontal stresses are significantly greater than the vertical stresses. Some of the measurements reported by Hast[75] in 1958 are included (points 86, 93, 94 and 95), and are certainly not in conflict with other measurements made in Scandinavia, Australia, South Africa and the U.S.A.

For depths in excess of 1 kilometre (3280 feet), the average horizontal stress and the vertical stress tend to equalise, as suggested by Heim's rule. This trend is not surprising since, as will be shown in the next chapter, rock is incapable of supporting very high stresses when there are large differences in the magnitudes of the three applied principal stresses. Hence, if very high horizontal stresses existed at depths in excess of 1 kilometre, these would have induced fracturing, plastic flow and time-dependent deformation in the rock, and all of these processes would tend to reduce the difference between horizontal and vertical stresses.

In considering the significance of figure 41, it must be remembered that the *average* horizontal stress has been plotted. In many cases there is a significant difference between the horizontal stresses in different directions and, as more reliable stress measurement results become

TABLE 8 - SUMMARY OF IN SITU STRESS MEASUREMENT RESULTS

Point	Location	Rock Type	Depth m	σ_v MPa	$\dfrac{\sigma_{h.av}}{\sigma_v}$	Ref.
AUSTRALIA						
1	CSA mine, Cobar, NSW	Siltstone, chloritic slate	360	16.6	1.46	78
2	CSA mine, Cobar, NSW	Siltstone, chloritic slate	360	8.0	1.30	78
3	CSA mine, Cobar, NSW	Siltstone, chloritic slate	540	15.2	1.70	78
4	CSA mine, Cobar, NSW	Siltstone, chloritic slate	330	10.0	1.40	78
5	CSA mine, Cobar, NSW	Siltstone, chloritic slate	455	11.0	1.90	78
6	CSA mine, Cobar, NSW	Siltstone, chloritic slate	245	8.4	2.10	78
7	CSA mine, Cobar, NSW	Siltstone, chloritic slate	633	13.7	2.00	78
8	NBHC mine, Broken Hill, NSW	Sillimanite gneiss	1022	6.2	1.66	78
9	NBHC mine, Broken Hill, NSW	Garnet quartzite	668	13.8	1.17	78
10	NBHC mine, Broken Hill, NSW	Garnet quartzite	668	4.8	2.73	78
11	NBHC mine, Broken Hill, NSW	Garnet quartzite	570	15.9	1.32	78
12	ZC mine, Broken Hill, NSW	Sillimanite gneiss	818	20.0	1.07	78
13	ZC mine, Broken Hill, NSW	Sillimanite gneiss	818	26.9	1.17	78
14	ZC mine, Broken Hill, NSW	Sillimanite gneiss	915	13.1	1.29	78
15	ZC mine, Broken Hill, NSW	Sillimanite gneiss	915	21.4	0.97	78
16	ZC mine, Broken Hill, NSW	Sillimanite gneiss	766	9.7	1.85	78
17	ZC mine, Broken Hill, NSW	Garnet quartzite	570	14.7	1.43	78
18	ZC mine, Broken Hill, NSW	Garnet quartzite	570	12.7	2.09	78
19	ZC mine, Broken Hill, NSW	Garnet quartzite	818	12.3	2.10	78
20	NBHC mine, Broken Hill, NSW	Gneiss and quartzite	670	13.0	2.40	78
21	NBHC mine, Broken Hill, NSW	Gneiss and quartzite	1277	19.2	1.60	78
22	NBHC mine, Broken Hill, NSW	Gneiss and quartzite	1140	6.9	2.40	78
23	NBHC mine, Broken Hill, NSW	Gneiss and quartzite	1094	25.5	0.82	78
24	NBHC mine, Broken Hill, NSW	Rhodonite	1094	15.9	1.81	78
25	NBHC mine, Broken Hill, NSW	Gneiss and quartzite	1094	18.6	1.62	78
26	NBHC mine, Broken Hill, NSW	Gneiss and quartzite	1094	26.9	1.34	78
27	NBHC mine, Broken Hill, NSW	Gneiss and quartzite	1140	29.7	1.43	78
28	NBHC mine, Broken Hill, NSW	Gneiss and quartzite	1423	24.2	1.51	78
29	Mount Isa Mine, Queensland	Silica dolomite	664	19.0	0.83	78
30	Mount Isa Mine, Queensland	Silica dolomite	1089	16.5	1.28	78
31	Mount Isa Mine, Queensland	Dolomite and shale	1025	28.5	0.87	78,79
32	Mount Isa Mine, Queensland	Shale	970	25.4	0.85	78
33	Warrego mine, Tennant Creek, NT	Magnetite	245	7.0	2.40	78
34	Warrego mine, Tennant Creek, NT	Chloritic slate, quartz	245	6.8	1.80	78
35	Warrego mine, Tennant Creek, NT	Magnetite	322	11.5	1.30	78
36	Kanmantoo, SA	Black garnet mica schist	58	2.5	3.30	78
37	Mount Charlotte mine, WA	Dolerite	92	11.2	1.45	78
38	Mount Charlotte mine, WA	Greenstone	152	10.4	1.42	78
39	Mount Charlotte mine, WA	Greenstone	152	7.9	1.43	78
40	Durkin mine, Kambalda, WA	Serpentine	87	7.4	2.20	78
41	Dolphin Mine, King Is., Tasmania	Marble and skarn	75	1.8	1.80	78
42	Poatina hydro. project, Tasmania	Mudstone	160	8.5	1.70	78,80
43	Cethana hydro. project, Tasmania	Quartzite conglomerate	90	14.0	1.35	78
44	Gordon River hydro. project, Tas.	Quartzite	200	11.0	2.10	78
45	Mount Lyell mine, Tasmania	Quartzite schist	105	11.3	2.95	78
46	Windy Creek, Snowy Mts., NSW	Diorite	300	12.4	1.07	78
47	Tumut 1 power stn., Snowy Mts., NSW	Granite and gneiss	335	11.0	1.20	78
48	Tumut 2 power stn., Snowy Mts., NSW	Granite and gneiss	215	18.4	1.20	78
49	Eucumbene Tunnel, Snowy Mts., NSW	Granite	365	9.5	2.60	78
CANADA						
50	G.W. MacLeod Mine, Wawa, Ontario	Siderite	370	16.1	1.29	81
51	G.W. MacLeod Mine, Wawa, Ontario	Tuff	370	15.1	2.54	81
52	G.W. MacLeod Mine, Wawa, Ontario	Tuff	575	21.5	1.23	81
53	G.W. MacLeod Mine, Wawa, Ontario	Tuff	575	14.6	1.25	81
54	G.W. MacLeod Mine, Wawa, Ontario	Meta-diorite	480	18.7	1.54	81
55	G.W. MacLeod Mine, Wawa, Ontario	Chert	575	26.6	1.52	81
56	Wawa, Ontario	Granite	345	20.0	2.50	82
57	Elliot Lake, Ontario	Sandstone	310	(11.0)*	2.56	83
58	Elliot Lake, Ontario	Quartzite	705	(17.2)	1.70	83
59	Elliot Lake, Ontario	Diabase dyke	400	17.2	1.90	84

* Vertical stresses in parentheses calculated from depth below surface.

Point	Location	Rock Type	Depth m	σ_v MPa	$\dfrac{\sigma_{h.av}}{\sigma_v}$	Ref.
60	Churchill Falls hydro., Labrador	Diorite gneiss	300	7.8	1.70	85
61	Portage Mountain hydro., BC	Sandstone and shale	137	6.8	1.42	86
62	Mica Dam, BC	Gneiss and quartzite	220	6.9	1.50	87

UNITED STATES

Point	Location	Rock Type	Depth m	σ_v MPa	$\dfrac{\sigma_{h.av}}{\sigma_v}$	Ref.
63	Rangeley oil field, Colorado	Sandstone	1910	(43.5)	1.04	88
64	Nevada Test Site, Nevada	Tuff	380	(7.0)	0.90	89
65	Helms hydro, Fresno, California	Granodiorite	300	(8.2)	0.91	90
66	Bad Creek hydro., South Carolina	Gneiss	230	(6.2)	3.12	90
67	Montello, Wisconsin	Granite	136	(3.5)	3.29	91
68	Alma, New York	Sandstone	500	(7.9)	1.61	92
69	Falls Township, Ohio	Sandstone	810	(14.1)	1.25	92
70	Winnfield, Louisiana	Salt	270	5.5	0.95	93
71	Barberton, Ohio	Limestone	830	24.0	1.94	93
72	Silver Summit Mine, Osburn, Idaho	Argillaceous quartzite	1670	56.7	1.26	94
73	Star Mine, Burke, Idaho	Quartzite	1720	37.9	0.60	95
74	Crescent Mine, Idaho	Quartzite	1620	40.3	1.17	96
75	Red Mountain, Colorado	Granite	625	18.1	0.56	97
76	Henderson Mine, Colorado	Granite	790	24.2	1.23	97
77	Henderson Mine, Colorado	Orebody	1130	29.6	0.98	97
78	Piceance Basin, Colorado	Oil shale	400	(9.8)	0.80	98
79	Gratiot Country, Michigan	Dolomite	2806	(63.1)	0.78	91

SCANDINAVIA

Point	Location	Rock Type	Depth m	σ_v MPa	$\dfrac{\sigma_{h.av}}{\sigma_v}$	Ref.
80	Bleikvassli Mine, N. Norway	Gneiss and mica schist	200	6.0	1.92	99
81	Bleikvassli Mine, N. Norway	Gneiss and mica schist	250	7.0	2.00	99
82	Bidjovagge Mine, N. Norway	Precambrian rocks	70	2.8	4.64	100
83	Bjornevan, N. Norway	Gneiss	100	(2.7)	5.56	99
84	Sulitjelma, N. Norway	Phyllite	850	10.0	0.99	100
85	Sulitjelma, N. Norway	Phyllite	900	11.0	0.55	100
86	Ställberg, Sweden	Precambrian rocks	915	(24.7)	1.56	75,99
87	Vingesbacke, Sweden	Granite and amphibolite	400	(10.8)	4.99	101
88	Laisvall, Sweden	Granite	220	(5.9)	3.72	101
89	Malmberget, Sweden	Granite	500	(13.4)	2.41	99
90	Grängesberg, Sweden	Gneiss	400	(10.8)	2.31	100
91	Kiruna, Sweden	Precambrian rocks	680	(18.4)	1.90	100
92	Stalldalen, Sweden	Precambrian rocks	690	(18.6)	2.58	101
93	Stalldalen, Sweden	Precambrian rocks	900	(24.3)	2.02	101
94	Hofors, Sweden	Precambrian rocks	470	(12.7)	2.74	101
95	Hofors, Sweden	Precambrian rocks	650	(17.6)	2.25	101

SOUTHERN AFRICA

Point	Location	Rock Type	Depth m	σ_v MPa	$\dfrac{\sigma_{h.av}}{\sigma_v}$	Ref.
96	Shabani Mine, Rhodesia	Dunite, serpentine	350	10.7	1.46	102
97	Kafue Gorge hydro., Zambia	Gneiss, amphibolite schist	160	7.5	1.57	103
98	Kafue Gorge hydro., Zambia	Gneiss, amphibolite schist	400	12.5	1.60	103
99	Ruacana hydro. project, S-W Africa	Granite gneiss	215	4.0	1.95	104
100	Drakensberg hydro. project, S.A.	Mudstone and sandstone	110	3.0	2.50	105
101	Braken Mine, Evander, S.A.	Quartzite	508	13.9	0.99	103
102	Winkelhaak Mine, Evander, S.A.	Quartzite	1226	38.4	0.82	103
103	Kinross Mine, Evander, S,A,	Quartzite	1577	49.5	0.64	103
104	Doornfontein Mine, Carltonville, SA	Quartzite	1320	39.0	0.48	103
105	Harmony Mine, Virginia, S.A.	Quartzite	1500	33.1	0.49	103
106	Durban Roodeport Deep Mine, S.A.	Quartzite	2300	68.5	0.67	103
107	Durban Roodeport Deep Mine, S.A.	Quartzite	2500	59.0	1.02	103
108	East Rand Proprietary Mine, S.A.	Quartzite and shale	2400	37.4	0.72	103
109	Prieska Mine, Copperton, S.A.	Quartz amphibolite schist	279	8.8	1.41	106
110	Prieska Mine, Copperton, S.A.	Quartz amphibolite schist	410	9.6	1.01	106
111	Western Deep Levels Mine, S.A.	Quartzite	1770	45.6	0.63	107
112	Doornfontein Mine, S.A.	Quartzite	2320	58.5	0.54	103

OTHER REGIONS

Point	Location	Rock Type	Depth m	σ_v MPa	$\dfrac{\sigma_{h.av}}{\sigma_v}$	Ref.
113	Dinorwic hydro. project, Wales, UK	Slate	250	9.0	1.28	108
114	Mont Blanc tunnel project, France	Gneiss-granite	1800	48.6	1.00	109
115	Cameron Highlands hydro., Malaysia	Granite	296	10.6	1.03	110
116	Idikki hydro. project, south India	Granite gneiss	360	8.3	1.96	111

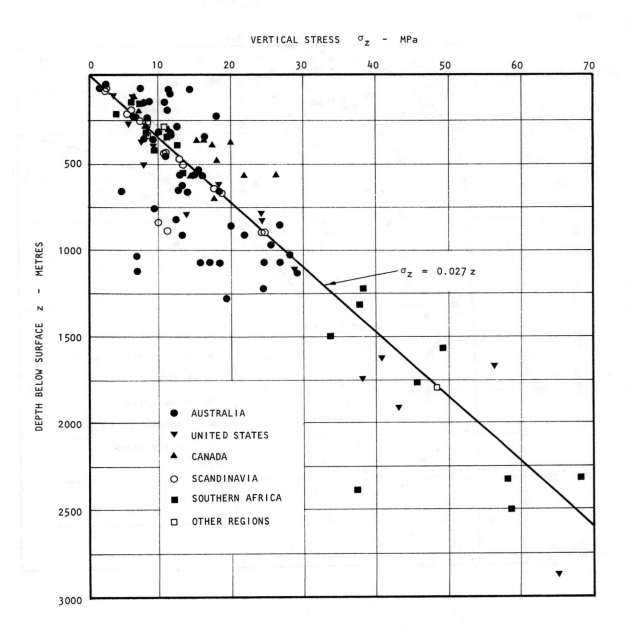

Figure 40 : Plot of vertical stresses against depth below surface.

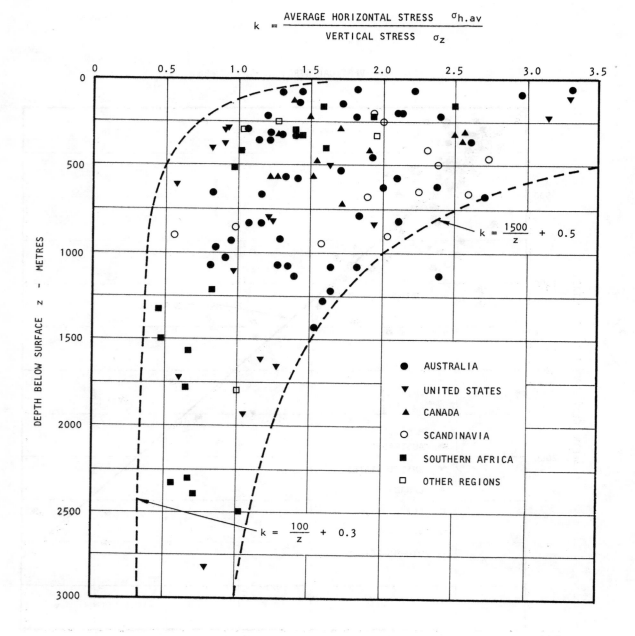

Figure 41 : Variation of ratio of average horizontal stress to vertical stress with depth below surface.

available, it may be useful to consider the significance of these individual stresses rather than their average. It must also be remembered that a number of stress measurements have been omitted from figure 41. In particular, a number of in situ stress measurements carried out in the Caledonian mountains in Norway[100] have not been included in this plot. These measurements show that the ratio of horizontal to vertical stress varies from zero to approximately 10 and that the horizontal stresses are aligned parallel to and normal to the trend of the mountain chain. These measurements were omitted from figure 41 because the influence of both topography and tectonic activity was considered to be such that the results could not be regarded as typical of in situ stresses in undisturbed rock masses. However, the wide variation in these measured stresses emphasises the uncertainty which is inherent in any attempt to predict in situ horizontal stresses on the basis of simple theoretical concepts and the necessity for in situ stress measurements. In fact, the authors believe that it is *essential* to carry out in situ stress measurements as part of the site investigation programme for any important underground excavation project.

Stress distributions around single excavations

The streamline analogy for principal stress trajectories

When an underground excavation is made in a rock mass, the stresses which previously existed in the rock are disturbed, and new stresses are induced in the rock in the immediate vicinity of the opening. One method of representing this new stress field is by means of *principal stress trajectories* which are imaginary lines in a stressed elastic body along which principal stresses act. Before considering in detail the distribution of stresses around single underground excavations of various cross-sections, it may be helpful to the reader to visualise the stress field by making use of the approximate analogy which exists between principal stress trajectories and the streamlines in a smoothly flowing stream of water.

Figure 42 shows the major and minor principal stress trajectories in the material surrounding a circular hole in a uniaxially stressed elastic plate. These principal stress trajectories may be regarded as dividing the material into elements on which the principal stresses act. On the right hand side of figure 42, two of the major principal stress trajectories are shown dashed and, at arbitrarily selected points along these trajectories, the principal stresses acting on imaginary elements are shown. In each case, the direction and magnitude of the principal stress is shown by an arrow, the length of which gives the magnitude of the principal stress to some specified scale. Note that the principal stresses depart significantly from being vertical and horizontal in the vicinity of the opening which deflects the stress trajectories.

When a cylindrical obstruction such as the pier of a bridge is introduced into a smoothly flowing stream, the water has to flow around this obstruction and the streamlines are deflected as shown in figure 43. Immediately upstream and downstream of the obstruction, the water flow is slowed down and the streamlines are spread outwards. This separation is

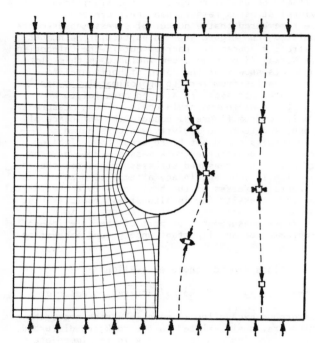

Figure 42 : Major and minor principal stress trajectories in the material surrounding a circular hole in a uniaxially stressed elastic plate.

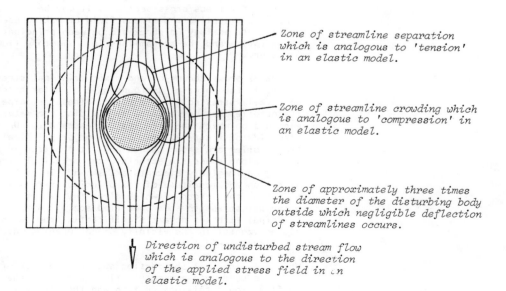

Zone of streamline separation which is analogous to 'tension' in an elastic model.

Zone of streamline crowding which is analogous to 'compression' in an elastic model.

Zone of approximately three times the diameter of the disturbing body outside which negligible deflection of streamlines occurs.

Direction of undisturbed stream flow which is analogous to the direction of the applied stress field in an elastic model.

Figure 43 : Deflection of streamlines around a cylindrical obstruction.

analogous to the separation of stress trajectories which occurs in zones of tensile stress and, as will be shown later in this chapter, such tensile stress zones occur in the roof and floor of a circular excavation subjected to uniaxial compressive applied stress.

In zones on either side of the obstruction the water flow has to speed up in order to catch up with the rest of the stream and the streamlines are crowded together as shown in figure 43. This is analogous to the crowding of stress trajectories which occurs in zones of increased compressive stress. Figure 43 shows that outside a zone of approximately three times the diameter of the obstruction, the streamlines are not deflected to any significant degree by the obstruction. The stream flowing outside this zone does not 'see' the obstruction which only creates a local disturbance. This effect is also found in stress fields. Points in the rock mass which are more than approximately three radii from the centre of the excavation are not significantly influenced by the presence of the excavation.

Stresses around a circular excavation

In order to calculate the stresses, strains and displacements induced around excavations in elastic materials, it is necessary to turn to the *mathematical theory of elasticity*. This requires that a set of equilibrium and displacement compatibility equations be solved for given boundary conditions and constitutive equations for the material. The process involved in obtaining the required solutions can become quite complex and tedious, and will not be described in this book. The reader interested in obtaining his own closed form solutions to problems should refer to the standard texts on the subject such as those by Love[70], Timoshenko and Goodier[73] and Jaeger and Cook[67].

One of the earliest solutions for the two-dimensional distribution of stresses around an opening in an elastic body was published in 1898 by Kirsch[112] for the simplest cross-sectional shape, the circular hole. A full discussion on the derivation of the *Kirsch equations*, as they are now known, is given by Jaeger and Cook[67] and no attempt will be made to reproduce this discussion here. The final equations are presented in figure 44, using a system of *polar co-ordinates* in which the stresses are defined in terms of the tractions acting on the faces of an element located by a radius r and a polar angle θ.

Some of the many interesting and important facts about stresses around openings are illustrated by this example and will be discussed in subsequent sections.

Stresses at the excavation boundary

The equations given in figure 44 show that the radial stress σ_r and the shear stress $\tau_{r\theta}$ are both zero at the boundary of the opening where $r = a$. The tangential stress on the boundary is given by

$$\sigma_\theta = p_z\{(1+k) - 2(1-k)\cos 2\theta\} \qquad (33)$$

In the roof and floor of the opening, θ = 0° and 180° respectively, and equation 33 reduces to

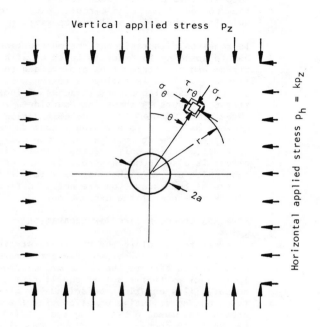

STRESS COMPONENTS AT POINT (r, θ)

Radial $\quad \sigma_r = \tfrac{1}{2}p_z\{(1+k)(1-a^2/r^2) + (1-k)(1-4a^2/r^2+3a^4/r^4)\cos 2\theta\}$

Tangential $\quad \sigma_\theta = \tfrac{1}{2}p_z\{(1+k)(1+a^2/r^2) - (1-k)(1+3a^4/r^4)\cos 2\theta\}$

Shear $\quad \tau_{r\theta} = \tfrac{1}{2}p_z\{-(1-k)(1+2a^2/r^2-3a^4/r^4)\sin 2\theta\}$

PRINCIPAL STRESSES IN PLANE OF PAPER AT POINT (r, θ)

Maximum $\quad \sigma_1 = \tfrac{1}{2}(\sigma_r + \sigma_\theta) + \left(\tfrac{1}{4}(\sigma_r - \sigma_\theta)^2 + \tau_{r\theta}^2\right)^{\tfrac{1}{2}}$

Minimum $\quad \sigma_2 = \tfrac{1}{2}(\sigma_r + \sigma_\theta) - \left(\tfrac{1}{4}(\sigma_r - \sigma_\theta)^2 + \tau_{r\theta}^2\right)^{\tfrac{1}{2}}$

Inclinations to radial direction $\quad \tan 2\alpha = 2\tau_{r\theta}/(\sigma_\theta - \sigma_r)$

Figure 44: Equations for the stresses in the material surrounding a circular hole in a stressed elastic body.

$$\sigma_\theta = p_z(3k - 1) \tag{34}$$

In the sidewalls of the opening, $\theta = 90°$ and $270°$ and equation 33 becomes

$$\sigma_\theta = p_z(3 - k) \tag{35}$$

Equations 34 and 35 are plotted in figure 45 which shows that, for $k = 0$, the stresses in the roof and floor of the opening are tensile. For $k = 0.33$, the stresses in the roof and floor are zero and, for higher k values, all stresses on the boundary of the opening are compressive. The sidewall stresses decrease from a maximum of $3p_z$ for $k = 0$ to a value of $2p_z$ for $k = 1$.

The condition that the only stresses which can exist at the boundary of an excavation are the stresses tangential to the boundary holds true for all excavation shapes which are free of internal loading. When the inside surface of the opening is loaded by means of water pressure, the reaction of a concrete lining or the loads applied through rockbolts, these internal stresses must be taken into account in calculating the stress distribution in the rock surrounding the opening.

Stresses remote from the excavation boundary

As the distance r from the hole increases, the influence of the opening upon the stresses in the rock decreases. A plot of the ratio of σ_θ/p_z against the distance r along the horizontal axis of the stressed model is given in figure 46. This plot shows that the stress concentrating effect of the hole dies away fairly rapidly and that, at $r = 3a$, the ratio of induced to applied stress is very close to unity. This means that, at this distance from the excavation boundary, the stresses in the rock do not 'see' the influence of the opening. This fact has been utilised by those concerned with model studies of stresses around underground excavations. The general rule is that the minimum size of the model should be 3 to 4 times the maximum dimension of the excavation in the model.

Axes of symmetry

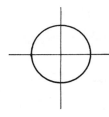

Close examination of the equations presented in figure 44 shows that, for a symmetrical excavation such as that considered, the stress pattern is repeated in each of the four quadrants. This means that a complete picture of the stresses surrounding the opening can be generated by solving the equations for values of θ from $0°$ to $90°$ and that the horizontal and vertical axes through the centre of the opening are *axes of symmetry*.

Axes of symmetry

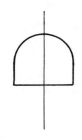

In the case of a horse-shoe shaped tunnel, the top and bottom halves of the opening are not symmetrical but the left and right hand sides are. Hence, the vertical axis is the only axis of symmetry and the complete stress picture can only be obtained by solving the appropriate stress equations for values of θ between $0°$ and $180°$. In some other cases, particularly those involving multiple openings, no axes of symmetry may exist and a complete solution of the stress equations in all parts of the model may be required in order to provide a complete solution of the stress distribution in the model.

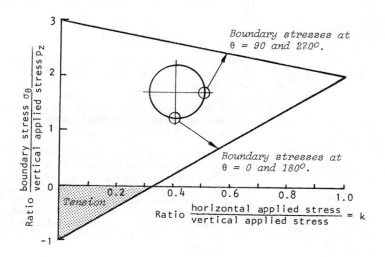

Figure 45 : Variation in boundary stresses in the roof and floor and sidewalls of a circular opening with variation in the ratio k of applied stresses.

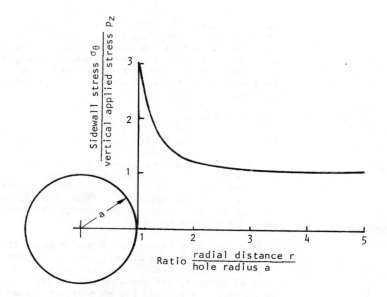

Figure 46 : Variation in ratio of tangential stress σ_θ to vertical applied stress p_z with radial distance r along horizontal axis for k = 0.

Stresses independent of elastic constants

The equations presented in figure 44 show that the stresses around the circular hole are dependent upon the magnitude of the applied stresses and the geometry or shape of the stressed body. The elastic constants E (Young's modulus) and ν (Poisson's ratio) do not appear in any of the equations and this means that the stress pattern is independent of the material used, provided that this is a linear elastic material.

This fact has been utilised by a number of researchers who have studied the distribution of stresses around excavations by means of *photoelasticity*[113]. This technique involves viewing a stressed glass or plastic model in polarised light. The stress pattern which appears under these conditions is related to the difference between the principal stresses σ_1 and σ_2 (or σ_3 if the smaller principal stress is tensile) in the plane of the model. Since these stresses do not depend upon the properties of the material, as discussed above, the photoelastic stress pattern can be used to calculate the stresses around an opening or openings of the same shape in hard rock. Photoelastic techniques are seldom used for this purpose today because stresses around underground excavations can be calculated more rapidly and more economically by means of the numerical techniques to be described later in this chapter.

Photoelastic stress pattern in a stressed glass plate containing several rectangular holes representing underground excavations in hard rock.

Stresses independent of size of excavation

It is important to note that the equations for the stresses around a circular hole in an infinite rock mass given in figure 44 do not include terms in the radius of the tunnel, a, but rather, include terms in the dimensionless parameter a/r. This means that the calculated stress levels at the boundaries of the excavation, for example, are independent of the absolute value of the radius. The same stress levels will be induced in the walls of a 1 metre diameter circular tunnel as in the walls of a 10 metre tunnel in the same elastic rock.

This fact has led to considerable confusion in the past. Some underground excavation designers have concluded that, because the stresses induced in the rock around an excavation are independent of the size of the excavation, the *stability* of the excavation is also independent of its size. If the rock mass were perfectly elastic and completely free of defects, this conclusion would be reasonably correct, but it is not valid for real rock masses which are already fractured. Even if the stresses are the same, the stability of an excavation in a fractured and jointed rock mass will be controlled by the ratio of excavation size to the size of the blocks in the rock mass. Consequently, increasing the size of an excavation in a typical jointed rock mass may not cause an increase in stress but it will almost certainly give rise to a decrease in stability.

The authors are aware of at least two mines where difficulties have been encountered when small scraper drifts have been enlarged to accommodate trackless mining equipment. The assumption was made that the stability of the excavations was independent of size and that a doubling of the span of the tunnels would have no significant influence upon their stability. This assumption has proved to be incorrect and serious stability problems have been encountered as a result of roof falls caused by the release of joints which had not been disturbed by the smaller scraper drifts.

Many early textbooks and papers on underground excavation design were based, almost entirely, upon elastic theory and ignored the influence of structural features such as joints, bedding planes and faults, discussed in the earlier chapters of this book. This over-simplification of the subject resulted in the sort of confusion discussed above. The reader should be aware of these historical facts when reading some of the earlier rock mechanics literature.

Principal stress contours

When considering the influence of the stresses in the rock surrounding an underground excavation upon the stability of that excavation, it is important that an assessment be made of the possible extent of the zone of fracturing around the excavation. As will be shown in the next chapter, the failure of a typical hard rock depends upon the magnitudes of the major and minor principal stresses acting at the point under consideration. Consequently, the most useful plot of the stresses surrounding an underground opening is a plot of the principal stress contours such as that reproduced in figure 47.

A set of 50 principal stress contour diagrams, for different excavation shapes and applied stress ratios, is presented in Appendix 3 at the end of this book. These plots were prepared by Dr El Sayed Ahmed Eissa, under the direction of Dr J.W. Bray, at Imperial College, London. The utilisation of these stress plots in assessing the extent of fracturing around underground excavations will be discussed in chapter 7.

Calculation of stresses around other excavation shapes

In the interests of simplicity, the discussion of stresses around underground excavations has been confined, to this point, to the case of the circular opening. Relatively few underground excavations are, in fact, circular in shape and it is important, therefore, that the stresses surrounding other excavation shapes be considered.

Available theoretical solutions have been reviewed by Jaeger and Cook[67], Obert and Duvall[68] and Denkhaus[114] who refer to the work of Muskhelishvili[71], Savin[72], Greenspan[115] and Heller et al[116]. These authors have published elegant solutions for stresses surrounding openings of various shapes. Such solutions played a very important role in the early development of rock mechanics, before the advent of the digital computer and the numerical techniques which are available today.

With the advent of the digital computer in the 1960's, a range of numerical stress analysis techniques was developed, and these have been refined to a high degree of effectiveness during the past fifteen years. The *Finite Element Method* was one of the first techniques for numerical stress analysis to be developed and it is still one of the most popular and powerful methods available. A very good review of this method has been published by Goodman[117] who gives a comprehensive list of references on the method and an illustrative finite element program for use by the reader. Goodman's

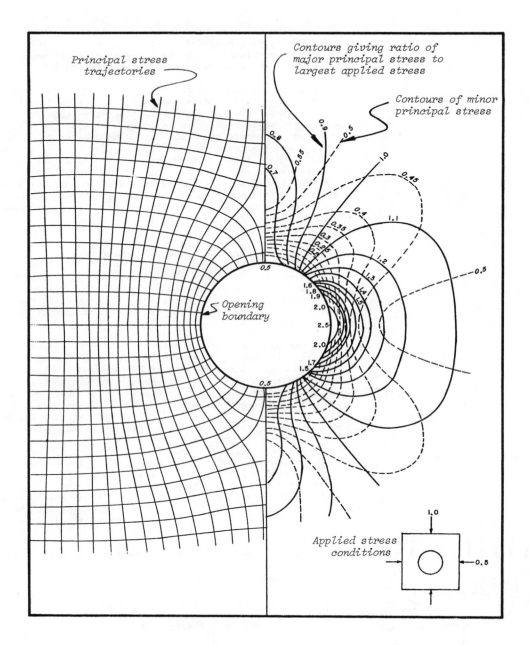

Figure 47 : Principal stress contours and principal stress trajectories in the material surrounding a circular hole in a stressed elastic body. As shown in the inset diagram, the ratio of applied stresses k = 0.5. Solid lines are major principal stress contours and dashed lines are minor principal stress contours. Contour values are ratios of principal stresses to the larger of the two applied stresses.

book is highly recommended for the interested engineer or geologist who wishes to obtain a clear introduction to the use of the finite element method in rock mechanics.

In spite of the power of the finite element method, the technique has certain disadvantages. When fine detail of the stresses around excavation boundaries is required or when large scale problems have to be analysed, the effort required to prepare the input data, the demands on computer storage and the costs of computer time can become considerable. Some of these problems are alleviated by the use of an alternative technique known as the *Boundary Element Method*. A discussion of this method is presented in Appendix 4 at the end of this book and a simple illustrative program is given for use by the interested reader. This program was used in the preparation of the stress contour plots given in Appendix 3.

Influence of excavation shape and orientation

An examination of the principal stress contour plots shown in Appendix 3 provides a number of useful guidelines for the designer of underground excavations by showing how adverse stress conditions can be induced for certain excavation shapes and orientations.

As an example, consider an excavation with an elliptical cross-section having axes with lengths in the ratio 2:1 oriented with its major axis at 0°, 45° and 90° to a uniaxial stress field, p_z, as shown in figure 48. The maximum compressive tangential stress in the sidewall of the elliptical opening increases from $2 p_z$ for the vertical ellipse to $3.62 p_z$ for the 45° ellipse to $5 p_z$ for the horizontal ellipse. The maximum tensile tangential stresses in the roof and floor of the excavation are $-p_z$ for the vertical and horizontal ellipses, and $-1.12 p_z$ for the 45° ellipse. Some qualitative sense of the way in which the boundary stresses might be expected to vary with orientation of the ellipse can be obtained by recognising that the sharper end of the ellipse will act as a more significant point of stress concentration than the flatter extremity. Clearly, the horizontal ellipse should be avoided for this particular stress field.

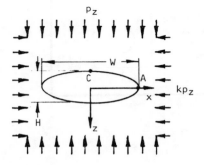

In a biaxial stress field, the tangential boundary stresses at the ends of the axes of an elliptical opening are given by the following equations:

$$\sigma_A = p_z\{1 + 2W/H - k\} \qquad (36)$$

$$= p_z\{1 + \sqrt{2W/\rho_A} - k\} \qquad (37)$$

$$\sigma_C = p_z\{k(1 + 2H/W) - 1\} \qquad (38)$$

$$= p_z\{k(1 + \sqrt{2H/\rho_C}) - 1\} \qquad (39)$$

where ρ_A and ρ_C are the radii of curvature at A and C, and the other symbols are defined in the margin sketch.

The form of equations 37 and 39 illustrates the influence of radius of curvature on the stress concentrations at the corners of the excavation; the smaller the radius of curvature, the higher the compressive stress concentration.

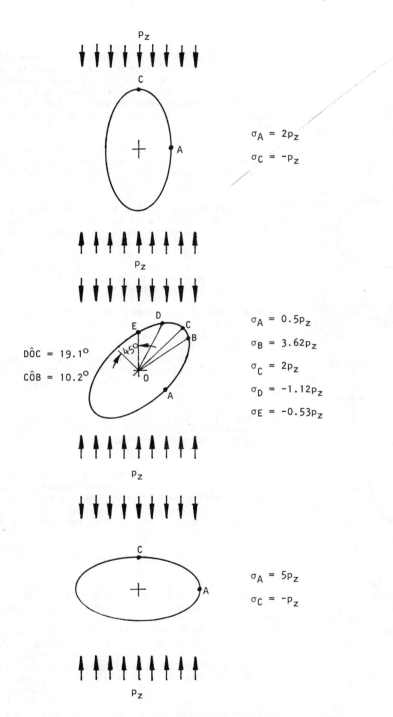

Figure 48 : Boundary stresses around elliptical excavations in a uniaxial stress field.

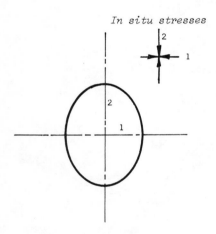

Lowest boundary stresses are given by an ovaloidal opening.

Minimum boundary stresses occur when the axis ratios of elliptical or ovaloidal openings are matched to the in situ stress ratio.

This general principle also applies to excavations with other cross-sections such as rectangles with rounded corners. The most favourable stress condition at the rounded corners of a rectangular opening is obtained when the radius of curvature takes the maximum possible value, half the excavation height, i.e. when the excavation has an ovaloidal cross-section.

Some of the main design principles which emerge from a consideration of the distribution of elastic stresses around excavations of various shapes and orientations in biaxial stress fields are :

- Critical stress concentrations increase as the relative radius of curvature of the boundary decreases. Openings with sharp corners should therefore be avoided.

- Since the lowest stresses on the boundary of the opening occur for the largest radius of curvature of that boundary, the optimum shape for an opening in a hydrostatic stress field (k = 1) is a circle.

- For stress fields other than hydrostatic (k ≠ 1), the lowest boundary stresses will be associated with an opening of ovaloidal shape. Hence, if a cavern with a height to width ratio of 1:2 has to be excavated in a stress field in which the horizontal stress is equal to half the vertical stress, the opening shape which will give the lowest boundary stresses is the ovaloid illustrated in the upper margin sketch.

- Boundary stresses in an elliptical opening can be reduced to a minimum if the axis ratio of the opening can be matched to the ratio between the in situ stresses.

- Under applied stress conditions in which the value of k is very low, tensile stresses occur on the boundaries of all excavation shapes. These tensile stresses are replaced by compressive stresses as the value of k increases above a value of approximately 1/3 as illustrated, for a circular excavation, in figure 45.

Stresses around multiple excavations

Consider the analogy of a smoothly flowing stream obstructed by three bridge piers as illustrated in figure 49. In order to accommodate the flow through the gaps between the piers, the streamlines are crowded together and the flow velocity increases through these gaps. The extent to which the flow velocity increases depends upon the ratio of the width of the stream to the sum of the distances between piers. The shape into which the streamlines are distorted depends upon the shape of the piers. Smoother streamlines will be associated with piers of circular or elliptical shape (with the major axis parallel to the flow direction) than with the square piers illustrated in figure 49.

A close analogy exists between this flow behaviour and the transmission of stress through the pillars between a series of parallel tunnels. This analogy gives rise to the term *Tributary theory* which is used by some authors[118,119] to describe the branching of stress trajectories and the

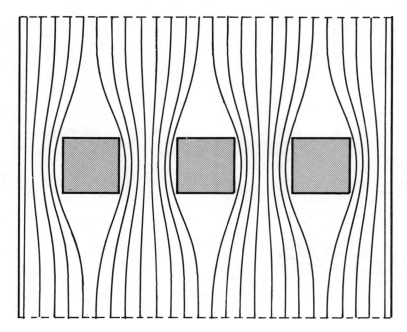

Figure 49 : Sketch of streamlines in a smoothly flowing stream obstructed by three bridge piers.

concentration of vertical stresses in the pillar between adjacent excavations.

The stresses at any point in a pillar depend upon

a. the *average* pillar stress which depends upon the ratio of the total area excavated to the total area remaining in the pillars, and

b. the *stress concentration* which is a function of the shape of the pillar between adjacent excavations.

It is convenient to treat these two effects separately and they will be dealt with in turn in the text which follows. In the interests of simplicity, this discussion will be confined to a set of uniform pillars in a single horizontal plane. The influence of multiple layers of pillars and of inclination of the entire system will be discussed later in this chapter.

Average pillar stresses

Figures 50 illustrates a typical square room and pillar layout used in mining horizontally bedded deposits of materials such as coal. Assuming that the pillars shown are part of a large array of pillars and that the rock load is uniformly distributed over these pillars, the average pillar stress is given by :

$$\sigma_p = p_z(1 + W_o/W_p)^2 = \gamma z(1 + W_o/W_p)^2 \qquad (40)$$

where γ is the unit weight of the rock, z is the depth below surface and W_o and W_p are the widths of the opening and the pillar respectively.

Average pillar stresses for different pillar layouts are summarised in figure 51 and, in all cases, the value of σ_p is given by the ratio of the weight of the rock column carried by an individual pillar to the plan area of the pillar.

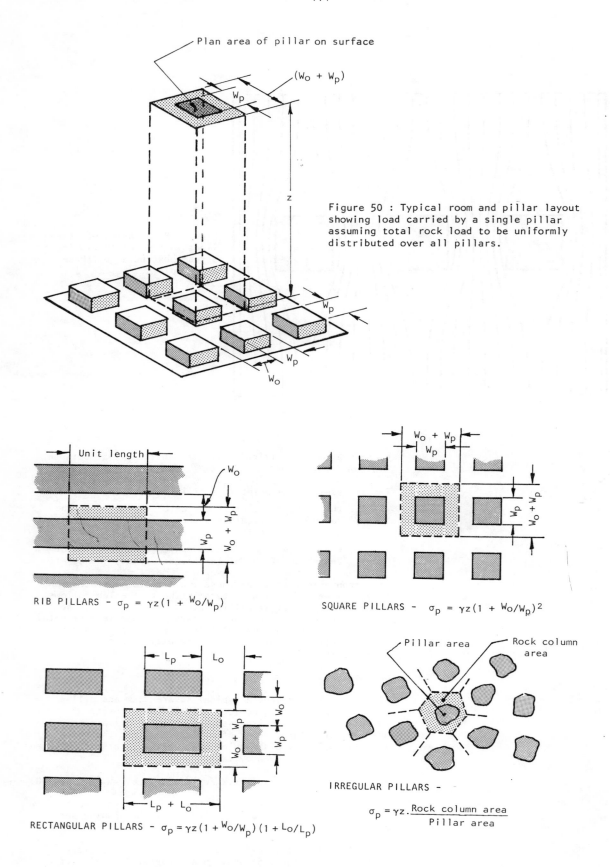

Figure 50 : Typical room and pillar layout showing load carried by a single pillar assuming total rock load to be uniformly distributed over all pillars.

RIB PILLARS - $\sigma_p = \gamma z (1 + W_o/W_p)$

SQUARE PILLARS - $\sigma_p = \gamma z (1 + W_o/W_p)^2$

RECTANGULAR PILLARS - $\sigma_p = \gamma z (1 + W_o/W_p)(1 + L_o/L_p)$

IRREGULAR PILLARS -

$\sigma_p = \gamma z \cdot \dfrac{\text{Rock column area}}{\text{Pillar area}}$

Figure 51 : Average vertical pillar stresses in typical pillar layouts. *Illustrations are all plan views.*

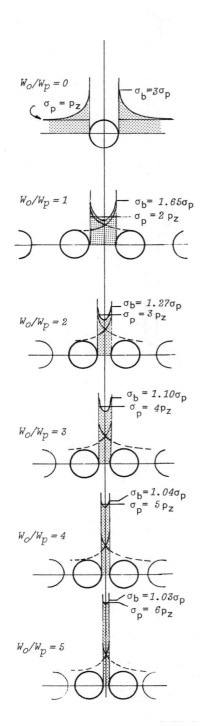

Stresses in rib pillars between parallel circular tunnels - after Obert and Duvall[68].

Influence of pillar shape

The shape of a pillar between two adjacent excavations depends upon the shape of the excavations and their distance apart. The shape of a pillar has a major influence upon the stress distribution within that pillar.

Obert and Duvall[68] report the results of photoelastic studies carried out to determine the stress distribution in rib pillars between a number of parallel circular tunnels. The type of plate model which could be used in such studies is illustrated in figure 52. Figure 53 shows that the average vertical stress at the mid height of the pillar is given by

$$\sigma_p = (1 + W_o/W_p)p_z \qquad (41)$$

The distribution of the maximum principal stress σ_1 across the mid height of the pillar can be approximated by super-imposing the two stress distributions surrounding the individual tunnels, as defined in figure 46 on page 106. Note that the *average* value of the maximum principal stress σ_1 across the pillar must be equal to the average pillar stress σ_p in order to satisfy the conditions of equilibrium in the model.

Obert and Duvall's results, for different ratios of W_o/W_p are summarised in the margin drawings. These show that the average pillar stress σ_p increases as the pillar becomes narrower. On the other hand, the maximum boundary stress concentration σ_b/σ_p decreases as the tunnels are moved closer together.

This trend is illustrated more clearly in the series of stress distributions reproduced in figures 54 to 58. These stress distributions* are for pillars between rectangular openings in which the width of the opening W_o is equal to the pillar width W_p. In all cases, the contour values are expressed as ratios of the major and minor principal stresses, σ_1 and σ_3 respectively, to the average pillar stress σ_p. These plots show that, as the pillar becomes taller and narrower, the stress distribution across the mid-height of the pillar becomes more uniform. In the case of the very slender pillar shown in figure 54, the stress conditions across the centre of the pillar are very close to uniaxial stress conditions in which $\sigma_1 = \sigma_p$ and $\sigma_3 = 0$. On the other hand, in the case of the squat pillar shown in figure 58, the distribution of stress across the pillar is anything but uniform. At the centre of the pillar, the maximum principal stress falls to a value below that of the average pillar stress but the minor principal stress increases to a level which is a significant proportion of the average pillar stress. As will be shown in a later chapter of this book, the triaxial stress conditions generated at the centre of squat pillars are very important in determining the stability of these pillars.

Up to this point, the discussion has been restricted to the distribution of stress in rib pillars between parallel excavations. In the case of square pillars, such as those illustrated in figure 50, it is necessary to consider the

---

* Computed by Dr. R.D. Hammett of Golder Associates Ltd., Vancouver, using a finite element program.

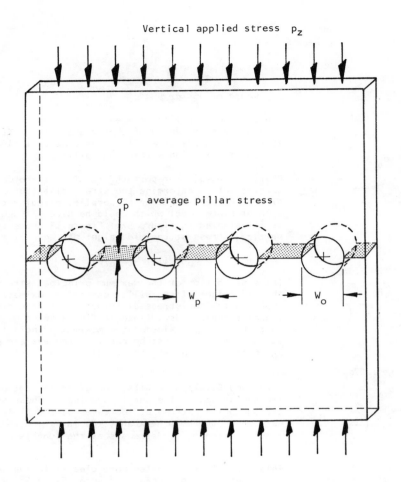

Figure 52 : Plate model containing a series of holes representing parallel circular tunnels.

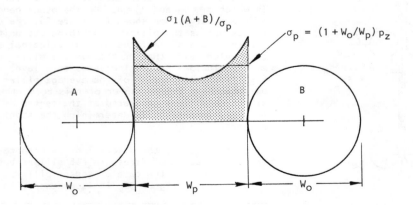

Figure 53 : Distribution of major principal stress in pillar depends upon average pillar stress and stress concentration around individual tunnels.

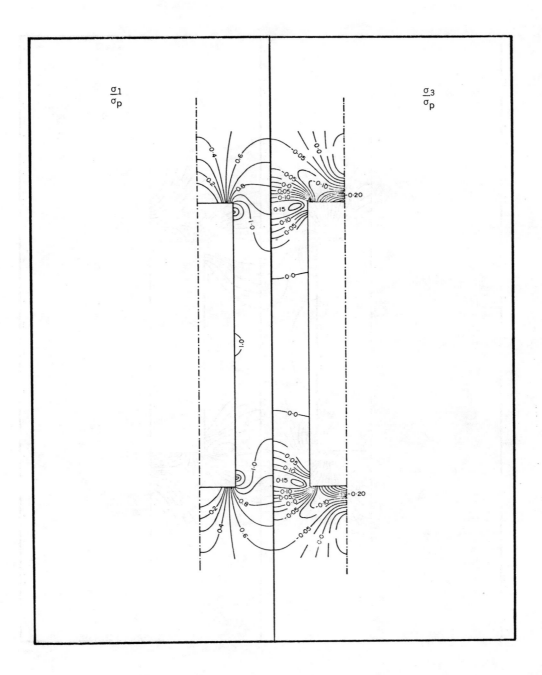

Figure 54 : Principal stress distributions in a rib pillar defined by a ratio of pillar height to pillar width of 4.0. The contour values are given by the ratio of major and minor principal stresses to the average pillar stress.
Plane strain analysis for uniformly distributed vertical applied stress with no horizontal stress.

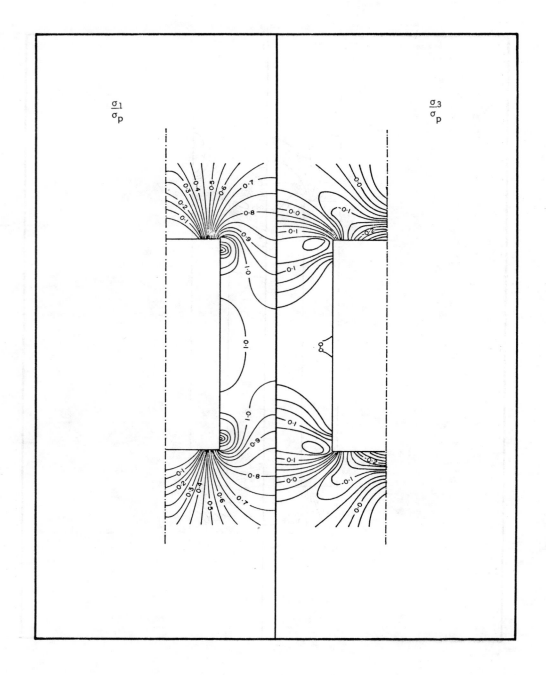

Figure 55 : Principal stress distributions in a rib pillar defined by a ratio of pillar height to pillar width of 2.0. The contour values are given by the ratio of major and minor principal stresses to the average pillar stress.
Plane strain analysis for uniformly distributed vertical applied stress with no horizontal stress.

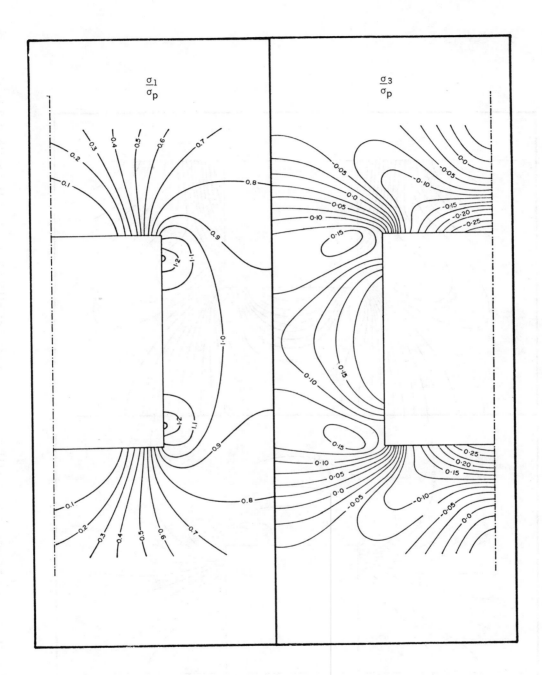

Figure 56 : Principal stress distributions in a rib pillar defined by a ratio of pillar height to pillar width of 1.0. The contour values are given by the ratio of major and minor principal stresses to the average pillar stress.
Plane strain analysis for uniformly distributed vertical applied stress with no horizontal stress.

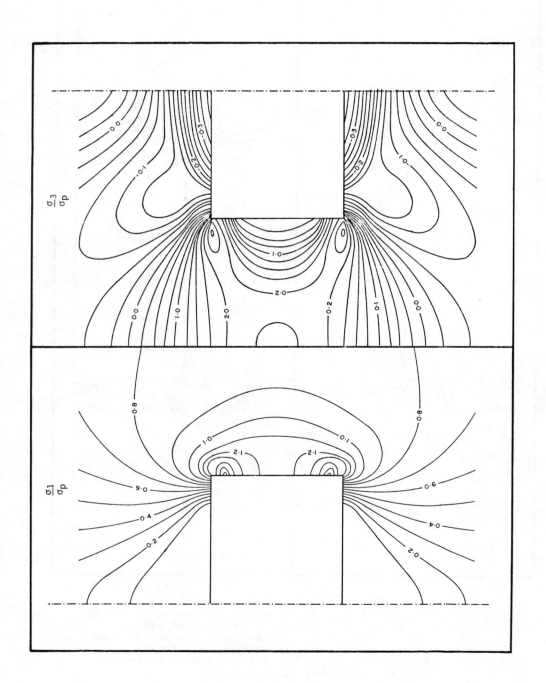

Figure 57 : Principal stress distributions in a rib pillar defined by a ratio of pillar height to pillar width of 0.5. The contour values are given by the ratio of major and minor principal stresses to the average pillar stress. Plane strain analysis for uniformly distributed vertical applied stress with no horizontal stress.

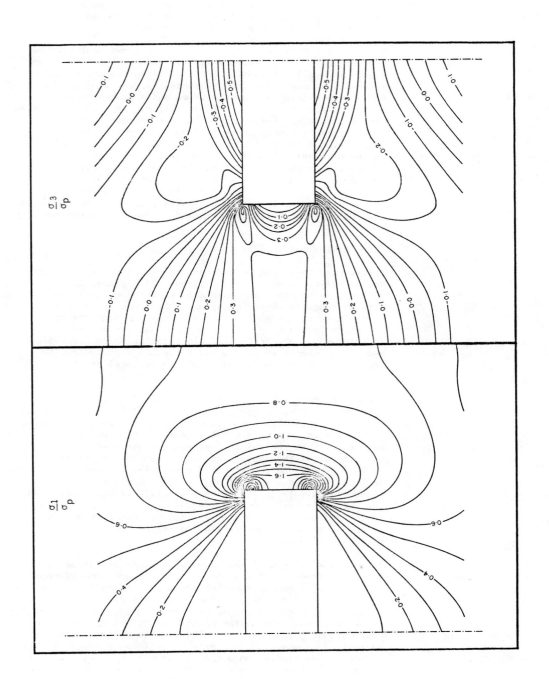

Figure 58 : Principal stress distributions in a rib pillar defined by a ratio of pillar height to pillar width of 0.25. The contour values are given by the ratio of major and minor principal stresses to the average pillar stress. Plane strain analysis for uniformly distributed vertical applied stress with no horizontal stress.

additional effects of the stress field due to the two openings running at right angles to the two openings on either side of the rib pillar. The superposition of these two stress fields is illustrated diagrammatically in figure 59.

Consider the example of the squat pillar shown in figure 58. The maximum principal stress at the mid-height of the excavation sidewall is $1.60\sigma_p$, where σ_p is the average pillar stress. This maximum principal stress is made up from the sum of the average pillar stress σ_p and a *stress increment* of $0.60\sigma_p$ due to the shape of the pillar. When super-imposing the second stress field due to two excavations running at right angles to those shown in figure 58, only the stress increment due to the second set of excavations is added to the original value of σ_1. Hence, at the corner of a square pillar with a height to width ratio of 0.25, the maximum principal stress at the mid-height of the pillar is given by $\sigma_1 = (1 + 0.60 + 0.60)\sigma_p = 2.20\sigma_p$. The addition of the stress increments to the average pillar stress ensures that the average pillar stress is only added into the sum once.

Three-dimensional pillar stress problems

In the case of a complex underground excavation layout such as that illustrated in figure 37 on page 85, the stress distribution in the rock mass can no longer be analysed with an adequate degree of accuracy by means of the two-dimensional stress analysis methods described on the previous pages. Unfortunately, very few practical and economical three-dimensional stress analysis techniques are available and those methods which are available are very tedious to apply and demand a high degree of experimental or theoretical skill.

One of the most powerful three-dimensional stress analysis techniques is *frozen stress photoelasticity*[120]. This method makes use of the fact that certain types of plastics, when slowly heated to a critical temperature while under load, will retain the photoelastic stress pattern after cooling and removal of the load. The models can then be sliced very carefully and the stress distributions in various sections through the model determined. The use of this method is only justified in very special circumstances since the extremely critical experimental techniques which have to be used coupled with the very tedious calculations required to separate the principal stresses make the technique very expensive to apply.

Recent developments of the boundary element technique have been described by Hocking, Brown and Watson[121] and by Brown and Hocking[122] and this method shows considerable promise for application to the solution of certain classes of three-dimensional stress problems. Figure 60 shows the distribution of major principal stresses on the boundaries of two intersecting underground caverns. This stress distribution was determined by means of the three-dimensional boundary element method during studies on a project involving large underground excavations where the stability of the intersections between the excavations was considered to be important enough to justify the use of this method. It is anticipated that future developments of numerical methods such as this will provide practical three-dimensional stress analysis tools in the years to come.

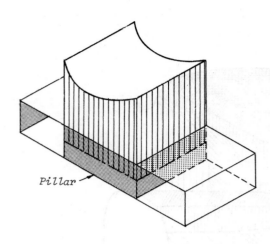

1. Stress distribution in north-south pillar due to interaction of the stress fields surrounding parallel north-south roadways.

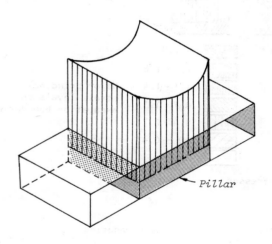

2. Stress distribution in east-west pillar due to interaction of the stress fields surrounding parallel east-west roadways.

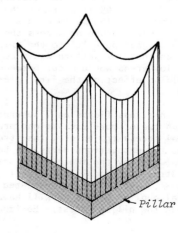

3. Stress distribution is a square or rectangular pillar obtained by superposition of the stresses in north-south and east-west pillars.

Figure 59 : Distribution of maximum principal stresses acting on a plane through the centre of a square or rectangular pillar surrounded by a large number of similar pillars.

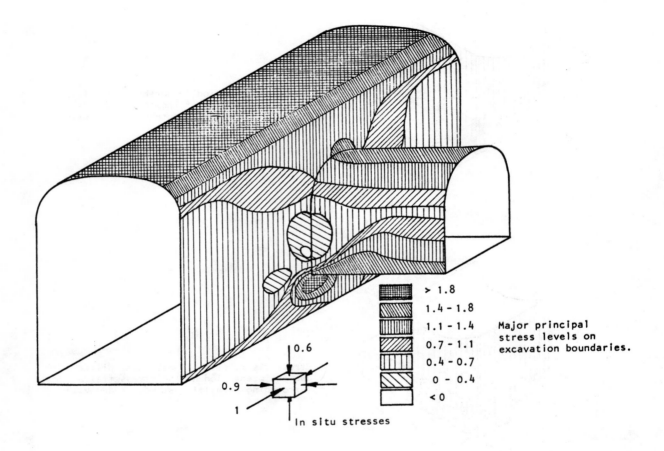

Figure 60 : Distribution of major principal stresses on the boundaries of two intersecting underground excavations.

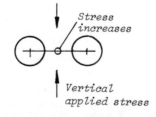

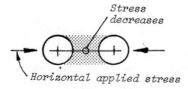

Stress shadows

Returning, for a moment, to the stream flow analogy discussed earlier in this chapter, figure 49 illustrates the effect of three bridge piers upon the flow pattern in a smoothly flowing stream. If, instead of running across the river, these piers were to be aligned parallel to the axis of the river, the effect on the flow pattern would be entirely different. Stagnant flow areas would occur between the piers due to the shielding effect of the first pier seen by the water flow.

Similar effects occur in stress fields as illustrated in the margin sketch. When two or more excavations are aligned along a major principal stress trajectory, the stress in the pillar between the two excavations decreases because it lies in the 'shadow' cast by the two excavations. Hence, when considering the stresses in a pillar which is part of a room and pillar layout, such as that illustrated in figure 50, the vertical applied stress p_z will have the most significant effect upon the pillar stress. Horizontal stresses (kp_z) will have very little effect on the stress distribution in pillars in the centre of the array since these pillars will be shielded from these stresses by the pillars near the edge of the panel.

Similar considerations apply when dealing with multi-seam mining in which the rock between excavations placed one above another will be shielded from vertical stresses by these excavations. This effect is illustrated in the photo-elastic pattern reproduced in the margin of page 107 . This shows that the horizontal pillars between the three excavations in a vertical line are stress relieved whereas the stress is intensified in the pillar between these three excavations and the single excavation to their left.

Influence of inclination upon pillar stresses

When mining in an inclined orebody, the stress field acting upon the excavations and the pillars between these excavations is no longer aligned normal to and parallel to the excavation boundaries as has been the case in all examples discussed thus far. The inclination of the stress field to the excavation boundaries results in a considerable change in the stress distributions induced in the rock surrounding the excavations, as illustrated in figures 61 and 62. These stress distributions are not particularly difficult to determine by means of either the finite element or boundary element techniques but care has to be taken in applying some of the approximate methods of pillar stress calculation or stress superposition to these inclined excavation problems.

Influence of gravity

In the preceding discussion it has been assumed that the applied stresses p_z and kp_z are uniform as illustrated in figure 44. These conditions are equivalent to those in a uniformly loaded plate model and they are based upon the assumption that the excavation under consideration is far enough below the ground surface that the influence of stress gradients due to gravitational loading can be ignored. Denkhaus[114] examined the errors associated with this assumption and concluded that these errors are less than 5% when the depth of the excavation below surface exceeds 10 times the span of the excavation.

Obviously, for shallow tunnels or for very large excavations at shallow depth, the influence of gravitational stresses must be taken into account in calculating the stresses induced in the rock around the excavations. A full discussion on this topic would exceed the scope of this chapter but allowance has been made for gravitational stresses in the boundary element program presented in Appendix 4 at the end of this book. An example of the stresses induced in the rock surrounding a shallow excavation has been included in this appendix.

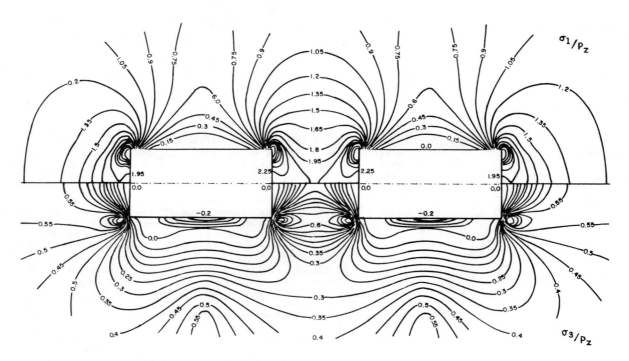

Figure 61 : Principal stress distributions in the rock surrounding two adjacent excavations aligned normal and parallel to the applied stress directions. (k = 0.5)

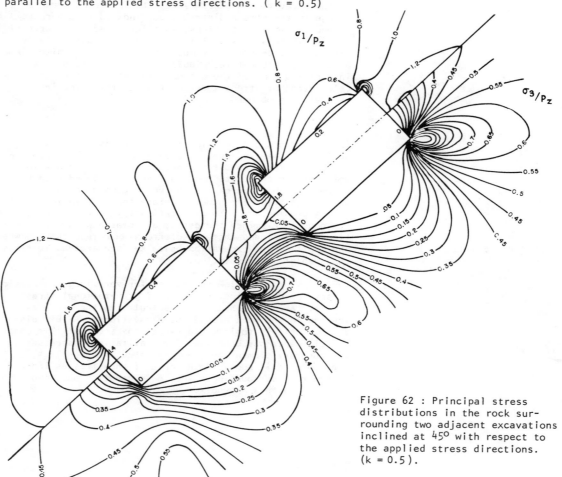

Figure 62 : Principal stress distributions in the rock surrounding two adjacent excavations inclined at 45° with respect to the applied stress directions. (k = 0.5).

Chapter 5 references

67. JAEGER, J.C. and COOK, N.G.W. *Fundamentals of Rock Mechanics*. Chapman and Hall, London, 1976, 585 pages.

68. OBERT, L. and DUVALL, W.I. *Rock Mechanics and the Design of Structures in Rock*. John Wiley & Sons, New York, 1967, 650 pages.

69. JAEGER, J.C. *Elasticity, Fracture and Flow*. Methuen & Co., London. 3rd edition, 1969, 268 pages.

70. LOVE, A.E.H. *A Treatise on the Mathematical Theory of Elasticity*. Dover Press, New York, 1927, 643 pages.

71. MUSKHELISHVILI, N.I. *Some Basic Problems of the Mathemetical Theory of Elasticity*. Translated from Russian by J.R.M. Radok, published by Noordhoff, Groningen, 4th edition, 1953.

72. SAVIN, G.N. *Stress Concentrations Around Holes*. Translated from Russian by E.Gros, published by Pergamon Press, Oxford, 1961, 430 pages.

73. TIMOSHENKO, S.P. and GOODIER, J.N. *Theory of Elasticity*. McGraw-Hill, New York, 2nd edition, 1951, 506 pages.

74. TERZAGHI, K. and RICHART, F.E. Stresses in rock about cavities. *Géotechnique*, Vol. 3, 1952, pages 57-90.

75. HAST, N. *The measurement of rock pressure in mines*. Sveriges Geol. Undersokn, Årsbok 52, no. 3, 1958.

76. TALOBRE, J. *La Méchanique des Roches*. Dunod, Paris, 1957, 444 pages.

77. HEIM, A. Zur Frage der Gebirgs- und Gesteinsfestigkeit. *Schweiz. Bauztg.* Vol. 50, February, 1912.

78. WOROTNICKI, G. and DENHAM, D. The state of stress in the upper part of the earth's crust in Australia according to measurements in mines and tunnels and from seismic observations. *Proc. ISRM Symposium on Investigation of Stress in Rock - Advances in Stress Measurement*, Sydney, Australia, August 1976, pages 71-82.

79. BRADY, B.H.G., FRIDAY, R.G. and ALEXANDER, L.G. Stress measurements in a bored raise at the Mount Isa Mine. *Proc. ISRM Symposium on Investigation of Stress in Rock - Advances in Stress Measurement*, Sydney, Australia, August 1976, pages 12-16.

80. ENDERSBEE, L.A. and HOFTO, E.O. Civil engineering design and studies in rock mechanics for Poatina underground power station, Tasmania. *Journal of the Institution of Engineers, Australia*, Vol. 35, 1963, pages 187-209.

81. HERGET, G. Variations of rock stress with depth at a Canadian iron mine. *Int. J. Rock Mechanics and Mining Sciences*, Vol. 10, 1973, pages 35-51.

82. BUCHBINDER, G.G.R., NYLAND, E. and BLANCHARD, J.E. Measurement of stress in boreholes. *Canadian Geol. Survey Paper* 66-13, 1966.

83. EISBACHER, G.H. and BIELENSTEIN, H.U. Elastic strain recovery in Proterozoic rocks near Elliot Lake, Ontario. *J. Geophys. Research*, Vol. 76, 1971, pages 2012-2021.

84. COATES, D.F. and GRANT, F. Stress measurements at Elliot Lake. *Canadian Min. Metall. Bulletin*, Vol. 59, 1966, pages 603-613.

85. BENSON, R.P., KIERANS, T.W. and SIGVALDSON, O.T. In situ and induced stresses at the Churchill Falls underground power house, Labrador. *Proc. 2nd. Congress Int. Soc. Rock Mech.*, Belgrade, Vol. 2, 1970, pages 821-832.

86. IMRIE, A.S. and JORY, L.T. Behaviour of the underground powerhouse arch at the W.A.C. Bennett dam during excavation. *Proc. 5th Canadian Rock Mech. Symp.*, Toronto, 1968, pages 19-39.

87. IMRIE, A.S. and CAMPBELL, D.D. Engineering geology at the Mica underground plant. *Proc. 1976 Rapid Excavn. & Tunn. Conf.*, R.J.Robbins and R.J. Conlon, eds., AIME, New York, 1976, pages 534-549.

88. HAIMSON, B.C. Earthquake related stresses at Rangely, Colorado. *Proc. 14th Rock Mech. Symp.*, H.R.Hardy and R.Stefanko, eds., ASCE, New York, 1973, pages 689-708.

89. HAIMSON, B.C., LACOMB, J., JONES, A.H. and GREEN, S.J. Deep stress measurements in tuff at the Nevada test site. *Proc. 3rd Congress, Int. Soc. Rock Mech.*, Denver, Vol. 2A, 1974, pages 557-561.

90. HAIMSON, B.C. Design of underground powerhouses and the importance of pre-excavation stress measurements. *Proc. 16th Rock Mech. Symp.*, C. Fairhurst and S.L. Crouch, eds., ASCE, New York, 1977, pages 197-204.

91. HAIMSON, B.C. The hydrofracturing stress measuring method and recent field results. *Intnl. J. Rock Mechanics and Mining Sciences*, Vol. 15, 1978, pages 167-178.

92. HAIMSON, B.C. and STAHL, E.J. Hydraulic fracturing and the excavation of minerals through wells. *Proc. 3rd Symp. on Salt*, Northern Ohio Geol. Soc., 1969, pages 421-432.

93. OBERT, L. In situ determination of stress in rock. *Min. Engng.*, Vol. 14, No. 8, 1962, pages 51-58.

94. CHAN, S.S.M. and CROCKER, T.J. A case study of in situ rock deformation behaviour in the Silver Summit mine, Coeur d'Alène mining district. *Proc. 7th Canadian Rock Mech. Symp.*, Edmonton, 1971, pages 135-160.

95. AGETON, R.W. Deep mine stress determination using flatjack and borehole deformation methods. *U.S. Bur. Mines Rept. Invn.* 6887, 1967.

96. CONWAY, J.P. Progress report on the Crescent Mine overcoring studies. *U.S. Bur. Mines*, Spokane, Wash. 1968.

97. HOOKER, V.E., BICKEL, D.L. and AGGSON, J.R. In situ determination of stresses in mountainous topography. *U.S. Bur. Mines Rept. Invn.* 7654, 1972.

98. BREDEHOFF, J.D., WOLFF, R.G., KEYS, W.S. and SHUTER, E. Hydraulic fracturing to determine the regional stress field, Piceance Basin, Colorado. *Bull. Geol. Soc. Am.* Vol. 87, 1976, pages 250-258.

99. MYRVANG, A.M. Practical use of rock stress measurements in Norway. *Proc. ISRM Symp. on Investigation of Stress in Rock - Advances in Stress Measurement*. Sydney, Australia, 1976, pages 92-99.

100. LI, B. Natural stress values obtained in different parts of the Fennoscandian rock mass. *Proc. 2nd Congr., Int. Soc. Rock Mech.*, Belgrade, 1970, Paper 1-28.

101. HAST, N. The state of stress in the upper part of the Earth's crust. *Engng. Geol.*, Vol. 2, 1967, pages 5-17.

102. DENKHAUS, H.G. The significance of stress in rock masses. *Proc. Int. Symp. Rock Mech.*, Madrid, 1968, pages 263-271.

103. GAY, N.C. In-situ stress measurements in Southern Africa. *Tectonophysics*, Vol. 29, 1975, pages 447-459.

104. VAN HEERDEN, W.L. Practical application of the C.S.I.R. triaxial strain cell for rock stress measurements. *Proc. ISRM Symp. on Investigation of Stress in Rock - Advances in Stress Measurement*, Sydney, Australia, 1976, pages 1-6.

105. BOWCOCK, J.B., BOYD, J.M., HOEK, E. and SHARP, J.C. Drakensberg Pumped Storage Scheme - rock engineering aspects. *Proc. Symp. Exploration for Rock Engineering*, Z.T.Bieniawski ed., A.A.Balkema, Rotterdam, Vol. 2, 1977, pages 121-139.

106. GAY, N.C. Principal horizontal stresses in Southern Africa. *Pure Appl. Geophys.*, Vol. 115, 1977, pages 1-10.

107. LEEMAN, E.R. The determination of the complete state of stress in rock in a single borehole - Laboratory and underground measurements. *Intnl. J. Rock Mechanics and Mining Sciences*, Vol. 5, 1968, pages 31-56.

108. DOUGLAS, T.H., RICHARDS, L.R. and O'NEILL, D. Site investigation for main underground complex - Dinorwic Pumped Storage Scheme. *Field Measurements in Rock Mechanics*, K. Kovari, ed., A.A.Balkema, Rotterdam, Vol. 2, 1977, pages 551-567.

109. HAST, N. Global measurements of absolute stress. *Phil. Trans. Royal Soc.*, 274A, 1973, pages 409-419.

110. KLUTH, D.J. Rock stress measurements in the Jor underground power station of the Cameron Highlands Hydroelectric scheme. *Trans. 8th Int. Conr. Large Dams*, Edinburgh, Vol. 1, 1964, pages 103-119.

111. LE FRANCOIS, P. In situ measurement of rock stress for the Idikki hydroelectric project. *Proc. 6th Canadian Rock Mech. Symp.*, Montreal, 1970, pages 65-90.

112. KIRSCH, G. Die theorie der elastizität und die bedürfnisse der festigkeitslehre. *Veit. Ver. Deut. Ing.*, Vol. 42, No. 28, 1898, pages 797-807.

113. HOEK, E. A photoelastic technique for the determination of potential fracture zones in rock structures. *Proc. 8th Rock. Mech. Symp.*, C. Fairhurst ed., AIME, New York, 1967, pages 94-112.

114. DENKHAUS, H.G. The application of the mathematical theory of elasticity to problems of stress in hard rock at great depth. *Papers and Discussions Assn. Mine Managers of South Africa*, 1958, pages 271-310.

115. GREENSPAN, M. Effect of a small hole on the stresses in a uniformly loaded plate. *Quart. Appl. Maths.*, Vol. 2, 1944, pages 60-71.

116. HELLER, S.R., BROCK, J.S. and BART, R. The stresses around a rectangular opening with rounded corners in a uniformly loaded plate. *Trans. 3rd US Cong. Appl. Mech.*, 1958, page 357.

117. GOODMAN, R.E. *Methods of Geological Engineering*. West Publishing Co., St. Paul, Minnesota, 1976, 472 pages.

118. MORRISON, R.G.K. *A Philosophy of Ground Control*. Published by Dept. Min. Metall. Engg., McGill Univ., 1976, 182 pages.

119. DUVALL, W.I. General principles of underground opening design in competent rock. *Proc. 17th Rock Mech. Symp.*, AIME, New York, 1977, pages 101-111.

120. DURELLI, A.J. and RILEY, W.F. *Introduction to Photomechanics*. Prentice-Hall, Englewood Cliffs, New Jersey, 1965.

121. HOCKING, G., BROWN, E.T. and WATSON, J.O. Three-dimensional elastic stress analysis of underground openings by the boundary integral equation method. *Proc. 3rd Symp. Engineering Applications of Solid Mechanics*, Toronto, 1976, pages 203-216.

122. BROWN, E.T. and HOCKING, G. The use of the three dimensional boundary integral equation method for determining stresses at tunnel intersections. *Proc. 2nd Australian Tunnelling Conf.*, Melbourne, 1976, pages 55-64.

Chapter 6: Strength of rock and rock masses

Introduction

The stability of an underground excavation depends upon the structural conditions in the rock mass, as discussed in chapter 2, and also upon the relationship between the stress in the rock and the strength of the rock. Shallow excavations such as most road and rail tunnels or the near surface workings in mines are most strongly influenced by the structural conditions and the degree of weathering of the rock mass. On the other hand, the stability of deep excavations depends more upon the response of the rock mass to the stress field induced around the excavations. The different types of instability which occur under these two extremes will be discussed in the next chapter.

In order to utilise the knowledge of stresses induced around underground excavations discussed in the previous chapter, it is necessary to have available a criterion or a set of rules which will predict the response of a rock mass to a given set of induced stresses. Such a need has long been recognised and a large proportion of rock mechanics literature is devoted to the search for a suitable failure criterion.

The difficulty of finding a realistic failure criterion for rock masses is emphasised in figure 63 which shows the transition from intact rock material to a heavily jointed rock mass. The underground excavation designer is concerned with all the stages in this transition. The processes of drilling and blasting or the use of tunnel boring machines or raise borers for the excavation of underground openings are strongly influenced by the strength of the intact rock material. The stability of the rock in the immediate vicinity of the underground openings and the behaviour of the rockbolts used to support this rock are related to the existing discontinuities and to fractures induced in the intact rock by blasting. The stability of the entire system of underground openings which make up a mine or an underground hydroelectric scheme depends upon the behaviour of the entire rock mass surrounding these openings. This rock mass may be so heavily jointed that it will tend to behave like an assemblage of tightly interlocking angular particles with no significant strength under unconfined conditions.

In considering the behaviour of the different systems in the transition between intact rock and a heavily jointed rock mass, it must be remembered that the quantity and quality of experimental data decrease rapidly as one moves from the intact rock sample to the rock mass. Because small samples of intact rock are easy to collect and to test under a variety of laboratory conditions, there is a vast amount of information on almost every aspect of intact rock behaviour. Experimental difficulties increase significantly in tests on specimens containing one set of discontinuities and become very serious when two or more sets of discontinuities are present. Full scale tests on heavily jointed rock masses are extremely difficult because of the experimental problems of preparing and loading the samples and are very expensive because of the scale of the operation. Consequently, test data for large scale rock mass behaviour will never be available in similar quantities to that for intact rock samples.

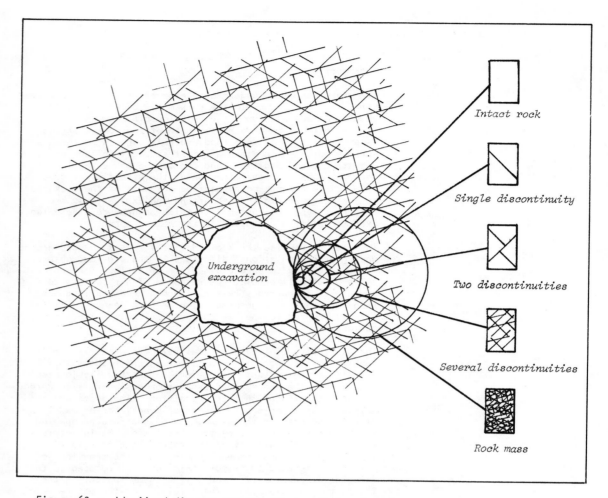

Figure 63 : Idealised diagram showing the transition from intact rock to a heavily jointed rock mass with increasing sample size.

Taking all of these factors into account, it can be seen that a rock failure criterion which will be of significant use to the underground excavation designer should satisfy the following requirements :

a. It should adequately describe the response of an intact rock sample to the full range of stress conditions likely to be encountered underground. These conditions range from uniaxial tensile stress to triaxial compressive stress.

b. It should be capable of predicting the influence of one or more sets of discontinuities upon the behaviour of a rock sample. This behaviour may be highly anisotropic, i.e. it will depend upon the inclination of the discontinuities to the applied stress direction.

c. It should provide some form of projection, even if approximate, for the behaviour of a full scale rock mass containing several sets of discontinuities.

While the authors do not claim that the failure criterion presented on the following pages meets all of these requirements, they do feel that it provides a simple empirical relationship which is sufficiently accurate for most underground excavation design processes which are dealt with in this book.

Brittle and ductile behaviour

Throughout this chapter, rock failure will be referred to as being brittle, ductile or at the brittle-ductile transition. It is important, therefore, that the meaning of these terms be defined at the outset.

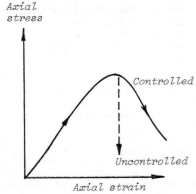

Stress-strain curves for brittle fracture in uniaxial compression

Brittle failure is said to occur when the ability of the rock to resist load decreases with increasing deformation. Brittle failure is often associated with little or no permanent deformation before failure and, depending upon the test conditions, may occur suddenly and catastrophically. Rock bursts in deep hard rock mines provide graphic illustrations of the phenomenon of explosive brittle fracture[123].

A material is said to be *ductile* when it can sustain permanent deformation without losing its ability to resist load. Most rocks will behave in a brittle rather than a ductile manner at the confining pressures and temperatures encountered in civil and mining engineering applications. Ductility increases with increased confining pressure and temperature, but can also occur in weathered rocks, heavily jointed rock masses and some weak rocks such as evaporites under normal engineering conditions.

As the confining pressure is increased it will reach the *brittle-ductile transition* value at which there is a transition from typically brittle to fully ductile behaviour. Byerlee[124] has defined the brittle-ductile transition pressure as the confining pressure at which the stress required to form a failure plane in a rock specimen is equal to the stress required to cause sliding on that plane.

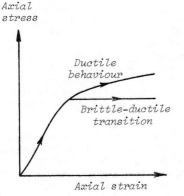

Stress-strain curves for ductile behaviour in compression

As noted above, the brittle failure occurring in rocks under either laboratory or field conditions is often of a violent or uncontrolled nature. In other situations, such as in mine pillars, the rock may be fractured and deformed past its peak load bearing capability in a controlled manner and reach equilibrium at some lower load. In the first case, explosive failure occurs at the peak stress, and the post-peak section of the stress-strain curve will not be recorded. In the second case, progressive fracture of the rock will be observed, and the post-peak section of the stress-strain curve will be recorded.

Which of these two general modes of behaviour occurs depends on the relative stiffnesses of the specimen being loaded and the loading system, be it a laboratory testing machine or the rock mass surrounding and overlying a volume of rock in-situ. In the laboratory, the likelihood of uncontrolled failure occurring can be reduced by using stiff or servo-controlled testing machines. A complete account of the mechanics of machine-specimen interaction and the use of stiff and servo-controlled machines is given by Hudson, Crouch and Fairhurst[125]. The interested reader is also referred to the discussion given by Jaeger and Cook[67], pages 177-183. In practice, the concept of the complete stress-strain or force-displacement curves for brittle rocks and rock masses is vital to the understanding and analysis of the behaviour of highly stressed rock in pillars or around underground excavations.

Laboratory testing of intact rock samples

Before entering into a discussion on the interpretation of strength data obtained from tests on intact samples, it is necessary briefly to review the range of laboratory tests which have been used in the generation of these data.

Uniaxial tensile tests

One extreme of the range of stress conditions of interest in this discussion is the state of uniaxial tension in which $\sigma_1 = \sigma_2 = 0$ and $\sigma_3 = -\sigma_t$ where σ_t is the uniaxial tensile strength of the specimen.

Figure 64 illustrates the uniaxial testing arrangement suggested by Hawkes and Mellor[126]*. This arrangement satisfies all the requirements which the authors of this book consider to be essential for the generation of meaningful tensile strength data. The use of a ball joint on the end of a non-twist cable ensures that the load will be applied along the axis of the specimen with an absence of torsion. Provided that the components are machined with a reasonable degree of precision and that care is taken in bonding the rock specimen into the collars, there will be a minimum of bending in the specimen. The use of aluminium for the collars and the use of epoxy resin for bonding ensures that the stress will be transmitted into the specimen without the severe stress concentrations which are usually associated with more rigid specimen gripping arrangements such as those used in testing steel specimens. Unless all of these conditions are satisfied, it is doubtful if much significance can be attached to tensile strength results.

Triaxial compression-tensile tests

In order to obtain rock fracture data under all the stress conditions of interest in this discussion, one of the authors used the testing arrangement illustrated in figure 65. The rubber sleeved "dog-bone" specimen is subjected to hydraulic pressure p which generates radial stresses $\sigma_1 = \sigma_2 = p$ and an axial tensile stress σ_3 which is given by :

$$\sigma_3 = -\frac{p(d_2^2 - d_1^2)}{d_1^2} \qquad (42)$$

where d_1 is the diameter of the centre of the specimen and d_2 is the diameter of the enlarged ends of the specimen.

Adjustment of the pressure applied to the rubber sealing rings can be achieved by tightening or loosening the

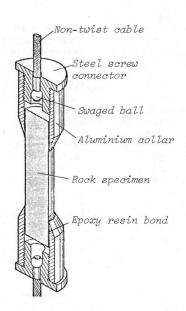

Figure 64 : Uniaxial tensile test arrangement suggested by Hawkes and Mellor[126].

---

* This paper, entitled "Uniaxial testing in rock mechanics laboratories" is one of the most comprehensive on this subject and is recommended reading for anyone seriously involved in or contemplating rock mechanics testing.

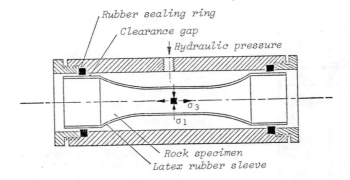

Figure 65 : Apparatus for inducing triaxial stress conditions in which one of the principal stresses is tensile. After Hoek[127,128].

threaded end pieces of the cell. This adjustment can be used to vary the clearance gap between the sleeved specimen and the sealing rings such that a small oil leakage occurs when the cell is pressurized. This leakage minimises end friction on the specimen and ensures that the specimen is not eccentrically loaded. Specimens for use in this apparatus are machined in a lathe by means of a diamond grinding wheel mounted on a tool-post grinding attachment actuated by a hydraulic or mechanical profile follower[127].

Uniaxial and triaxial testing

In the apparatus illustrated in figure 66, the specimen is subjected to an axial stress σ_1 and to radial confinement giving $\sigma_2 = \sigma_3 = p$, where p is the hydraulic pressure in the cell. The normal test conditions are arranged so that σ_1, the major principal stress, acts along the axis of the specimen.

A variation of this test, known as the *extension* test, involves the application of a hydraulic cell pressure which is higher than the axial stress in the specimen. This gives rise to the situation in which $\sigma_1 = \sigma_2 = p > \sigma_3$ where σ_3 is the axial stress in the specimen. Although this axial stress is compressive, the axial *strain* ε_3 is tensile when σ_3 is small (see equation 10 on page 91) and this causes the length of the specimen to increase.

The importance of these two variations of the triaxial test is that they represent the upper and lower bound conditions for the intermediate principal stress σ_2 and they can be used to test the influence of this stress upon the failure of rock.

A further variation of the triaxial test is the simple uniaxial compression test in which $\sigma_2 = \sigma_3 = 0$ and, at failure, $\sigma_1 = \sigma_c$, the uniaxial compressive strength of the rock.

In all cases, the end conditions of the specimen are critical if a uniformly distributed axial stress is to be induced in the specimen. Hawkes and Mellor[126] have discussed these end conditions in considerable detail and any reader who may be involved in rock testing should pay particular attention to these conditions since careless preparation or load-

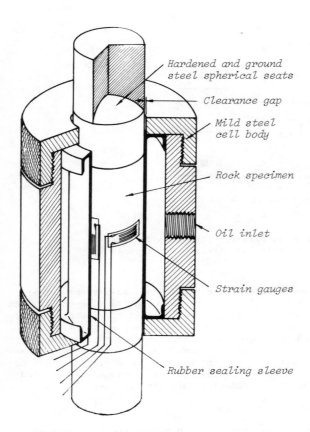

Figure 66: Cutaway drawing of triaxial cell for testing cylindrical rock specimens at confining pressures of up to 70 MPa (10,000 lb/in^2). Cell does not require drainage between tests. Designed by Hoek and Franklin[129,130].

ing of the specimen can result in the production of very poor test results.

The authors favour the use of a diamond impregnated wheel, mounted on a lathe tool-post grinding attachment as shown in the margin photograph, for the preparation of the ends of the specimen for uniaxial and triaxial testing. By mounting the core specimen in the lathe chuck and rotating it at approximately 20 revolutions per minute in the opposite direction to the direction of rotation of the diamond wheel (running at about 3000 revolutions per minute with water cooling), a clean square specimen edge is formed. If the wheel is run past the centre of the specimen, a flat end with a very good surface finish can be produced quickly and economically. The use of spherical seats at either end of the specimen, as shown in figure 66, will minimise loading eccentricity due to the fact that the ends of the specimen may not be absolutely parallel.

Use of a diamond impregnated cutting wheel mounted on the tool-post grinder of a lathe for the preparation of the ends of specimens for uniaxial and triaxial testing.

A modification to the loading platens of the apparatus illustrated in figure 66 permits the testing of rock specimens with internal pore pressure. If the domed spherical seats (shown in contact with the rock specimen) are drilled to accept high pressure couplings and porous discs are placed between these seats and the ends of the specimen, water under pressure can be introduced into the specimen under test. The significance of this water pressure will become obvious later in this chapter when the principle of effective stress is discussed.

An empirical failure criterion for rock

In spite of the excellent research carried out by many workers in this field, the authors are not aware of any failure criterion which meets all the requirements set out on page 132. Many of the available failure theories offer an excellent explanation for some aspects of rock behaviour but fail to explain others or cannot be extended beyond a limited range of stress conditions. Consequently, faced with the task of providing a failure criterion which will be of practical value to the underground excavation designer, the authors had no alternative but to seek a new criterion which would meet at least most of the requirements listed on page 132.

Jaeger and Cook[67] give a comprehensive discussion on the various failure theories which have been proposed to explain observed rock failure phenomena and no attempt will be made to reproduce this discussion here. These theories, particularly the theory proposed by Griffith[131,132] and modified by McClintock and Walsh[133,134], formed the basis for the development of the empirical failure theory presented in this book.

The authors have drawn on their experience in both theoretical and experimental aspects of rock behaviour to develop, by a process of trial and error, the following empirical relationship between the principal stresses associated with the failure of rock:

$$\sigma_1 = \sigma_3 + \sqrt{m\sigma_c\sigma_3 + s\sigma_c^2} \qquad (43)$$

where σ_1 is the major principal stress at failure,
    σ_3 is the minor principal stress applied to the specimen,
    σ_c is the uniaxial compressive strength of the intact rock material in the specimen,
    m and s are constants which depend upon the properties of the rock and upon the extent to which it has been broken before being subjected to the stresses σ_1 and σ_3.

This relationship can be represented graphically by means of a diagram such as that presented in the upper left hand portion of figure 67.

The uniaxial compressive strength of the specimen is given by substituting $\sigma_3 = 0$ in equation 43, giving:

$$\sigma_{cs} = \sqrt{s\sigma_c^2} \qquad (44)$$

For intact rock, $\sigma_{cs} = \sigma_c$ and $s = 1$. For previously broken rock, $s < 1$ and the strength at zero confining pressure is given by equation 44, where σ_c is the uniaxial compressive strength of the pieces of *intact* rock material making up the specimen.

The uniaxial tensile strength of the specimen is given by substitution of $\sigma_1 = 0$ in equation 43 and by solving the resulting quadratic equation for σ_3:

$$\sigma_t = \tfrac{1}{2}\sigma_c(m - \sqrt{m^2 + 4s}) \qquad (45)$$

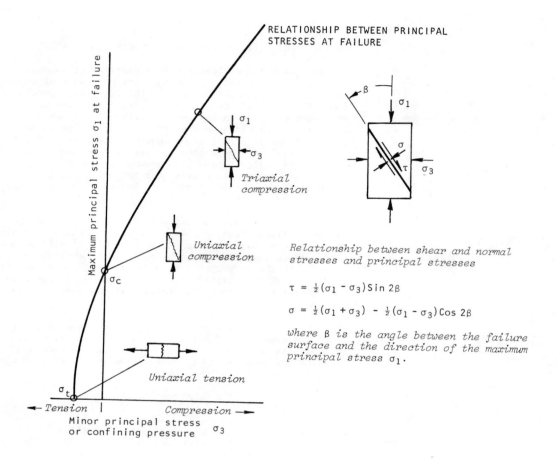
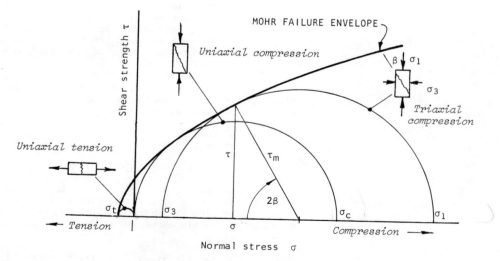

Figure 67 : Graphical representation of stress conditions for failure of intact rock.

In addition to the relationship between the major and minor principal stresses at failure, it is sometimes convenient to express the failure criterion in terms of the shear and normal stresses acting on a plane inclined at an angle β to the major principal stress direction, as illustrated in the upper right hand portion of figure 67. When the inclination β of the failure surface is known, the shear and normal stresses τ and σ can be determined directly from the equations presented in figure 67.

When an isotropic specimen is tested, it is usually assumed that the relationship between shear strength τ and normal stress σ is defined by the envelope to a set of Mohr circles representing the principal stresses at failure (see page 94). Under these conditions, it is assumed that the inclination β of the failure surface is defined by the normal to the Mohr envelope as illustrated in the lower diagram in figure 67. As will be shown later in this chapter, this assumption is probably an over-simplification. It has been included in this discussion because of its historical importance in rock mechanics literature and also because it does provide a rough guide to the inclination of the failure surface or surfaces under some stress conditions.

Balmer[135] derived a general relationship between the shear and normal stresses and the principal stresses at which failure of an isotropic rock specimen occurs. Substitution of equation 43 into Balmer's equations gives :

$$\sigma = \sigma_3 + \frac{\tau_m^2}{\tau_m + m\sigma_c/8} \qquad (46)$$

$$\tau = (\sigma - \sigma_3)\sqrt{1 + m\sigma_c/4\tau_m} \qquad (47)$$

where $\tau_m = \tfrac{1}{2}(\sigma_1 - \sigma_3)$

The angle β is defined by :

$$\text{Sin } 2\beta = \frac{\tau}{\tau_m} \qquad (48)$$

A detailed discussion on the derivation of these equations is given in part 3 of Appendix 5 at the end of this book.

Brittle-ductile transition

A practical limitation has to be placed on the use of equations 43, 46 and 47 because, as discussed on page 133, the behaviour of most rocks changes from brittle to ductile at high confining pressures. Mogi[136] investigated this transition and found that, for most rocks, it is defined by:

$$\sigma_1 = 3.4\,\sigma_3 \qquad (49)$$

This transition is illustrated in the margin drawing in which the results obtained by Schwartz[137] from a series of triaxial tests on Indiana limestone are presented.

The authors have used equation 49 in their analysis of

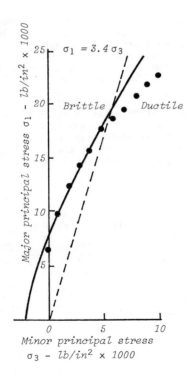

Transition from brittle to ductile failure illustrated by test data obtained by Schwartz[137] for Indiana limestone.

triaxial test data on intact rock specimens, presented later in this chapter. Hence, equation 43 has only been applied to data where $\sigma_1 > 3.4\sigma_3$.

In triaxial tests on specimens containing planes of weakness or specimens which have previously been broken, the transition from brittle to ductile failure is not as clearly defined as in the case for tests on specimens of intact homogeneous rock. In addition, this transition, when it can be determined, appears to occur at a smaller ratio of σ_1 to σ_3 than for intact rock. In the absence of any published guidelines on this transition, the authors have arbitrarily used $\sigma_1 = 2\sigma_3$ as the limit of applicability of equation 43 in analysing the results of triaxial tests on rock specimens.

Survey of triaxial test data on intact rock specimens

In order to check the applicability of the empirical failure criterion described on the previous pages and to provide a starting point for the analysis of rock failure around underground excavations, the authors have analysed published triaxial test data for a wide range of rock types. The sources of the information included in this survey are listed in table 9 and the results of the study are presented graphically on pages 143 to 149.

In selecting these data, care has been taken to ensure that the test conditions used in the generation of the data were comparable to those described earlier in this chapter. Wherever possible, the authors have gone back to the raw experimental data in order to avoid using information which may already have been adjusted to fit some other failure criterion.

In order to compare the results of triaxial tests on different rock types and on different samples of the same rock type, all the data has been reduced to dimensionless form by dividing the principal stresses at failure by the uniaxial compressive strength of each sample. Hence, equation 43, with s = 1 for intact rock, becomes

$$\sigma_{1n} = \sigma_{3n} + \sqrt{m\sigma_{3n} + 1} \qquad (50)$$

where σ_{1n} and σ_{3n} are normalized principal stresses σ_1/σ_c and σ_3/σ_c respectively.

The corresponding normal and shear stresses, from equations 46 and 47, become :

$$\sigma_n = \sigma_{3n} + \frac{\tau_{mn}^2}{\tau_{mn} + m/8} \qquad (51)$$

$$\tau_n = (\sigma_n - \sigma_{3n})\sqrt{1 + m/4\tau_{mn}} \qquad (52)$$

The values of σ_c and m for a given data set are determined by means of the linear regression analysis set out part 1 of Appendix 5. This analysis has only been applied to data sets containing more than 5 experimental points well spaced in the stress space defined by $\sigma_t < \sigma_3 < \sigma_1/3.4$.

TABLE 9 - ANALYSIS OF ROCK FRACTURE DATA

Rock type	Name/location	Tested by	Symbol	Number tested	Uniax. compr. strength σ_c lb/in²	Uniax. compr. strength σ_c MPa	Each sample m	Each sample r²	Rock type m	Rock type r²
AMPHIBOLITE	Norway (// to foliation)	Broch[138]	●	5	21290	146.8	24.8	0.98	25.1	0.98
	Norway (⊥ to foliation)	Broch[138]	■	5	29240	201.6	21.8	0.99		
CHERT	Chert dyke, South Africa	Hoek[139]	●	24	84040	579.5	20.3	0.93	20.3	0.93
DOLERITE	Frederick diabase, USA	Brace[140]	▲	7	82970	572.1	15.1	0.97	15.2	0.97
	Witwatersrand, South Africa	Hoek[139]	■	6	48660	335.5	10.7	0.90		
	Northumberland, UK	Franklin & Hoek[130]	●	38	42580	293.6	13.4	0.92		
DOLOMITE	Webatuck, USA	Brace[140]	●	6	*	*	7.9	0.95	6.8	0.90
	Blair, USA	Brace[140]	■	10	73430	506.3	5.9	0.84		
	Dunham, USA	Mogi[141]	▲	9	42290	291.6	8.0	0.99		
GABBRO	Norway (dry)	Broch[138]	●	5	50880	350.8	17.3	0.98	23.9	0.97
	Norway (saturated)	Broch[138]	■	5	29730	205.0	22.9	1.00		
GNEISS	Norway (// to foliation)	Broch[138]	●	5	36820	253.9	29.8	0.98	24.5	0.91
	Norway (⊥ to foliation)	Broch[138]	■	5	34010	234.5	21.2	0.96		
GRANITE	Westerley, USA	Heard et al[142]	■	17	31040	214.0	26.7	1.00	29.2	0.99
	Westerley, USA	Wawersik & Brace[143]	◀	7	43310	298.6	27.0	1.00		
	Westerley, USA	Brace[140]	▶	7	49820	343.5	28.3	0.98		
	Westerley, USA	Mogi[141]	▲	6	32440	223.7	32.8	0.99		
	Stone Mountain, USA	Schwartz[137]	▼	14	16850	116.2	28.9	0.93		
	Blackingstone, UK	Franklin & Hoek[130]	●	48	30410	209.7	20.8	0.91		
	Mount Sorrel, UK	Misra[144]	●	5	39910	275.2	26.5	0.99		
	Carnmarth, Redruth, UK	Misra[144]	◆	5	23540	162.3	27.7	0.99		
LIMESTONE	Portland, UK	Franklin & Hoek[130]	◆	30	13300	91.7	7.5	0.72	5.4	0.68
	Indiana, USA	Schwartz[137]	■	6	7090	48.9	3.2	0.95		
	Bath, UK	Misra[144]	●	7	6830	47.1	5.5	0.97		
	Grindling Stubbs, UK	Misra[144]	●	6	19450	134.1	8.8	0.97		
	Kirbymoorside, UK	Misra[144]	●	6	23830	164.3	12.3	0.98		
	Blackwell, UK	Misra[144]	●	5	29211	201.4	10.0	0.92		
	Foster Yeaman, UK	Misra[144]	●	5	24265	167.3	14.1	0.95		
	Gigglewick, UK	Misra[144]	●	5	22423	154.6	8.8	0.97		
	Kelmac, UK	Misra[144]	●	5	16897	116.5	7.3	1.00		
	Threshfield, UK	Misra[144]	●	5	21423	147.7	6.9	0.98		
	Swinden Cracoe, UK	Misra[144]	●	5	16027	110.5	8.4	0.96		

TABLE 9 - ANALYSIS OF ROCK FRACTURE DATA - CONTINUED

Rock type	Name/location	Tested by	Symbol	Number tested	Uniax. compr. strength σ_c lb/in²	Uniax. compr. strength σ_c MPa	Each sample m	Each sample r²	Rock type m	Rock type r²
MARBLE	Tennessee II, USA	Wawersik & Fairhurst[145]	●	44	19330	133.3	5.9	0.99		
	Norwegian	Broch[138]	■	5	8370	57.7	8.2	1.00		
	Carrara, Italy	Kovari & Tisa[146]	●	26	15050	103.8	6.6	0.99	10.6	0.90
	Carrara, Italy	Franklin & Hoek[130]	◆	14	13590	93.7	7.7	0.99		
	Carthage, USA	Bredthauer[147]	◀	7	9060	62.5	11.7	0.95		
	Georgia, USA	Schwartz[137]	▼	9	7210	49.7	7.1	0.98		
MUDSTONE	South Africa	Bieniawski[148]	●	29	*	*	7.3	0.85	7.3	0.82
	Horton-in-Ribblesdale, UK	Misra[144]	■	5	18910	130.4	11.1	1.00		
NORITE	South Africa	Bieniawski[148]	●	17	*	*	23.3	0.97	23.2	0.97
QUARTZDIORITE	Norway (dry)	Broch[138]	●	5	35210	242.8	20.2	1.00	23.4	0.98
	Norway (saturated)	Broch[138]	■	5	27160	187.3	23.8	0.99		
QUARTZITE	Witwatersrand, South Africa	Hoek[139]	●	19	32860	226.6	14.1	0.99		
	Witwatersrand, South Africa	Bieniawski[148]	■	35	*	*	18.5	0.82	16.8	0.84
	Boons Nuneaton, UK	Misra[144]	▶	5	47450	327.2	23.3	1.00		
SANDSTONE	Berea, USA	Aldrich[149]	●	27	10540	72.7	15.0	0.99		
	Pottsville, USA	Schwartz[137]	■	44	12970	89.4	19.3	0.87		
	South Africa	Bieniawski[148]	◆	56	7820	53.9	18.6	0.90		
	Derbyshire, UK	Franklin & Hoek[130]	▶	33	8950	61.7	15.6	0.93		
	Darley Dale, UK	Franklin & Hoek[130]	◀	27	11660	80.4	15.9	0.99		
	Pennant, UK	Franklin & Hoek[130]	▼	31	30310	209.0	11.9	0.98		
	Nugget, USA	Schock et al[150]	▲	13	57780	398.4	15.3	0.96		
	Buchberg, Switzerland	Kovari & Tisa[146]	●	47	10070	69.4	14.3	0.87		
	Mév, Hungary	Bodonyi[151]	○	14	12110	83.5	22.5	0.84	14.3	0.87
	Elland Edge, UK	Misra[144]	□	6	14910	102.8	12.8	0.99		
	Darley Dale, UK	Misra[144]	▷	11	6630	45.7	12.5	1.00		
	Bretton Blue, UK	Misra[144]	◁	6	25950	178.9	7.4	0.99		
	Horsforth, UK	Misra[144]	▽	5	7006	48.3	16.7	1.00		
	Ramsbottom Wild, UK	Misra[144]	△	5	15288	105.4	17.7	1.00		
	Buckstone, UK	Misra[144]	◇	5	14359	99.0	27.3	1.00		
	St Bees, UK	Misra[144]	◆	7	9780	67.4	9.2	0.98		
	Darley Dale, UK	Ramez[152]	●	28	14780	101.9	8.8	0.97		
	Darley Dale, UK	Price[154]	●	5	5780	39.9	6.4	0.85		
	Gosford, Australia	Jaeger[153]	■	5	9010	62.1	15.7	0.99		

* Quoted in dimensionless form in original paper.

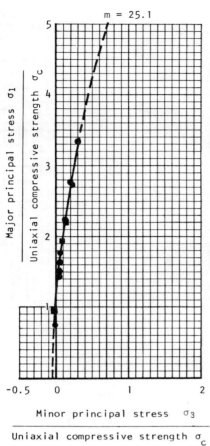
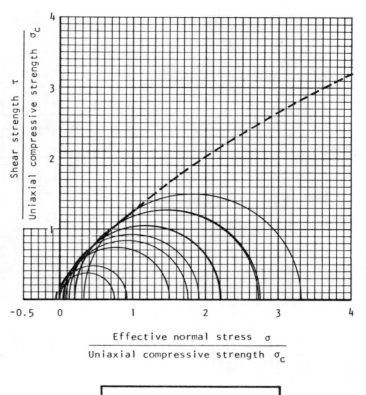

AMPHIBOLITE

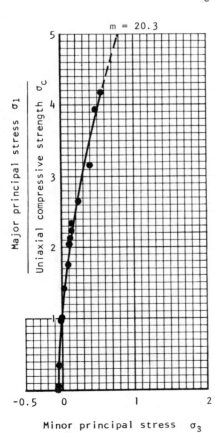
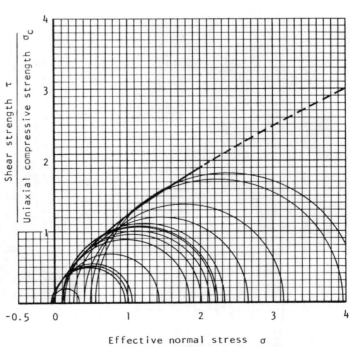

CHERT

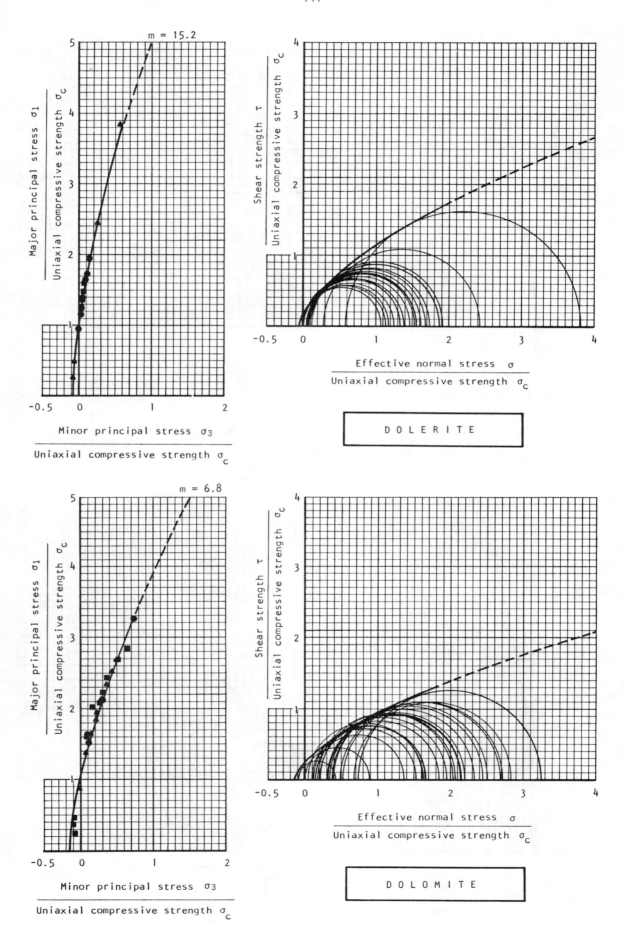

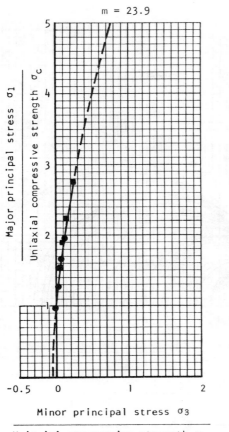
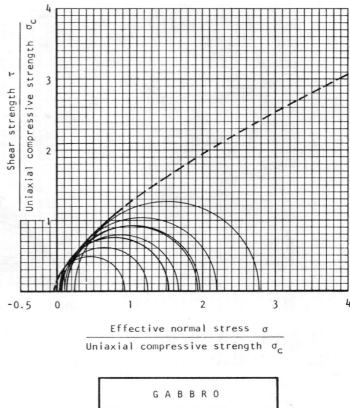

GABBRO

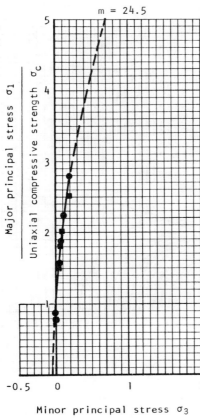
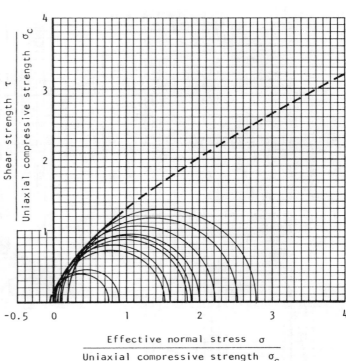

GNEISS

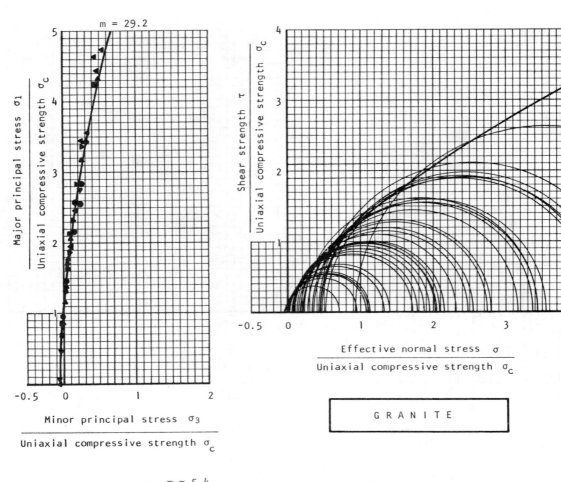

GRANITE

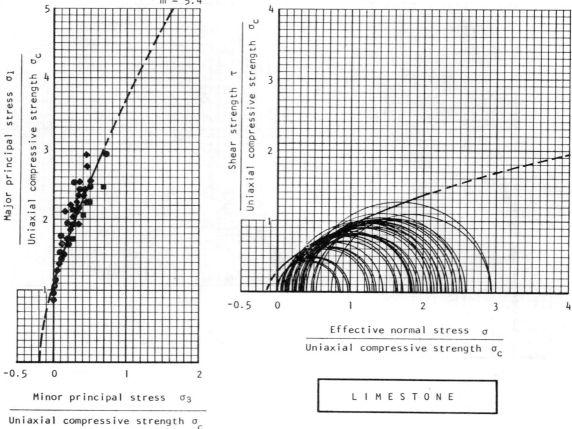

LIMESTONE

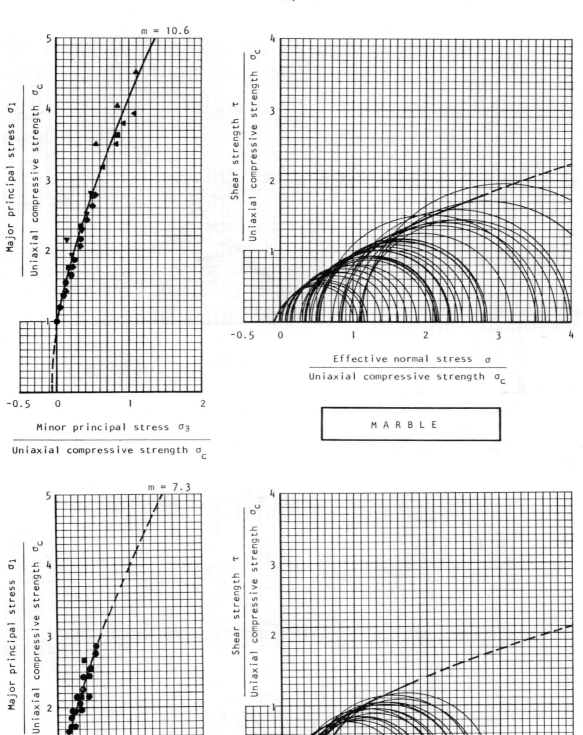

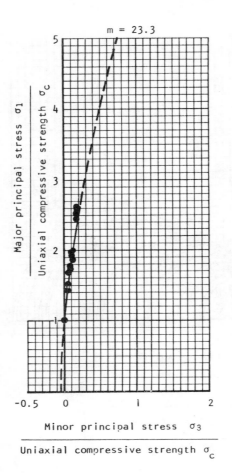
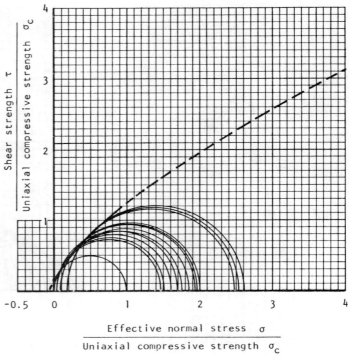

NORITE

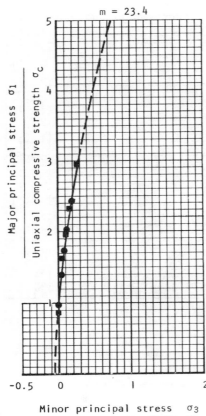
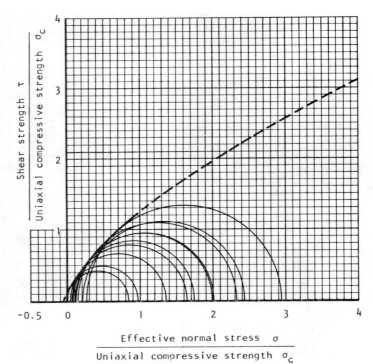

QUARTZDIORITE

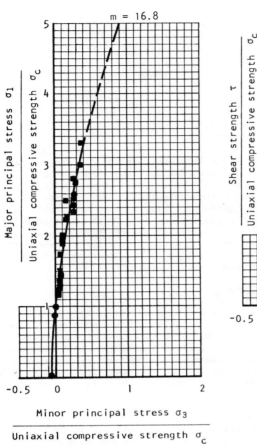
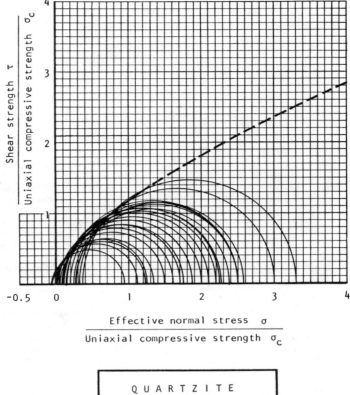

QUARTZITE

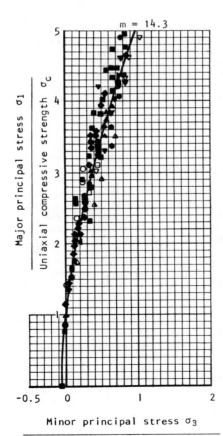
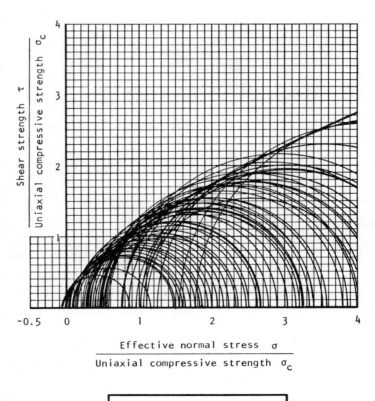

SANDSTONE

Simplifying assumptions

Anyone familiar with the literature on the failure of rock will realise that a number of simplifications have been made in developing the failure criterion presented on the previous pages. While these simplifications are necessary in order to achieve a practical solution to the problem, it is important that the assumptions upon which these simplifications are based are examined to ensure that no basic laws are violated. The following discussion briefly reviews the main simplifying assumptions which have been made in the development of the empirical failure criterion.

Definition of failure

The authors have adopted the maximum stress carried by the specimen as their definition of the "failure" stress. The justification for this choice is that most of the problems which will be discussed in later chapters of this book are concerned with the "failure" of underground excavations, i.e. the relationship between the induced stresses around excavations and the stress levels at which failure occurs in the rock.

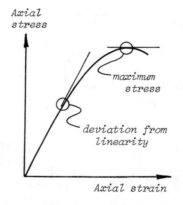

Several alternative definitions of failure will be found in rock mechanics literature but the authors believe that the definition adopted should always depend upon the application for which the failure criterion is intended. Hence, in designing the foundations for equipment or buildings which cannot tolerate differential displacements, an appropriate definition of failure might be the stress level at which significant deviation from linearity of the stress-strain curve occurs. For underground excavation engineering in which deformation can be tolerated but collapse of the excavation is to be avoided, the maximum stress carried by the specimen is considered to be a more appropriate definition of failure.

Grouping of data

In compiling the rock strength data presented in table 9, an arbitrary decision was made to group the data under traditional rock names. In addition to satisfying the obvious need to condense the data onto a reasonable number of pages, the authors believe that this grouping also serves a practical purpose. In many cases, the starting point in the design of an underground excavation, to which no access is available, is the knowledge that the rock surrounding the proposed excavation is granite or quartzite or sandstone. Consequently, the grouping of rock strength data under these rock types can serve a useful practical purpose, provided that there is some relationship between the failure characteristics of a rock and the name applied to that rock.

Detailed studies on rock fracture suggest that factors such as mineral composition, grain size and angularity, grain packing patterns and the nature of the cementing materials between grains all contribute to the manner in which fracture initiates and propagates in a rock specimen. In the case of a rock type such as granite, these factors are reasonably uniform, irrespective of the source of the granite. Hence, the fracture characteristics of a Westerley granite from the USA are similar to those of a Mount Sorrel granite from the

United Kingdom.

On the other hand, the term limestone is applied to a range of carbonate rocks, formed by either organic or inorganic processes, and it is hardly surprising that the materials grouped under this rock type exhibit a relatively wide range of failure characteristics. The difference between the rocks grouped under the name granite and those described as limestones is evident in the two plots on page 146.

The manner in which fracture initiates and propagates determines the *shape* of the curve relating the principal stresses at failure and this is reflected in the value of m in equation 43. On the other hand, the *strength* of the rock depends upon the strength of the individual particles making up the specimen and this is defined by the uniaxial compressive strength σ_c in equation 43. (Note that s = 1.00 for intact rock). Hence, Westerley granite tested by Brace[140] has a value of m which is very close to that of the Stone Mountain granite tested by Schwartz[137] but the strengths of these materials differ by a factor of three. On a dimensionless plot such as that given on page 146, the triaxial test results for these two materials fall on the same line but, on a true stress plot the results are widely separated.

Inclination of failure surfaces

In deriving the Mohr failure envelope for brittle rock, equations 46 and 47 on page 139, it was assumed that the surface or zone along which failure of the specimen occurs is inclined at an angle β to the major principal stress direction. This angle is defined by equation 48 and is illustrated in figure 67.

Wawersik and Brace[143] carried out a very detailed study of the initiation and propagation of fracture in specimens of granite and of diabase. Specimens of these materials were tested in a stiff testing machine (see page 133) and complete stress-strain curves were obtained. A typical stress-strain curve for Westerley granite tested under uniaxial compressive stress conditions is given in the margin figure. This stress-strain curve is divided into eight regions and Wawersik and Brace give the following description of the initiation and propagation of fractures in these regions :

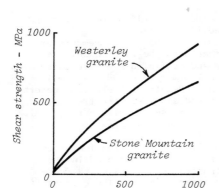

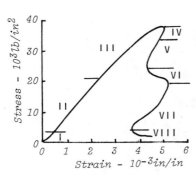

Complete uniaxial stress-strain curve for Westerley granite tested by Wawersik and Brace[143].

" Regions I to III are marked by the closure of pre-existing cracks, approximately linear elastic behaviour, and a combination of random crack formation, crack growth and sliding on existing crack interfaces. (Note, this cracking is on a microscopic scale). In region IV, which includes the stress maximum, a large number of small fractures are formed predominantly parallel to the direction of loading. This local cracking continues throughout the entire subsequent failure history with very little change of the angular crack distribution. In addition, relatively large cracks appear towards the end of region IV near and approximately parallel to the free sides of the sample. These cracks develop at about midheight of all samples and lead to the onset of spalling at the beginning of region V. Because spalling causes an area reduction, the onset of region V is associated with an increase in the true average stress in the sample. Spalling proceeds during regions V and VI due to buckling instability of thin near surface layers of rock. In region

VI small, steeply inclined shear fractures are formed; these grow into an open fault in region VII. In region VIII the deformation results in a loose mass of broken material held together by friction."

This description emphasises the complexity of the processes of fracture initiation and propagation in rock and the difficulties which have been associated with attempts to derive failure criteria based on simple crack models[67].

Wawersik and Brace[143] measured the inclination of the shear fractures and faults formed in Westerley granite in regions VI, VII and VIII. The results of these measurements are plotted in figure 68 together with the relationship between fracture plane inclination and confining pressure calculated by means of equation 48 (using m = 27.0 and s = 1.00). In spite of the large amount of scatter, there is a general similarity between the measured and predicted trends which suggests that equation 48 can be used to give a rough guide to the inclination of shear fractures and faults in rock.

The formation of cracks parallel to the direction of the major principal stress (i.e $\beta = 0$) is mentioned in the description of fracture formation by Wawersik and Brace[143] and has been noted by many other authors. These types of fractures occur most frequently under conditions of low confining stress, for example, near the free surfaces of underground excavations. Fairhurst and Cook[155] suggest that rock splitting or slabbing is an important fracture process in the rock surrounding underground excavations and this suggestion can be confirmed by many practical miners with experience in hard rock mines.

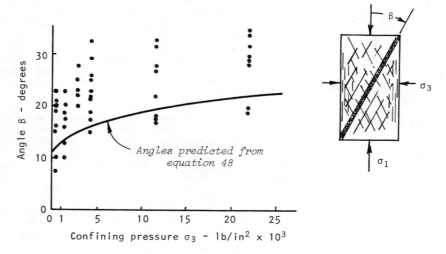

Figure 68 : Inclination of shear fractures and faults in specimens of Westerley granite tested by Wawersik and Brace[143].

Effective stress

The influence of internal fluid pressure or *pore pressure* on the strength of rock has been a controversial topic for many years. In 1923, Terzaghi[156] revolutionised soil mechanics by the introduction of his principle of effective stress which stated that the strength and volume change behaviour of saturated soils is governed not by the total or applied stress but by the effective stress σ' given by the difference between the applied stress σ and the pore pressure u. Terzaghi himself later suggested that this definition of effective stress may be incorrect for low porosity materials such as rocks and concrete, and that the effective stress governing the behaviour of the material should be written

$$\sigma' = \sigma - (1 - n_b) u \qquad (53)$$

where n_b was termed the boundary porosity.

Many authors have argued strongly that the original expression for effective stress ($\sigma' = \sigma - u$) is the correct one for rock while others argue, equally strongly, that it does not apply. This controversy has been discussed by Jaeger and Cook[67] (pages 219-225) and need not be reviewed here. On balance, the authors take the view that the original definition of effective stress is satisfactory for practical purposes. It will be applicable to rock provided that the pore structure of the rock is sufficiently interconnected and the loading rate is sufficiently slow to permit internal fluid pressure to equalise during testing. In the case of very low permeability rocks, this may require extremely slow loading rates as found necessary by Brace and Martin[157] in their tests on crystalline silicate rocks. On the other hand, in the case of porous rocks such as sandstones, the loading rate required to satisfy effective stress conditions appears to be within the range of normal laboratory tests and the original definition of effective stress is readily shown to apply (see, for example, Handin et al[158]).

For the stress components used throughout this chapter, the corresponding effective stresses may be calculated from :

$$\begin{aligned} \sigma_1' &= \sigma_1 - u \\ \sigma_3' &= \sigma_3 - u \\ \sigma' &= \sigma - u \end{aligned} \qquad (54)$$

Note that the shear stress τ is not affected since, as shown in figure 67 on page 138, it is a function of the difference between the major and minor principal stresses and substitution of the effective stresses in place of the applied stresses does not give rise to any difference in the shear stress.

Influence of pore fluid on strength

In addition to the effects of internal fluid pressure in the pores of the rock, discussed above, there is strong evidence to suggest that the presence of pore fluid without pressure can have a significant influence upon the strength of rock. Colback and Wiid[159] showed that the presence of water caused the strength of samples of shale and sandstone to reduce by a factor of 2 from oven dried to saturated specimens. The following ratios between the uniaxial compressive strengths of dry and saturated specimens were found by Broch[138] :

    Quartzdiorite   1.5
    Gabbro   1.7
    Gneiss ($\perp$ foliation)   2.1
    Gneiss (// foliation)   1.6

Broch's results are particularly interesting because they were obtained as part of a triaxial testing programme on these rocks and the results of his tests have been included in table 9 and in the plots on pages 148 and 145. The dimensionless plots of Broch's results show that there is not a great deal of difference between the fracture characteristics of dry and saturated samples (given by the value of m) and that the principal change is in the uniaxial compressive strength σ_c.

Within the limits of accuracy of the empirical failure criterion presented earlier in this chapter, it can be assumed that the presence of water in the pores of a rock will cause a reduction in its uniaxial compressive strength but not in the value of the material constant m.

It is important, when comparing the strengths of rock specimens taken from a particular rock mass, to ensure that the moisture content of all the specimens is the same. If the specimens are left standing around in the laboratory for various periods of time before testing, as is frequently the case, a significant scatter in experimental results can occur, particularly when testing sedimentary rocks which appear to be the most sensitive to changes in moisture content. Ideally, the moisture content of the specimens should be the same as that of the rock mass with which the underground excavation designer is concerned and the authors recommend that, when in doubt, the specimens should be tested saturated rather than dry. Where a moist room for curing concrete specimens is available, rock cores can be stored in a damp atmosphere for several weeks before testing to ensure that their moisture contents are uniform and close to saturation.

Influence of intermediate principal stress

In all the discussions up to this point, it has been assumed that the failure of rock is controlled by the major and minor principal stresses σ_1 and σ_3 and that the intermediate principal stress σ_2 has no effect upon the failure process. This is certainly an over-simplification but the authors consider that it is justifiable in the interests of keeping the failure criterion as simple as possible, in view of the fact that it still has to be extended to accommodate the behaviour of jointed and fractured rock.

The triaxial strength data on dolerite, dolomite, granite and quartzite published by Brace[140] contain both triaxial compression and triaxial extension test results. As discussed on page 135, this means that his results include the maximum possible variation of the intermediate principal stress σ_2 which can lie between the major and the minor principal stress values. Brace found no significant variation between the results obtained when $\sigma_2 = \sigma_3$ and when $\sigma_2 = \sigma_1$ and he

concluded that the intermediate principal stress has a
negligible influence upon the failure of the rocks which
he tested. He did emphasise that he considered that this
subject requires further study.

Hojem and Cook[160] carried out tests in a "polyaxial" cell
in which small flat jacks were used to subject rectangular
specimens to three independent principal stresses. In discussing the results of these tests, Jaeger and Cook[67] conclude that the strength of rock increases with increasing
intermediate principal stress level but that this increase
is small enough to ignore for most practical applications.

Mogi[161,162] tested a number of rocks under polyaxial stress
conditions and found more significant strength differences
than those reported by Brace or by Hojem and Cook. However,
close examination of Mogi's results suggests that many of
his tests involved brittle/ductile transitions and it is
not clear to what extent this may have complicated the
interpretation of his results.

On the basis of available evidence, the authors feel that
it is admissible to ignore the influence of the intermediate
principal stress upon the failure of brittle rock. This
assumption is important in keeping the failure criterion
as simple as possible in order that it can be extended to
include the effects of joints and pre-existing fractures.

Influence of loading rate

Vutukuri, Lama and Saluja[163] have compiled information on
the influence of loading rate upon the strength of Westerley
granite under triaxial test conditions and have shown that
the strength can be halved by decreasing the loading rate
by four orders of magnitude. In general, the faster the
rate of load application, the stronger will be the rock
specimen tested. In terms of this discussion, the loading
rate encountered in underground excavation engineering (with
the exception of blasting) does not differ to such an extent
that this effect need be taken into account.

Influence of specimen size

Rock mechanics literature contains a number of conflicting
observations on the effect of specimen size upon the strength
of rock. Hodgson and Cook[164] and Obert et al[165] report no
change in rock strength with change in specimen size. On the
other hand, significant strength reductions with increasing
specimen size have been reported by Mogi[166], Bieniawski[167],
Pratt et al[168], Protodiakonov and Koifman[169] and Hoskins and
Horino[170].

On balance, the authors believe that there is a strength
reduction with increasing specimen size for most rocks
and a collection of typical experimental results is presented
in figure 69. These data have been reduced to dimensionless
form by dividing individual strength values by the strength
of a specimen 50mm in diameter, the average size of a laboratory specimen. This process not only makes it possible to
compare the experimental results but it also eliminates
differences due to variations in moisture content, specimen
shape, loading rate etc since these factors are generally
the same for a given data set.

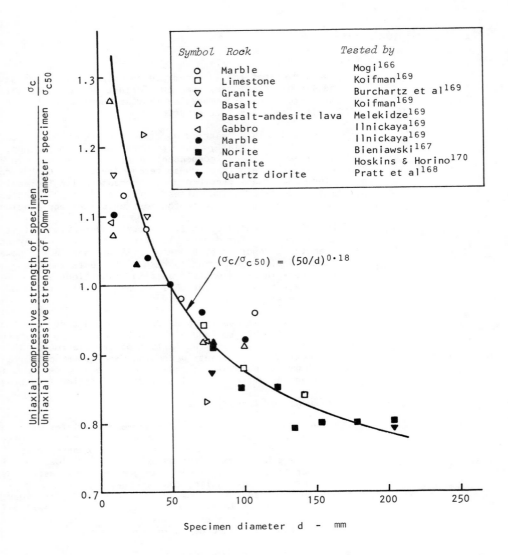

Figure 69 : Influence of specimen size upon the strength of intact rock.

An approximate relationship between uniaxial compressive strength and specimen diameter (for specimens between 10 and 200 mm diameter) is given by

$$\sigma_c = \sigma_{c50}(50/d)^{0.18} \qquad (55)$$

where σ_{c50} is the uniaxial compressive strength of a specimen of 50mm diameter and
d is the diameter of the specimen in mm.

Note that the data presented in figure 69 are for laboratory tests on unjointed intact rock only. In the case of jointed rock masses, the variation in strength with size is related to discontinuity spacing and is taken into account in using rock mass classifications to predict values of m and s as discussed on page 171.

Anisotropic rock strength

As shown in figure 63, the next stage in this process of developing a failure criterion which will be of practical value to the underground excavation designer is to consider the effect of a pre-existing discontinuity on the strength of a rock specimen.

A considerable amount of experimental and theoretical work has been done in this field and the papers by Jaeger[171], McLamore and Gray[172] and Donath[173] are essential reading for anyone wishing to become familiar with the background to this subject.

The single plane of weakness theory developed by Jaeger[171] (see also Jaeger and Cook[67], pages 65-68) provides a useful starting point for considerations of the effect of pre-existing discontinuities on rock strength. Jaeger determined the conditions under which, for the situation shown in the margin sketch, slip would occur on the discontinuity AB. If the discontinuity has a shear strength given by

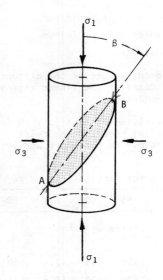

$$\tau = c + \sigma \operatorname{Tan} \phi \qquad (56)$$

where c is the cohesive strength of the surface and
ϕ is the angle of friction,

then slip will occur when

$$\sigma_1 \geqq \sigma_3 + \frac{2(c + \sigma_3 \operatorname{Tan} \phi)}{(1 - \operatorname{Tan}\phi \operatorname{Tan}\beta)\operatorname{Sin} 2\beta} \qquad (57)$$

For those combinations of c, ϕ, σ_3 and β for which the inequality of equation 57 is not satisfied, slip on the discontinuity cannot occur and the only alternative is fracture of the rock material independent of the presence of the discontinuity. These two types of failure, slip on the discontinuity and fracture of the intact rock, are represented graphically in the margin sketch.

Experience has shown that this simple relationship, while suitable for cases in which a single well defined discontinuity is present in a rock specimen, does not adequately describe the behaviour of naturally occurring anisotropic rocks such as slates. Modifications to equation 57 have been proposed by McLamore and Gray[172] but these modifications obscure the simple logic used by Jaeger in deriving the original equation and the end result is an empirical equation.

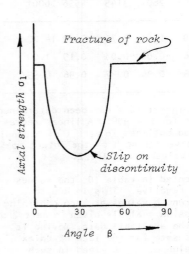

Rather than follow McLamore and Gray's approach, the authors have adopted an empirical equation* which is used to modify the material constants m and s which have been used in the earlier part of this chapter. The resulting equations for m and s are :

$$m = m_i(1 - Ae^{-\theta^4}) \qquad (58)$$

$$s = 1 - Pe^{-\zeta^4} \qquad (59)$$

---

* Suggested by Mr I. Miller of Golder Associates, Vancouver.

where m_i is the value of m for the intact rock,
A and P are constants,

$$\theta = \frac{\beta - \xi_m}{A_2 + A_3\beta}$$

$$\zeta = \frac{\beta - \xi_s}{P_2 + P_3\beta}$$
(60)

ξ_m is the value of β at which m is a minimum,

ξ_s is the value of β at which s is a minimum,

A_2, A_3, P_2 and P_3 are constants.

The values of m and s for different values of the discontinuity angle β are calculated by means of the linear regression analysis given in part 2 of Appendix 5.

In order to illustrate the application of equations 58 and 59 to a typical set of experimental results, the tests carried out by Donath[174] on Martinsburg slate are analysed on the following pages.

TABLE 10 - DONATH'S TEST RESULTS FOR MARTINSBURG SLATE*.

Angle β	0°	15°	30°	45°	60°	75°	90°
Confining pressure σ_3 - bars	\multicolumn{7}{l}{Axial failure stress σ_1 - bars}						
35	1275	510	221	425	748	1275	1938
105	1615	816	442	629	1020	1598	2414
350	2720	1343	867	1071	1496	2159	3349
500	3553	1717	1292	1496	1938	2839	4097
1000	5304	2856	2295	2601	3145	4556	6001
m	11.54	2.12	1.10	1.61	2.71	7.40	14.22
s	0.14	0.01	-0.06	-0.04	0.08	0.19	1.00
r^2	0.99	0.98	0.96	0.96	0.97	0.96	0.99

Note that the strength of the intact rock has been determined from the values given by tests for β = 90°. A linear regression analysis, using the method outlined in part 1 of Appendix 5, gives the uniaxial compressive strength of the intact material σ_c = 1551 bars and the value of m = 14.22.

In analysing the results presented in table 10, the values of s for β = 30° and 45° are negative. This is due to the fact that there are no strength values at zero confining pressure and the linear regression analysis given in part 2 of Appendix 5 fits a curve to the available data, moving it slightly to the right to compensate for the lack of uniaxial test results. The negative value of s obtained in such cases has no physical significance and is set to zero in order to avoid mathematical complications in the subsequent analysis (see part 2, Appendix 5).

* Donath's results for σ_3 = 2000 bars have been omitted because the stress-strain curves indicate that these specimens behaved plastically. Note: 1 bar = 0.1 MPa.

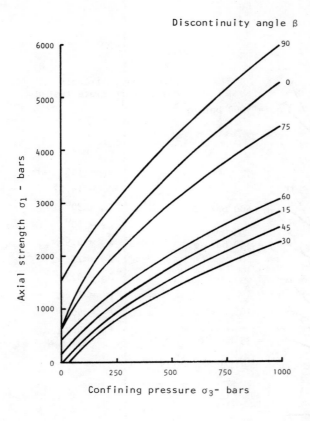

Figure 70 : Relationship between principal stresses at failure for Martinsburg slate tested by Donath[174].

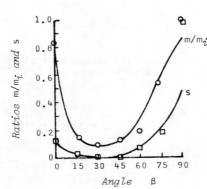

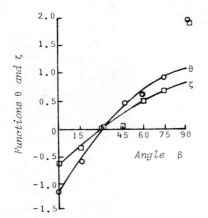

From equations 58 and 59 :

$$\theta = \left(\text{Log}_e \left(\frac{A}{1-m/m_i} \right) \right)^{\frac{1}{4}} \qquad (61)$$

$$\zeta = \left(\text{Log}_e \left(\frac{P}{1-s} \right) \right)^{\frac{1}{4}} \qquad (62)$$

where $A = (m_i - m_{min}/m_i)$

$P = (1 - s_{min})$

In solving equations 62 and 63 it should be noted that the values of θ and ζ are negative for values of β less than ξ_m and ξ_s.

Plots of m/m_i and s versus β and of θ and ζ versus β are given in the margin. Equations 60 and 61 have been fitted to the values of θ and ζ, calculated from equations 62 and 63, by a process of trial and error. The values of θ and ζ from these fitted curves have then been substituted into equations 58 and 59 to obtain the curves of m/m_i and s versus β given in the upper margin drawing. These values of m and s were then substituted into equation 43 (page 137) to give the curves of axial strength σ_1 versus discontinuity angle β presented in figure 71.

The results of similar analyses on a variety of anisotropic rocks are presented in figures 72 to 76. In all cases the authors feel that equation 43, with values of m and s calculated from equations 58 and 59, describes the influence of a single discontinuity with sufficient accuracy for most practical purposes.

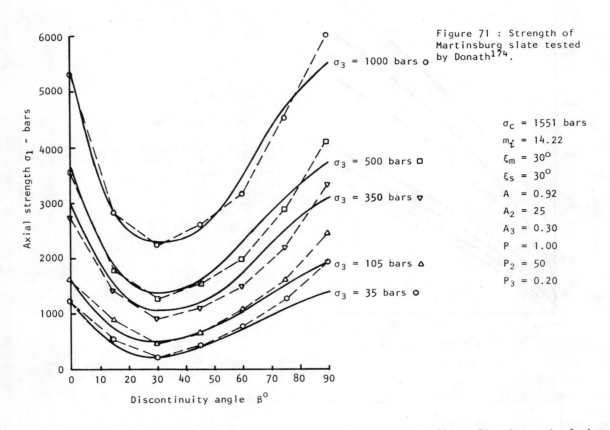

Figure 71 : Strength of Martinsburg slate tested by Donath[174].

σ_c = 1551 bars
m_i = 14.22
ξ_m = 30°
ξ_s = 30°
A = 0.92
A_2 = 25
A_3 = 0.30
P = 1.00
P_2 = 50
P_3 = 0.20

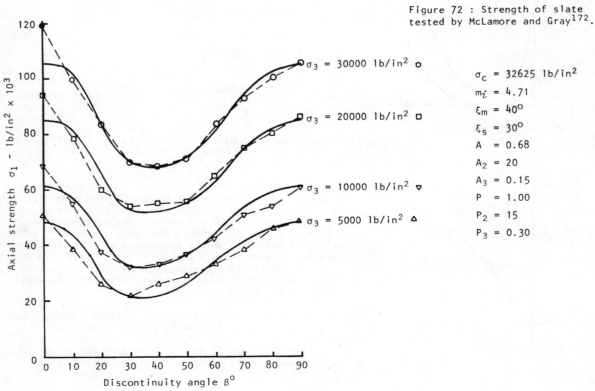

Figure 72 : Strength of slate tested by McLamore and Gray[172].

σ_c = 32625 lb/in²
m_i = 4.71
ξ_m = 40°
ξ_s = 30°
A = 0.68
A_2 = 20
A_3 = 0.15
P = 1.00
P_2 = 15
P_3 = 0.30

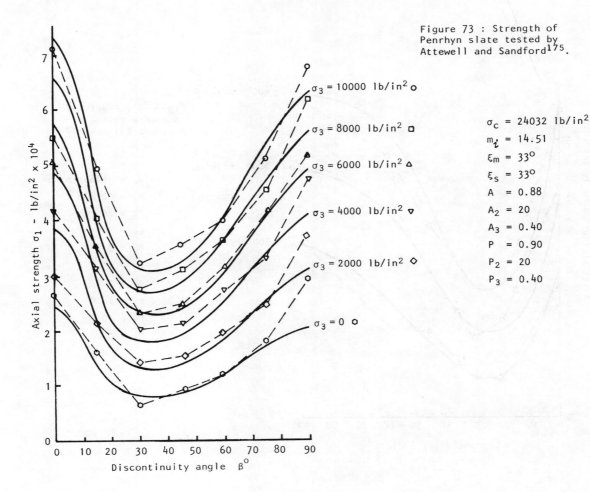

Figure 73 : Strength of Penrhyn slate tested by Attewell and Sandford[175].

σ_c = 24032 lb/in²
m_i = 14.51
ξ_m = 33°
ξ_s = 33°
A = 0.88
A_2 = 20
A_3 = 0.40
P = 0.90
P_2 = 20
P_3 = 0.40

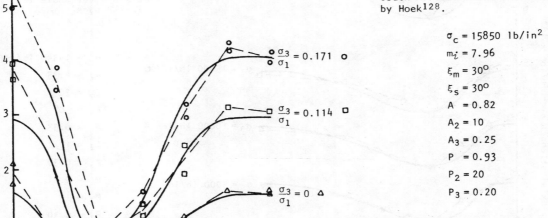

Figure 74 : Strength of a South African slate tested by Hoek[128].

σ_c = 15850 lb/in²
m_i = 7.96
ξ_m = 30°
ξ_s = 30°
A = 0.82
A_2 = 10
A_3 = 0.25
P = 0.93
P_2 = 20
P_3 = 0.20

Note that these specimens were tested at a constant stress ratio and equation 43 has to be modified to :

$$\sigma_1 = \sigma_c \left[\frac{m\sigma_3/\sigma_1 + \sqrt{(m\sigma_3/\sigma_1)^2 + 4s(1 - \sigma_3/\sigma_1)^2}}{2(1 - \sigma_3/\sigma_1)^2} \right]$$

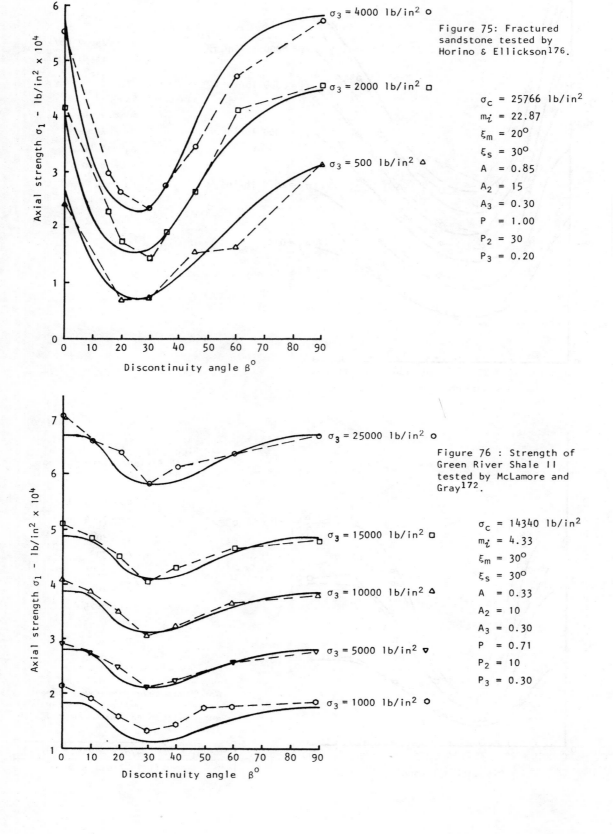

Figure 75: Fractured sandstone tested by Horino & Ellickson[176].

σ_c = 25766 lb/in²
m_i = 22.87
ξ_m = 20°
ξ_s = 30°
A = 0.85
A_2 = 15
A_3 = 0.30
P = 1.00
P_2 = 30
P_3 = 0.20

Figure 76: Strength of Green River Shale II tested by McLamore and Gray[172].

σ_c = 14340 lb/in²
m_i = 4.33
ξ_m = 30°
ξ_s = 30°
A = 0.33
A_2 = 10
A_3 = 0.30
P = 0.71
P_2 = 10
P_3 = 0.30

Strength of rock with multiple discontinuities

Consider a rock specimen containing two pre-existing discontinuities, as illustrated in the margin sketch. Failure of the specimen associated with discontinuity AB is defined by equations 43, 58, 59 and 60 as discussed in the previous section. Plotting the axial strength σ_1, calculated from these equations, against the discontinuity angle β in a polar diagram results in the type of strength curve shown by the full line in figure 77. Note that the curve in each quadrant is a mirror image of the curve in the preceding quadrant.

The influence of the second discontinuity CD on the failure of the specimen is indicated by the dashed strength curve in figure 77. Note that this curve is identical to that associated with discontinuity AB but it is rotated by the angle α which is the included angle between the two discontinuities, measured in the same direction as the angle β.

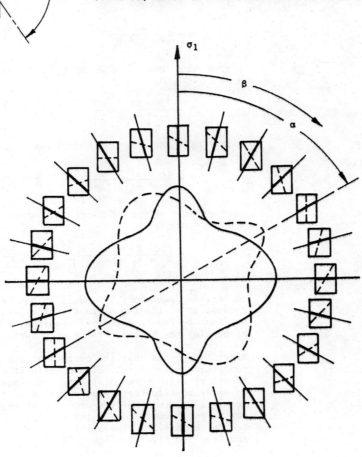

Figure 77 : Polar plot of axial strength σ_1 against discontinuity inclination β for a rock specimen containing two pre-existing discontinuities inclined at an angle α to one another.

The full set of equations defining the behaviour of a rock specimen containing two pre-existing discontinuities is listed below. For convenience, equations 43, 58 and 59 are repeated here.

Axial strength of specimen :

$$\sigma_1 = \sigma_3 + \sqrt{m\sigma_c\sigma_3 + s\sigma_c^2} \qquad (43)$$

Material constants :

$$m = m_i(1 - Ae^{-\theta^4}) \qquad (58)$$

$$s = 1 - Pe^{-\zeta^4} \qquad (59)$$

Values of θ and ζ for discontinuity AB in 1st & 3rd quadrants

$$\theta = \frac{\beta - \xi_m}{A_2 + A_3\beta} \qquad \zeta = \frac{\beta - \xi_s}{P_2 + P_3\beta} \qquad (60)$$

Mirror image values of θ and ζ for discontinuity AB in 2nd & 4th quadrants

$$\theta = \frac{|90 - \beta| - \xi_m}{A_2 + A_3|90 - \beta|} \qquad \zeta = \frac{|90 - \beta| - \xi_s}{P_2 + P_3|90 - \beta|} \qquad (63)$$

Values of θ and ζ for discontinuity CD in 1st & 3rd quadrants

$$\theta = \frac{|\beta - \alpha| - \xi_m}{A_2 + A_3|\beta - \alpha|} \qquad \zeta = \frac{|\beta - \alpha| - \xi_s}{P_2 + P_3|\beta - \alpha|} \qquad (64)$$

Mirror image values of θ and ζ for discontinuity CD in 2nd & 4th quadrants

$$\theta = \frac{|90 - \beta - \alpha| - \xi_m}{A_2 + A_2|90 - \beta - \alpha|} \qquad \zeta = \frac{|90 - \beta - \alpha| - \xi_s}{P_2 + P_3|90 - \beta - \alpha|} \qquad (65)$$

In deriving equations 64 and 65 it has been assumed that the properties of the two discontinuities are identical. This need not be the case and, if required, a different set of properties may be assigned to each surface.

Bray[177] suggests that the overall strength of a rock mass containing several sets of discontinuities is given by the lowest strength envelope to the individual strength curves. In order to illustrate the influence of several sets of identical discontinuities upon the strength of a rock mass, the strength curves presented in figure 78 have been constructed on the basis of the strength of slate tested by McLamore and Gray[172], given in figure 72.

It is clear, from figure 78, that as the number of discontinuities in a rock mass increases, the overall strength behaviour of the mass tends to become more and more isotropic. In the context of this text on the design of underground excavations, the authors feel that it is justifiable to treat rock masses containing four or more discontinuity sets as isotropic.

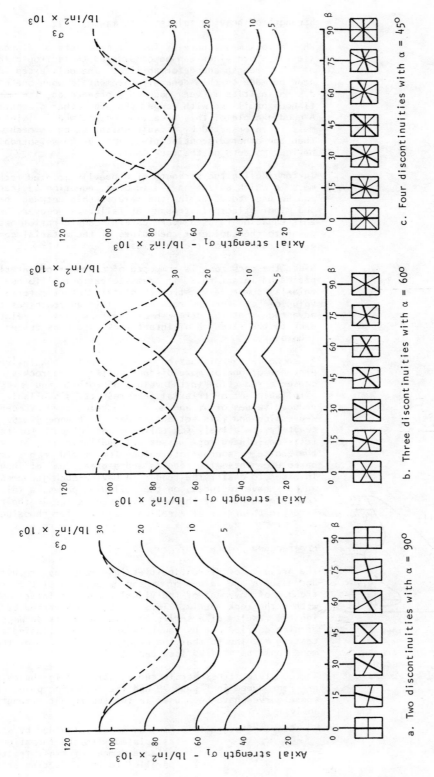

Figure 78 : Strength curves for rock specimens with multiple discontinuities.

Strength of heavily jointed rock masses

When a rock mass contains four or more sets of discontinuities, its behaviour can be considered as isotropic in terms of its strength and deformability. The only exception which need be made to this general statement is when one of the discontinuities is very much more pronounced, i.e more continuous or filled with gouge, than the other discontinuities. A good example of this exception occurs when a jointed rock mass is intersected by a fault which, being so much weaker than the other discontinuities, imposes an anisotropic behaviour pattern on the rock mass.

In considering the strength of a heavily jointed rock mass in which anisotropy is ignored, equation 43 (page 137) can be used to determine the relationship between the major and minor principal stresses at failure. However, before this equation is used, very careful consideration has to be given to the choice of the values of the material constants σ_c, m, and s for the rock mass.

Since the rock mass is composed of a number of interlocking pieces of intact rock, it seems to be logical to use the uniaxial compressive strength of this intact material as the value of σ_c for the rock mass. This approach has the advantage that the strength of the rock mass is related back to the strength of intact rock specimens tested in the laboratory.

As shown in the discussion on anisotropic rock failure, the presence of one or more discontinuities in a rock specimen causes a reduction in the values of both m and s. Where a reliable set of triaxial test results is available, the reduced values of m and s can be calculated by means of the regression analysis set out in part 2 of Appendix 5. Unfortunately, relatively few reliable sets of data for triaxial tests on jointed rock masses are available and, at present, the choice of appropriate values for m and s has to be based upon these few results and a great deal of judgement. In order to assist the reader in making such judgements, the authors have included, on the following pages, a fairly complete account of the most important steps in their own struggle to arrive at a rational process for choosing m and s.

Progressive fracture of granite

The presence of discontinuities in a rock mass results in a decrease in the values of m and s because of the greater freedom of movement of individual pieces of intact material within the rock matrix. This can be demonstrated by analysing the results of a series of triaxial tests on Westerley granite carried out by Wawersik and Brace[143], using a stiff testing machine so that the progressive failure of the specimens could be studied (see page 133).

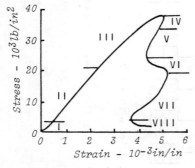

Complete uniaxial stress-strain curve for Westerley granite tested by Wawersik and Brace[143].

The complete stress-strain curves obtained in these tests are reproduced in figure 79 and the following points on these curves have been used in the subsequent strength analysis:

    Start of stage IV - maximum stress attained by specimen associated with the formation of a large number of small fractures parallel to the direction of loading.

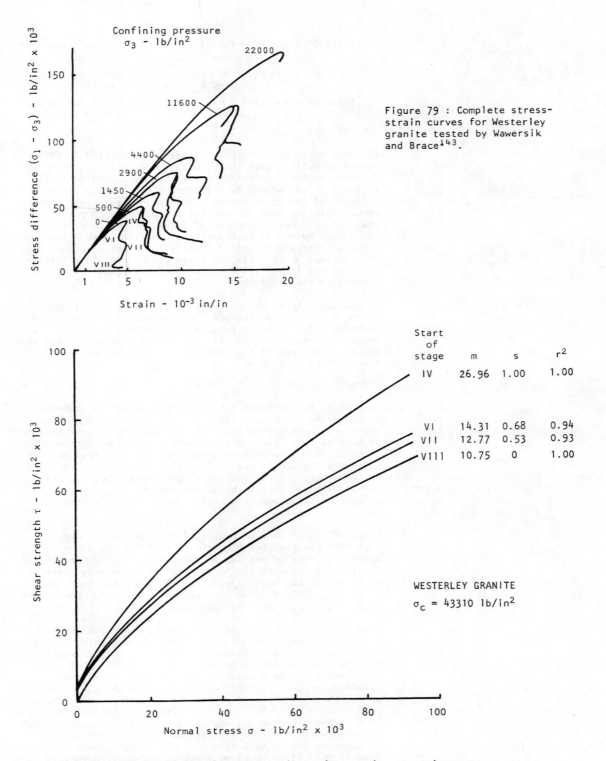

Figure 79 : Complete stress-strain curves for Westerley granite tested by Wawersik and Brace[143].

Figure 80 : Mohr envelopes for tests on Westerley granite at various stages.

Start of stage VI - formation of small, steeply inclined shear fractures.

Start of stage VII - growth of small steeply inclined fractures into an open fault.

Start of stage VIII - ultimate strength of loose broken material held together by friction between particles.

The Mohr envelopes for these stages are presented in figure 80 and it will be seen that the values of both m and s decrease as the fracturing of the specimens progress. The value of m is reduced by a factor of approximately 2 as the strength of the rock specimens progresses from peak to ultimate. This decrease is relatively modest, as compared with that for rock masses considered later in this chapter, because the individual rock particles have not been allowed to move very far and are still tightly interlocking. Note that the constant m reflects the curvature of the Mohr failure envelope and is sensitive to the degree of particle interlocking in the specimen.

On the other hand, the value of s decreases from 1.00 to 0 as the fracturing of the specimen progresses from stage IV to stage VIII. This is because the constant s reflects the tensile component of the strength of the rock matrix and this decreases to zero as the rock matrix is broken up by the formation of fractures.

Strength of "granulated" marble

An elegant set of experiments was carried out on Wombeyan marble by Rosengren and Jaeger[178] and the results of these tests are presented in figure 81 in the form of Mohr envelopes. By heating this coarse grained marble to approximately 600°C, these authors were able to obtain "granulated" specimens in which the individual grains were almost completely separated but were still in their original positions. The resulting low-porosity tightly interlocking aggregate is interesting to consider as a possible small scale model of a heavily jointed rock mass.

Figure 81 shows that the value of the constant s drops from 1.00 for the intact marble to 0.19 for the granulated material. This is not unexpected in view of the fact that the tensile strength of the material has been very largely destroyed by the heating process. On the other hand, the value of the constant m increases from 4.00, for the intact material, to 5.24 for the aggregate.

This increase in the value of m is interesting in that it goes against the trend of most of the other results analysed by the authors. It is considered likely that this result was obtained because the value of σ_c for the intact rock unit was changed by the process of granulation. A value of σ_c obtained for the original rock material which is, in fact, an aggregate of crystals bonded across grain boundaries, was used in this analysis. However, after granulation, the basic unit of the interlocking aggregate becomes a calcite crystal or grain which would be expected to have a higher value of σ_c than the intact rock material. It is considered likely that if this value could be measured and applied in this analysis, the value of m obtained would reduce below 4.00.

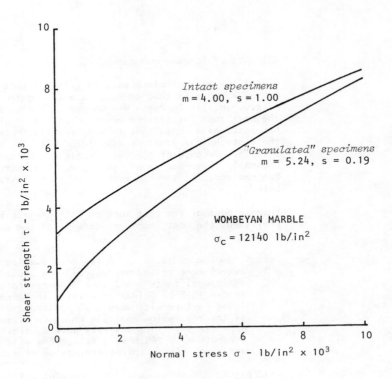

Figure 81 : Results of triaxial tests on intact and "granulated" Wombeyan marble by Rosengren and Jaeger[178].

It is also likely that, because no shearing or displacement was involved in the breakdown of the marble, the grain boundaries were rough and tightly interlocking so that under anything other than low confining stress conditions, slip would not be permitted without fracture of the intact material. In fact, under high confining pressures, it could be expected that the behaviour of the granulated material would be very similar to that of the intact marble.

Because of these reasons, the authors feel that the granulated marble is not an adequate model for the study of rock mass behaviour. It is, as Rosengren and Jaeger described it, simply a low porosity interlocking aggregate, and not a jointed rock mass. Just as the production of granulated marble does not adequately simulate either the tectonic processes by which the joints were formed or the behaviour of the joints themselves, so too might the plaster models widely used in research studies[179-184] give misleading results. In particular, the dilatant behaviour and stiffness of joints are rarely adequately modelled. This statement is not intended as a criticism nor is there any desire on the part of the authors to discount the excellent work and the many contributions to the science of rock mechanics which have come from model studies. Indeed, the authors have made use of and will continue to use models in their own work but, in future, will pay more attention to the techniques used to simulate 'joints' in these models.

Strength of Panguna andesite

One of the most complete sets of triaxial test data available to the authors is that on Panguna andesite from the island of Bougainville in Papua New Guinea. This material comprises the host rock for a large copper deposit and the senior author has acted as a consultant to Bougainville Copper Limited on the geotechnical problems associated with the mining of this deposit. Because of the large scale of the operation, an unusually large amount of testing has been carried out on the Panguna andesite and the results of these tests are analysed below.

Because most of the test work on this project has not been published, the most important details of the tests are listed below.

Intact Panguna andesite - A few core samples (1 inch diameter) were tested by Jaeger[185] in about 1968. Some additional tests (on 2 inch diameter cores) were carried out in 1978 by Golder Associates in Vancouver. All of these test results were combined and analysed by means of the regression analysis which is given in part 1 of Appendix 5 . This analysis gave a uniaxial compressive strength σ_c = 265 MPa, a value of m = 18.9, s = 1.00 and a coefficient of determination r^2 = 0.85.

Undisturbed core samples - Very careful drilling, using 6 inch diameter triple tube drilling equipment, produced a number of undisturbed core samples of jointed Panguna andesite. These samples were transported, in the inner core barrels, to Canberra where they were prepared and tested triaxially by Jaeger[185]. Using the regression analysis presented in part 2 of Appendix 5, assuming a value of σ_c = 265 MPa, the following values were obtained : m = 0.278, s = 0.0002, r^2 = 0.99.

Recompacted graded samples - Samples were obtained from bench faces in the mine and typical grading curves were established for these samples. These grading curves were scaled down, as suggested by Marsal[186], and the samples were compacted to as near the in situ density as possible before testing in a 6 inch diameter triaxial cell in the mine laboratory. A preliminary analysis of these data gave a small negative value for s and the method given at the end of part 2 of Appendix 5 was used to calculate the value of m for s = 0. This analysis gave m = 0.116 for s = 0. The same calculation technique was used for all the other samples listed below.

Fresh to slightly weathered Panguna andesite - Substantial quantities of this material were shipped to Cooma in Australia where samples were tested in a 22½ inch diameter triaxial cell by the Snowy Mountains Engineering Corporation. These samples were compacted to densities ranging between 1.94 and 2.07 tonnes/m³ (intact rock γ = 2.55 tonnes/m³) before testing. A regression analysis of the results of these tests gave m = 0.040 for s = 0.

Moderately weathered Panguna andesite - Tested by the Snowy Mountains Engineering Corporation in their 22½ inch diameter triaxial cell after compaction to 1.97 tonnes/m^3, these samples gave m = 0.030 for s = 0.

Highly weathered Panguna andesite – Tested by the Snowy Mountains Engineering Corporation in a 6 inch diameter triaxial cell after compaction to 1.97 tonnes/m^3, this material gave m = 0.012 for s = 0.

The Mohr failure envelopes for these samples of Panguna andesite are plotted in figures 82 and 83 and it is evident that there is a systematic decrease in the values of m and s with the degree of jointing and weathering of the samples. This trend is consistent with qualitative behaviour patterns which are apparent from published discussions on rock mass strength (e.g by Manev and Avramova-Tacheva[187]) and on rock fill (e.g by Marsal[186] and Marachi, Chan and Seed[188]). Unfortunately, very few publications on this subject contain sufficient detail to permit the same type of analysis as that presented above for Panguna andesite. The authors have only been able to carry out very crude studies on published test data on rock mass and rock fill strength but these studies have reinforced their view that the Panguna andesite results can be taken as a reasonable model for the in situ strength of heavily jointed hard rock masses.

Use of rock mass classifications for rock strength prediction

In view of the scarcity of reliable information on the strength of rock masses and of the very high cost of obtaining such information, the authors consider it unlikely that a comprehensive quantitative analysis of rock mass strength will ever be possible. Since this is one of the key questions in rock engineering, it is clear that some attempt should be made to use whatever information is available to provide some form of general guidance on reasonable trends in rock mass strength.

Having considered several possible alternatives, the authors have turned to the rock mass classification schemes, presented in chapter 2, for the prediction of rock mass strength. Table 11 on page 173 gives the details of the ratings which have been chosen for the various samples of Panguna andesite described earlier in this chapter. The classifications proposed by Bieniawski of the South African Council for Scientific and Industrial Research[25,26] and by Barton, Lien and Lunde[1] of the Norwegian Geotechnical Institute have been used in this table and individual ratings have been listed in order to provide the reader with a guide in using these classifications for rock mass strength predictions.

Figure 84 gives a plot of the ratio of m/m_i and the value of s against the CSIR and the NGI classification ratings for Panguna andesite. Bieniawski[26] proposed the following relationship between the CSIR and the NGI systems :

$$RMR = 9 \, Log_e Q + 44 \qquad (66)$$

where RMR is the rock mass rating obtained from the CSIR classification and Q is the quality index obtained from the NGI classification. This relationship has been used to position the scales in figure 84.

In spite of the very low density of experimental data, the

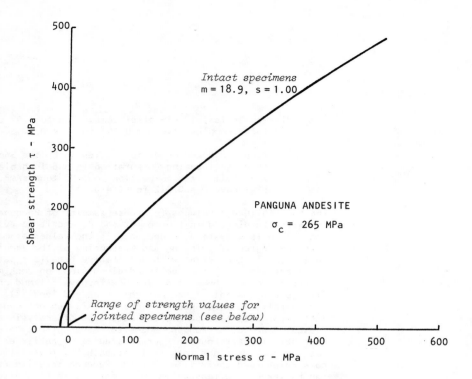

Figure 82 : Mohr failure envelope for intact specimens of Panguna andesite from Bougainville, Papua New Guinea.

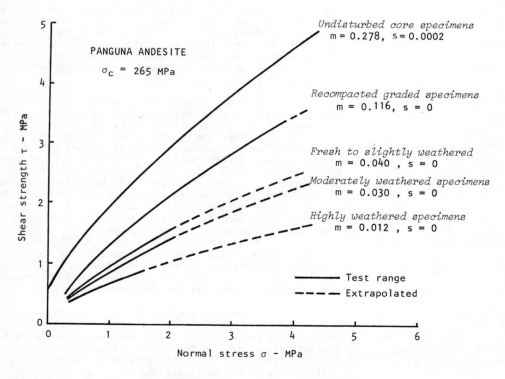

Figure 83: Triaxial test results for jointed Panguna andesite.

TABLE 11 - CSIR AND NGI CLASSIFICATIONS OF PANGUNA ANDESITE SAMPLES

		Intact rock specimens	Undisturbed core samples	Recompacted, graded samples	Fresh to slightly weathered samples	Moderately weathered samples	Highly weathered samples
	Ratio m/m_i	1.00	0.0147	0.0061	0.0021	0.0016	0.0006
	Value of constant s	1.00	0.002	0	0	0	0
CSIR CLASSIFICATION	Intact strength rating	15	15	15	15	15	15
	RQD rating	20	3	3	3	3	3
	Joint spacing rating	30	5	5	5	5	5
	Joint condition rating	25	20	10	10	5	0
	Groundwater rating	10	8	5	5	5	5
	Joint orientation rating	0	-5	-10	-12	-12	-12
	CSIR total rating	100	46	28	26	18	8
NGI CLASSIFICATION	RQD	100	10	10	10	10	10
	Joint set number J_n	1	9	12	15	18	20
	Joint roughness number J_r	4	3	1.5	1	1	1
	Joint alteration number J_a	0.75	2	4	4	6	8
	Joint water reduction J_w	1	1	1	1	1	1
	Stress reduction factor SRF	1	2.5	5	7.5	10	10
	NGI quality index Q	533	0.67	0.06	0.02	0.009	0.006

authors have taken it upon themselves to draw in lines giving approximate relationships between the values of m/m_i and s and the classification ratings. It is suggested that these relationships may be used as a very crude guide in estimating rock mass strength.

Deformability of rock masses

A further application of rock mass classifications was proposed by Bieniawski[189] in a paper dealing with the deformability of rock masses. A number of in situ deformation modulus measurements were reviewed and the rock masses in which these measurements were carried out were classified using the CSIR system.

Figure 85 has been constructed from the information presented in Bieniawski's paper and the authors consider the relationship illustrated to be a useful guide to choosing the modulus of

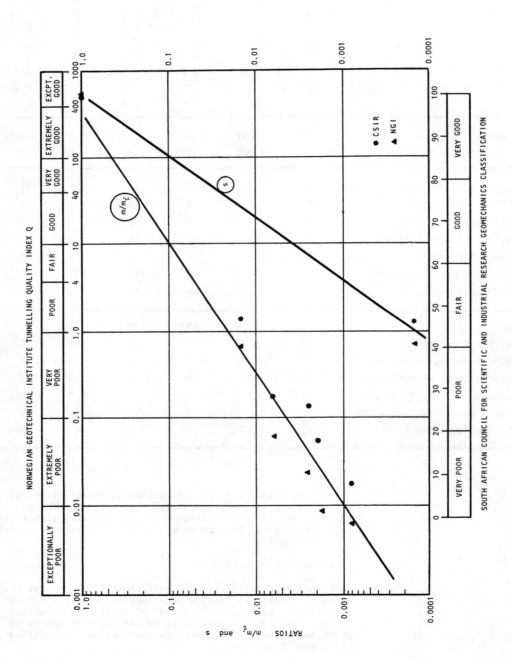

Figure 84 : Plot of the ratio m/m_i and s for Panguna andesite against rock mass classifications for this material.

deformation of an in situ rock mass. This modulus is required for numerical studies of the stress and displacement distribution around underground excavations.

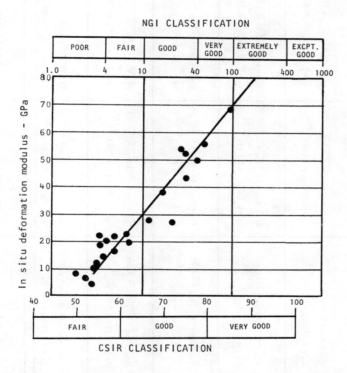

Figure 85 : Relationship between in situ deformation modulus of rock masses and rock mass classifications. After Bieniawski[189].

Approximate equations defining the strength of intact rock and heavily jointed rock masses

On the basis of the discussion presented in this chapter, the authors have attemped to summarise their conclusions in table 12 on the next page. This table gives a set of approximate equations defining the relationships between principal stresses and the Mohr envelopes for the failure of intact rock specimens and of heavily jointed rock masses.

The principal stress relationships are presented in the form:

$$\sigma_{1n} = \sigma_{3n} + \sqrt{m\sigma_{3n} + s} \qquad (67)$$

where m and s are the material constants defined on page 137 and σ_{1n} and σ_{3n} are the normalised principal stresses σ_1/σ_c and σ_3/σ_c, σ_c being the uniaxial compressive strength of the intact rock pieces in the rock mass.

TABLE 12 - APPROXIMATE EQUATIONS FOR PRINCIPAL STRESS RELATIONSHIPS AND MOHR ENVELOPES FOR INTACT ROCK AND JOINTED ROCK MASSES

	CARBONATE ROCKS WITH WELL DEVELOPED CRYSTAL CLEAVAGE *dolomite, limestone and marble*	LITHIFIED ARGILLACEOUS ROCKS *mudstone, siltstone, shale and slate (normal to cleavage)*	ARENACEOUS ROCKS WITH STRONG CRYSTALS AND POORLY DEVELOPED CRYSTAL CLEAVAGE *sandstone and quartzite*	FINE GRAINED POLYMINERALLIC IGNEOUS CRYSTALLINE ROCKS *andesite, dolerite, diabase and rhyolite*	COARSE GRAINED POLYMINERALLIC IGNEOUS AND METAMORPHIC CRYSTALLINE ROCKS *amphibolite, gabbro, gneiss, granite, norite and quartz-diorite*
INTACT ROCK SAMPLES *Laboratory size rock specimens free from structural defects* *CSIR rating 100+, NGI rating 500*	$\sigma_{1n} = \sigma_{3n} + \sqrt{7\sigma_{3n} + 1.0}$ $\tau_n = 0.816(\sigma_n + 0.140)^{0.658}$	$\sigma_{1n} = \sigma_{3n} + \sqrt{10\sigma_{3n} + 1.0}$ $\tau_n = 0.918(\sigma_n + 0.099)^{0.677}$	$\sigma_{1n} = \sigma_{3n} + \sqrt{15\sigma_{3n} + 1.0}$ $\tau_n = 1.044(\sigma_n + 0.067)^{0.692}$	$\sigma_{1n} = \sigma_{3n} + \sqrt{17\sigma_{3n} + 1.0}$ $\tau_n = 1.086(\sigma_n + 0.059)^{0.696}$	$\sigma_{1n} = \sigma_{3n} + \sqrt{25\sigma_{3n} + 1.0}$ $\tau_n = 1.220(\sigma_n + 0.040)^{0.705}$
VERY GOOD QUALITY ROCK MASS *Tightly interlocking undisturbed rock with unweathered joints spaced at ±3 metres* *CSIR rating 85, NGI rating 100*	$\sigma_{1n} = \sigma_{3n} + \sqrt{3.5\sigma_{3n} + 0.1}$ $\tau_n = 0.651(\sigma_n + 0.028)^{0.679}$	$\sigma_{1n} = \sigma_{3n} + \sqrt{5\sigma_{3n} + 0.1}$ $\tau_n = 0.739(\sigma_n + 0.020)^{0.692}$	$\sigma_{1n} = \sigma_{3n} + \sqrt{7.5\sigma_{3n} + 0.1}$ $\tau_n = 0.848(\sigma_n + 0.013)^{0.702}$	$\sigma_{1n} = \sigma_{3n} + \sqrt{8.5\sigma_{3n} + 0.1}$ $\tau_n = 0.883(\sigma_n + 0.012)^{0.705}$	$\sigma_{1n} = \sigma_{3n} + \sqrt{12.5\sigma_{3n} + 0.1}$ $\tau_n = 0.998(\sigma_n + 0.008)^{0.712}$
GOOD QUALITY ROCK MASS *Fresh to slightly weathered rock, slightly disturbed with joints spaced at 1 to 3 metres.* *CSIR rating 65, NGI rating 10*	$\sigma_{1n} = \sigma_{3n} + \sqrt{0.7\sigma_{3n} + 0.004}$ $\tau_n = 0.369(\sigma_n + 0.006)^{0.669}$	$\sigma_{1n} = \sigma_{3n} + \sqrt{1.0\sigma_{3n} + 0.004}$ $\tau_n = 0.427(\sigma_n + 0.004)^{0.683}$	$\sigma_{1n} = \sigma_{3n} + \sqrt{1.5\sigma_{3n} + 0.004}$ $\tau_n = 0.501(\sigma_n + 0.003)^{0.695}$	$\sigma_{1n} = \sigma_{3n} + \sqrt{1.7\sigma_{3n} + 0.004}$ $\tau_n = 0.525(\sigma_n + 0.002)^{0.698}$	$\sigma_{1n} = \sigma_{3n} + \sqrt{2.5\sigma_{3n} + 0.004}$ $\tau_n = 0.603(\sigma_n + 0.002)^{0.707}$
FAIR QUALITY ROCK MASS *Several sets of moderately weathered joints spaced at 0.3 to 1 metre.* *CSIR rating 44, NGI rating 1.0*	$\sigma_{1n} = \sigma_{3n} + \sqrt{0.14\sigma_{3n} + 0.0001}$ $\tau_n = 0.198(\sigma_n + 0.0007)^{0.662}$	$\sigma_{1n} = \sigma_{3n} + \sqrt{0.20\sigma_{3n} + 0.0001}$ $\tau_n = 0.234(\sigma_n + 0.0005)^{0.675}$	$\sigma_{1n} = \sigma_{3n} + \sqrt{0.30\sigma_{3n} + 0.0001}$ $\tau_n = 0.280(\sigma_n + 0.0003)^{0.688}$	$\sigma_{1n} = \sigma_{3n} + \sqrt{0.34\sigma_{3n} + 0.0001}$ $\tau_n = 0.295(\sigma_n + 0.0003)^{0.691}$	$\sigma_{1n} = \sigma_{3n} + \sqrt{0.50\sigma_{3n} + 0.0001}$ $\tau_n = 0.346(\sigma_n + 0.0002)^{0.700}$
POOR QUALITY ROCK MASS *Numerous weathered joints spaced at 30 to 500mm with some gouge filling / clean waste rock* *CSIR rating 23, NGI rating 0.1*	$\sigma_{1n} = \sigma_{3n} + \sqrt{0.04\sigma_{3n} + 0.00001}$ $\tau_n = 0.115(\sigma_n + 0.0002)^{0.646}$	$\sigma_{1n} = \sigma_{3n} + \sqrt{0.05\sigma_{3n} + 0.00001}$ $\tau_n = 0.129(\sigma_n + 0.0002)^{0.655}$	$\sigma_{1n} = \sigma_{3n} + \sqrt{0.08\sigma_{3n} + 0.00001}$ $\tau_n = 0.162(\sigma_n + 0.0001)^{0.672}$	$\sigma_{1n} = \sigma_{3n} + \sqrt{0.09\sigma_{3n} + 0.00001}$ $\tau_n = 0.172(\sigma_n + 0.0001)^{0.676}$	$\sigma_{1n} = \sigma_{3n} + \sqrt{0.13\sigma_{3n} + 0.00001}$ $\tau_n = 0.203(\sigma_n + 0.0001)^{0.686}$
VERY POOR QUALITY ROCK MASS *Numerous heavily weathered joints spaced less than 50mm with gouge filling / waste rock with fines* *CSIR rating 3, NGI rating 0.01*	$\sigma_{1n} = \sigma_{3n} + \sqrt{0.007\sigma_{3n} + 0}$ $\tau_n = 0.042(\sigma_n)^{0.534}$	$\sigma_{1n} = \sigma_{3n} + \sqrt{0.010\sigma_{3n} + 0}$ $\tau_n = 0.050(\sigma_n)^{0.539}$	$\sigma_{1n} = \sigma_{3n} + \sqrt{0.015\sigma_{3n} + 0}$ $\tau_n = 0.061(\sigma_n)^{0.546}$	$\sigma_{1n} = \sigma_{3n} + \sqrt{0.017\sigma_{3n} + 0}$ $\tau_n = 0.065(\sigma_n)^{0.548}$	$\sigma_{1n} = \sigma_{3n} + \sqrt{0.025\sigma_{3n} + 0}$ $\tau_n = 0.078(\sigma_n)^{0.556}$

The Mohr envelopes are defined by equations having the form:

$$\tau_n = A(\sigma_n - \sigma_{tn})^B \qquad (69)$$

where τ_n and σ_n are normalised shear and normal stresses τ/σ_c and σ/σ_c and σ_{tn} is the normalised uniaxial tensile strength which is defined by

$$\sigma_{tn} = \sigma_t/\sigma_c = \tfrac{1}{2}(m - \sqrt{m^2 + 4s}) \qquad (70)$$

The constants A and B in equation 69 are determined by generating a series of values for σ_n and τ_n for given values of m and s (using equations 51 and 52 on page 140) and then fitting the best curve defined by equation 69 by means of a regression analysis. This process is discussed in greater detail in Appendix 5 at the end of this book.

In order to use table 12 for the approximate analysis of rock failure, match the descriptions of the intact rock components and the rock mass quality and use the indicated equations to calculate the normalised major principal stress or the normalised shear strength as required. Multiplication of the calculated normalised stresses and strengths by the uniaxial compressive strength of the intact rock pieces will give the rock mass strength in appropriate stress units. Note that, in all cases, the uniaxial compressive strength of the intact rock pieces which make up the rock mass is used in this analysis and that, if necessary, this strength may be estimated by indirect tests such as the point load test (see page 52).

The equations presented in table 12 should only be used for preliminary analyses of underground excavation or rock slope designs and they should be used to establish the sensitivity of the design to changes in rock mass behaviour. In the case of critical designs, a more accurate analysis should be carried out on the basis of test data such as those presented for Panguna andesite on pages 170 to 172.

Chapter 6 references

123. COOK, N.G.W., HOEK, E., PRETORIUS, J.P.G., ORTLEPP, W.D. and SALAMON, M.D.G. Rock mechanics applied to the study of rockbursts. *J. South African Inst. Min. Metall.*, Vol. 66, 1966, pages 425-528.

124. BYERLEE, J.D. Brittle-ductile transition in rocks. *J. Geophys. Res.*, Vol. 73, No. 14, 1968, pages 4741-4750.

125. HUDSON, J.A., CROUCH, S.L. and FAIRHURST, C. Soft, stiff and servo-controlled testing machines; a review with reference to rock failure. *Engineering Geology*, Vol. 6, No. 3, 1972, pages 155-189.

126. HAWKES, I. and MELLOR, M. Uniaxial testing in rock mechanics laboratories. *Engineering Geology*, Vol. 4, 1970, pages 177-285.

127. HOEK, E. Rock mechanics - an introduction for the practical engineer. Part II. *Mining Magazine*, London, Vol. 114, No. 6, 1966, pages 13-23.

128. HOEK, E. Fracture of anisotropic rock. *J. South African Inst. Min. Metall.*, Vol. 64, No. 10, 1964, pages 510-518.

129. HOEK, E. and FRANKLIN, J.A. A simple triaxial cell for field and laboratory testing of rock. *Trans. Inst. Min. Metall.*, London, Section A, Vol. 77, 1968, pages 22-26.

130. FRANKLIN, J.A. and HOEK, E. Developments in triaxial testing equipment. *Rock Mechanics*, Vol. 2, 1970, pages 223-228.

131. GRIFFITH, A.A. The phenomena of rupture and flow in solids. *Phil. Trans. Royal Soc.*, London, Series A, Vol. 221, 1921, pages 163-198.

132. GRIFFITH, A.A. Theory of rupture. *Proc. Intnl. Gongress Appl. Mech.*, Delft, 1924, pages 55-63.

133. McCLINTOCK, F.A. and WALSH, J.B. Friction on Griffith cracks under pressure. *Proc. 4th National Congress Appl. Mech.*, 1962, pages 1015-1021.

134. HOEK, E. Brittle failure of rock. In *Rock Mechanics in Engineering Practice*, K.G.Stagg and O.C.Zienkiewicz, eds., J. Wiley and Sons, London, 1968, pages 99-124.

135. BALMER, G. A general analytical solution for Mohr's envelope. *Amer. Soc. Testing Materials*, Vol. 52, 1952, pages 1260-1271.

136. MOGI, K. Pressure dependence of rock strength and transition from brittle fracture to ductile flow. *Bull. Earthquake Res. Inst.*, Japan, Vol. 44, 1966, pages 215-232.

137. SCHWARTZ, A.E. Failure of rock in the triaxial shear test. *Proc. 6th Rock Mechanics Symp.*, Rolla, Missouri, 1964, pages 109-135.

138. BROCH, E. The influence of water on some rock properties. In *Advances in Rock Mechanics, Proc. 3rd Congr., Intnl. Soc. Rock Mech.*, Denver, 1974, Vol. 2, Part A, pages 33-38.

139. HOEK, E. Rock fracture under static stress conditions. *Nat. Mech. Engg. Res. Inst. Report* MEG 383, CSIR, S.Africa, 1965, 200 pages.

140. BRACE, W.F. Brittle fracture of rocks. In *State of Stress in the Earth's Crust*, W.R.Judd, ed., Elsevier, New York, 1964, pages 111-174.

141. MOGI, K. Effect of the intermediate principal stress on rock failure. *J. Geophys. Res.*, Vol. 72, No. 20, 1967, pages 5117-5131.

142. HEARD, H.C., ABEY, A.E., BONNER, B.P. and SCHOCK, R.N. Mechanical behaviour of dry Westerley granite at high confining pressure. *Lawrence Livermore Laboratory Report*, UCRL 51642, 1974, 14 pages.

143. WAWERSIK, W.R. and BRACE, W.F. Post failure behaviour of a granite and a diabase. *Rock Mechanics*, Vol. 3, No. 2, 1971, pages 61-85.

144. MISRA, B. Correlation of rock properties with machine performance. *Ph.D Thesis*, Leeds University, 1972.

145. WAWERSIK, W.R. and FAIRHURST, C. A study of brittle rock fracture in laboratory compression experiments. *Intnl. J. Rock Mech. Min. Sci.*, Vol. 7, No. 5, 1970, pages 561-575.

146. KOVARI, K. and TISA, A. Multiple failure state and strain controlled triaxial tests. *Rock Mechanics*, Vol. 7, 1975, pages 17-33.

147. BREDTHAUER, R.O. Strength characteristics of rock samples under hydrostatic pressure. *Amer. Soc. Mech. Engrs. Trans.*, Vol. 79, 1957, pages 695-708.

148. BIENIAWSKI, Z.T. Estimating the strength of rock materials. *J. South African Inst. Min. Metall.*, Vol. 74, No. 8, 1974, pages 312-320.

149. ALDRICH, M.J. Pore pressure effects on Berea sandstone subjected to experimental deformation. *Geol. Soc. Amer. Bull.*, Vol. 80, No. 8, 1969, pages 1577-1586.

150. SCHOCK, R.N., ABEY, A.E., BONNER, B.P., DUBA, A. and HEARD, H.C. Mechanical properties of Nugget sandstone. *Lawrence Livermore Laboratory Report*, UCRL 51447, 1973.

151. BODONYI, J. Laboratory tests on certain rocks under axially symmetrical loading conditions. *Proc. 2nd. Congress Intnl. Soc. Rock Mech.*, Belgrade, Vol. 1, 1970, Paper 2-17.

152. RAMEZ, M.R.H. Fractures and the strength of a sandstone under triaxial compression. *Intnl. J. Rock Mech. Min. Sci.*, Vol. 4, 1967, pages 257-268.

153. JAEGER, J.C. Rock failure at low confining pressures. *Engineering*, Vol. 189, 1960, pages 283-284.

154. PRICE, N.J. The strength of coal measure rocks in triaxial compression. *UK Nat. Coal Board MRE Report*, No. 2159, 1960.

155. FAIRHURST, C. and COOK, N.G.W. The phenonenon of rock splitting parallel to a free surface under compressive stress. *Proc. 1st Congress Intnl. Soc. Rock Mech.*, Lisbon, Vol. 1, 1966, pages 687-692.

156. TERZAGHI, K. Stress conditions for the failure of saturated concrete and rock. *Proc. Amer. Soc. Testing Materials*, Vol. 45, 1945, pages 777-801.

157. BRACE, W.F. and MARTIN, R.J. A test of the law of effective stress for crystalline rocks of low porosity. *Intnl. J. Rock Mech. Min Sci.*, Vol. 5, No. 5, 1968, pages 415-426.

158. HANDIN, J., HAGER, R.V., FRIEDMAN, M. and FEATHER, J.N. Experimental deformation of sedimentary rocks under confining pressure; pore pressure tests. *Bull. Amer. Ass. Petrol. Geol.*, Vol. 47, 1963, pages 717-755.

159. COLBACK, P.S.B. and WIID, B.L. The influence of moisture content on the compressive strength of rock. *Proc. 3rd Canadian Rock Mech. Symp.*, Toronto, 1965, pages 65-83.

160. HOJEM, J.M.P. and COOK, N.G.W. The design and construction of a triaxial and polyaxial cell for testing rock specimens. *South African Mech. Engr.*, Vol. 18, 1968, pages 57-61.

161. MOGI, K. Effect of triaxial stress system on rock failure. *Rock Mech. in Japan*, Vol. 1, 1970, pages 53-55.

162. MOGI, K. Fracture and flow of rocks under high triaxial compression. *J. Geophys. Res.*, Vol. 76, No. 5, 1971, pages 1255-1269.

163. VUTUKURI, V.S., LAMA, R.D. and SALUJA, S.S. *Handbook on Mechanical Properties of Rocks*. Vol. 1, Trans Tech Publications, Clausthal, 1974, 280 pages.

164. HODGSON, K. and COOK, N.G.W. The effects of size and stress gradient on the strength of rock. *Proc. 2nd Congress Intnl. Soc. Rock Mech.*, Belgrade, Vol. 2, 1970, Paper 3-5.

165. OBERT, L., WINDES, S.L. and DUVALL, W.I. Standardized tests for determining the physical properties of mine rock. *U.S. Bureau of Mines Report of Investigations*, 3891, 1946.

166. MOGI, K. The influence of the dimensions of specimens on the fracture strength of rocks. *Bull. Earthquake Res. Inst. Tokyo Univ.*, Vol. 40, 1962, pages 175-185.

167. BIENIAWSKI, Z.T. Propagation of brittle fracture in rock. *Proc. 10th Symp. Rock Mech.*, AIME, New York, 1972, pages 409-427.

168. PRATT, H.R., BLACK, A.D., BROWN, W.S. and BRACE, W.R. The effect of specimen size on the mechanical properties of unjointed diorite. *Intnl. J. Rock Mech. Min. Sci.*, Vol. 9, 1972, pages 513-529.

169. PROTODIAKONOV, M.M. and KOIFMAN, M.I. The scale effect in investigations of rock and coal. *Proc. 5th Congress Intnl. Bureau Rock Mech.*, Leipzig, 1963.

170. HOSKINS, J.R. and HORINO, F.G. The influence of spherical head size and specimen diameter on the uniaxial compressive strength of rocks. *U.S. Bureau of Mines Report of Investigations*, 7234, 1969, 16 pages.

171. JAEGER, J.C. Shear failure of anisotropic rock. *Geol. Mag.*, Vol. 97, 1960, pages 65-72.

172. McLAMORE, R. and GRAY, K.E. The mechanical behaviour of anisotropic sedimentary rocks. *Amer. Soc. Mech. Engrs. Trans.*, Series B, 1967, pages 62-76.

173. DONATH, F.A. Effects of cohesion and granularity on the deformational behaviour of anisotropic rock. *Geol. Soc. Amer.*, Memoir 135, 1972, pages 95-128.

174. DONATH, F.A. Strength variations and deformational behaviour in anisotropic rock. In *State of Stress in the Earth's Crust*, W.R.Judd, ed., Elsevier, New York, 1964, pages 281-297.

175. ATTEWELL, P.B. and SANDFORD, M.R. Intrinsic shear strength of a brittle, anisotropic rock - I, Experimental and mechanical interpretation. *Intnl. J. Rock Mech. Min. Sci.*, Vol. 11, 1974, pages 423-430.

176. HORINO, F.G. and ELLICKSON, M.L. A method of estimating strength of rock containing planes of weakness. *U.S. Bureau of Mines Report of Investigations*, 7449, 1970.

177. BRAY, J.W. A study of jointed and fractured rock. *Rock Mech. and Engr. Geol.*, Vol. 5, Nos. 2 and 3, 1967, pages 119-136 and 197-216.

178. ROSENGREN, K.J. and JAEGER, J.C. The mechanical properties of a low porosity interlocked aggregate. *Géotechnique*, Vol. 18, 1968, pages 317-326.

179. BROWN, E.T. Strength of models of rock with intermittent joints. *J. Soil Mechs. Foundns. Div., ASCE*, Vol. 96, No. SM6, 1970, pages 1935-1949.

180. BROWN, E.T. and TROLLOPE, D.H. Strength of a model of jointed rock. *J. Soil Mechs. Foundns. Div., ASCE*, Vol. 96, No. SM2, 1970, pages 685-704.

181. EINSTEIN, H.H. and HIRSCHFELD, R.C. Model studies on mechanics of jointed rock. *J. Soil Mechs. Foundns. Div., ASCE*, Vol. 99, No. SM3, 1973, pages 229-248.

182. JOHN, K.W. Festigkeit und Verformbarkeit von drukfesten, regelmassig gefügten Diskontinuen. *Veroffentlichungen des Institutes für Bodenmechanic und Felsmechanik der Universität Fredericiana in Karlsruhe*, Heft 37, 1969, 99 pages.

183. LADANYI, B. and ARCHAMBAULT, G. Simulation of the shear behaviour of a jointed rock mass. *Proc. 11th Symposium on Rock Mech*, published by AIME, New York, 1970, pages 105-125.

184. MULLER, L. and PACHER, P. Modellversuche zur Klarung der Bruchefahr geklufteter Medien. *Rock Mech. and Engr. Geol.*, Supplement 2, 1965, pages 7-24.

185. JAEGER, J.C. The behaviour of closely jointed rock. *Proc. 11th Symposium on Rock Mech.*, Berkeley, published by AIME, New York, 1970, pages 57-68.

186. MARSAL, R.J. Mechanical properties of rockfill. in *Embankment Dam Engineering - Casagrande Volume*, Edited by R.C.Hirschfeld and S.J.Poulos, published by J.Wiley & Sons, New York, 1973, pages 109-200.

187. MANEV, G. and AVRAMOVA-TACHEVA, E. On the valuation of strength and resistance condition of the rocks in natural rock massif. *Proc. 2nd Intnl. Congress on Rock Mech.*, Belgrade, Vol. 1, 1970, pages 59-65.

188. MARACHI, N.D., CHAN, C.K. and SEED, H.B. Evaluation of the properties of rockfill materials. *J. Soil Mechs. Foundns. Div., ASCE*, Vol. 98, No. SM1, 1972, pages 95-114.

189. BIENIAWSKI, Z.T. Determining rock mass deformability : experience from case histories. *Intnl. J. Rock Mechanics and Mining Sciences*, Vol. 15, 1978, pages 237-247.

Chapter 7: Underground excavation failure mechanisms

Overburden soil and heavily weathered rock - squeezing and flowing ground, short stand-up time

Blocky jointed rock partially weathered - gravity falls of blocks from roof and sidewalls

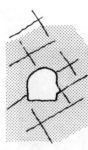

Massive rock with few unweathered joints - no serious stability problems

Massive rock at great depth - stress induced failures, spalling and popping with possible rockbursts

Introduction

The left hand margin sketch gives a simplified picture of the underground excavation stability problems which are encountered with increasing depth below surface.

At shallow depth in overburden soil or heavily weathered poor quality rock, excavation problems are generally associated with squeezing or flowing ground and very short stand-up times. This means that cut and cover or soft ground tunnelling techniques have to be used and adequate support has to be provided immediately behind the advancing face. The stability of underground excavations in very poor quality rock or soil will not be discussed in detail in this book and the interested reader is referred to books such as those by Pequignot[190], Hewett and Johannesson[191] and Szechy[192] and to journals such as Tunnels and Tunnelling for further details.

Stability problems in blocky jointed rock are generally associated with gravity falls of blocks from the roof and sidewalls. Rock stresses at shallow depth are generally low enough that they do not have a significant effect upon this failure process which is controlled by the three-dimensional geometry of the excavation and of the rock structure.

Excavations in unweathered massive rock with few joints do not usually suffer from serious stability problems when the stresses in the rock surrounding the excavations are less than approximately one fifth of the uniaxial compressive strength of the rock. These are generally the most ideal conditions for the creation of large unsupported excavations in rock.

As the depth below surface increases or as a number of excavations are mined close to one another, as in room and pillar mining, the rock stress increases to a level at which failure is induced in the rock surrounding the excavations. This failure can range from minor spalling or slabbing in the surface rock to major rockbursts in which explosive failure of significant volumes of rock can occur.

Obviously, there are many underground situations in which two or more of these failure processes can occur simultaneously. Such cases can only be dealt with on an individual basis and the discussion which follows is intended to cover the basic failure processes and to give the reader sufficient background to tackle the more complex failure mechanisms encountered underground.

Structurally controlled failure

Figure 86 shows a tunnel through a bench in an old slate quarry in Wales. This tunnel was constructed approximately 100 years ago without any form of support and the tunnel shape has stabilised to conform to the structural pattern in the slate. Another example of structurally controlled failure is illustrated in figure 87 which shows blocky roof conditions in jointed hard rock in a mine in Australia.

In order that a block of rock should be free to fall from the roof or the sidewalls of an excavation, it is necessary that this block should be separated from the surrounding rock mass by at least three intersecting structural discontinuities.

Figure 86 : Structurally controlled excavation shape in an unsupported tunnel in slate.

Figure 87 : Blocky roof conditions in an underground mine excavation in jointed hard rock.

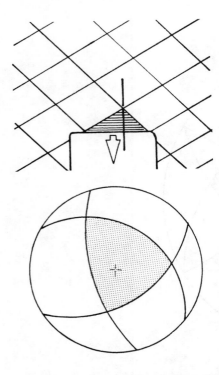

Conditions for gravity falls of roof wedges

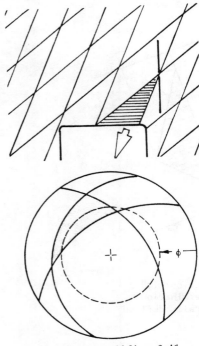

Conditions for sliding failure of roof wedges

Structurally controlled failure can be analysed by means of the stereographic projection technique described in chapter 4 of this book. A simple example of the application of this method is illustrated in the margin sketch which shows a wedge of rock falling from the roof of an excavation in jointed rock. A vertical line drawn through the apex of the wedge must fall within the base of the wedge for failure to occur without sliding on at least one of the joint planes.

In the stereographic plot, the vertical line through the apex of the wedge is represented by the centre point of the net and the conditions stated above are satisfied if the great circles representing the joint planes form a closed figure which *surrounds* the centre of the net.

This very simple kinematic check is useful for evaluating the potential for roof falls during preliminary studies of structural geology data which have been collected for the design of an underground excavation. The stereographic method can also be used for a much more detailed evaluation of the shape and volume of potentially unstable wedges as illustrated in figure 88.

Three planes are represented by their great circles, marked A, B and C in figure 88. The strike lines of these planes are marked a, b and c and the traces of the vertical planes through the centre of the net and the great circle intersections are marked ab, ac and bc. Suppose that a square tunnel with a span of S runs in a direction from 290° to 110° as shown in the lower part of figure 88. The directions of the strike lines correspond to the traces of the planes A, B and C on the horizontal roof of the tunnel. These strike lines can be combined to give the maximum size of the triangular figure which can be accommodated within the tunnel roof span, as shown in figure 88.

In the plan view, the apex of the wedge is defined by finding the intersection point of the lines ab, ac and bc, projected from the corners of the triangular wedge base as shown. The height h of the apex of the wedge above the horizontal tunnel roof is found by taking a section through the wedge apex and normal to the tunnel axis. This section, marked XX in figure 88, intersects the traces a and c at the points shown and these points define the base of the triangle as seen in view XX. The apparent dips of the planes C and A are given by the angles α and β which are measured on the stereographic projection along the line XX through the centre of the net.

The volume of the wedge is given by $1/3.h$ x the base area of the wedge as determined from the plan view in figure 88.

If three joints intersect to form a wedge in the roof of an underground excavation but the vertical line through the apex of the wedge does not fall within the base of the wedge, failure can only occur by sliding on one of the joint surfaces or along one of the lines of intersection. This condition is represented stereographically if the intersection figure formed by the three great circles falls to one side of the centre of the net as illustrated in the lower margin drawing.

An additional condition which must be satisfied for sliding

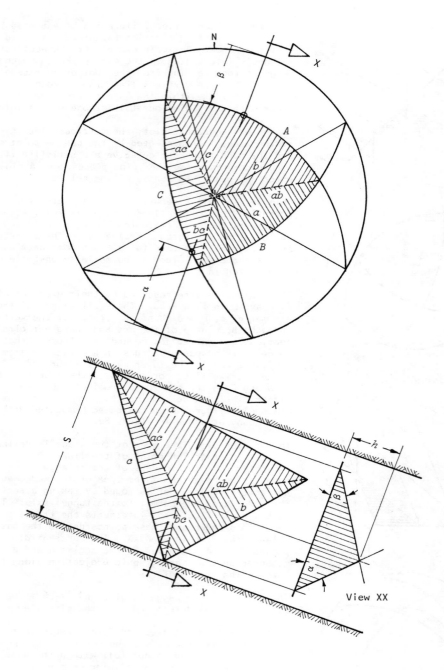

Figure 88: Supplementary construction in conjunction with a stereographic projection for the determination of the shape and volume of a structurally defined wedge in the roof of a tunnel.

of the wedge to occur is that the plane or the line of intersection along which sliding is to occur should be steeper than the angle of friction ϕ. This condition is satisfied if at least part of the intersection figure falls within a circle defined by counting off the number of degree divisions corresponding to the angle of friction from the outer circumference of the net.

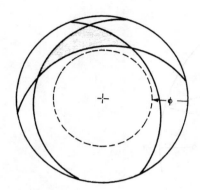
Stable wedge conditions

The construction of the true plan view of the wedge follows the same principles as used in figure 88 and the construction for the case under consideration is illustrated in figure 89. In this example, the strike length of the trace c of plane C is defined by the dimension L.

In determining the height h of the wedge, the view XX has to be taken at right angles to the line ab which passes through the centre of the net and the intersection of the great circles representing planes A and B. The angle α is the true dip of the line of intersection of these two planes.

When the entire intersection figure falls outside the friction circle, as shown by the drawing in the margin, the gravitational weight of the wedge is not high enough to overcome the frictional resistance of the plane or planes on which sliding would take place. Under these conditions, the wedge is stable against sliding.

In the sidewall of an excavation in jointed rock, failure of wedges can occur in much the same way as in the roof except that falls are not possible and all sidewall failures involve sliding on a plane or along the line of intersection of two planes. Two methods for analysing sidewall failure are presented below.

Sidewall failure analysis - method 1

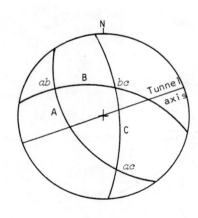

Consider a square tunnel running in a direction from 250° to 70° through a rock mass in which three joint sets occur. These joints are represented by the great circles marked A, B and C in the stereographic projection given in the margin. The traces of the great circles in this drawing have been obtained by projection onto a *horizontal* plane through the centre of the reference sphere. In order to find the shape of the wedge in the tunnel sidewall, it is necessary to determine the shape of the intersection figure projected onto a *vertical* plane.

This intersection figure is obtained by *rotation* of the great circle intersections ab, bc and ac through 90° about the tunnel axis. This rotation is carried out stereographically as follows :

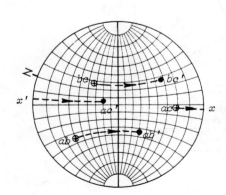

Trace the points ab, bc and ac onto a clean piece of tracing paper. Mark the centre point and north point and also the tunnel axis on this tracing.

Locate the tracing on the meridional net by means of the centre pin so that the *tunnel axis* coincides with the *north-south* axis of the net.

Rotate each of the three intersections onto a vertical plane by counting off 90° along the *small circles* passing through the points ab, bc and ac.

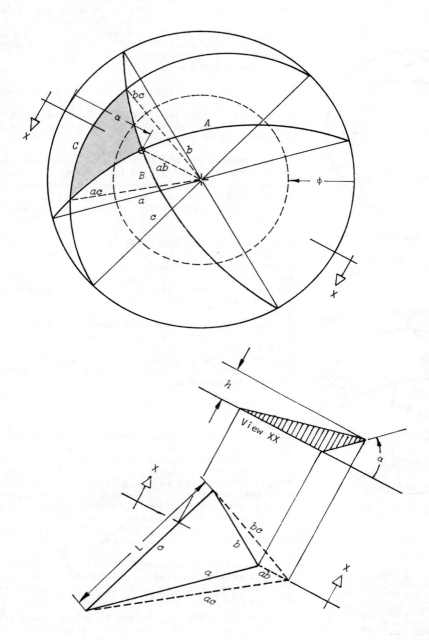

Figure 89: Construction of true plan view and determination of the height of a wedge where failure takes place as a result of sliding along the line of intersection of planes A and B.

It should be noted that the rotation of all the points must be in the same direction. It should also be noted that the small circle through ac passes out of the net circumference at x and re-enters it at a diametrically opposite point x'. This procedure ensures that all intersection points lie within the same hemisphere and that the projection onto the vertical plane is meaningful.

Mark the rotated intersections ab', bc' and ac' and find the great circles which pass through pairs of intersection points. The strike lines of these great circles represent the *traces* of the joint planes on the vertical sidewalls of the tunnel.

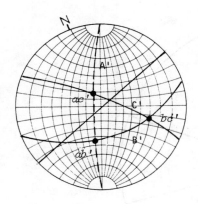

The complete construction is shown in the upper part of figure 90 which gives the stereographic projection of the planes and their intersections in a *vertical plane* parallel to the tunnel sidewalls.

Construction of the true view of the wedge in the sidewall follows the same procedure as that which was used for the roof (figures 88 and 89). The traces a', b' and c' of the joints in the sidewall are parallel to the strike lines of the great circles in the vertical stereographic projection. The lines of intersection ab', bc' and ac' as seen in the vertical sidewall are also parallel to the lines from the centre of the vertical projection to the points ab', bc' and ac'.

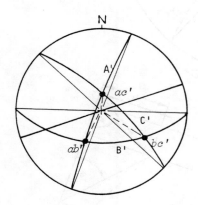

Note that the views in the lower parts of figures 90 and 91 represent the joint traces seen in the *northern* sidewall from the *inside* of the tunnel or in the *southern* sidewall from the *outside* of the tunnel, looking towards 340°. This can be checked by comparing the dips α, β and ξ of the traces of the planes A, B and C in the vertical sidewall, obtained from the stereographic projections, with the corresponding traces in the views of the tunnel sidewall. A *mirror image* of the view given in the lower parts of figures 90 and 91 represents the joint traces in the *southern* sidewall seen from the *inside* or in the *northern* sidewall seen from the *outside* of the tunnel, looking in a direction of 160°.

It is very important that these views should be fully understood since an error could result in an incorrect assessment of stability and in the application of the incorrect remedial measures.

The height h of the wedge shown in figure 90 is found by taking a section XX through the apex of the wedge and finding the apparent dips κ and θ of the planes A' and B' as seen in the vertical projection. This construction is identical to that used in figure 88 to find the height of the wedge in the tunnel roof.

Sidewall failure analysis - method 2

In this method, the traces a, b and c of the joints in the sidewall of the tunnel are found by determining the *apparent* dips α, β and ξ of the planes A, B and C in a vertical plane parallel to the tunnel axis. The determination of these apparent dips is illustrated in figure 91.

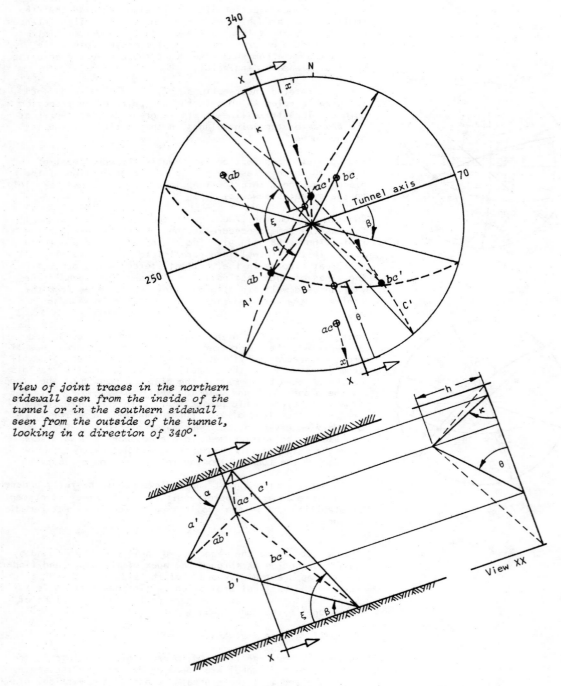

View of joint traces in the northern sidewall seen from the inside of the tunnel or in the southern sidewall seen from the outside of the tunnel, looking in a direction of 340°.

Figure 90: Construction of the true view of a wedge in the sidewall of a tunnel by Method 1.

The appearance of the traces ab, bc and ac in the sidewall is established by finding the dips ψ_{abt}, ψ_{bct} and ψ_{act} of the projections of these lines of intersection onto the vertical sidewall. The angle ψ_{abt} is given by

$$\text{Tan } \psi_{abt} = \frac{\text{Tan } \psi_{ab}}{\text{Cos } \theta_{ab}}$$

where θ_{ab} is the angle between the tunnel axis and the projection of the line of intersection ab on the horizontal plane and ψ_{ab} is the true dip of the line of intersection ab.

The angles ψ_{bct} and ψ_{act} are found in the same way.

The height h of the wedge is found by determining the angles ψ_{bct}' and ψ_{act}' which represent the dips of the lines of intersection as seen in a vertical plane at right angles to the tunnel axis. The angle ψ_{bct}' is given by

$$\text{Tan } \psi_{bct}' = \frac{\text{Tan } \psi_{bc}}{\text{Sin } \theta_{bc}}$$

The other angles are determined in a similar way.

Method of orthographic projections

In addition to the two methods described above, the method of orthographic projections described by Goodman[117] (page 77) can be used for sidewall stability analysis.

Computer analysis of structurally controlled instability

Stereographic techniques are useful for gaining an understanding of structurally controlled failure and for checking the stability of isolated wedges in underground excavations. When designing major excavations in well jointed rock, the use of these manual techniques is too time-consuming and computer techniques can be used for structural stability analysis.

A good example of the use of computer techniques for structural stability analysis has been discussed in papers by Cartney[194] and by Croney, Legge and Dhalla[195] and brief details of this example are summarised on the following pages*.

Figure 92 shows an isometric drawing of the underground excavation complex of the Dinorwic Pumped Storage Scheme which is located in north Wales[196]. The machine hall, which will house six 300 MW pump-turbines, is 180 metres long, 24.5 metres in span and has a maximum height of 52.2 metres. This hall, together with the other caverns shown in figure 92, has been excavated at a depth of approximately 300 metres below an abandoned slate quarry. This slate is of high quality but is well jointed as illustrated in the photograph reproduced in figure 93 showing a surface exposure of slate.

---

* Details of this analysis are published with permission from the Central Electricity Generating Board and from James Williamson and Partners.

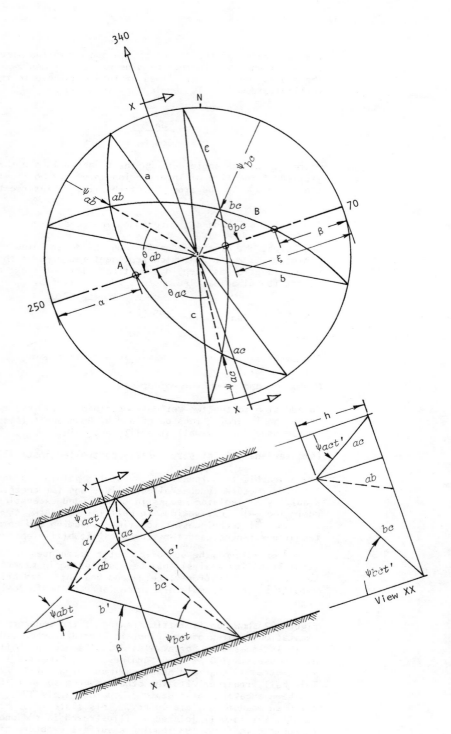

Figure 91 : Construction of the true view of a wedge in the sidewall of a tunnel by Method 2.

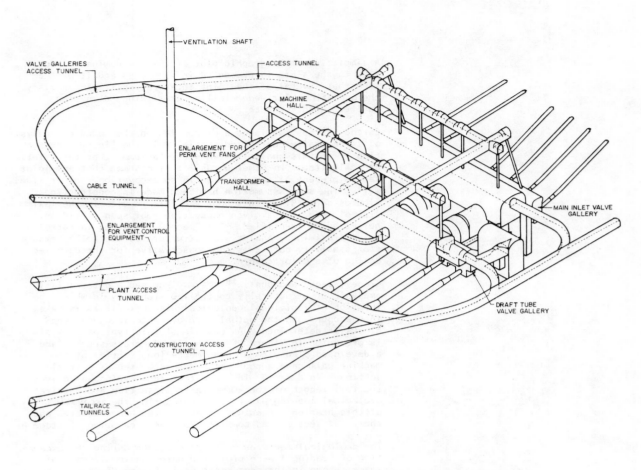

Figure 92 : Isometric drawing showing the underground excavation complex for the Dinorwic Pumped Storage Scheme in north Wales. (Drawing prepared by James Williamson and Partners, Glasgow)

Figure 93 : Surface excavation showing well developed jointing in the slate.

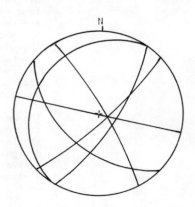

Figure 94 : Stereographic plot of the six main joints in the slate in which the caverns shown in figure 92 have been excavated.

A simplified stereographic plot of the six dominant joint sets in the slate is presented in figure 94 and it will be seen that several of the joint intersection figures satisfy the conditions for wedge falling or sliding, discussed earlier in this chapter.

Cartney[194] has published a discussion on the computer analysis of structurally controlled stability in the Dinorwic excavations. In this analysis it has been assumed that the joints are ubiquitous. In other words, it is assumed that any joint can be present at any location in an excavation roof or sidewall. The same assumption has been made in constructing figures 88 to 91 in this book and this results in the maximum size of wedge for a given excavation roof span or sidewall height being identified. This method is useful in carrying out a preliminary analysis of structurally controlled excavation stability but it is too conservative for the detailed design of a major cavern.

Croney, Legge and Dhalla[195] have discussed a deterministic analysis of structurally controlled stability in which the location of individual geological discontinuities is taken into account and the actual size of unstable wedges and blocks is determined. An example of this type of analysis is presented in figure 95 which gives a cross-section and a developed plan view of a 40 metre long section of the machine hall cavern roof which was excavated as a trial section. An 8 metre wide, 6 metre high heading was driven the full length of the 40 metre trial section and detailed geological mapping was carried out during the excavation of this heading. Headings 2 and 3 were then excavated and the trial section was completed by the excavation of stage 4.

The geological features which were mapped during the excavation of heading 1 were projected onto the idealised roof shape (shown in the cross-section in figure 95) and this information was then used as input for the computer program. This input consists of the location and orientation of all relevant structural discontinuities and the program generates a developed plan view of the excavation showing all of these discontinuities. All wedges and blocks made up of three or four discontinuities and the excavation surface are located and the kinematic possibility of falling or sliding checked for each one. For kinematically unstable wedges or blocks, the coordinates of all corners are printed out together with the base area, volume and height of the wedge or block. This information can then be used to calculate the reinforcing load required to give a specified factor of safety for each wedge or block.

The developed plan view of the roof of the machine hall trial section, presented in figure 95, shows the potentially unstable wedges and blocks which were taken into account in designing the rockbolt reinforcement for the machine hall excavation. The appearance of the machine hall during an advanced stage of the excavation process is illustrated in figure 96.

Optimum orientation and shape of excavations in jointed rock

It will be obvious, from the discussion presented on the previous pages, that the optimum orientation and shape of an underground excavation in jointed rock will be those which give the smallest volume of potentially unstable wedges.

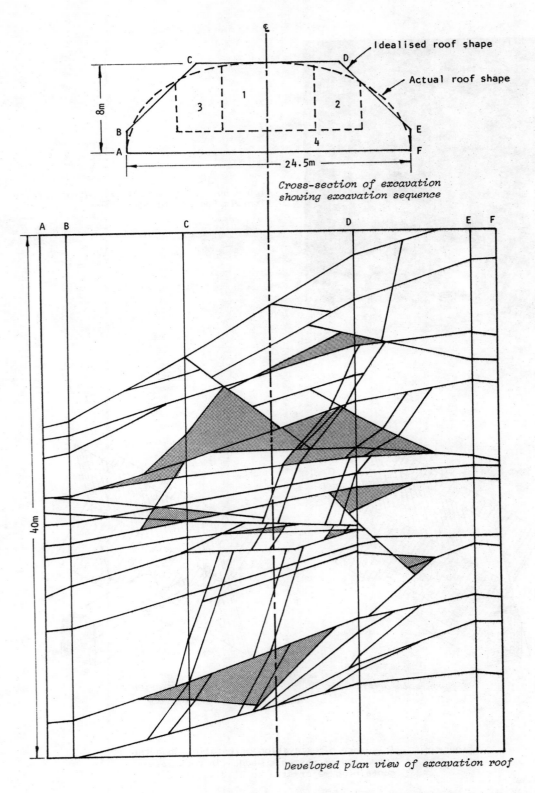

Figure 95 : Cross-section and developed plan view of a trial section of the machine hall roof at Dinorwic. Potentially unstable wedges and blocks are shaded. (After Croney, Legge and Dhalla[195]).

Figure 96: Machine hall excavation for the Dinorwic Pumped Storage Scheme.

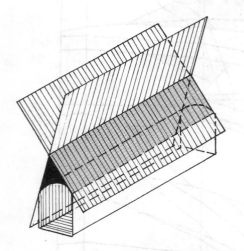
Unfavourable orientation

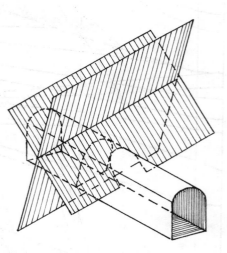
Optimum orientation

Figure 97 : Influence of excavation orientation upon the formation of unstable wedges in rock masses containing major structural discontinuities.

A very simple example is illustrated in figure 97 which shows two alternative orientations for an excavation in relationship to the strike direction of two major discontinuities. The most unfavourable orientation is that in which the excavation axis is aligned parallel to the strike of the line of intersection of the discontinuities. This orientation can result in the formation of a large wedge, running parallel to the excavation axis as illustrated in figure 98.

The optimum orientation for the excavation illustrated in figure 97 is at right angles to the strike direction of the line of intersection of the two discontinuities. This orientation gives the minimum volume of unstable material in the excavation roof.

In the case of a rock mass containing a number of intersecting discontinuities, all having similar strength characteristics, the choice of an optimum direction becomes more difficult. In critical cases it may be necessary to carry out an analysis of potential failures for a range of possible excavation orientations. Once again, the optimum orientation of the excavation is that which gives the minimum volume of unstable material.

In some cases it may not be possible to alter the orientation of the excavation with respect to the rock structure. An example of such a case is illustrated in figure 99 which shows a cut and fill stope in the Mount Isa Mine in Australia. Here the ore body lies between slaty hanging and footwalls and, since the object of mining is to recover this orebody, the most economical excavation is parallel to the strike of the bedding system. Figure 99 shows that the excavation shape has been chosen to give the most favourable hangingwall and back stability conditions.

Influence of excavation size upon structurally controlled failure

Consider the very simple example of an excavation, such as that illustrated in figure 98, in which the axis of the excavation runs parallel to the strike direction of the line of intersection of two sets of joints. Assume that these joints are uniformly spaced at intervals of 1 foot as shown in figure 100. A square shaped scraper drive is excavated in this rock mass and the shaded areas adjacent to the smaller excavation in figure 100 show that a 6 foot by 6 foot drive will release potentially unstable wedges having a volume of approximately 12 cubic feet per foot of drive length. Suppose that it is decided to mechanise this area of the mine and to open the drive to 12 feet by 12 feet to accommodate a front-end loader. The shaded areas adjacent to the larger excavation shown in figure 100 show that the creation of this larger drive will release potential wedges having a volume of approximately 70 cubic feet per foot of drive length. Hence, the increase in the volume of unstable material released by the increase in size of the excavation is approximately proportional to the increase in cross-sectional area of the excavation. Since support costs are roughly proportional to the volume of unstable material to be supported, it could be anticipated that support costs in this example would be proportional to the square of the excavation size.

Unfortunately, the lesson to be learned from this simple example is frequently ignored or overlooked and there are

Figure 98 : Sidewall failure in a mine haulage aligned parallel to the strike direction of two major discontinuities.

Figure 99 : Choice of excavation shape to give the most favourable stability conditions in cut and fill stopes in Mount Isa Mine, Australia.
(Photograph reproduced with permission from Mount Isa Mines Limited).

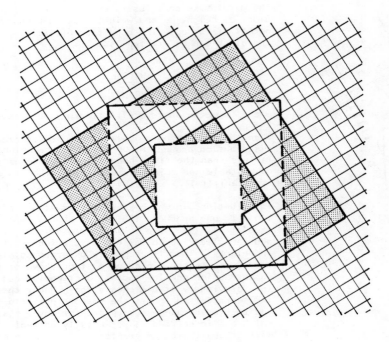

Figure 100 : Increase in unstable rock volume with increase in excavation size in example in which the excavation axis is parallel to the strike of the line of intersection of the joints.

many mines where the engineers have been surprised at the significant increase in stability problems and associated support costs involved in converting from scraper to mechanised operations. In some cases this problem is further compounded by lack of adequate control of blasting which results in serious overbreak and a further deterioration of excavation stability.

Influence of in situ stress on structurally controlled instability

In the preceding discussion, the influence of in situ rock stress has been ignored and it has been assumed that the kinematically unstable wedges and blocks are acted upon by gravity only. This is clearly an over-simplification, particularly in the case of excavations at considerable depth or in rock masses in which the horizontal stresses are exceptionally high.

Unfortunately the current state of the art in rock mechanics does not extend to a satisfactory solution of this interactive problem and it is only possible to discuss the influence of in situ stresses in very simple terms.

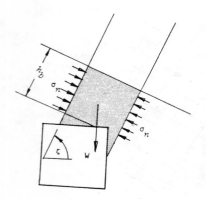

The margin sketch illustrates the case of a parallel sided block of height h_b and weight W acted upon by an average normal stress σ_n. The condition of limiting equilibrium for the block can be written as

$$\sigma_n = \frac{W . \sin \zeta}{2 h_b . \tan \phi} \qquad (71)$$

where ϕ is the angle of friction of the discontinuity surfaces.

When the average normal stress σ_n, calculated from the stress distribution around the opening and the weight of the block, is less than the right hand side of the equation, the block will be unstable and reinforcement is required in order to restore its stability.

In situations in which several excavations are to be mined close to one another, the stress shadows which occur between excavations (see page 124) can cause local stress reductions and can give rise to failure of previously stable roof wedges and blocks. This is common in underground mining where stress-relief due to the mining of a stope above or adjacent to an extraction opening such as a drawpoint can cause failure in a previously stable excavation. Under these circumstances it is prudent to ignore the support provided by rock stress and to provide adequate support for any wedges or blocks which are identified as potentially unstable. This subject will be dealt with in greater detail in the next chapter.

Pillar failure

The simplest example of stress induced instability in underground excavations is that of pillars which are crushed as a result of over-loading. It is useful to consider this example in some detail since there are many valuable practical lessons to be learned from it.

Stress analysis approach to pillar design

Figure 101 shows a pillar in a large metal mine and, in the following hypothetical example, it will be assumed that it is required to carry out a study of the stability of a series of such pillars. It is assumed that these pillars form part of a large panel in a flat lying part of the orebody and that the overburden load is uniformly distributed over the pillars. The following information is provided :

Pillar width W_p = 1.5 metres (pillar is square in plan),
Pillar height h = 3 metres,
Excavation width W_o = 4.2 metres,
Depth below surface z = 100 metres,
Unit weight of rock γ = 0.028 MN/m^3,
Uniaxial compressive strength of intact rock σ_c = 150 MPa,

Rock mass is of very good quality and it is assumed that its triaxial strength is defined by

$$\sigma_1 = \sigma_3 + \sqrt{8.5 \sigma_3 \sigma_c + 0.1 \sigma_c^2} \quad \text{(see page 176)}$$

$$\text{or} \quad \sigma_1 = \sigma_3 + \sqrt{1275 \sigma_3 + 2250}$$

Figure 101 - Typical pillar in a large metal mine.

The average pillar stress σ_p can be calculated from the equation for square pillars given on page 114:

$$\sigma_p = \gamma z (1 + W_o/W_p)^2 = 0.028 \times 100 (1 + 4.2/1.5)^2 = 40 \text{ MPa}.$$

In order to evaluate the strength of the pillar it will be assumed that the stress distribution given in figure 55 on page 118 is applicable to this pillar. This is an approximation since this stress distribution was derived for a rib pillar (ie a two-dimensional pillar) while the pillar under consideration is square in plan (ie a three-dimensional pillar). As shown in figure 59 on page 123, the stress distribution in a three-dimensional pillar is more complex than that in a rib pillar. However, for the purposes of this analysis it will be assumed that the error introduced by this difference is within the overall accuracy of the analysis.

The left hand side of figure 102 gives superimposed contours of the major and minor principal stresses. The values of σ_1 and σ_3 have been obtained by multiplying the values of σ_1/σ_p and σ_3/σ_p, from figure 55, by the average pillar stress $\sigma_p = 40$ MPa, calculated above.

Figure 103 is a plot of the rock mass strength and it shows the strength at a minor principal stress value of $\sigma_3 = 2$ MPa as $\sigma_{1s} = 71.3$ MPa. If the stress conditions at a point are defined by $\sigma_1 = 40$ MPa and $\sigma_3 = 2$ MPa, then the *strength/stress ratio* at this point is $\sigma_{1s}/\sigma_1 = 71.3/40 = 1.78$. Contours of equal strength/stress ratios are shown on the right hand side of figure 102.

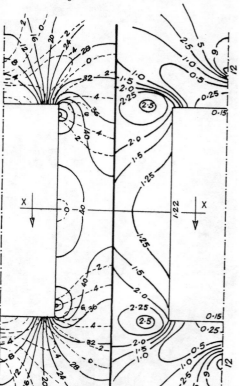

Figure 102 : Distribution of major and minor principal stresses and of strength/stress ratio in pillar subjected to an average pillar stress of 40 MPa.

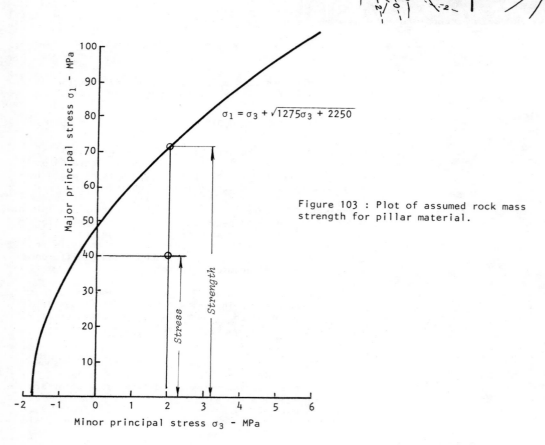

$$\sigma_1 = \sigma_3 + \sqrt{1275\sigma_3 + 2250}$$

Figure 103 : Plot of assumed rock mass strength for pillar material.

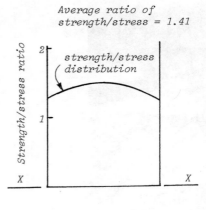

Average ratio of strength/stress = 1.41

strength/stress distribution

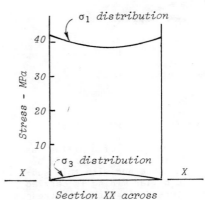

σ_1 distribution

σ_3 distribution

Section XX across pillar in figure 102

The distributions of the stresses σ_1 and σ_3 across the centre of the pillar (section XX in figure 102) are plotted in the margin drawing together with a plot of the distribution of the strength/stress ratio across the pillar. The average strength/stress ratio is 1.41 and the authors suggest that this average value is equivalent to the *factor of safety* of the pillar.

It is important to emphasise the difference between the ratio of strength to stress at a point and the factor of safety of the entire pillar. When the strength/stress ratio at a point falls below 1.00, failure will initiate at this point. As will be shown later in this chapter, the propagation of failure from this initiation point can be a very complex process and it does not necessarily lead to failure of the entire pillar.

In the case of the pillar analysed in figure 102, the lowest strength/stress ratio occurs in the roof and floor of the opening adjacent to the pillar. The strength/stress value of 0.15 will lead to the formation of vertical cracks in the roof and floor and, while this may lead to some instability in the roof, it does not have a significant influence on the stress distribution in the pillar. This tensile failure in the roof and floor will be discussed in greater detail in a later example in this chapter.

The next lowest strength/stress ratio value of 1.22 occurs in the centre of the pillar sidewall. As shown in figure 59 on page 123, the highest stress concentrations occur at the corners of a square pillar and hence the value of 1.22 given in figure 102 may be higher than that at the corners. Suppose the the strength/stress ratio on the pillar corners is approximately 1, this means that failure will initiate at these points. Spalling or slabbing along pillar corners is a fairly common sight in underground mines and it is not usually a cause for concern unless it propagates a long way into the core of the pillar.

In a highly stressed pillar, the failure which initiates at the corners and in the centre of the pillar sidewalls will give rise to some load transfer from the failed material onto the core of the pillar. In extreme cases, the magnitude of this load transfer may be large enough that the strength/stress ratio of the material forming the central core of the pillar falls below unity. Under these circumstances, collapse of the entire pillar can occur.

Bearing this progressive failure and stress transfer process in mind, the authors suggest that overall pillar instability can occur when the *average strength/stress ratio* across the centre of the pillar falls below 1.00. Hence, this average strength/stress ratio is equivalent to the *factor of safety* which has been used by other authors.

One of the unfortunate consequences of pillar failure is that it can give rise to a domino effect. If all the pillars in a panel are highly stressed and their individual factors of safety are all approximately unity, the collapse of one pillar will cause load transfer onto the surrounding pillars which may, in turn, collapse as illustrated in figure 104. Whether or not pillar collapse will be sudden and total or gradual and incomplete will depend upon the relationship between the stiffness of the pillar and that of the surrounding rock. This relationship will be discussed at the end of this chapter.

Figure 104 : Surface damage resulting from the collapse of a large number of pillars in a small metal mine.

Influence of width to height ratio on pillar strength

It has long been recognised that the *shape* of a pillar has a significant influence upon its strength and there is a large body of literature which is concerned with this subject[197-210]. Since most room and pillar mining is carried out in coal, most of this literature deals with the strength of coal pillars in horizontal seams.

The stress distributions presented in figures 54 to 58 together with the failure criterion represented by equation 43 provide a basis for examining the influence of pillar shape upon pillar strength. If it is assumed that the overall strength of a pillar is approximately equal to the average strength $\sigma_{1s.av.}$ across the centre of the pillar, as suggested on page 203, then the strength of a variety of pillar shapes can be calculated for a range of material properties. Figure 105 gives the results of a series of such calculations in which the average pillar strength is represented by $\sigma_{1s.av.}/\sigma_c$, where σ_c is the uniaxial compressive strength of the intact material.

In order to calculate the approximate factor of safety of a pillar, the procedure illustrated in the following example can be followed :

Assume: Pillar width W_p = 4 metres
Pillar height h = 3 metres
Roadway width W_o = 5 metres
Depth below surface z = 180 metres
Unit weight of rock mass γ = 0.028 MN/m^3
Uniaxial compressive strength of
intact rock material σ_c = 100 MPa.

Assume a good quality rock mass defined by material constants

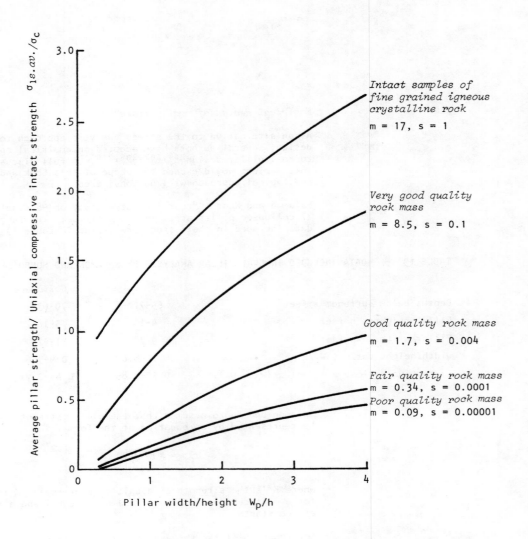

Figure 105 : Influence of pillar width to height ratio on average pillar strength.

$m = 1.7$ and $s = 0.004$. The value of $\sigma_{1s.av.}/\sigma_c$, from figure 105 for $W_p/h = 1.33$, is 0.4 and hence, $\sigma_{1s.av.} = 0.4 \times 100 = 40$ MPa.

The average pillar stress is given by the appropriate equation on page 114 and, assuming that the pillar is square in plan,

$$\sigma_p = \gamma z (1 + W_o/W_p)^2 = 0.028 \times 180 \times (1 + 5/4)^2 = 25.5 \text{ MPa}.$$

The approximate factor of safety of the pillar is given by

$$F = \sigma_{1s.av.}/\sigma_p = 40/25.5 = 1.57.$$

A factor of safety of 1.0 or less implies that the pillar is theoretically unstable and that failure could propagate across the entire pillar, resulting in its collapse. As will be shown in the next section, the authors consider that a factor of safety in excess of 1.5 should be used for pillars which are required to provide permanent support in an underground mine.

Empirical design of coal pillars

As an alternative to the stress analysis approach to pillar design, several authors have adopted an empirical approach to coal pillar design[204,207,208,210]. A full discussion on this subject would exceed the scope of this book and only one typical approach will be considered.

Salamon and Munro[200] carried out a study on 98 stable and 27 collapsed pillar areas in South Africa. The ranges of data included in their study are listed in table 13 below.

TABLE 13 - DATA INCLUDED IN COAL PILLAR ANALYSIS BY SALAMON AND MUNRO[200]

	Stable	*Collapsed*
Depths below surface, z feet	65-720	70-630
Pillar heights, h feet	4-16	5-18
Pillar widths, W_p feet	9-70	11-52
Width/height ratios, W_p/h	1.2-8.8	0.9-3.6
Extraction ratios, $e = 1 - (W_p/(W_o + W_p))^2$	0.37-0.89	0.45-0.91

Salamon and Munro assumed that the pillar strength could be represented by an equation of the form :

$$\sigma_{ps} = K h^a W_p^b \qquad (71)$$

where K is the strength of a unit cube of coal (K in lb/in² for a 1 ft cube or in kPa for a 1 metre cube) and a and b are constants.

For square pillars, the average pillar stress σ_p is given by

$$\sigma_p = \gamma z (1 + W_o/W_p)^2 = \gamma z/(1 - e) \qquad (72)$$

The factor of safety of a pillar is given by

$$F.S. = \frac{\sigma_{ps}}{\sigma_p} = \frac{K h^a W_p^b (1 - e)}{\gamma z} \qquad (73)$$

In order to determine the values of K, a and b, Salamon and Munro carried out a statistical study on the 27 collapsed pillar cases and adjusted the values of K, a and b until a mean factor of safety of 1.0 was obtained for these cases. A histogram of factors of safety obtained by Salamon and Munro is reproduced in figure 106. The values used in calculating this histogram were K = 1320 lb/in² (for dimensions in feet) or K = 7176 kPa (for dimensions in metres), a = -0.66 and b = 0.46.

Also included in figure 106 is a histogram of the factors of safety for the 98 stable pillar cases studied by Salamon and Munro. Because of the wide range of factors of safety included in this study (no generally accepted pillar design

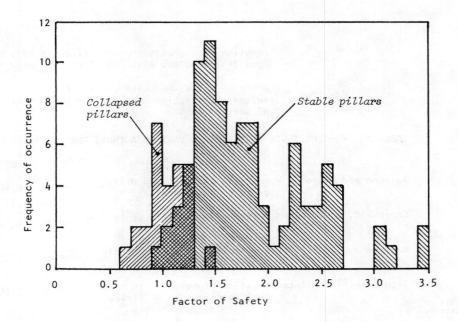

Figure 106 : Histogram of factors of safety for coal pillars in South Africa analysed by Salamon and Munro[200].

rules had been used in South Africa up to that time), Salamon and Munro decided to consider only that 50% of the stable pillar population which fell in the densest cluster between factors of safety of 1.31 and 1.88. The mean factor of safety for these 49 cases was 1.57 and Salamon and Munro suggested that a factor of safety of 1.6 is an appropriate design value for pillars similar to those studied.

In rock slope engineering[2], factors of safety range from about 1.2 for temporary mine slopes to about 1.5 for slopes in which failure could have serious economic and safety consequences. In view of the potential for a "domino effect" failure in pillars, the authors consider that the factor of safety should be in the same range as that for critical slopes. Hence, the factor of safety of 1.6 suggested by Salamon and Munro is considered to be a reasonable value for permanent pillar design.

Influence of shape and size on pillar strength

The volume V of a pillar which is square in plan can be expressed as

$$V = W_p^2 h \qquad (74)$$

Substitution of this expression into equation 71 and rearrangement of the resulting equation gives

$$\sigma_{ps} = K(W_p/h)^c V^d \qquad (75)$$

where

$$c = (b - 2a)/3$$
$$d = (a + b)/3$$

Equation 75 suggests that the strength of a pillar depends upon its *shape* and also upon its *size* (or volume).

Table 14 lists the values of the constants a, b, c and d suggested by various authors who have written on the design of coal pillars.

TABLE 14 - CONSTANTS SUGGESTED BY VARIOUS AUTHORS FOR EMPIRICAL PILLAR DESIGN

	a	b	c	d
Salamon and Munro[200] - analysis of collapsed pillar areas in South Africa	-0.66	0.46	0.59	-0.067
Greenwald, Howarth and Hartman[198] - in situ tests on small pillars in USA	-0.85	0.50	0.73	-0.117
Holland and Gaddy[199] - extrapolation of small scale laboratory tests	-1.00	0.50	0.83	-0.167
Bieniawski[211] - interpretation of tests on in situ coal specimens in S. Africa	-0.55	0.16	0.42	-0.130

At first glance it appears that there is a very wide scatter in the values of the constants suggested by various authors but the significance of this scatter can only be evaluated by comparing typical coal pillar strengths. Figures 107 and 108 show the influence of pillar shape and pillar volume on the strength of typical pillars. It will be seen that the general trends are similar and that an "average" pair of curves have been calculated for c = 0.60 and d = -0.10.

On the basis of these "average" values, pillar strengths have been calculated for a range of pillar shapes and sizes and the results are plotted in figure 109. The three dashed curves presented in figure 109 have been traced from figure 105 on page 205 and it is interesting that the trends predicted by these "theoretical" curves are very similar to those suggested by the empirical approach. In fact, figure 109 suggests that equations 71 and 75 can be used for the design of pillars in materials other than coal, provided that appropriate values for the constants K, a, b, c and d can be determined. Alternatively, the approach outlined on pages 204 and 205 can be used if values of σ_c, m and s are available or can be estimated.

It is interesting to note that the size effect included in equation 75 has already been taken care of by the inclusion of the rock mass characteristics in the determination of the values of the constants m and s which have been used in calculating the values plotted in figure 105. This emphasises once again the concept illustrated in figure 63 on page 132, namely that the scale of the problem under consideration must be taken into account when determining the rock mass characteristics which are most appropriate to the solution of that problem. This is particularly important when estimating the values of m and s from rock mass classifications, as suggested on page 171 and in figure 84.

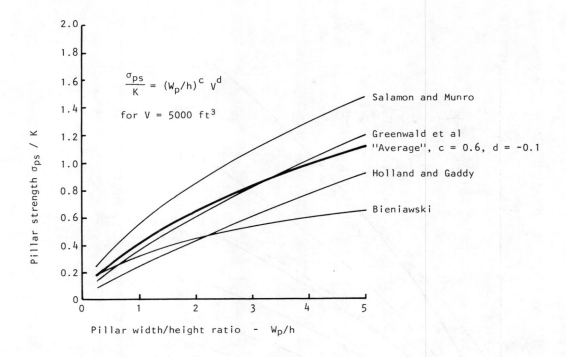

Figure 107 : Relationship between pillar shape and pillar strength for constants suggested by various authors.

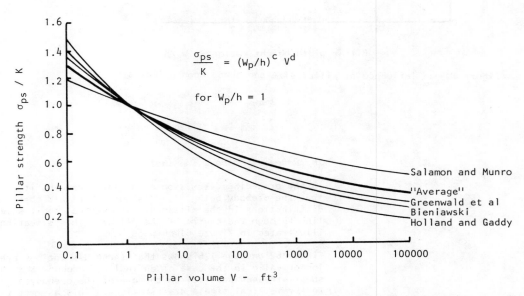

Figure 108 : Relationship between pillar volume and pillar strength for constants suggested by various authors.

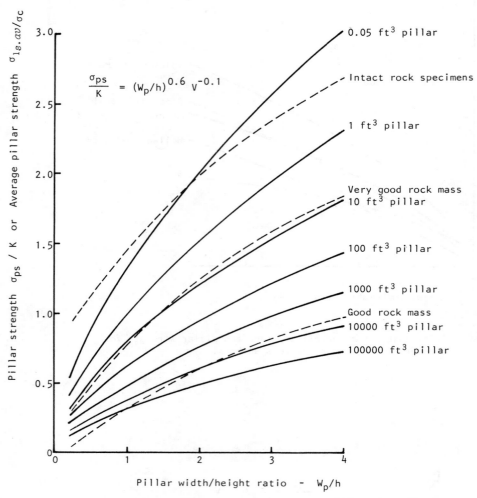

Figure 109 : Influence of pillar size and shape upon pillar strength

Influence of orebody inclination

In the preceding discussion on pillar strength, it is assumed that the orebody being mined is horizontal and that the stress distributions in the pillar are symmetrical about a vertical line through the centre of the pillar. This situation is illustrated in figure 61 on page 126.

Figure 62 on page 126 shows that these assumptions are no longer valid in the case of an inclined orebody and that the shear stresses parallel to the dip of the orebody give rise to asymmetrical stress distributions. This asymmetry is even more pronounced when the excavations are close to surface and are influenced by stress gradients due to gravitational loading. Under these circumstances, it can no longer be assumed that pillar failure follows the same sequence as it does in a horizontal pillar, namely that failure initiates at the pillar corners or edges and propagates uniformly towards the core of the pillar. The pattern of failure propagation in a pillar in an inclined orebody is unknown at the present time but it can be stated that the pillar design methods presented earlier in this chapter are not applicable in cases where the orebody inclination exceeds about 20°.

The question of how fracture propagates in stressed rock around an underground excavation has considerable practical significance in relation to overall excavation stability. While currently available rock mechanics knowledge does not provide a complete solution to this problem, useful conclusions can be drawn from studies of specific cases such as that discussed in the next section of this chapter.

Fracture propagation in rock surrounding a circular tunnel

One of the commonest underground excavation shapes is that of a circular horizontal tunnel. The propagation of failure in a homogeneous brittle rock surrounding such a tunnel has been studied by Hoek[139] by means of small scale laboratory models.

Details of one of these models, illustrated in figure 110a, are listed below :

   Model size : 5 in x 5 in x 0.125 in thick,
   Model material : Chert dyke material from East Rand
                    Proprietary Mines Ltd, South Africa,
      Uniaxial compressive strength σ_c = 84040 lb/in^2,
      Material constant m = 20.3 (see table 9, page 141),
   Ratio horizontal/vertical applied stress k = 0.15.

The model was biaxially loaded as shown in figure 110a and no load or restraint was applied normal to the surface of the model. Hence, the stress conditions are closer to those of plane stress (see discussion on pages 90 and 91) than to plane strain conditions which are believed to occur in the rock surrounding an actual tunnel.

In order to observe the fracture propagation in the rock surrounding the tunnel, a thin layer of photoelastic plastic was bonded onto the rock surface by means of a reflective cement. A series of photographs was taken at 0.2 second intervals using a high intensity electronic flash unit with a flash duration of approximately one microsecond. Three photographs selected from one of the model study sequences are reproduced in figure 110 together with strength/stress contours and fracture trajectories. Note that the strength/stress ratios are expressed in terms of the uniaxial compressive strength σ_c of the rock.

Determination of the strength/stress ratios and fracture trajectories was carried out on the basis of the principal stresses σ_1 and σ_3 determined from photoelastic and electrical analog models [139]. The two types of failure considered are discussed below.

Tensile failure occurs when $\sigma_3 < \sigma_t = \frac{1}{2}\sigma_c(m - \sqrt{m^2 + 4s})$

The strength/stress ratio, expressed in terms of the uniaxial compressive strength σ_c is given by equation 76 on the next page.

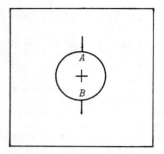

$p_z = 9000\ lb/in^2$

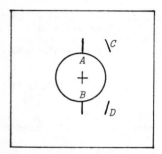

$p_z = 30000\ lb/in^2$

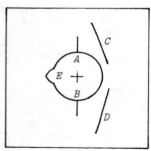

$p_z = 34000\ lb/in^2$

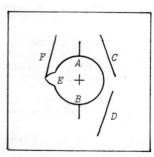

$p_z = 36000\ lb/in^2$

$$\text{Strength/stress ratio (tension)} = \frac{\sigma_c(m - \sqrt{m^2 + 4s})}{2\sigma_3} \quad (76)$$

The failure angle β (defined in figure 67 on page 138) is zero for tensile failure and hence a crack will propagate parallel to the direction of the major principal stress σ_1.

Because of the definition of strength/stress ratio adopted here, a step may occur between contours of tensile strength/stress ratio and those for shear failure (discussed below). This step can be minimised by moving the transition into the range $\sigma_t < \sigma_3 < 0$.

Shear failure occurs when $\sigma_3 > \sigma_t$ and the strength/stress ratio is defined by

$$\text{Strength/stress ratio (shear)} = \frac{\sigma_3 + \sqrt{m\sigma_c\sigma_3 + s\sigma_c^2}}{\sigma_1} \quad (77)$$

By putting the vertical applied stress $p_z = \sigma_c$, the strength/stress ratio for shear is expressed in terms of the uniaxial compressive strength σ_c.

The failure angle β is defined by

$$\beta = \tfrac{1}{2} \text{Arcsin} \frac{\sqrt{1 + m\sigma_c/4\tau_{ms}}}{1 + m\sigma_c/8\tau_{ms}} \quad (78)$$

where $\tau_{ms} = \tfrac{1}{2}\sqrt{m\sigma_c\sigma_3 + s\sigma_c^2}$

The directions of shear failure propagation are assumed to be at $+\beta$ and $-\beta$ to the direction of the major principal stress σ_1.

The following commentary is offered on the observed and predicted fracture patterns.

$p_z = 9000\ lb/in^2$ - Vertical tensile cracks formed at points A and B in the roof and floor of the tunnel (see margin sketch for locations). These cracks propagated almost instantaneously to a length of approximately one third of the tunnel diameter and remained stable for the remainder of the loading process. This stable crack formation has been observed in other models[139] and it appears that, having relieved the tensile stresses in the roof and floor, these cracks play no further part in the fracture process.

The predicted stress level at which tensile failure should have occurred at points A and B is $0.09\ \sigma_c = 7500\ lb/in^2$ which is about 17% lower that the stress level at which these cracks were observed.

$p_z = 30000\ lb/in^2$ - Cracks initiated at points C and D remote from the tunnel boundary. These cracks propagated in a direction parallel to the direction of the major principal stress σ_1 and it is therefore presumed that these cracks are tensile failures.

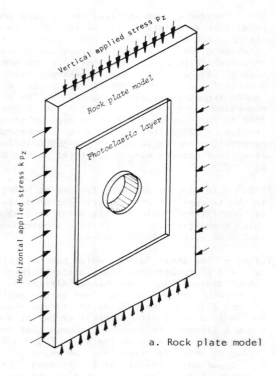

a. Rock plate model

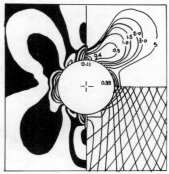

b. Formation of vertical cracks in the roof and floor of the tunnel.

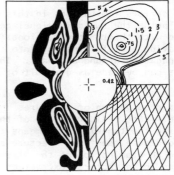

c. Fracture initiation remote from the excavation boundary.

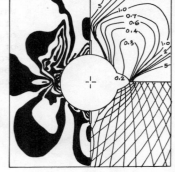

d. Shear failure in the tunnel sidewall with subsequent change in fracture direction.

Figure 110 : Model study of failure in the rock surrounding a circular tunnel.

The formation of the vertical cracks at A and B results in a redistribution of the stresses in the rock and an analysis of these redistributed stresses gives the strength/stress ratios and fracture trajectories shown in the right hand side of figure 110c. It will be seen that a zone of low strength/stress ratios occurs remote from the excavation boundary. The minimum value in this region is $0.38\sigma_c$ which suggests that failure should occur at an applied stress level $p_z = 32000$ lb/in^2. This is in reasonable agreement with the value of $p_z = 30000$ lb/in^2 at which fractures were observed at points C and D.

$p_z = 34000$ lb/in^2 - Initiation of shear failure at point E in the sidewall of the tunnel. The predicted strength/stress ratio at this point is $0.42\ \sigma_c$ which suggests that shear failure should have initiated at $p_z = 35000$ lb/in^2.

$p_z = 36000$ lb/in^2 - The shear failure which had initiated at point E propagated a small distance into the tunnel sidewall and then changed direction as shown by crack F in the lower margin sketch on page 212. An analysis of the stress distribution in a model with vertical cracks at A and B and a sidewall failure at E showed that the strength/stress ratio in the area of crack F is $0.3\ \sigma_c$ which is lower than the value of $0.42\ \sigma_c$ at which failure at E initiated. This suggests that the crack propagation process is unstable and that, once shear failure has initiated at E, the formation of the crack at F is inevitable. In fact, a violent collapse of the model occurred very shortly after the formation of the crack at F and no further analysis was possible.

While there are several serious limitations in this model study, the most serious being the two-dimensional geometry of the model as compared with the three-dimensional geometry of a tunnel, it does provide a useful insight into the process of fracture propagation around a tunnel. The following general observations can be made as a result of this study :

1. The initiation and propagation of fractures in the rock causes a significant redistribution of the stresses surrounding the tunnel. This means that it is not possible to deduce the final fracture zone configuration from an examination of the stress distribution around an unfractured excavation.

2. The formation of tensile cracks in the roof and floor, which only occurs for $k < 0.33$ as shown in figure 45 on page 106, results in stress relief but not necessarily in instability. In unjointed rock, these cracks propagate a short distance into the roof and floor rock and then become stable, playing very little further role in the fracture process.

3. This study has shown that it is theoretically possible for fractures to initiate in rock remote from the

boundary of an excavation when tensile stresses occur as a result of stress redistribution resulting from fracturing elsewhere in the rock surrounding the excavation. This prediction has been confirmed in this model study but it is not known to what extent this type of cracking occurs under different stress conditions or how significant it is if it does occur.

4. The shear failure which was predicted and observed in the tunnel sidewall is regarded, by these authors, as the most important type of failure in the rock around the opening. This is because, as shown in the model study, it can initiate a complex process of progressive failure which can lead, under certain circumstances, to collapse of the excavation. This progressive failure process is very poorly understood at the present time and it constitutes a challenging problem for rock mechanics research workers. In discussions on excavation stability later in this chapter and on excavation support in the next chapter, sidewall failure can only be dealt with in very simplified or even qualitative terms.

The initial stages of sidewall failure in a bored raise through highly stressed massive brittle rock are shown in the photograph reproduced in figure 111. Figure 112 shows a more advanced stage of sidewall failure in a mine tunnel in very highly stressed rock. A final stage of sidewall failure under severe rockburst conditions is illustrated in figure 113.

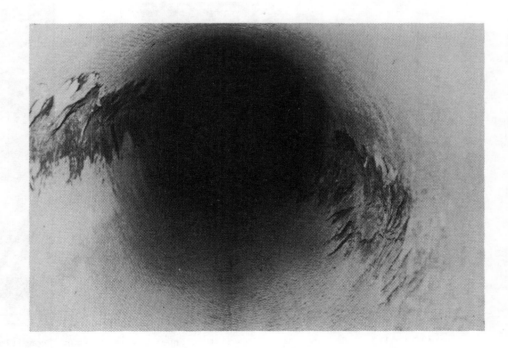

Figure 111 : Sidewall failure in a bored raise in highly stressed massive brittle rock.

Figure 112 : Sidewall failure in a mine tunnel in very highly stressed rock in a mine in Zambia.

Figure 113 : Collapse of a mine haulage under severe rockburst conditions in a deep level gold mine in South Africa.

Sidewall failure in square tunnels

Wilson[213] and Ortlepp, More O'Ferrall and Wilson[214] have reported observations on the failure of rock surrounding square tunnels in massive quartzite in deep level gold mines in South Africa. Their observations are summarised in table 15 and in figure 114. The points plotted in figure 114 represent values of the vertical in situ stress p_z at which sidewall slabbing of the tunnels was judged to have occurred.

TABLE 15 - OBSERVED SIDEWALL FAILURE IN SQUARE TUNNELS IN MASSIVE QUARTZITE IN SOUTH AFRICA.

Data point, figure 114	Material description	Uniaxial compressive strength σ_c MPa	Vertical in situ stress at failure p_z MPa	Ratio p_z/σ_c
	Unsupported tunnels			
1	Gritty quartzite	326	58	0.18
2	Gritty quartzite	252	56	0.22
3	Hard clean quartzite	324	55	0.17
4	Weak sandstone with shale partings	207	43	0.21
5	Hard clean gritty quartzite	298	55	0.18
6	Hard clean gritty quartzite	336	55	0.16
7	Well bedded weak quartzite with frequent shale layers	259	39	0.15
8	Waxy quartzite with some shale	170	52	0.31
9	Extremely variable quartzite	220	50	0.23
10	Hard gritty quartzite	322	61	0.19
11	Weak argillaceous quartzite	254	47	0.19
12	Silicious quartzite with pyrite bands	304	56	0.18
13	Argillaceous quartzite	242	54	0.22
	Timber set support			
14	Orange Free State footwall quartzite	240	62	0.26
15	Klerksdorp quartzite	183	59	0.32
	Steel girders on steel pipe columns			
16	Orange Free State footwall quartzite	240	83	0.35
17	Klerksdorp quartzite	183	73	0.40
	Steel girders on concrete walls			
18	Orange Free State footwall quartzite	240	69	0.29
19	Klerksdorp quartzite	183	79	0.43
	Steel arches			
20	Orange Free State footwall quartzite	240	100	0.42
21	Klerksdorp quartzite	183	88	0.48

$p_z/\sigma_c = 0.1$ - stable unsupported tunnel.

$p_z/\sigma_c = 0.2$ - minor sidewall spalling.

$p_z/\sigma_c = 0.3$ - severe sidewall spalling.

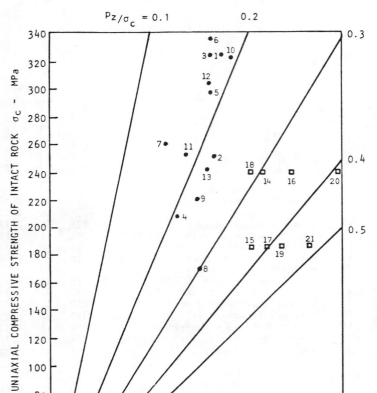

$p_z/\sigma_c = 0.4$ - heavy support required.

$p_z/\sigma_c > 0.5$ - possible rockburst conditions.

Figure 114 : Stability of square tunnels in very good quality quartzite with increasing vertical applied stress level.

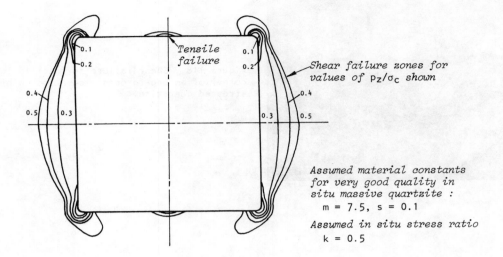

Figure 115: Zones of overstressed rock for different ratios of p_z/σ_c predicted from elastic theory.

Assuming that the ratio of horizontal to vertical in situ stress in the deep level mines in which these observations were made is k = 0.5 (see figure 41 on page 100), and that the properties of very good quality massive quartzite can be represented by m = 7.5 and s = 0.1 (see table 12 on page 176), the strength/stress contours reproduced in figure 115 were calculated by means of the boundary element program given in appendix 4.

These contours show that tensile failure can occur in the roof and floor of the square tunnel but, as demonstrated in the model study analysis discussed on page 212, these tensile cracks are not expected to have any significant effect upon the stability of the tunnels.

Shear failure initiates in the sharp corners of the tunnels at relatively low applied stress levels but, because of the rapid decrease in both major and minor principal stress values with distance from these sharp corners, the zone of overstressed rock is very limited in extent. The appearance of shear failure planes in a sharp corner in a deep level underground excavation in quartzite is illustrated in the photograph reproduced in figure 116. Because of the very high stress gradients and the confining influence of the surrounding rock, these shear failure zones do not usually give rise to major stability problems in excavations.

For values of p_z/σ_c in excess of 0.2, the zone of overstressed rock extends over the whole sidewall and, as shown by the observations recorded in figure 114, failure of the sidewall occurs. The severity of the sidewall damage appears to be

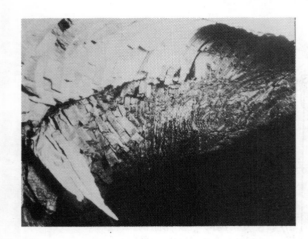

Figure 116 : Shear failure surfaces in the corner of an underground excavation in highly stressed quartzite.

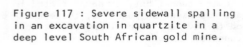

Figure 117 : Severe sidewall spalling in an excavation in quartzite in a deep level South African gold mine.

approximately proportional to the volume of overstressed rock adjacent to the sidewall of the excavation.

While it is obviously dangerous to draw too many conclusions from this limited amount of evidence, qualitative evidence from practical experience does tend to suggest that there is a reasonable correlation between the zone of overstressed rock predicted from elastic theory and the stability and support requirements of underground excavations. In fact, many underground excavation designers use these elastic theory predictions as rough guides in judging excavation stability and details such as the length of rockbolts. Examples of such applications will be discussed in later sections of this chapter.

Influence of excavation shape and in situ stress ratio

In order to extend the observations and conclusions discussed in the previous section to other excavation shapes and to different in situ stress ratios, it is necessary to consider the influence of the excavation shape upon the stresses induced in the surrounding rock. Stress distributions around several typical excavation shapes are presented in appendix 3 and these stress plots have been used to compile the graph, presented in figure 118, showing the values of the maximum boundary stresses in the roof and sidewalls of excavations for different stress ratios. Note that, in compiling this summary, the stress concentrations in sharp corners have been ignored because, as shown in the previous section, these zones of very high stress gradient do not appear to have a major influence upon overall excavation stability.

In order to demonstrate the use of figure 118, consider the influence of different in situ stress ratios upon the stresses surrounding the square tunnel discussed in the previous section. The shape constants A and B for this excavation shape are both equal to 1.9 as shown in the table in figure 118. The maximum roof stress is given by

$$\sigma_r/p_z = (1.9k - 1)$$

while the maximum sidewall stress is given by

$$\sigma_s/p_z = (1.9 - k)$$

For $k = 0.5$, $\sigma_r/p_z = -0.05$ (tension) and $\sigma_s/p_z = 1.40$. As shown in figure 115, these boundary stresses result in tensile failure in the roof and floor and shear failure in the sidewalls when the applied stress p_z is high enough.

For $k = 1.0$, $\sigma_r/p_z = \sigma_s/p_z = 0.9$. In this case the roof and sidewall stresses are equal and are both slightly lower than the applied stress level p_z. This situation would be more favourable for stability than that for $k = 0.5$.

For $k = 2.0$, $\sigma_r/p_z = 2.8$ and $\sigma_s/p_z = -0.1$. This shows that the stress field has been rotated through 90° as compared with that for $k = 0.5$ and that the stress values are twice as high. For the same depth below surface, this situation would result in much less favourable roof stability conditions than that for $k = 0.5$ but the sidewalls may only suffer minor tensile cracking.

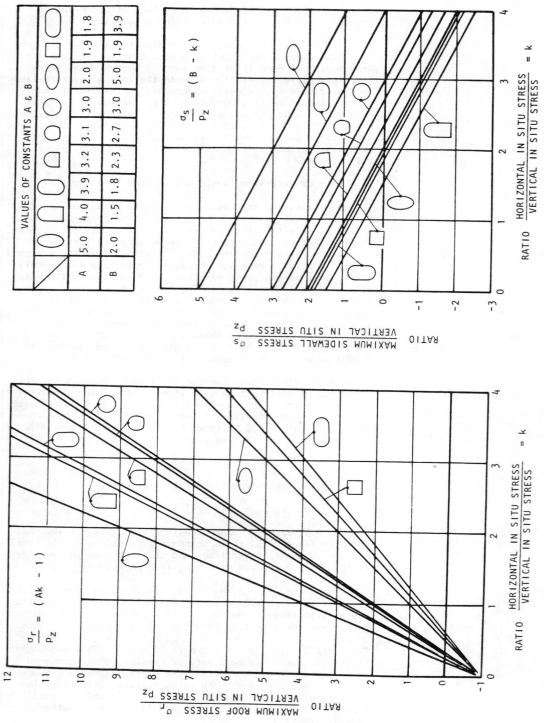

Figure 118 : Influence of excavation shape and ratio of applied stresses upon maximum excavation boundary stress.

In choosing an excavation shape for a given stress field, one of the objectives should be to try to achieve a uniform compressive stress distribution around the excavation. This question has been discussed in some detail by Richards and Bjorkman[215] who refer to excavation shapes around which the stresses are uniformly distributed as "harmonic holes".

A very rapid method for approximating the optimum excavation shape for a given stress field is to superimpose the two graphs presented in figure 118. The excavation shape which gives the same compressive roof and sidewall stresses will be the optimum shape for that stress field. As discussed on page 112, the excavation shape which gives the most uniform compressive stress distribution is usually an ovaloid or an ellipse with the same axis ratio as the ratio of the in situ principal stresses.

The concept of "harmonic holes" has two serious practical limitations when applied to underground excavation design.

1. The underground excavation designer seldom has complete freedom in the choice of the excavation which is usually being created in order to house a particular piece of equipment or to recover an orebody of a given shape. Hence, it is usually necessary to compromise between the optimum "harmonic hole" shape and the shape required in order to fulfil the practical requirements for which the excavation is being created.

2. When the boundary stress values are relatively low as compared with the strength of the rock mass, the concept of equalising the compressive stresses by choosing the optimum excavation shape is acceptable. However, when the in situ stresses are high enough to induce a significant amount of failure in the rock surrounding the excavation, the elastic stress distribution may no longer be a reasonable basis for excavation design. The extent to which extensive fracturing can alter the stress field has been demonstrated in figure 110 on page 213. Under these conditions, Fairhurst[216] suggests that the choice of a "harmonic hole" excavation shape may not lead to the best stability conditions. He suggests that consideration should be given to choosing an excavation shape in which high compressive stresses are concentrated at sharp corners in order that the zone of overstressed rock is limited in extent and confined by the surrounding rock mass. The authors consider that this recommendation has considerable merit when applied to very highly stressed rock and that the concept deserves serious consideration by rock mechanics research workers.

An example of excavation shape optimisation

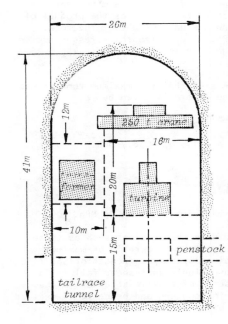

Suppose that it is required to design an underground powerhouse cavern which is to be excavated at a depth of 300 metres below surface in massive gneiss. The owner, after discussions with equipment manufacturers and on the basis of an examination of some papers on precedent cases of underground powerhouses, has proposed the general layout illustrated in the margin sketch. This layout is modelled on the Kemano powerhouse in Canada[217] and allows for the turbines and transformers to be housed in the same cavern.

The reader has been appointed geotechnical consultant on the project and his first task is to confirm and extend the preliminary geological information provided by the owner. An exploration adit is available and this is extended and several new boreholes are drilled from the adit and from the surface. The extended exploration adit is carefully mapped and all borehole core is geotechnically logged. In situ stress measurements are carried out from the exploration adit which has been extended to a depth of 250 metres below surface.

During the early part of the geotechnical site investigation programme, the following information is summarised as a basis for the preliminary design :

1. The rock mass in which the powerhouse cavern is to be located is projected to be a good quality granitic gneiss with clean rough joints spaced at 1 to 2 metres. The orientation of the cavern, dictated by the penstock and tailrace alignment, is considered to be favourable in relation to the joint orientations and it is concluded that structural failures will not play a significant part in the stability of the cavern.

2. Classification of the rock mass in the exploration adit and from the borehole core gives a CSIR rock mass rating of 65 (good quality rock) and an NGI quality index of 12 (good quality rock).

3. The average uniaxial compressive strength of the intact gneiss is found to be 150 MPa (22000 lb/in^2).

4. The average ratio of horizontal to vertical in situ stress is found to be k = 2.0.

5. The unit weight of the rock mass is γ = 0.027 MN/m^3.

On the basis of this information it is assumed that the stability of the cavern will be controlled by the stresses in the rock surrounding the excavation. Assuming that the vertical in situ stress can be calculated as the product of the depth below surface and the unit weight of the rock, p_z = 0.027 x 300 = 8.10 MPa. The horizontal stresses are assumed to be of equal magnitude and are given by kp_z = 16.2 MPa.

From figure 118 it will be seen that the maximum roof stress for a tall horseshoe shaped cavern is given by

$$\sigma_r = (4.0k - 1)p_z$$

Substituting k = 2 and p_z = 8.10, σ_r = 56.7 MPa.

The maximum sidewall stress is given by

$$\sigma_s = (1.5 - k)p_z$$

For k = 2 and p_z = 8.10, this gives σ_s = -4.05 MPa.

Practical experience based upon observations such as those presented in figure 114 suggests that a ratio of maximum boundary stress to uniaxial compressive strength of 56.7/150 = 0.38 is high enough to cause significant shear failure in the cavern roof. The presence of tensile stresses in the sidewalls is also a cause for concern and hence it is decided to investigate the stress distribution around the cavern in greater detail.

The geotechnical consultant does not have access to a finite element or boundary element program at this stage in his investigations but, being familiar with this book, he decided to attempt to use the stress distributions presented in appendix 3 for a further study of the cavern stability. In order to utilise the appropriate stress distribution directly, it is necessary to determine a failure criterion for the rock mass which is expressed in the same terms as those used in plotting the contours. Note that for k values in excess of 1.0, the stress contours presented in appendix 3 are given as ratios σ_1/kp_z and σ_3/kp_z. The corresponding failure criterion, derived from equation 43 on page 137, is

$$\frac{\sigma_1}{kp_z} = \frac{\sigma_3}{kp_z} + \sqrt{\frac{m\sigma_c}{kp_z}\cdot\frac{\sigma_3}{kp_z} + s\left(\frac{\sigma_c}{kp_z}\right)^2} \qquad (79)$$

Substituting σ_c = 150 MPa, m = 2.5 and s = 0.004 (from table 12 on page 176 for good quality gneiss) and kp_z = 16.2 MPa, gives the curve plotted in figure 119. This curve is used to define the zone of overstressed rock shown in figure 120.

Figure 120 shows that the zone of potential tensile failure extends approximately 6 metres into the sidewall of the cavern and that a substantial volume of material is included in the overstressed zone. There is also a significant zone of potential shear failure in both the roof and floor of the cavern. At this stage in his investigations the geotechnical consultant decides that a different approach is required and he approaches the owner for permission to investigate a rearrangement of the principal components which are to be housed in the cavern.

The purpose of this rearrangement is to attempt to achieve a better stress distribution while, at the same time, meeting the hydraulic and mechanical requirements of the project. After considerable discussion with the equipment manufacturers and the hydraulic engineers, a compromise solution is arrived at which involves placing the transformers and valves in separate galleries and reducing the size of the machine hall as illustrated in figure 121.

At this stage, the geotechnical consultant has installed the boundary element program given in appendix 4 on the computer of a local university and he has been able to investigate a number of possible excavation shapes and spacings. Since setting up and running the program for this problem takes about one half hour, the consultant has no hesitation in conducting a number of trial runs and in making adjustments after each run to improve the stress distributions.

The excavation layout finally selected is given in figures 122 and 123 which show the principal stress distributions, the principal stress trajectories and the strength/stress ratios. The maximum principal stress distribution shows that all boundary stresses are now compressive and that the maximum roof stress in the cavern has been reduced to 34 MPa as compared with 56.7 MPa for the original tall horseshoe cavern shape.

In order to achieve the stress distribution illustrated in figure 122, the three caverns are positioned in such a way that the two smaller caverns deflect the stress trajectories around the main cavern to give a principal stress flow path

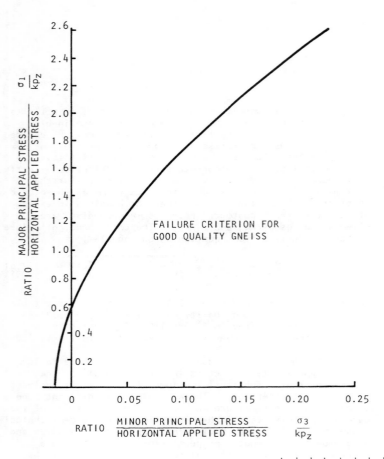

Figure 119 : Failure criterion for good quality gneiss expressed in terms of the ratios of principal stresses to applied horizontal stress kp_z.

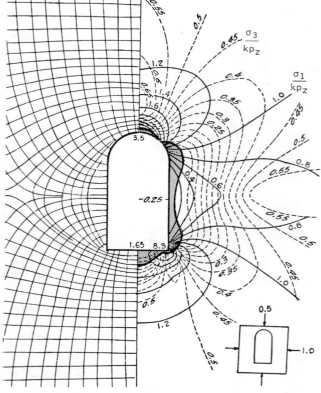

Figure 120 : Distribution of principal stresses around a tall horse-shoe shaped cavern subjected to a horizontal stress field of twice the vertical in situ stress. Potentially overstressed rock is shaded.

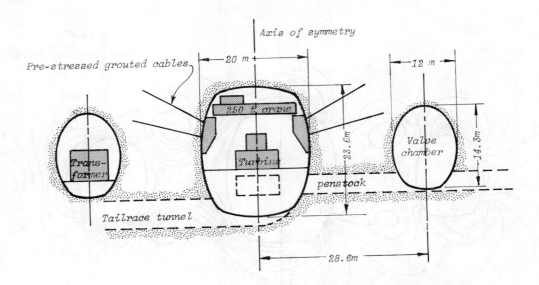

Figure 121 : Alternative arrangement of equipment with provision of separate transformer and valve caverns in order to improve the stress distribution and hence the stability of the excavations.

approximately similar to that around an ellipse with its major axis parallel to the largest in situ stress, in this case horizontal. At the same time, the separation distance between the three caverns is chosen to ensure that the pillars between the caverns are uniformly stressed in compression. By curving the sidewalls and roof of the excavations, in the style of recent German cavern designs[218], the stress distributions around the individual caverns are made as uniform as possible.

The strength/stress ratio contours given in figure 123 show that the zone of overstressed rock extends approximately 1.5 metres into the rock surrounding the excavations and that it is fairly uniformly distributed around the excavation boundaries. At this stage the geotechnical consultant concludes that the stability of the excavations can be maintained by means of a systematic pattern of relatively short rockbolts, say 4 m long, placed in the roofs and sidewalls of all the caverns.

As a final refinement, the use of cast in situ concrete crane beams, anchored to the cavern walls by pre-stressed grouted cables, is proposed in place of the column supported crane beams planned in the original excavation. This arrangement has the advantage that the crane can be installed very early in the construction programme and it can be used to gain access to the roof and to assist in the excavation of the turbine pits. The use of anchored crane beams is becoming a fairly common feature in large underground powerhouse excavations.

In comparing the original cavern design, shown in the margin drawing on page 223, to the revised layout given in figure 121, it will be evident that the excavation volume for the

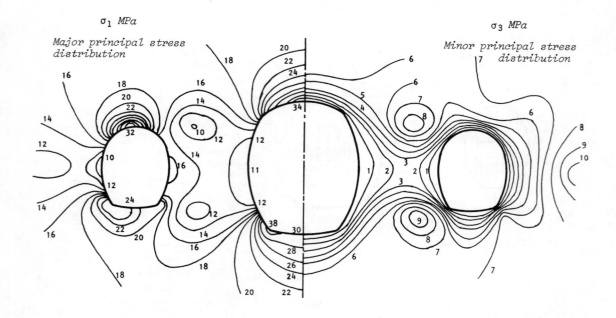

Figure 122 : Distribution of principal stresses around the revised excavation layout.

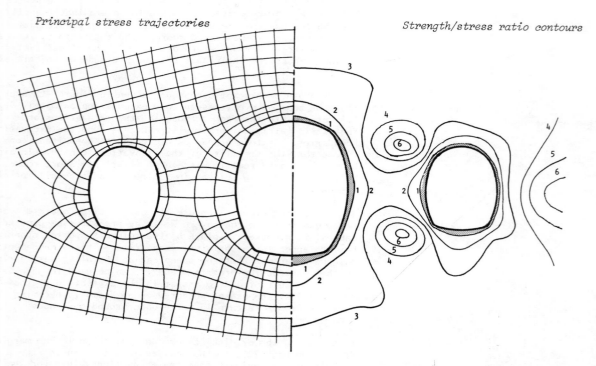

Figure 123 : Principal stress trajectories and strength/stress contours around the revised excavation layout.

revised excavation layout could be greater than that
for the original design. Unless the increased excavation
cost is at least matched by the savings in support costs,
the owner will be justifiably reluctant to authorise a change
from the original to the revised design. Under such circumstances, the geotechnical consultant may have to settle for
more modest changes and he may have to be prepared to deal
with more difficult stability problems which arise during
excavation of the cavern.

The inexperienced reader may assume that, having carried out
the rather sophisticated analysis presented on the preceding
pages and having persuaded the owner to pay for the changes,
his task as a geotechnical consultant is now complete. This
is far from the case since any real underground excavation
design is and must be an interactive process and the design
is only complete when the last load of broken rock has been
removed and the last rockbolt installed.

The analysis presented in this example assumes that the
final excavation shape is the only one which is important.
In fact, a careless choice of the excavation stages for
each cavern and of the interrelation between the excavation
sequence of the three caverns could easily give rise to
severe stress induced damage to the rock at some intermediate
excavation stage. Consequently, a task which remains to be
carried out is an analysis of the stage by stage excavation
process. This analysis can be done utilising the boundary
element program and its purpose is to check for adverse
stress conditions during excavation. Such adverse conditions
can usually be remedied by relatively minor changes in excavation sequence or by providing additional support to
protect a critical rock area until the unfavourable stress
condition has passed.

It has also been assumed that the entire stress analysis
can be treated two dimensionally. While this assumption
may be reasonable for most of the excavation length, there
may be regions at the ends of the cavern or adjacent to
the intersections between the cavern and major access tunnels
and connecting galleries which require special consideration.
Depending upon the size of the intersections and the excavation sequence to be adopted, a three-dimensional stress
analysis similar to that described on page 122 and illustrated
in figure 60 may be necessary.

There are many other problems which will face the geotechnical
consultant before the project is completed. Some will be
solved by the application of engineering common sense but
others will require detailed study and careful monitoring
of the actual excavation performance to ensure that reasonable
solutions have been achieved. In many cases it may be necessary to draw up contingency plans to be implemented if the
monitored behaviour indicates that the installed support
is inadequate or that some element of the design is not
performing as anticipated.

This raises an important question of how the contract has
been negotiated to accommodate changes during construction.
Since the geotechnical engineer never has sufficient information at his disposal and since many of his design
methods are less than perfect, he would generally prefer
to maintain as much flexibility as possible in the design
and he may request significant design changes during con-

struction. On the other hand, the contractor has a critical construction path to follow and is usually working under a tight budget control and, consequently, the last thing that he needs is a "design-as-you-go" consultant.

A full discussion of the contractual aspects of underground excavation engineering exceeds the scope of this book but it will be evident to the reader that all possible attempts must be made to anticipate major problems and to keep the contractor fully informed of likely design changes and the reasons why they may be requested. It will also be obvious that a contract which is written on the assumption that all design details should be absolutely fixed before construction commences will lead to difficulties in all but the most ideal excavation conditions.

Excavation shape changes to improve stability

The major changes in excavation layout described in the previous section may not be acceptable to the owner or the contractor on a project because of cost or of timing. The following example is intended to demonstrate that significant improvements can still be achieved in some circumstances by much more modest changes.

Consider the case of a powerhouse cavern which is to be excavated at a depth of 370 metres in massive gneiss which is characterised by the following material properties: uniaxial compressive strength σ_c = 150 MPa, material constants m = 2.5, s = 0.004 (see table 12 on page 176), unit weight of rock mass γ = 0.027 MN/m^3. The ratio of horizontal to vertical in situ stress in the rock mass is k = 0.5.

The excavation designer has based his initial cavern design upon the traditional method used in the 1950s and 1960s in which the cavern roof was supported by means of a full concrete arch. The arch reaction was taken by notched haunches cut into the cavern sidewalls resulting in the cavern shape illustrated in figure 124a. For the assumed in situ stress conditions and the given rock mass properties, the zone of overstressed rock is indicated by the shaded region in the drawing.

Most excavation designers would regard the zone of overstressed rock shown in figure 124a to be unacceptably large and two alternatives would have to be considered. Long rockbolts could be used to improve the stability of the sidewalls or, alternatively, the shape of the excavation could be changed to improve the induced stress distribution.

By using rockbolts rather than a concrete arch to support the roof and by supporting the crane beams on columns rather than on the haunches below the arch, the notch in the cavern sidewall can be eliminated as shown in figure 124b. This results in a significant improvement in the stress distribution and in a reduction of the volume of overstressed rock adjacent to the excavation sidewall.

A further improvement can be achieved by slightly curving the excavation sidewalls as illustrated in figure 124 c. This curvature reduces the zone of overstressed rock to a relatively narrow strip which can be supported by means of short rockbolts.

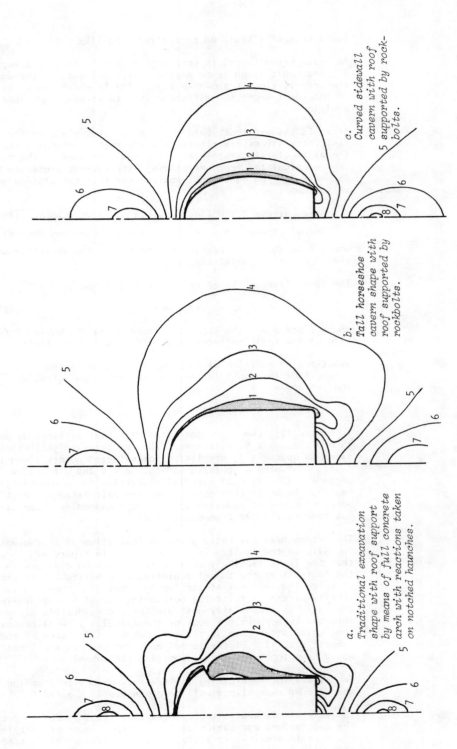

Figure 124 : Strength/stress ratio contours in the rock surrounding powerhouse caverns of different shapes.

Inlfluence of a fault on excavation stability

An unsupported tunnel in very good rock intersects a steeply dipping fault at an acute angle. A typical section through the intersection is shown in figure 125 in which the stability of the fault under two different in situ stress conditions is analysed.

If it is assumed that, before any slip takes place on the fault, the stress distribution in the rock surrounding the tunnel is elastic, the shear and normal stresses on the fault can be calculated from the stress distributions presented in Appendix 3. The equations required for this calculation are as follows :

$$\text{Shear stress} : \tau = \tfrac{1}{2}(\sigma_1 - \sigma_3)\sin 2\beta \qquad (80)$$

$$\text{Normal stress} : \sigma = \tfrac{1}{2}\{(\sigma_1 + \sigma_3) - (\sigma_1 - \sigma_3)\cos 2\beta\} \qquad (81)$$

where β is the angle between the fault and the direction of the major principal stress σ_1.

The shear strength τ_s of the fault is defined by

$$\tau_s = c + \sigma \tan \phi \qquad (82)$$

where c and ϕ are the cohesive strength and the angle of friction of the fault surfaces or of the gouge in the fault.

Substituting equation 81 into equation 82 gives the shear strength of the fault in terms of the principal stresses and the angle β :

$$\tau_s = c + \tfrac{1}{2}\{(\sigma_1 + \sigma_3) - (\sigma_1 - \sigma_3)\cos 2\beta\}\tan\phi \qquad (83)$$

In figure 125, the major and minor principal stresses σ_1 and σ_3 and the angle β, determined from the stress distributions given in appendix 3, are listed for selected points along the fault. Assuming a cohesive strength $c = 0$ and a friction angle $\phi = 20°$ (typical material properties for a gouge-filled fault), the shear strength τ_s has been calculated for each point by means of equation 83. The corresponding shear stress has been calculated from equation 80.

The strength/stress ratio gives an indication of the potential for slip on the fault. The values given in figure 125 show that the fault is likely to slip for a limited distance in the footwall of the tunnel subjected to a horizontal stress of twice the vertical stress (upper drawing). When the vertical in situ stress is twice the horizontal stress (lower drawing) the fault is potentially unstable for a considerable distance into the rock above the roof of the tunnel. It will be obvious that this latter case is the more dangerous in terms of the overall stability of the tunnel and that serious consideration would have to be given to supporting the tunnel roof.

Techniques for reinforcing faulted rock masses surrounding underground excavations will be discussed in chapter 8.

The very simple analysis presented above is intended to provide the underground excavation designer with a means of carrying out a rapid preliminary analysis of stability problems associated with faults. This analysis should be used in conjunction with the structural analysis presented on pages 183 to 194 to ensure that wedges which are free to fall or slide are not

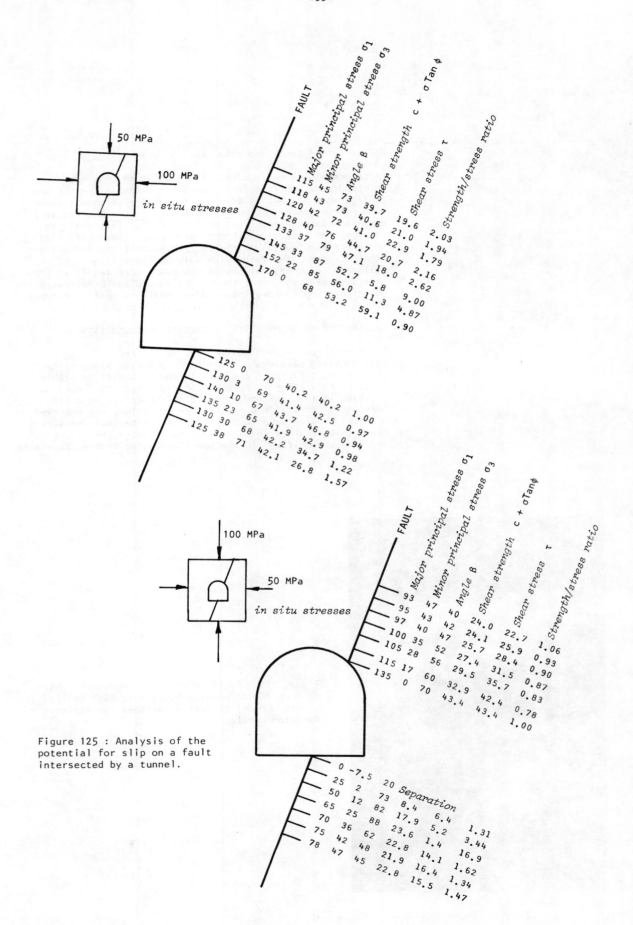

Figure 125 : Analysis of the potential for slip on a fault intersected by a tunnel.

formed by the fault and other faults or joints.

In the case of a major excavation such as an underground powerhouse cavern, the simple analysis discussed on the preceding pages may not be adequate for design purposes. Under such circumstances, it may be justified to use the finite element technique described by Goodman[117]. This analysis incorporates both strength and deformation characteristics of joints (or faults) and takes into account the redistribution of stress associated with slip on these structural features. Such an analysis is obviously more realistic than that discussed above but the reader should not under-estimate the effort and expense involved in obtaining realistic input data and in setting up the finite element mesh.

Buckling of slabs parallel to excavation boundaries

The photographs reproduced in figures 113 and 117 suggest that buckling of slabs or plates of rock play a significant role in the failure of the rock in the sidewalls of highly stressed excavations. Figure 126 illustrates the buckling of slabs in the roof and floor of an excavation in a high horizontal stress field. This type of failure was observed in model studies conducted by the Australian Coal Industries Research Laboratory in an attempt to simulate the structural and stress conditions in the coalfields near Sydney, Australia.

Figure 126 : Buckling of roof and floor slabs in a coal mine model subjected to high horizontal stress. (Model by Australian Coal Industries Research Laboratory).

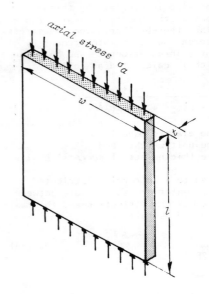

Fairhurst and Cook[155] have postulated that slabs can form in massive rock under the conditions of nearly uniaxial compressive stress which occur adjacent to the boundaries of underground excavations (see discussion on page 152). In jointed or bedded rock masses, the presence of structural features parallel to the excavation surfaces will result in the formation of plates and slabs. Whatever the reason for the presence of these slabs, it takes very little imagination to visualise that they are susceptible to buckling under axial stress.

Referring to the margin sketch, the axial stress σ_a at which a plate will buckle is given by

$$\sigma_a = \frac{\pi^2 E}{12 q^2 (l/t)^2} \qquad (84)$$

where E is the modulus of elasticity of the rock,
l/t is the *slenderness ratio* of the plate and
q is a constant which depends upon the end conditions of the plate. The constant q has the following values :

Both ends pin-jointed $q = 1$
Both ends clamped $q = \frac{1}{2}$
One end clamped, one free $q = 2$
One end clamped, one pin-jointed $q = 1/\sqrt{2}$

Equation 84 shows that the axial stress which can be carried by the plate before it buckles is inversely proportional to the square of the slenderness ratio. Consequently, thin plates will buckle more easily than thick plates. This suggests that an effective method for reinforcing an underground excavation in which slab buckling is considered to be a problem is to pin the slabs together by means of short rockbolts.

Excavations in horizontally bedded rock

As illustrated in figure 126, the roof and floor of an opening in horizontally bedded rock can suffer from buckling failure when the rock plates are relatively thin and when the in situ horizontal stress is high. When the in situ horizontal stress is low, the roof slabs in a similar opening can fail as a result of the tensile stresses induced by bending of the slabs under their own weight.

This problem has been studied in some detail by Obert and Duvall[68] who give the following equations for the maximum vertical deflection and the maximum tensile stress in a thin roof slab overlain by thicker slabs :

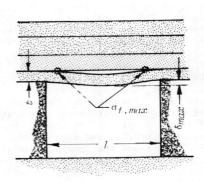

$$\delta_{max} = \frac{\gamma l^4}{32 E t^2} \qquad (85)$$

$$\sigma_{t.max} = \frac{\gamma l^2}{2t} \qquad (86)$$

where δ_{max} is the maximum vertical deflection at the centre of the roof span,
$\sigma_{t.max}$ is the maximum tensile stress in the top of the slab near the pillars,

γ is the unit weight of the rock,
E is the modulus of elasticity of the rock,
l is the span of the opening and
t is the thickness of the roof slab.

When a relatively thick roof slab is overlain by thinner slabs, the larger deflections of the upper slabs will cause them to rest on the roof slab, thereby increasing the stress in this slab. This increase can be taken into account by by replacing the unit weight γ in equations 85 and 86 by an adjusted unit weight γ_a which is calculated as follows :

$$\gamma_a = \frac{E_1 t_1^2 (\gamma_1 t_1 + \gamma_2 t_2 +..+ \gamma_n t_n)}{E_1 t_1^3 + E_2 t_2^3 +..+ E_n t_n^3}$$

where E_1, E_2 ... E_n are the elastic modulii
γ_1, γ_2 ... γ_n are the unit weights and
t_1, t_2 ... t_n are the thicknesses of successive slabs.

If the roof slab is subjected to a uniformly distributed pressure due, for example, to water or gas pressure between it and the upper layers, the maximum deflection and the maximum tensile stress are increased as follows :

$$\delta_{max} = \frac{\gamma l^4}{32 E t^2} + \frac{p l^4}{32 E t^3} \tag{88}$$

$$\sigma_{t.max} = \frac{\gamma l^2}{2t} + \frac{p l^2}{2 t^2} \tag{89}$$

A few trial calculations will soon convince the reader that the increase in tensile stress due to the presence of water or gas pressure above the roof slab can be significant. This suggests that, if this problem is suspected, drainage holes should be drilled through the roof slab in order to relieve pressures which could build up behind the slab.

Equations 85 and 86 show that the slenderness ratio of the roof slab plays an important role in determining the stability of gravity loaded roof slabs, as it does in the case of slabs which buckle under axial loads. This means that the installation of short rockbolts which pin the roof slabs together to decrease the slenderness ratio can be effective in stabilizing openings in horizontally bedded rock.

Stiffness, energy and stability

The discussions presented so far in this chapter have been in terms of peak strengths and strength/stress ratios. However, factors other than induced stress levels and rock mass strengths may influence the behaviour of an underground excavation. As was noted on page 203, the violence and the completeness of the collapse of a component such as a pillar once the peak strength of the rock has been exceeded, will depend upon the relationship between the stiffness of the pillar and that of the surrounding rock. This important aspect of the stability of underground excavations has been discussed in detail by authors such as Fairhurst[216], Brady[219], Salamon[204], Petukhov and Linkov[220], and Starfield and Fairhurst[221] and only the basic concepts will be outlined here.

The possibility of local instability or sudden collapse in

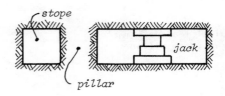

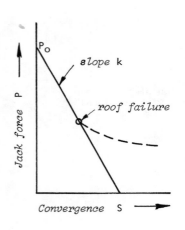

a mine structure arises when the strain energy stored locally in the structure exceeds the total energy required to crush the pillar support. The same basic concept governs the phenomena of controlled or uncontrolled failure of brittle rock specimens in laboratory compression tests referred to on page 133. Techniques for the determination of the required stiffness and the assessment of pillar stability have been developed by Starfield and Fairhurst[221], Salamon[204] and Brady[219] and the following discussion is based upon these contributions. It is assumed that the country rock is continuous and linearly elastic and that only the pillars exhibit non-linear behaviour.

Consider the situation shown in the upper margin sketch in which one of the pillars in a simple mining lay-out is replaced by a jack. Say the initial load on the jack is P_O and that it is then slowly retracted to simulate a pillar collapse. As the jack is retracted, the load P will fall and the roof and floor will converge. Provided that the roof remains intact, the curve relating jack force and convergence, S, will be as shown in the lower margin sketch. If the roof fails at some stage, the jack will be required to support a gravitational load, and in this case, the force-convergence curve may follow the dashed line. The negative slope of the P-S curve will depend upon the mechanical properties of the roof and floor, the widths of the openings, the sizes of adjacent pillars and abutments and the location of the jack. This slope is called the *local mine-stiffness*.

Now replace the jack by a pillar. The deformation of the pillar, which corresponds to the contraction of the jack, will depend on the force exerted on it by the roof and floor. A typical force-deformation curve for a pillar is shown in figure 127a on page 238. As the pillar is loaded, it compresses along the line OA until its peak load bearing capability, P_{max}, is reached. At this point the pillar may be internally fractured and some slabbing of the sides may be in evidence. However, the pillar still has the capacity to support loads less than P_{max} if it is deformed along the post-peak curve AB. It is this post-peak behaviour that is influenced by the local mine-stiffness.

Two different local mine-stiffnesses AE (low) and AG (high) are shown superposed on the pillar load-deformation curve in figures 127b and 127c. In the case of the example in figure 127b, an increase in convergence of Δs beyond P_{max} would result in a force P_H being exerted on the pillar by the roof. Since the pillar can only sustain a force P_J at this new deformation, the situation is unstable and the pillar will collapse as soon as P_{max} is reached. In this case, the energy released by the mine (the area AHDC) is greater than the energy required to deform the pillar (the area AJDC), and excess energy represented by the shaded area in the sketch is available to crush the pillar.

If, on the other hand, the pillar is in a region of high local mine-stiffness, the roof is unable to supply the force necessary to deform the pillar beyond A and the situation is stable. That is to say, the elastic strain energy released locally by the mine during the convergence increment Δs, is less than that required to deform the pillar along the curve AK. Controlled failure of the pillar, but not violent collapse, may occur. In practice, the local stiffness will vary over the life of the pillar so that the pillar may exhibit both stable and unstable behaviour during its history. The pillar stiffness will not be

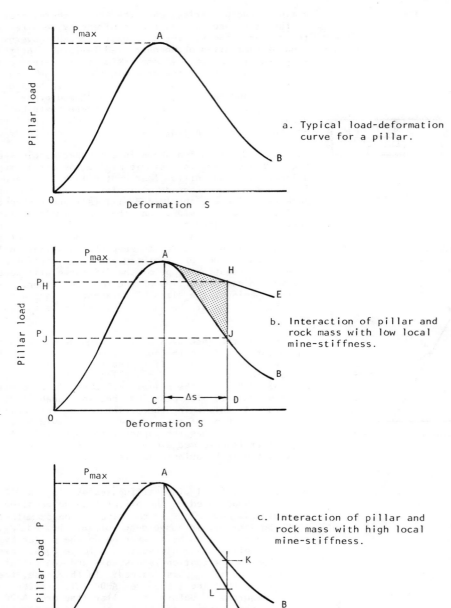

Figure 127 : Interaction between post-peak pillar deformation and local mine-stiffness.

constant but will vary with convergence.

Salamon[204] has shown that the condition for pillar stability may be expressed mathematically as

$$k + \lambda > 0$$

where k is the local mine-stiffness (a positive number) and $\lambda = f(S)$ is the slope of the force-deformation curve for the pillar (negative in the post-peak region). The system becomes unstable at the point at which $k + \lambda = 0$.

As the width to height ratio of the pillar increases, the post-peak portions of the force-deformation curves become flatter and the likelihood of sudden collapse of any given pillar decreases. The determination of this post-peak behaviour is no simple matter. Some indication of the types of behaviour to be expected can be obtained from the results of the large-scale tests on coal pillars conducted in South Africa by Bieniawski[211] and Van Heerden[222], and from laboratory model tests such as those described by Starfield and Wawersik[223].

As noted above, the local mine-stiffness, k, varies with the local mine geometry and with the mechanical properties of the rock mass. Analyses which permit the calculation of local stresses and displacements at points of interest in the mine structure are required in order to determine appropriate values of k. Salamon[204], Starfield and Fairhurst[221] and Brady[219] describe procedures for doing this for specific mine configurations.

Figure 128 shows the results of an analysis for pre-failure conditions carried out by Brady[219] using the boundary element method. In this problem the extraction has been examined of an 8m thick horizontal orebody using long rooms and rib pillars. The orebody and the country rock were assumed to behave elastically and to both have a Young's modulus of 50 GPa and a Poisson's ratio of 0.25. The pre-mining principal stresses were taken as p_x = 9 MPa (normal to the long axis of the excavation), p_y = 6 MPa (parallel to the long axis of the excavation) and p_z = 12 MPa (vertical). Two stopes of equal span, S_s, were excavated to generate a 12m wide pillar. In order to determine the local mine-stiffness characteristics, the pillar was replaced by a series of uniformly distributed loads of various magnitudes and the displacement distributions calculated. For figure 128, the pillar performance characteristic was obtained by plotting pillar load against convergence across the centre line of the pillar, while the country rock performance characteristics for various stope spans were obtained by plotting the applied strip load magnitudes against average displacements under the loaded strip. In any given case, the intersection of the two characteristics gives the load-convergence equilibrium condition. Clearly, any calculation involving post-peak pillar behaviour will be more complex than that required for this example.

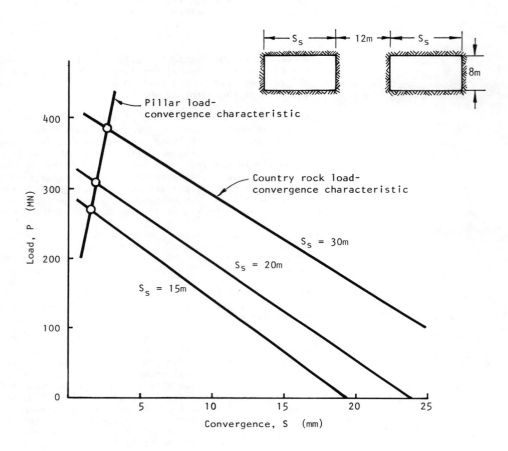

Figure 128 : Load-convergence characteristics for hypothetical example computed using boundary element method. After Brady[219].

Chapter 7 references

190. PEQUIGNOT, G.A. *Tunnels and Tunnelling*. Hutchinson & Co., London, 1963, 555 pages.

191. HEWETT, B.H.M. and JOHANNESSON, S. *Shields and Compressed Air Tunnelling*. McGraw-Hill, New York, 1922.

192. SZECHY, K. *The Art of Tunnelling*. Akademiai Kiado, Budapest, 1970, 1097 pages.

193. HOEK, E. Structurally controlled instability in underground excavations. *Proc. 18th Rock Mechanics Symposium*, Keystone, Colorado, 1977.

194. CARTNEY, S.A. The ubiquitous joint method, cavern design at Dinorwic power station. *Tunnels and Tunnelling*, Vol. 9, No. 3, 1977, pages 54-57.

195. CRONEY, P., LEGGE, T.F. and DHALLA, A. Location of block release mechanisms in tunnels from geological data and the design of associated support. *Computer Methods in Tunnel Design*, The Institution of Civil Engineers, London, 1978, pages 97-119.

196. DOUGLAS, T. and MARKLAND, J.T. Dinorwic Pumped Storage power station. *Tunnels and Tunnelling*, Vol. 9, No. 6, 1977, pages 59-66.

197. BUNTING, G. Chamber pillars in deep anthracite mines. *Trans. AIME*, Vol. XLII, 1911, pages 236-245.

198. GREENWALD, H.P., HOWARTH, H.C. and HARTMAN, I. Experiments on the strength of small pillars of coal in the Attsburg bed. *U.S. Bureau of Mines Tech. Rep.* No. 605, 1939, 22 pages.

199. HOLLAND, C.T. and GADDY, F.L. Some aspects of permanent support of overburden on coal beds. *Proc. W. Virginia Coal Mining Institute*, 1957, pages 43-66.

200. SALAMON, M.D.G. and MUNRO, A.H. A study of the strength of coal pillars. *J. South African Institute of Mining and Metallurgy*, Vol. 68, No. 2, 1967, pages 55-67.

201. BIENIAWSKI, Z.T. In situ strength and deformation characteristics of coal. *Engineering Geology*, Vol. 2, 1968, pages 325-340.

202. BIENIAWSKI, Z.T. and VAN HEERDEN, W.L. The significance of in situ tests on large rock specimens. *Intnl. J. Rock Mechanics and Mining Sciences*, Vol. 12, 1975, pages 101-113.

203. DENKHAUS, H.G. A critical review of the present state of scientific knowledge related to the strength of mine pillars. *J. South African Institute of Mining and Metallurgy*, 1962, pages 59-75.

204. SALAMON, M.D.G. Stability, instability and design of pillar workings. *Intnl. J. Rock Mechanics and Mining Sciences*, Vol. 7, 1970, pages 613-631.

205. BORECKI, M. and KIDYBINSKI, A. Coal strength and bearing capacity of coal pillars. *Proc. 2nd Congress Intnl. Soc. Rock Mechanics*, Belgrade, Vol.2, pages 145-152.

206. GROBBELAAR, C. The theoretical strength of mine pillars, part II. *The University of Witwatersrand*, Department of Mining Engineering, Report 113, 1968.

207. WILSON, A.H. Research into the determination of pillar size - Part 1. *The Mining Engineer*, Vol.131, 1971-72, pages 409-417.

208. HOLLAND, C.T. Pillar design for permanent and semi-permanent support of the overburden in coal mines. *Proc. 9th Canadian Rock Mechanics*, Montreal, 1973, pages 113-139.

209. PARISEAU, W.G. Limit design of mine pillars under uncertainty. *Proc. 16th Symposium on Rock Mechanics*, Minneapolis, 1975, S.L. Crouch & C. Fairhurst, eds., pages 183-187.

210. HARDY, M.P. and AGAPITO, J.F.T. Pillar design in underground oil shale mines. *Proc. 16th Symposium on Rock Mechanics*, Minneapolis, 1975, S.L. Crouch & C. Fairhurst, eds., pages 257-266.

211. BIENIAWSKI, Z.T. The effect of specimen size on the compressive strength of coal. *Intnl. J. Rock Mechanics and Mining Sciences*, Vol. 5, 1968, pages 325-335.

212. BIENIAWSKI, Z.T. In situ large scale testing of coal. *Proc. Conf. In Situ Investigations in Soils and Rocks*, British Geotechnical Society, London, 1969, pages 67-74.

213. WILSON, J.W. The design and support of underground excavations in deep level, hard rock mines. *Ph.D Thesis*, University of the Witwatersrand, 1971.

214. ORTLEPP, W.D., MORE O'FERRALL, R.C. and WILSON, J.W. Support methods in tunnels. *South African Association of Mine Managers*, Circular No. 2/73, 1973, pages 1-19.

215. RICHARDS, R. and BJORKMAN, G.S. Optimum shapes for unlined tunnels and cavities. *Engineering Geology*, Vol. 12, No. 2, 1978, pages 171-179.

216. FAIRHURST, C. The application of mechanics to rock engineering. *Proc. Symposium on Exploration for Rock Engineering*, Johannesburg, 1976, Vol. 2, pages 1-22.

217. WISE, L.L. World's largest underground power plant. *Engineering News Record*, Vol. 149, 1952, page 31.

218. WITTKE, W. A new design concept for underground openings in jointed rock. *Proc. Intnl. Symposium on Numerical Methods in Soil and Rock Mechanics*, Karlsruhe, 1975

219. BRADY, B.H.G. Boundary element methods for mine design. *Ph.D. Thesis*, University of London, 1979.

220. PETUKHOV, I.M. and LINKOV, A.M. The theory of post-failure deformations and the problem of stability in rock mechanics. *Int. J. Rock Mech. Min. Sci.*, Vol. 16, No. 2, 1979, pages 57-76.

221. STARFIELD, A.M. and FAIRHURST, C. How high-speed computers advance design of practical mine pillar systems. *Engineering and Mining Journal*, Vol. 169, No. 5, 1968, pages 78-84.

222. VAN HEERDEN, W.L. In-situ complete stress-strain characteristics of large coal pillars. *J. South African Inst. Min. Metall.*, Vol. 78, No. 8, 1975, pages 207-217.

223. STARFIELD, A.M. and WAWERSIK, W.R. Pillars as structural components in room and pillar mine design. *Proc. 10th Symp. Rock Mech.*, AIME, New York, 1972, pages 793-809.

Chapter 8: Underground excavation support design

Introduction

The principal objective in the design of underground excavation support is to help the rock mass to support itself.

Consider the example illustrated in figure 129 which shows a tunnel being driven by full face drill and blast methods with steel set supports being installed after each mucking cycle. The horizontal and vertical in situ stresses are assumed to be equal and to have a magnitude p_o.

In step 1, the tunnel face has not yet reached section X-X which defines the tunnel section under consideration. The rock mass inside the proposed tunnel profile, shown dotted in the step 1 cross-section drawing, is in equilibrium with the rock mass surrounding the tunnel. The internal support pressure p_i acting across the proposed excavation profile is equal to the in situ stress p_o (point A, figure 129).

In step 2, the tunnel face has been advanced beyond section X-X and the support pressure p_i, previously provided by the rock inside the tunnel, has dropped to zero. However, the tunnel will not collapse because the radial deformation u is limited by the proximity of the tunnel face which provides a significant amount of restraint. If this restraint provided by the face were not available, an internal support pressure p_i, given by points B and C in the graph in figure 129, would be required to limit the radial deformation u to the same value. Note that the support pressure p_i which would be required to limit the deformation of the roof is higher than that required to limit the sidewall deformation because the weight of the zone of loosened rock above the tunnel roof must be added to the support pressure required to limit the stress-induced displacement in the roof.

In step 3, the tunnel has been mucked out and steel sets have been installed close to the face. At this stage, the supports carry no load, as shown by point D on the graph in figure 129, because no further deformation of the tunnel has taken place. Assuming that the rock mass does not exhibit time-dependent deformation characteristics, the radial deformations of the tunnel are still those defined by points B and C.

In step 4, the tunnel face has been advanced about 1½ tunnel diameters beyond section X-X and the restraint provided by the proximity of the face is now considerably reduced. This causes further radial deformation of the tunnel sidewalls and roof as indicated by the curves CEG and BFH in figure 129. This inward radial deformation or convergence of the tunnel induces load in the support system which acts like a stiff spring. The support pressure p_i available from the blocked steel sets increases with radial deformation of the tunnel as indicated by the line DEF in figure 129.

In step 5, the tunnel face has advanced so far beyond section X-X that it no longer provides any restraint for the rock mass at section X-X. If no support had been installed, the radial deformations in the tunnel would increase as indicated by the dashed curves marked EG and FH in figure 129. In the case of the sidewalls, the pressure required to limit further deformation drops to zero at point G and, in this case, the sidewalls would be stable since there is no remaining driving force to induce further deformation. On the other hand, the support

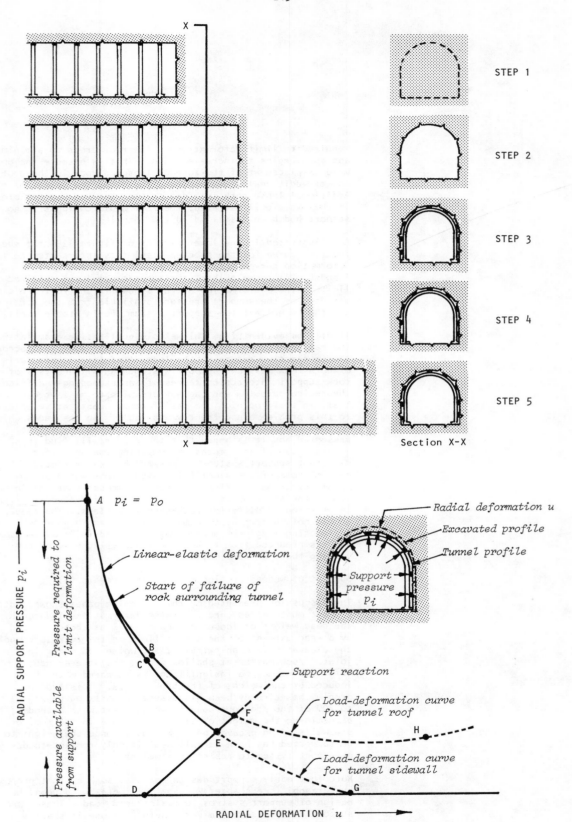

Figure 129 : Hypothetical example of a tunnel being advanced by full face drill and blast methods with blocked steel sets being installed after each mucking cycle. The load-deformation curves for the rock mass and the support system are given in the lower part of the figure. (After Daeman[224]).

required to limit deformation of the roof drops to a minimum and then begins to increase again. This is because the downward displacement of the zone of loosened rock in the roof causes additional rock to become loose and the weight of this additional loose rock is added to the required support pressure. In the example illustrated, the roof would collapse if no support had been installed in the tunnel.

As illustrated in the lower portion of figure 129, the support reaction curve for the blocked steel sets intersects the load-deformation curves for the tunnel sidewalls and roof at points E and F. At these points, the support pressures required to limit further deformation of the sidewalls and roof are exactly balanced by the support pressure available from the steel sets and the tunnel and the support system are in stable equilibrium.

It will be evident from this simple qualitative example that the rational design of support systems must take into account the interactive nature of the load-deformation characteristics of both rock mass and support system. A full analysis of rock-support interaction represents an exceedingly difficult theoretical problem and the authors wish to make it clear that, in their opinion, no satisfactory quantitative solutions to this problem are currently available. In the pages which follow, an approximate solution to this problem will be presented and it is hoped that the reader will find this solution useful as a means of comparing the behaviour of different support systems. However, the reader should take careful note of the simplifying assumptions which are made in deriving the solution and should remember that the intended use of the solution is limited to sensitivity studies of the interaction of different rock-support systems. The reader should not, under any circumstances, abandon his or her engineering judgement and common sense and rely solely upon the answers produced from this type of analysis, however convincing those answers may appear.

Support of wedges or blocks which are free to fall

The discussion presented on the preceding pages shows that support must be designed to resist deformations induced by the dead weight of loosened rock as well as those induced by a readjustment of the stress field in the rock surrounding the excavation. In an extreme case, which can occur in jointed rock masses at shallow depth, the stress-induced deformations may be insignificant as compared with those induced by the weight of the broken rock. This type of problem has already been discussed, in part, on pages 183 to 200 of the previous chapter which dealt with methods for determining the weight of blocks or wedges in the roof or sidewalls of an excavation. Obviously, once the weight to be supported has been established, it only remains to design a support system to resist this weight.

Note that early support design methods, such as that proposed by Terzaghi (see chapter 2, page 14), were based upon the design of support systems to resist dead load. These *passive* support designs served a very useful purpose in shallow tunnels where the dead weight of the loosened rock plays a dominant role but their use is limited in designing deeper excavations in which stress-induced deformations are important.

In general, the authors recommend the use of rockbolts or cables to support potentially unstable wedges or blocks which are free to fall or slide under their own weight. This is because these blocks or wedges move independently of the remainder of the rock mass and hence apply concentrated or eccentric loading to the support system. Rockbolts and cables are better able to resist these eccentric loads than steel sets or concrete linings and the latter should therefore be avoided if possible.

In designing rockbolt or cable support for blocks or wedges, the authors recommend that a liberal allowance be made for variations in the strength of the installed bolt or cable and for time-dependent strength reduction due to corrosion or creep at the anchor points. Typically, one hundred one inch diameter mechanically anchored rockbolts with an average failure load of 30 tons (determined from pull-out tests) would be used to support a 1500 ton block. This gives a factor of safety of 2 for the system. When the bolts or cables are grouted, as they should be for all permanent installations, this factor of safety can be reduced to about 1.5.

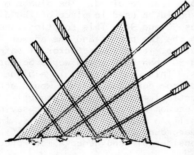

Use of bolts or cables to suspend a wedge which is free to fall from the roof of an underground excavation.

The length of the rockbolts or cables should be chosen to ensure that they are anchored well beyond the boundaries of the block. The choice of whether to use one hundred 30 ton rockbolts or twenty 150 ton cables depends upon the nature of the jointing in the block and in the rock mass surrounding the block. In the case of massive hard rock with widely spaced joints, the use of cables may prove to be the best and most economical solution. On the other hand, if the rock mass is closely jointed, it may be necessary to use bolts to ensure that small wedges do not fall out between the bolts. As a general rule, the maximum spacing between the bolts or cables should not exceed three times the average joint spacing in the rock mass when support is required for dead weight loading.

Support of wedges or blocks which are free to slide

Another special case in which the stress-induced deformations in the rock mass are not significant is that of wedges or blocks which are free to slide. This case is similar to that discussed above except that the frictional resistance of the sliding surfaces should be taken into account in designing the support system. Once again, the use of rockbolts or cables is preferred to steel sets or concrete lining.

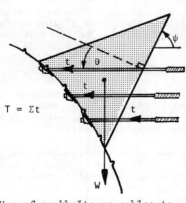

Use of rockbolts or cables to reinforce a wedge against sliding on a single plane.

The use of stereographic techniques to determine the weight and the sliding direction of wedges in the roof or the sidewalls of an excavation has been discussed in chapter 7, pages 183 to 192. The factor of safety of a block or a wedge which is reinforced against sliding on a single plane is given by

$$F = \frac{cA + (W\cos\psi + T\cos\theta)\tan\phi}{W\sin\psi - T\sin\theta} \qquad (90)$$

where W is the weight of the wedge or block,
   T is the load in the bolts or cables,
   A is the base area of the sliding surface,
   ψ is the dip of the sliding surface,
   θ is the angle between the plunge of the bolt or cable and the normal to the sliding surface,
   c is the cohesive strength of the sliding surface
and ϕ is the friction angle of the sliding surface.

Hence, the total bolt or cable load required is

$$T = \frac{W(F.\sin\psi - \cos\psi \tan\phi) - cA}{\cos\theta \tan\phi + F.\sin\theta} \qquad (91)$$

A factor of safety of 1.5 to 2 should be used, depending upon the damage which would result from sliding of the block or wedge and upon whether or not the bolts or cables are to be grouted.

When the geometry of the wedge or block is such that sliding would occur along the line of intersection of two planes, the analysis presented above can be used to give a first approximation of the support load required. The plunge of the line of intersection should be used in place of the dip ψ of the plane in equations 90 and 91. This solution ignores the wedging action between the two planes and the answer obtained would be conservative, ie. a lower factor of safety would be given by equation 90 than that which would be obtained from a full wedge analysis. This would result in a higher bolt load being calculated from equation 91 than would actually be required. In many practical applications, the quality of the input data and the economic importance of the saving in rockbolts would not justify a more refined analysis. In the case of very large underground caverns, the sizes of wedges or blocks can be considerable and hence a more precise analysis may be justified.

Hoek and Bray[2] have dealt with the problem of the reinforcement of sliding wedges in rock slopes and the problem is identical to that which occurs in underground excavations. Computer programs, such as that written by Croney, Legge and Dhalla[195], are available to identify wedges or blocks which slide or fall and to calculate the support loads required to achieve given factors of safety. A simplified analysis, suitable for use on a programmable calculator, is presented in Appendix 6 at the end of this book.

Before leaving this topic, it is worth noting that the plunge and trend of a bolt or cable which will give the highest factor of safety when used to reinforce a wedge which can slide along the line of intersection of two planes are defined in the margin sketch. The plunge of the bolt is at an angle equal to the friction angle of the sliding surfaces, measured from the line of intersection. The trend of the bolt is parallel to the line of intersection.

Rock-support interaction analysis

Having dealt, albeit rather briefly, with the most important types of dead weight loading in underground excavation support design, we can now turn to the problem of designing support systems to resist stress-induced deformations.

As pointed out in the introduction to this chapter, the analysis of rock-support interaction is a formidable theoretical problem because of the large number of factors which have to be taken into account in order to derive meaningful solutions. A number of simplifying assumptions have to be made in order to reduce this problem to manageable proportions and these assumptions will be discussed in detail in the following text. The interested reader who wishes to pursue the subject of rock-support interaction further is referred to the papers by Daemen[224,225], Rabcewicz[227], Ladanyi[228],

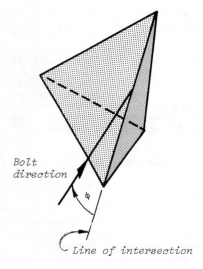

Optimum bolting direction to reinforce a wedge which slides along the line of intersection of two planes.

Lombardi[229,230], Egger[231] and Panet[232]. The solution presented below is based upon that derived by Ladanyi[228] but it utilises the rock strength criterion discussed in chapter 6.

Basic assumptions

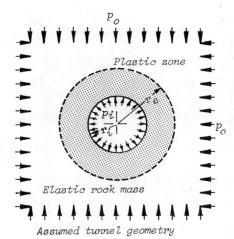

Assumed tunnel geometry

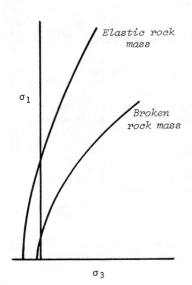

Assumed failure criteria for original elastic and broken rock masses

- *Tunnel geometry:* The analysis assumes a circular tunnel of initial radius r_i. The length of the tunnel is such that the problem can be treated two-dimensionally.

- *In situ stress field:* The horizontal and vertical in situ stresses are assumed to be equal and to have a magnitude p_o.

- *Support pressure:* The installed support is assumed to exert a uniform radial support pressure p_i on the walls of the tunnel.

- *Material properties of original rock mass:* The original rock mass is assumed to be linear-elastic and to be characterised by a Young's modulus E and a Poisson's ratio ν. The failure characteristics of this material are defined by the equation

$$\sigma_1 = \sigma_3 + (m\sigma_c.\sigma_3 + s\sigma_c^2)^{\frac{1}{2}} \qquad (92)$$

- *Material properties of broken rock mass:* The broken rock mass surrounding the tunnel is assumed to be perfectly plastic and to satisfy the following failure criterion

$$\sigma_1 = \sigma_3 + (m_r\sigma_c.\sigma_3 + s_r\sigma_c^2)^{\frac{1}{2}} \qquad (93)$$

Note that, in the interests of simplicity, it is assumed that the strength reduces suddenly from that defined by equation 92 to that defined by equation 93. Daemen[224,225], Egger[231] and Panet[232] have allowed for progressive strain-softening post-failure behaviour in their analyses.

- *Volumetric strains:* In the elastic region these are governed by the elastic constants E and ν. At failure, the rock will dilate (increase in volume) and the strains are calculated using the associated flow rule of the theory of plasticity.

- *Time-dependent behaviour:* In is assumed that both the original and broken rock masses do not exhibit time-dependent behaviour. Ladanyi[228] allowed for both short- and long-term material behaviour in his solution.

- *Extent of plastic zone:* It is assumed that the plastic zone extends to a radius r_e which depends upon the in situ stress p_o, the support pressure p_i and the material characteristics of both the elastic and the broken rock mass.

- *Radial symmetry:* Note that, in all details, the problem being analysed is symmetrical about the tunnel axis. If the weight of the rock in the broken zone were included in this analysis, this simplifying symmetry would be lost. Since the weight of this broken rock is extremely important in the support design, an allowance for this weight is added after the basic analysis has been completed.

Notation for stresses around tunnel

Analysis of stresses

For the case of cylindrical symmetry, the differential equation of equilibrium is

$$\frac{d\sigma_r}{dr} + \frac{(\sigma_r - \sigma_\theta)}{r} = 0 \qquad (94)$$

Satisfying this equation for linear-elastic behaviour and the boundary conditions $\sigma_r = \sigma_{re}$ at $r = r_e$ and $\sigma_r = p_o$ at $r = \infty$ gives the following equations for the stresses in the elastic region:

$$\sigma_r = p_o - (p_o - \sigma_{re})(r_e/r)^2 \qquad (95)$$

$$\sigma_\theta = p_o + (p_o - \sigma_{re})(r_e/r)^2 \qquad (96)$$

Within the broken zone, the failure criterion defined by equation 93 must be satisfied. Recognising that, in this problem $\sigma_\theta = \sigma_1$ and $\sigma_r = \sigma_3$, equation 93 may be re-written as

$$\sigma_\theta = \sigma_r + (m_r \sigma_c \cdot \sigma_r + s_r \sigma_c^2)^{\frac{1}{2}} \qquad (97)$$

Integration of equation 94 and substitution of the boundary condition $\sigma_r = p_i$ at $r = r_i$ gives the following equation for the radial stress in the broken rock:

$$\sigma_r = \frac{m_r \sigma_c}{4}\left(\ln(r/r_i)\right)^2 + \ln(r/r_i)(m_r \sigma_c p_i + s_r \sigma_c^2)^{\frac{1}{2}} + p_i \qquad (98)$$

In order to find the value of σ_{re} and the radius r_e of the broken zone, use is made of the fact that the failure criterion of the original rock mass must be satisfied at the internal boundary of the elastic region, ie at $r = r_e$ where, from equations 95 and 96, the principal stress difference is

$$\sigma_{\theta e} - \sigma_{re} = 2(p_o - \sigma_{re}) \qquad (99)$$

The failure criterion for the original rock mass is given by equation 92 which may be re-expressed as

$$\sigma_1 - \sigma_3 = \sigma_c(m\sigma_3/\sigma_c + s)^{\frac{1}{2}} \qquad (100)$$

Substitution of $\sigma_1 = \sigma_{\theta e}$ and $\sigma_3 = \sigma_{re}$ in equation 100 and then equating the right hand sides of equations 99 and 100 leads to the result

$$\sigma_{re} = p_o - M\sigma_c \qquad (101)$$

where

$$M = \frac{1}{2}\left(\left(\frac{m}{4}\right)^2 + mp_o/\sigma_c + s\right)^{\frac{1}{2}} - \frac{m}{8} \qquad (102)$$

The failure criterion for the broken rock must also be satisfied at $r = r_e$ and hence, from equation 98

$$\sigma_{re} = \frac{m_r \sigma_c}{4}\left(\ln(r_e/r_i)\right)^2 + \ln(r_e/r_i)(m_r \sigma_c p_i + s_r \sigma_c^2)^{\frac{1}{2}} + p_i \qquad (103)$$

Equating the values of σ_{re} given by equations 101 and 103 results in the following equation for the plastic zone radius

$$r_e = r_i \cdot e^{\left(N - \frac{2}{m_r \sigma_c}(m_r \sigma_c p_i + s_r \sigma_c^2)^{\frac{1}{2}}\right)} \tag{104}$$

where

$$N = \frac{2}{m_r \sigma_c}(m_r \sigma_c p_o + s_r \sigma_c^2 - m_r \sigma_c^2 M)^{\frac{1}{2}} \tag{105}$$

It will be seen, from equation 101, that the zone of broken rock will exist only if the internal support pressure p_i is lower than a critical value given by

$$p_i < p_{icr} = p_o - M\sigma_c \tag{106}$$

Analysis of deformations

The radial displacement of the elastic boundary u_e produced by the reduction of σ_r from its initial value of p_o to σ_{re} is found from the theory of elasticity to be

$$u_e = \frac{(1 + \nu)}{E}(p_o - \sigma_{re}) \cdot r_e \tag{107}$$

or, using equation 101

$$u_e = \frac{(1 + \nu)}{E} M\sigma_c \cdot r_e \tag{108}$$

Let e_{av} be the average plastic volumetric strain (positive for volume decrease) associated with the passage of the rock from the original to the broken state. By comparing the volumes of the broken zone before and after its formation we obtain

$$\pi(r_e^2 - r_i^2) = \pi\left((r_e + u_e)^2 - (r_i + u_i)^2\right)(1 - e_{av}) \tag{109}$$

Simplification gives

$$u_i = r_{io} \cdot \left(1 - \left[\frac{1 - e_{av}}{1 + A}\right]^{\frac{1}{2}}\right) \tag{110}$$

where

$$A = (2^{u_e/r_e} - e_{av})(r_e/r_i)^2 \tag{111}$$

Substitution for the terms r_e/r_i and u_e/r_e from equations 104 and 108 gives

$$A = \left(\frac{2(1+\nu)}{E} M\sigma_c - e_{av}\right) e^{2N - \frac{4}{m_r \sigma_c}(m_r \sigma_c p_i + s_r \sigma_c^2)^{\frac{1}{2}}} \tag{112}$$

The derivation of the expression for e_{av} is beyond the scope of this book but it can be found in the paper by Ladanyi[228].

$$e_{av} \approx \frac{2(u_e/r_e)(r_e/r_i)^2}{((r_e/r_i)^2 - 1)(1 + 1/R)} \tag{113}$$

The value of R depends upon the thickness of the broken zone.

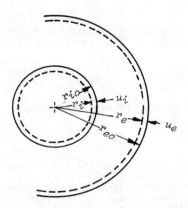

Notation for displacements around tunnel

For a relatively thin broken zone, defined by $r_e/r_i < \sqrt{3}$,

$$R = 2D \ln r_e/r_i \qquad (114)$$

For a thick broken zone, where $r_e/r_i > \sqrt{3}$

$$R = 1.1 D \qquad (115)$$

where

$$D = \frac{-m}{m + 4(m\sigma r_e/\sigma_c + s)^{\frac{1}{2}}} \qquad (116)$$

Equation for the required support line

For $p_{icr} < p_i < p_o$, the response of the rock mass is elastic and the equation for the required support line is given by

$$\frac{u_i}{r_{io}} = \frac{(1+\nu)}{E}(p_o - p_i) \qquad (117)$$

For $p_i < p_{icr}$, a broken zone exists and the required support line is given by equation 110.

Allowance for the dead weight of broken rock

The required support line defined by equations 110 and 117 can be considered to represent the behaviour of the sidewalls of the tunnel since the stresses and deformations in these regions are not influenced, to any significant extent, by the dead weight of the broken rock surrounding the tunnel. In order to allow for the dead weight of the broken rock in the roof and in the floor of the tunnel, the support pressure p_i can be increased or decreased by the amount $\gamma_r(r_e - r_i)$ where γ_r is the unit weight of the broken rock. Note that this correction can only be done *after* the required support line for the weightless conditions has been calculated by means of equations 110 and 117.

This correction must be recognised as a gross simplification but, within the overall accuracy of the analysis presented above, it gives a reasonable estimate of the effect of the dead weight of the broken rock.

Analysis of available support

As illustrated in figure 129, support is usually installed after a certain amount of convergence has already taken place in the tunnel. This initial convergence, denoted by u_{io}, is shown in the margin sketch.

The stiffness of the support installed within the tunnel is characterised by a stiffness constant k. The radial support pressure p_i provided by the support is given by

$$p_i = k \cdot u_{ie}/r_i \qquad (118)$$

where u_{ie} is the elastic part of the total deformation u_i. Hence

$$u_i = u_{io} + \frac{p_i r_i}{k} \qquad (119)$$

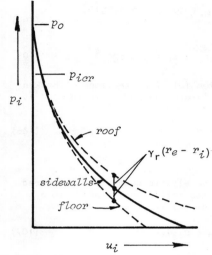

Required support lines for the rock surrounding the tunnel

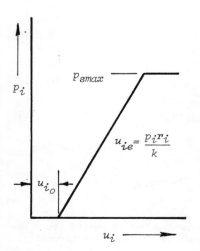

Available support curve

Equation 119 will apply up to a point at which the strength of the support system is reached. In the case of concrete or shotcrete lining, blocked steel sets or grouted bolts or cables, it will be assumed that plastic failure of the support system occurs at this point and that further deformation occurs at a constant support pressure as illustrated in the margin sketch on page 252. The maximum support pressure is defined by $Psmax$.

In the case of ungrouted mechanically anchored bolts, sudden failure of the bolt system can occur when the maximum support pressure $Psmax$ is exceeded. This is a situation which should be avoided since it can have unpredictable consequences.

Available support for concrete or shotcrete lining

A cast in situ concrete or a shotcrete lining of thickness t_c is placed inside a tunnel of radius r_i. The support pressure generated by this lining in response to convergence of the tunnel is given by equation 118 where

$$k_c = \frac{E_c(r_i^2 - (r_i - t_c)^2)}{(1 + \nu_c)((1 - 2\nu_c)r_i^2 + (r_i - t_c)^2)} \quad (120)$$

where

E_c = elastic modulus of concrete,
ν_c = Poisson's ratio of concrete,
r_i = tunnel radius,
t_c = concrete or shotcrete thickness.

Note that the influence of light reinforcing in the lining is not taken into accout in this stiffness calculation. Reinforcement such as mesh in shotcrete or light rebars in concrete plays a very important role in controlling and distributing stresses and cracking in the lining but it does not significantly increase the stiffness.

When very heavy reinforcing is included in the lining, for example when concrete is cast over steel sets, the contribution of both systems should be taken into account. The action of combined support systems will be discussed later in this chapter.

It should also be noted that the concrete or shotcrete lining is assumed to be permeable so that any internal or external water pressures do not influence the support pressure p_i. In the case of hydraulic tunnels in which the lining is impermeable, the additional stresses induced by water pressure must be taken into account. It is not particularly difficult to extend the analysis presented here to include this effect and the dedicated reader is left to carry out this modification. Alternatively, reference is made to textbooks such as those by Szechy[192] or Pequignot[190] for detailed discussions on this subject.

The maximum support pressure which can be generated by a shotcrete or concrete lining can be calculated from the theory of hollow cylinders under external pressure [67] and is given by:

$$Pscmax = \tfrac{1}{2}\sigma_{c.conc.}\left[1 - \frac{(r_i - t_c)^2}{r_i^2}\right] \quad (121)$$

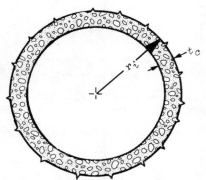

Concrete lining

Note : The analysis presented opposite assumes a closed ring of concrete or shotcrete and much of the support stiffness derives from the continuity of this ring. When applying this analysis to non-circular or to partial concrete or shotcrete linings, care must be taken that the available support is not over-estimated.

Blocked steel set

where $\sigma_{c.conc.}$ is the uniaxial compressive strength of shotcrete or concrete.

Note that equation 121 can only be applied when the lining is circular and the amount of overbreak is limited.

Available support for blocked steel sets

The stiffness of a blocked steel set is defined by

$$\frac{1}{k_s} = \frac{S \cdot r_i}{E_s A_s} + \frac{S \cdot r_i^3}{E_s I_s}\left[\frac{\theta(\theta + \sin\theta \cos\theta)}{2\sin^2\theta} - 1\right] + \frac{2S \cdot \theta \cdot t_B}{E_B W^2} \quad (122)$$

where
- r_i = tunnel radius,
- S = set spacing along length of tunnel,
- θ = half angle between blocking points (radians),
- W = flange width of set,
- A_s = cross-sectional area of steel section,
- I_s = moment of inertia of steel section,
- E_s = Young's modulus of steel,
- t_B = thickness of block,
- E_B = modulus of elasticity of block material.

The block is assumed to be square in plan and to have a side length equal to W, the flange width of the steel set.

The maximum support pressure which can be accommodated by the steel set is

$$p_{ssmax} = \frac{3 A_s I_s \sigma_{ys}}{2S \cdot r_i \cdot \theta \left[3 I_s + X A_s \left(r_i - (t_B + \tfrac{1}{2}X)\right)(1 - \cos\theta)\right]} \quad (123)$$

where
- σ_{ys} = yield strength of steel,
- X = depth of steel section.

Available support for ungrouted rockbolts

The available support for an ungrouted mechanically or chemically anchored rockbolt depends upon the deformation characteristics of the anchor, washer plate and bolt head as well as upon the deformation of the bolt shank. The results of a typical pull-out test on a mechanically anchored bolt are presented in figure 130. The displacement u_{eb} due to the elastic strain in the bolt shank is given by

$$u_{eb} = \frac{4 l T_b}{\pi d_b^2 E_b} \quad (124)$$

where
- l = free length of bolt between anchor and head,
- d_b = bolt diameter,
- E_b = Young's modulus of bolt material,
- T_b = load in bolt.

To this elastic displacement must be added a quantity

$$u_{ab} = Q T_b \quad (125)$$

where Q is a quantity related to the load-deformation characteristics of the anchor, washer plate and bolt head.

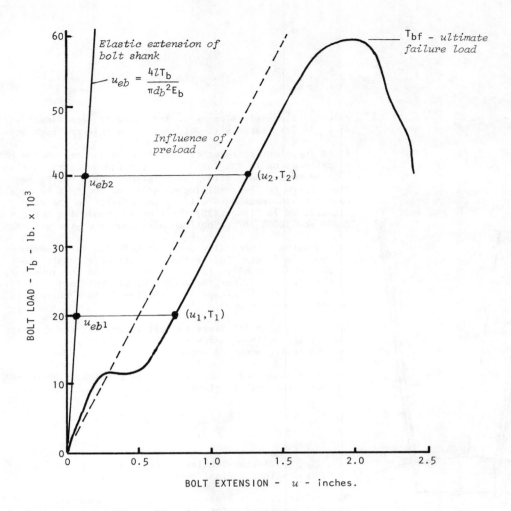

Figure 130 : Typical bolt load-extension curve determined by means of a pull-out test on a 1 inch diameter, 6 foot long bolt anchored by means of a 4 leaf Rawlplug expansion shell. After Franklin and Woodfield[226].

The value of Q can be determined from the load-extension curve obtained from a pull out test :

$$Q = \frac{(u_2 - u_{eb2}) - (u_1 - u_{eb1})}{T_2 - T_1} \qquad (126)$$

where (u_1, T_1) and (u_2, T_2) are two points on the linear portion of the load-extension plot as shown in figure 130.

In order to eliminate the non-linear response of the bolt system resulting from initial bedding in of the components of the mechanical anchor and washer, mechanically anchored bolts are normally preloaded immediately after installation. In the example illustrated in figure 130, a preload of 20000 lb would normally be used and this would have the effect, in terms of the rock-support interaction analysis, of moving the load-extension curve to the position shown by the dashed line. If no preload were applied to the bolt, it would follow the

Ungrouted mechanically anchored rockbolts

original load-extension curve and would be significantly less stiff in its response to deformations within the rock mass.

It is important to remember that load is induced in the bolt by deformation in the rock mass. Consequently, the preload applied to the bolt after installation should not be too high otherwise the remaining capacity of the bolt to accept load from the rock mass will be too small. In the case illustrated in figure 130, the preload should lie between 20000 and 30000 lb for optimum performance of the system.

An ungrouted mechanically anchored bolt can fail suddenly, as shown in figure 130, if the strength of the bolt shank is exceeded. Usually, failure occurs in the threaded portion of the shank at either the anchor or the head end. If failure occurs as a result of anchor slip, the failure process is generally more gradual.

Table 13 on page 257 lists typical values for the quantity Q and for the pull-out strength T_{bf} of a variety of mechanical and chemically anchored rockbolts.

The stiffness k_b of a mechanically or chemically anchored ungrouted rockbolt is given by

$$\frac{1}{k_b} = \frac{s_c s_l}{r_i} \left[\frac{4l}{\pi d_b^2 E_b} + Q \right] \qquad (127)$$

where
s_c = circumferential rockbolt spacing,
s_l = longitudinal rockbolt spacing.

The maximum support pressure which can be generated in a rockbolt system by deformation in the rock mass is given by

$$p_{sbmax} = \frac{T_{bf}}{s_c s_l} \qquad (128)$$

where
T_{bf} = ultimate strength of bolt system from pull-out test on a similar rock mass to that for which the rockbolt system is being designed.

Support provided by grouted rockbolts or cables

The rock-support interaction concepts, applied to the support systems discussed on the preceding pages, cannot be applied to grouted rockbolts or cables. This is because they do not act independently of the rock mass and hence the deformations which occur in both rock mass and support system cannot be separated.

Alternative theotetical solutions to this problem have not been explored in detail by the authors of this book and, hence, the support mechanism of grouted reinforcing elements will be discussed in a qualitative manner.

In the opinion of the authors, the support action of grouted rockbolts or cables arises from internal reinforcement of the rock mass in much the same way as the presence of reinforcing steel acts in reinforced concrete. By knitting the rock mass together and by limiting the separation of individual blocks

TABLE 13 - ULTIMATE STRENGTH AND LOAD-DEFORMATION CHARACTERISTICS OF TYPICAL ROCKBOLTS.

Bolt diameter in.	Bolt diameter mm	Bolt length ft.	Bolt length m	Anchor type	Rock type	T_{bf}^* lb.	Q^{**} in/lb	T_{bf} MN	Q m/MN
0.63	16.0	4.0	1.22	Expansion shell	Good rock	11 000	2.8×10^{-5}	0.049	0.160
0.63	16.0	6.0	1.83	Expansion shell	Shale	13 000	4.2×10^{-5}	0.058	0.241
0.63	16.0	4.0	1.22	Expansion shell	Unknown	9 000	9.0×10^{-6}	0.040	0.053
0.63	16.0	4.0	1.22	Expansion shell and resin	Unknown	14 000	5.0×10^{-6}	0.062	0.030
0.75	19.0	4.0	1.22	Expansion shell	Good rock	11 500	2.2×10^{-5}	0.051	0.126
0.75	19.0	6.0	1.83	Expansion shell	Unknown	20 000	4.0×10^{-6}	0.089	0.024
0.75	19.0	6.0	1.83	Expansion shell and resin	Unknown	22 000	4.0×10^{-6}	0.098	0.029
0.75	19.0	10.0	3.0	Slotted bolt and wedge	Unknown	22 000	1.3×10^{-5}	0.098	0.074
0.87	22.0	10.0	3.0	Expansion shell	Gneiss	48 000	5.5×10^{-6}	0.214	0.032
0.87	22.0	10.0	3.0	Expansion shell	Sandstone	44 000	7.3×10^{-6}	0.196	0.042
0.87	22.0	10.0	3.0	Expansion shell	Sandy shale	28 500	1.2×10^{-5}	0.127	0.069
0.87	22.0	10.0	3.0	Expansion shell	Shale	13 000	2.2×10^{-5}	0.058	0.126
1.00	25.4	19.7	6.0	Expansion shell	Massive gneiss	72 600	8.9×10^{-6}	0.323	0.051
1.00	25.4	6.0	1.83	Expansion shell	Granite	57 000	2.5×10^{-5}	0.254	0.143
1.00	25.4	6.0	1.93	Resin anchor	Granite	64 000	3.2×10^{-6}	0.285	0.018
1.00	25.4	4.0	1.22	Slotted bolt and wedge	Good rock	20 000	1.1×10^{-5}	0.089	0.064
1.00	24.5	6.0	1.83	Resin anchor	Shale	36 000	3.5×10^{-6}	0.160	0.020

* Ultimate pull out load determined in a field test.
** Defined by equation 126 on page 255.

Note : The values listed in this table have been determined from published test data and the authors cannot guarantee the accuracy of the results. For critical applications it is strongly recommended that pull-out and load-deformation characteristics be determined from field tests on the bolts to be used.

the grouted reinforcing elements limit the dilation in the rock mass immediately surrounding the tunnel. This has the effect of limiting the extent to which the original rock mass material constants m and s reduce to m_r and s_r. Sensitivity studies of the influence of the values of m_r and s_r upon the required support line for the rock surrounding a tunnel (see page 252) show that the deformation u_i is sharply reduced for relatively modest increases in m_r and s_r.

Unfortunately, no direct evidence is available on the strength of reinforced rock masses and hence the mechanism discussed above cannot be quantified. A few sample calculations will be presented later in this chapter to demonstrate the importance of this reinforcing but the practical design of grouted reinforcing systems is currently a matter of engineering judgement.

Reaction of combined support systems

When two support systems, for example rockbolts and shotcrete lining, are combined in a single application, it is assumed that the stiffness of the combined support system is equal to the sum of the stiffnesses of the individual components:

$$k' = k_1 + k_2 \qquad (129)$$

where
k_1 = stiffness of first system and
k_2 = stiffness of second system.

Note that the two support systems are assumed to be installed at the same time.

The available support curve for the combined system is defined by

$$u_i = u_{io} + \frac{p_i r_i}{k'} \qquad (130)$$

(see equation 119 on page 252).

Equation 131 will apply until the maximum deformation which can be tolerated by one of the systems is reached. At this point the remaining support system will be required to carry most of the load but its response will probably be unpredictable. Consequently, the failure of the first system is regarded as failure of the overall support system.

The maximum deformation which can be tolerated by each support system is determined by substituting the appropriate value of the maximum support pressure p_{smax} (from equations 121, 122, 124 and 129) in equation 119.

An example of the use of combined support will be discussed later in this chapter.

Summary of rock-support interaction equations

For the convenience of the reader, the rock-support interaction equations which have been derived on the preceding pages are summarised in table 14. These equations are listed in the sequence in which they would be used for calculation on a programmable calculator.

TABLE 14 - CALCULATION SEQUENCE FOR ROCK-SUPPORT INTERACTION ANALYSIS

1. Required support line for rock mass

Input data required

- σ_c = uniaxial compressive strength of intact rock pieces
- $m \atop s$ } material constants for original rock mass (see table 12 on page 176)
- E = modulus of elasticity of original rock mass
- ν = Poisson's ratio of original rock mass
- $m_r \atop s_r$ } material constants for broken rock mass (see table 12 on page 176)
- γ_r = unit weight of broken rock mass
- p_o = in situ stress magnitude
- r_i = radius of tunnel

Calculation sequence

a. $M = \frac{1}{2}\left(\left(\frac{m}{4}\right)^2 + m\, p_o/\sigma_c + s\right)^{\frac{1}{2}} - \frac{m}{8}$ (102)

b. $D = \dfrac{-m}{m + 4\left(m/\sigma_c(p_o - M\sigma_c) + s\right)^{\frac{1}{2}}}$ (101,116)

c. $N = 2\left[\dfrac{p_o - M\sigma_c}{m_r \sigma_c} + \dfrac{s_r}{m_r^2}\right]^{\frac{1}{2}}$ (105)

[Input p_i]

d. For $p_i > p_o - M\sigma_c$, deformation around tunnel is elastic

$\dfrac{u_i}{r_{io}} = \dfrac{(1+\nu)}{E}(p_o - p_i)$ (117)

e. For $p_i < p_o - M\sigma_c$, plastic failure occurs around tunnel

$\dfrac{u_e}{r_e} = \dfrac{(1+\nu)}{E} M\sigma_c$ (108)

f. $\dfrac{r_e}{r_i} = e^{N - 2\left(\frac{p_i}{m_r\sigma_c} + \frac{s_r}{m_r^2}\right)^{\frac{1}{2}}}$ (104)

g. For $r_e/r_i < \sqrt{3}$: $R = 2D \ln r_e/r_i$ (114)

h. For $r_e/r_i > \sqrt{3}$: $R = 1.1 D$ (115)

i. $e_{av} = \dfrac{2(u_e/r_e)(r_e/r_i)^2}{\left((r_e/r_i)^2 - 1\right)\left(1 + 1/R\right)}$ (113)

j. $A = (2u_e/r_e - e_{av})(r_e/r_i)^2$ (111)

k. $\dfrac{u_i}{r_{io}} = 1 - \left[\dfrac{1 - e_{av}}{1 + A}\right]^{\frac{1}{2}}$ (110)

l. For roof of tunnel, plot u_i/r_{io} against $\dfrac{p_i + \gamma_r(r_e - r_i)}{p_o}$

m. For sidewalls of tunnel, plot u_i/r_{io} against p_i/p_o

n. For floor of tunnel, plot u_i/r_{io} against $\dfrac{p_i - \gamma_r(r_e - r_i)}{p_o}$

TABLE 14 - CALCULATION SEQUENCE FOR ROCK-SUPPORT INTERACTION ANALYSIS

2. Support stiffness and maximum support pressure for concrete or shotcrete lining

Input data required

E_c = modulus of elasticity of concrete or shotcrete
ν_c = Poisson's ratio of concrete or shotcrete
t_c = thickness of lining
r_i = tunnel radius
$\sigma_{c.conc.}$ = uniaxial compressive strength of concrete or shotcrete

Support stiffness and maximum support pressure

a. $$k_c = \frac{E_c\left(r_i^2 - (r_i - t_c)^2\right)}{(1 + \nu_c)\left((1 - 2\nu_c)r_i^2 + (r_i - t_c)^2\right)} \qquad (120)$$

b. $$p_{scmax} = \tfrac{1}{2}\sigma_{c.conc.}\left[1 - \frac{(r_i - t_c)^2}{r_i^2}\right] \qquad (121)$$

3. Support stiffness and maximum support pressure for blocked steel sets

Input data required

W = flange width of steel set
X = depth of section of steel set
A_s = cross-sectional area of steel set
I_s = moment of inertia of steel section
E_s = modulus of elasticity of steel section
σ_{ys} = yield strength of steel
r_i = tunnel radius
S = steel set spacing along tunnel axis
θ = half angle between blocking points (radians)
t_B = thickness of block
E_B = modulus of elasticity of block material

Support stiffness and maximum support pressure

a. $$\frac{1}{k_s} = \frac{S.r_i}{E_s A_s} + \frac{S.r_i^3}{E_s I_s}\left[\frac{\theta(\theta + \sin\theta\cos\theta)}{2\sin^2\theta} - 1\right] + \frac{2S.\theta.t_B}{E_B W^2} \qquad (122)$$

b. $$p_{ssmax} = \frac{3 A_s I_s \sigma_{ys}}{2S.r_i.\theta\left[3I_s + XA_s\left(r_i - (t_B + \tfrac{1}{2}X)\right)(1 - \cos\theta)\right]} \qquad (123)$$

4. Support stiffness and maximum support pressure for ungrouted mechanically or chemically anchored rockbolts or cables

Input data required

l = free bolt or cable length
d_b = bolt diameter or equivalent cable diameter
E_b = elastic modulus of bolt or cable material
Q = load-deformation constant for anchor and head
T_{bf} = ultimate failure load from pull-out test
r_i = tunnel radius
s_c = circumferential bolt spacing
s_l = longitudinal bolt spacing

Support stiffness and maximum support pressure

a. $$\frac{1}{k_b} = \frac{s_c s_l}{r_i}\left[\frac{4l}{\pi d_b^2 E_b} + Q\right] \qquad (127)$$

b. $$p_{sbmax} = T_{bf}/s_c s_l \qquad (128)$$

TABLE 14 - CALCULATION SEQUENCE FOR ROCK-SUPPORT INTERACTION ANALYSIS

5. Available support curve for a single support system

Input data required

- k = stiffness of support system under consideration
- p_{smax} = maximum support pressure which can be accommodated
- u_{io} = initial tunnel deformation before installation of support

Available support curve

For $p_i < p_{smax}$: $\quad \dfrac{u_i}{r_i} = \dfrac{u_{io}}{r_i} + \dfrac{p_i}{k}$ \hfill (119)

6. Available support curve for a combined support system

Input data required

- k_1 = support stiffness of system 1
- p_{smax_1} = maximum support pressure for system 1
- k_2 = stiffness for support system 2
- p_{smax_2} = maximum support pressure for system 2
- u_{io} = initial tunnel deformation before installation of support
  (Note that the two support systems are assumed to be installed at the same time and to start responding to tunnel deformation simultaneously)

Calculation sequence for available support curve

a. $u_{max_1} = r_i \cdot p_{smax_1}/k_1$

b. $u_{max_2} = r_i \cdot p_{smax_2}/k_2$

c. $u_{12} = r_i \cdot p_i /(k_1 + k_2)$

d. For $u_{12} < u_{max_1} < u_{max_2}$

$$\dfrac{u_i}{r_i} = \dfrac{u_{io}}{r_i} + \dfrac{p_i}{(k_1 + k_2)} \hfill (130)$$

e. For $u_{12} > u_{max_1} < u_{max_2}$

$$p_{max_{12}} = u_{max_1}(k_1 + k_2)/r_i$$

f. For $u_{12} > u_{max_2} < u_{max_1}$

$$p_{max_{12}} = u_{max_2}(k_1 + k_2)/r_i$$

TABLE 15 - TYPICAL VALUES FOR MODULUS OF DEFORMATION AND MODULUS OF ELASTICITY FOR USE IN ROCK-SUPPORT INTERACTION ANALYSIS. After Lama and Vutukuri[233].

Rock type and description	Test location	Test description	Modulus of deformation GPa*	Modulus of elasticity ** GPa
Amphibolite	Czechoslovakian dams	0.71m Ø jack	2.5-15.6	13.5-35.9
Amphibolite granodiorite	Koshibu dam, Japan	pressure chamber	23.1	
Amphibolite	Oroville power plant, USA.	jack test cavern deformn.		60.3 56.5
Andesite	Japanese hydroelectric projects	0.18m² jack	4.9-10.1	30.3-32.4
Agglomerate and tuff with some andesite	Cachi dam	pressure chamber		2.1-15.8
Arkose sandstone	Japanese hydroelectric projects	0.18m² jack		1.2-3.2
Biotite gneiss	Amsteg pressure tunnel, Switzerland	pressure chamber		31.7
Biotite gneiss	Czechoslovakian dams	0.71m Ø jack	2.5-15.6	13.5-35.9
Biotite gneiss	Kariba dam, Rhodesia	jack test cavern deformn.	3.4-6.9	6.2
Biotite gneiss and amphibolites unweathered - partly weathered- highly weathered -	South Moravia, Czechoslovakia	0.71m x 0.71m plate jacking	2.5-10.0 1.0-2.5 0.25-1.0	6.0-15.0 2.5-4.0 1.0-2.5
Biotite shale	Mica project, Canada	0.06m² jack	1.0-6.2	3.6-26.1
Biotite shale parallel bedding- normal to bedding-	Andermatt, Switzerland	plate jacking	28.0 8.0	
Calcareous siltstone	Bhakra dam, India	plate jacking	2.7-7.6	
Calcareous marl	Iznajar dam, Spain	plate jacking	1.0-33.8	6.0-50.0
Calcareous schist	Kaunertal, Austria	radial jacking	3.0-6.0	5.0-10.7
Calcareous shale	Mauvoisin dam, Switzerland	plate jacking	4.0	28.0
Calcareous shale, sandy, faulted	Rothenbrunnen press. shaft, Switzerland	pressure chamber		7.9-11.3
Calcareous schist	Sabbione dam, Italy	overall deformn.	34.5	
Chlorite schist	Grocio dam, Italy	jack test	22.6	36.9
Coal	Witbank, S. Africa	in situ compression	3.6	
Conglomerate sandstone	Agri River, Italy	jack test pressure chamber		11.4 1.2-1.8
Conglomerate	Dez dam, Iran	jack test pressure chamber		1.8-50.0 4.9-16.2
Conglomerate vertical - horizontal -	Grado dam, Spain	jack test		8.5-19.1 6.9-8.4
Conglomerate, fissured	Nagase dam, Japan	0.25m Ø jack		2.3
Conglomerate sandstone	Nagase dam, Japan	0.25m Ø jack		1.7-9.4
Conglomerate	Not known	pressure chamber	2.8-3.1	
Conglomerate	Nuclear power plant, Japan	jack test		3.7-4.4
Cretaceous sandstone	Slavia, Czechoslovakia	0.5m² plate jack 2m radial jack		0.08-0.1 0.07

* 1 GPa = 1.02×10^4 kgf/cm² = 1.45×10^5 lbf/in². ** Modulus of Elasticity determined from linear portion of load-deformation curve, Modulus of Deformation from overall load versus displacement.

TABLE 15 continued.

Rock type and description	Test location	Test description	Modulus of deformation GPa	Modulus of elasticity GPa
Dacite	Pongolapoort dam, South Africa	jack test		43.2
Diabase, weathered, jointed	Kloof gold mine, South Africa	0.5m Ø jack	2.4	3.5
Diabase	Zillierback dam, Germany	deformation analysis		1.0
Diabase	Nuclear power plant, Japan	jack test		2.0
Diorite	Nuclear power plant, Japan	jack test		1.8
Diorite gneiss	Tehachapi, USA	jack test	0.7-5.4	2.6-11.7
Diorite	Karadj, Iran	jack test		7.4-17.1
Diorite gneiss and granodiorite gneiss	Edmonston pump plant, USA	plate loading relaxation	3.8 4.0	
Dolomite, stratified and fractured	Glagno dam, Italy	jack test pressure chamber	0.4-6.2	2.4-11.0 3.7-5.1
Dolomite & limestone	Mis dam, Italy	jack test		6.6
Dolomite	Sylvenstein dam, Germany	jack test		6.7-14.6
Dolomite	Val Vestino, Italy	jack test	3.3-4.5	3.7-16.5
Dolerite, Triassic intrusion, massive fractured	Khantaika hydro-plant, USSR	2.2m Ø pressure tunnel test	10.4-37.0 5.9	
Gabbro	Nuclear power plant, Japan	jack test		1.1
Gneiss	Bort dam, France	pressure chamber		15.0-20.0
Gneiss	Besserve dam, France	jack test	1.1	2.2
Gneiss	Funil, Brazil	jack test	7.3-30.3	
Gneiss, altered dry wet	Irongate dam, Romania	1.5-2.0m Ø jack	1.0-9.5 0.4-4.5	2.8-18.5 0.6-9.5
Gneiss	Kaunertal, Austria	radial jack	35.2	36.2
Gneiss, fine grained	Lake Delio, Italy	0.25m Ø jack	9.5	14.9
Gneiss	La Bathie, France	pressuremeter	34.7	
Gneiss, fine grained, foliated	Lago Debo	pressure chamber jack test deformation analysis		9.6-26.2 6.9-20.7 6.6
Gneiss pegmatite	Morrow Point, USA	0.86m Ø jack	6.2-11.7	9.0-15.7
Gneiss	Malpasset dam, France	jack test	0.5	1.0
Gneiss	Manapouri power proj. New Zealand	plate loading flat jack		4.8 22.1
Gneiss, slightly laminated	Sufers dam, Switzerland	overall deformn.	10.0	20.0
Gneiss	St. Cassein, France	jack test	4.8-32.0	
Gneiss, schistose	St. Cassein, France	pressure chamber		1.0-52.5
Gneiss, altered	St. Jean du Gard., France	jack test	1.0-2.8	2.3-4.1
Granite parallel bedding normal to bedding	Andermatt, Switzerland	jack test		40.0 25.0

TABLE 15 continued.

Rock type and description	Test location	Test description	Modulus of deformation GPa	Modulus of elasticity GPa

Granite
  weathered
  highly weathered
	Alto Rabago dam, Portugal			6.6
				1.0

Granite	Cabril dam, Portugal	jack test pressure chamber		1.0-11.5 11.5-23.0

Granite
  slightly weathered
  highly altered
	Canicada dam, Portugal	jack test		10.9-12.8
				0.6-14.0

Granite	Candes dam, France	jack test	1.5	3.1
Granite gneiss	Dworshak dam, USA	jack test	4.1-50.3	7.6-68.9
Granite	Grimsel dam, Switzerland	deformation analysis		15.0
Granite, massive unweathered	Inner Kirchen, Switzerland	pressure chamber		20.0

Granite
  weathered
  jointed
  unweathered
  weathered, joint filling
	Kurobe dam, Japan	pressure chamber
		jack test
		jack test
		jack test
		jack test
			
			2.3-3.3
			2.0-3.0
			3.1-6.4
			1.5-1.6
				1.6-3.2

Granite
  weathered, jointed
  weathered, highly jointed
  unweathered, jointed
	Kariba dam, Rhodesia	triaxial	0.8-1.3
			
			1.1-1.5
			0.3-1.6
				1.1-1.3
				
				1.1-1.2
				1.6-3.4

Granite gneiss	Lyse pressure shaft, Norway	pressure chamber		8.0-15.0
Granite gneiss	Mica project, Canada	$0.06m^2$ jack	7.5-84.0	15.2-86.0
Granite	Nagawado dam, Japan	$2m^2$ shear test		3.0-10.0
Granite, biotite and feldspar	Salamonde dam, Portugal	$1m^2$ jack pressure chamber	0.6-23.0 0.8-6.3	
Granite, massive	Schwarzenbach tunnel, Germany	pressure chamber	24.6	37.5
Granite	Tsruga dam, Japan	jack test		2.6-7.2
Granite, granite gneiss	Tumut 1, Australia	jack test pressure chamber deformation analysis		34.5 13.8 4.8-13.8
Granodiorite	Tumut 2, Australia	jack test pressure chamber plate loading deformation analysis	 5.5-48.3	41.4 20.7 6.9
Granite	Villefort dam, France	jack test	1.9	6.6
Granite	Valdecanas, Portugal	jack test	1.5-10.0	
	Japanese hydro. proj.	jack test	1.5-8.6	
Granodiorite	Pantano D'Avio dam, Italy	pressure chamber dam displacement overall structure	 11.7-16.5	12.0-24.0 17.0-32.0
Hematite ore	Germany	$2m^2$ compression		70.0
Hornfels	Emosson dam, Switzerland	jack test		46.0
Kiesengite	Place Moulin dam, Italy	pressure chamber		15.0-25.0
Limestone	Ambiesta, Italy	pressure chamber		12.5

TABLE 15 continued.

Rock type and description	Test location	Test description	Modulus of deformation GPa	Modulus of elasticity GPa
Limestone	Adiguzel dam, Turkey	0.58m Ø jack	0.3-3.4	2.7-8.4
Limestone, massive	Achensee pressure shaft, Austria			10.0-30.0
Limestone	Arrens tunnel, France	jack test	13.5	40.0
Limestone & dolomite	Dubrovnik power plant, Yugoslavia	pressuremeter	4.3-6.0	6.2
Limestone	Fedaia dam, Italy	pressure chamber	7.5	38.5
Limestone, massive	Finodal pressure shaft, Yugoslavia	jack test		22.3
Limestone	Greoux dam, France	jack test pressure chamber	3.2	5.1 5.7-12.3
Limestone upper Cretaceous fissured with clay highly fissured	Hydro. power projects, USSR	0.8m Ø jack	57.5 166.3 3.3	81.1 173.0
Limestone jointed	Kastenbell pressure shaft, Italy	pressure chamber		27.0 0.2
Limestone, laminated	Limberg, Austria	jack test		4.0-15.0
Limestone, massive Triassic	Mratje dam, Yugoslavia	jack test	0.002-0.016	0.002-0.03
Limestone dolomite	Mae dam, Italy	pressure chamber	6.6-8.3	25.4-30.3
Limestone dolomite	Mis dam, Italy	jack test	4.5-7.4	3.9-11.5
Limestone	Mequinenze dam, Spain	jack test		18.2-27.6
Limestone with shale and sandstone	Niagra tunnel, USA	overall structure	17.2	
Limestone, highly fissured	Pieve di Cadore, Italy	pressure chamber	3.0	4.0
Limestone	Prutz pressure shaft, Austria	radial jack	6.0-6.9	10.3-12.5
Limestone	Tena Termini, Italy	jack & seismic	1.5-41.6	4.7-46.1
Limestone, bituminous with fractured gneiss	Hydro. projects, USSR	jack test	0.5-15.2	
Limestone dolomite	Vouglans dam, France	0.28m Ø jack	3.2-10.0	7.0-160.0
Limestone, fractured	Val Gallina, Italy	pressure chamber	2.7-3.9	2.8-4.1
Limestone	Vaiont dam, Italy	pressure chamber	4.0-12.0	30.7-46.0
Limestone	Yellow Tail, USA	jack test overall structure	11.0-33.1 2.8-9.7	16.5-40.7
Limestone	Yugoslavian hydro. Yugoslavia		2.9-41.6	4.7-63.6
Limestone, massive	Not known	1m² flat jack	52.5-56.5	
Liparite fault, hard	Kawamata dam, Japan	0.8m Ø jack	0.4-7.2	
Liparite	Ikari dam, Japan		2.3	
Marl	Afourer tunnel, Morocco	jack test	4.0-5.0	6.0-7.0
Marl	Iznajar, Spain	jack test	2.3-5.0	
Mica schist	Beuregard, Italy	pressure chamber	0.69-14.0	1.0-24.0
Micaceous gneiss	Giovaretto, Italy	pressure chamber		21.5
Mica schist	Morasco dam, Italy	dilatometer dam displacement pressuremeter overall structure		17.5 210.0 17.2 18.6-20.7

TABLE 15 continued.

Rock type and description	Test location	Test description	Modulus of deformation GPa	Modulus of elasticity GPa
Mica schist	Roujanel dam, France	pressure chamber		1.8-7.0
Mudstone	Nuclear power plant, Japan	jack test	0.8-2.0	4.9-7.1
Mudstone	Poatina power plant, Tasmania	overall structure jack test	11.7	16.6-22.1
Mudstone	Kameyama dam, Japan	0.3m Ø jack	2.4	3.2
Orthogneiss, massive	Telessio dam, Italy	jack test		30.0-40.0
Paragneiss	Teisnach dam, Germany	jack test	3.1-8.5	6.1-13.4
Quartzite	Alvito Sussidenga, Portugal	jack test	1.4-7.3	
Quartziferous phyllitic gneiss	Careser dam, Italy	pressure chamber pressuremeter		24.0-35.0 23.4
Quartz-feldspar-biotite-granite-gneiss	Fallone dam, Italy	0.6m Ø jack	0.4	1.0
Quartzite with phyllite and shale	Fontana dam, USA		41.4	
Quartziferous phyllite	Frera dam, Italy			25.5-43.0
Quartz-porphyry	Forte Buso, Italy	pressure chamber		6.0
Quartz-phyllite	Gerlos pressure shaft, Austria	overall structure		3.5-5.0
Quartzite schist	D'Avene dam, France	jack test	2.8-5.4	5.5-12.5
Quartzite & phyllite	Gordon dam, Tasmania	jack test		5.6-28.0
Quartz-feldspar-biotite-granite-gneiss	Jassa dam, Italy	0.6m Ø jack	0.6	1.0
Quartzite, jointed	Kariba dam, Rhodesia	jack test	0.9-6.1	2.4-9.1
Quartzite sandstone	Latiyan dam, Iran	jack test	1.3-1.8	3.8-4.6
Quartz-seritic shale	Lovero tunnel, Italy	jack test	2.5-3.0	4.1-5.3
Quartz diorite	Malga Bissona, Italy	pressuremeter overall structure	16.5	21.4
Quartzite, micaceous	Morrow Point dam, USA	overall structure	3.5-14.5	
Quartzite	Srisailam dam, India	jack test	1.1-34.9	
Quartz-porphyry	Sudegai dam, Japan			5.0-10.0
Quartz-porphyry	St. Antonio pressure shaft, Italy	pressure chamber		40.0
Quartz-diorite	Cedar City, USA			82.7
Quartzite	Tignes dam, France			10.0-35.0
Quartzite with shale	Tachian dam, Taiwan	0.3m Ø jack	7.0	15.0
Quartzite, massive	Zinnoun, Morocco	jack test	38.0-45.0	
Quartzite	Not known	pressure chamber		10.8-47.9
Rhyolite & gneiss	Davis dam, USA	jack test overall structure	0.1-0.8 2.4-8.3	
Sandstone	Barbellino dam, Italy	overall structure		46.2
Sandstone	Bhakra dam, India	jack test	2.5-16.9	
Sandstone, compact	Caprile dam, Italy	pressure chamber	7.8	9.8-10.6
Sandstone	Sydney, Australia	jack test	0.6-3.2	
Sandstone	Glen Canyon dam, USA	jack test	1.1-8.9	7.4-9.6
Sandstone, arkose	Cambambe dam, Angloa	jack test		12.3-24.0
Sandstone	Antioch, USA	jack test		7.3
Sandstone, clay & shale	Hitotsuse dam, Japan	jack test	2.4-4.4	5.2-16.3

TABLE 15 continued.

Rock type and description	Test location	Test description	Modulus of deformation GPa	Modulus of elasticity GPa
Sandstone & conglomerate	Inferno dam, Italy			14.0
Sandstone & slate	Japanese hydro. proj.	0.18m² jack	1.4-5.3	5.1-25.4
Sandstone	Kamishiba dam, Japan			5.0-10.0
Sandstone	Latiyan dam, Iran	jack test	1.5-5.9	1.9-14.7
Sandstone and shale	Latiyan dam, Iran	jack test	0.8-1.7	1.9
Sandy limestone	Marmorera tunnel, Switzerland	pressure chamber	21.5	31.0-34.4
Sandstone	Ogochi dam, Japan			60.0-70.0
Sandstone	Paltinul dam, Romaina	jack test	1.0-2.0	1.8-5.0
Sandstone & schist	Poiana Uzului dam, Romania	1.5m Ø jack	1.0-3.0	1.7-7.0
Sandstone			4.8	9.0
Sandstone & conglomerate	Pietra del Pertusillo, Italy			2.0
Sandstone	Rossens dam, Switzerland	plate loading pressure chamber	0.8-2.0	1.3-3.6 3.8-7.0
Sandstone & mudstone	Silisian Beskids power plant, Poland	0.8m Ø jack	0.5-8.2	
Sandstone	Tonoyama dam, Japan			5.3-13.0
Sandstone & shale	Tawa dam, India	0.3m Ø jack	0.02	0.02
Sandstone & conglomerate	Trona dam, Italy	pressure chamber pressuremeter overall structure	17.9	13.3-18.0 25.5
Sandstone	Waldshut pressure tunnel, Germany	jack test		3.0-19.5
Sandstone	Nuclear power plant, Japan		0.5-1.7	13.0-16.0
Schist	Adiguzel dam, Turkey	0.6m Ø jack	0.36	1.0-2.4
Schist	Arrens tunnel, France	jack test	5.0	12.5-19.2
Schist	Barbellino dam, Italy	pressure chamber		48.0
Schist	Beauregard, Italy	pressure chamber	23.5	46.1
Schist, micaceous	Grandval dam, France	jack test		9.0-12.0
Schist & gneiss	Morrow Point, USA	jack test	5.9-11.3	15.5-18.3
Schist, Silurian	Tihange nuclear power plant, Belgium	dilatometer	0.6-3.0	0.9-5.0
Schist	Uzbekistan dam, USSR	jack test	1.2-4.6	2.4-36.3
Serpentine schist	Alpe Gera, Italy	pressure chamber jack test dam displacement	7.3-23.1	6.7 8.9-25.6 32.0
Seritic shale	Amsteg, Switzerland	pressure chamber	2.7	3.2-3.5
Shale	Benposta & Miranda, Portugal	jack test	3.2-29.4	
Shale & quartzite	Burgin mine, USA	jack test		3.1-3.8
Shale	Jablanica tunnel, Yugoslavia	pressure chamber		12.5
Shale	Naruko dam, Japan	jack test dam settlement	4.0 4.1	
Shale	Neudaz, Switzerland	pressure chamber		5.1
Slate	Wallsee dam, Austria	pressure chamber		0.1
Slate	Nuclear power plant, Japan	jack test		9.0-25.0

TABLE 16 - TYPICAL INPUT DATA FOR ROCK-SUPPORT INTERACTION ANALYSIS.

Rock mass

Uniaxial compressive strength of intact rock	- σ_c -	see table 9 on page 141.
Material constant for original rock mass	- m -	see table 12 on page 176.
Material constant for original rock mass	- s -	see table 12 on page 176.
Modulus of deformation for rock mass	- E -	see table 15 on page 262.
Poisson's ratio for rock mass	- ν -	0.15 to 0.30.
Material constant for broken rock mass	- m_r -	see table 12 on page 176.
Material constant for broken rock mass	- s_r -	see table 12 on page 176.
Unit weight of broken rock mass	- γ_r -	see below

Sedimentary rocks γ_r - 0.025±0.003 MN/m³ - 155±19 lb/ft³
Igneous and metamorphic rocks γ_r - 0.030±0.003 MN/m³ - 187±19 lb/ft³
Monomineralic aggregates (ores) γ_r - 0.034±0.012 MN/m³ - 210±75 lb/ft³

Depending upon the degree to which the rock mass is jointed or broken, these values may be decreased by up to approximately 20%.

Concrete or shotcrete linings

Modulus of elasticity of shotcrete or concrete	- E_c -	21±7 GPa - 3 × 10⁶ ± 1 × 10⁶ lb/in²
Poisson's ratio of shotcrete or concrete	- ν_c -	0.25
Compressive strength of shotcrete or concrete	- $\sigma_{c.conc.}$ -	35±20 MPa - 5000±3000 lb/in²

(Depending upon age and quality)

Blocked steel sets

		Light section 6 I 12	*Medium section 8 I 23*	*Heavy section 12 W 65*
Flange width	W	0.0762m 2.0 inches	0.1059m 4.16 inches	0.3048m 12.0 in.
Section depth	X	0.1524m 6.0 inches	0.2023m 8.0 inches	0.3048m 12.0 in.
Section area	A_s	0.00228m² 3.52 in²	0.00433m² 6.71 in²	0.01233m² 19.12 in²
Moment of Inertia	I_s	8.74×10⁻⁶m⁴ 21 in⁴	2.67×10⁻⁵m⁴ 64 in⁴	2.22×10⁻⁴m⁴ 534 in⁴

Young's modulus of steel	- E_s -	207 GPa - 30 × 10⁶ lb/in²
Yield strength of steel	- σ_{ys} -	245 MPa - 36 000 lb/in²
Young's modulus of blocking material	- E_B -	*Stiff blocking* 10 000 MPa - 1.5 × 10⁶ lb/in²
		Soft blocking 500 MPa - 72 000 lb/in²

Rockbolts

Bolt diameter	- d_b -	16mm/⅝ in. - 19mm/¾ in. - 25mm/1 in. - 34mm/1⅜ in.
Young's modulus of bolts	- E_b -	207 GPa - 30 × 10⁶ lb/in²
Anchor stiffness	- Q -	see table 13 on page 257.
Pull out strength	- T_{bf} -	see table 13 on page 257.

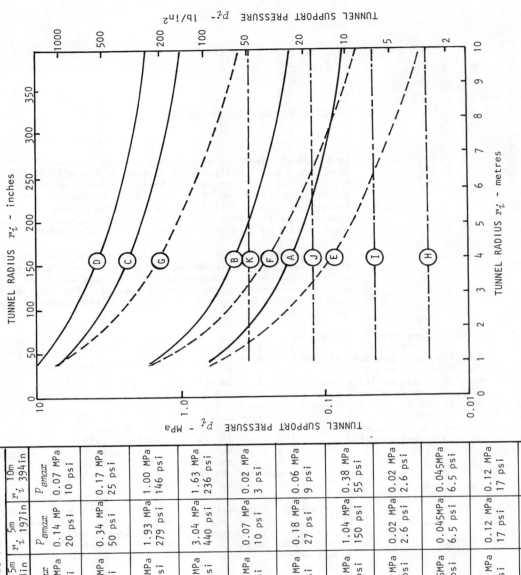

TABLE 17 - MAXIMUM SUPPORT PRESSURES FOR VARIOUS SYSTEMS.

Support system \ Tunnel radius	r_i 1m 39in p_{smax}	r_i 2.5m 98in p_{smax}	r_i 5m 197in p_{smax}	r_i 10m 394in p_{smax}
A - SHOTCRETE - 5cm (0.05m)/ 2 inches thick shotcrete. $\sigma_{c.conc.}$ = 14 MPa/2000 psi after 1 day.	0.65 MPa 95 psi	0.27 MPa 39 psi	0.14 MP 20 psi	0.07 MPa 10 psi
B - SHOTCRETE - 5cm(0.05m)/ 2 inches thick shotcrete. $\sigma_{c.conc.}$ = 35 MPa/5000 psi after 28 days.	1.63 MPa 236 psi	0.68 MPa 99 psi	0.34 MPa 50 psi	0.17 MPa 25 psi
C - CONCRETE - 30cm(0.30m)/ 12 inches thick concrete. $\sigma_{c.conc.}$ = 35 MPa/5000 psi after 28 days.	7.14 MPa 1036 psi	3.55 MPa 515 psi	1.93 MPa 279 psi	1.00 MPa 146 psi
D - CONCRETE - 50cm(0.50m)/ 19.5 inches thick concrete. $\sigma_{c.conc.}$ = 35 MPa/5000 psi after 28 days.	9.72 MPa 1410 psi	5.35 MPa 775 psi	3.04 MPa 440 psi	1.63 MPa 236 psi
E - STEEL SETS - (6 I 12) space 2m/79 in.. Blocked $2\theta=22\tfrac{1}{2}°$, σ_{ys} = 248MPa/36 000 psi.	0.61 MPa 88 psi	0.18 MPa 27 psi	0.07 MPa 10 psi	0.02 MPa 3 psi
F - STEEL SETS - (8 I 23) space 1.5m/59 in.Blocked $2\theta=22\tfrac{1}{2}°$, σ_{ys} = 248MPa/36 000 psi.	1.59 MPa 230 psi	0.50 MPa 72 psi	0.18 MPa 27 psi	0.06 MPa 9 psi
G - STEEL SETS - (12 W 65) at 1m/39 in. Blocked $2\theta=22\tfrac{1}{2}°$, σ_{ys} = 248MPa/36 000 psi.	7.28 MPa 1055 psi	2.53 MPa 366 psi	1.04 MPa 150 psi	0.38 MPa 55 psi
H - VERY LIGHT ROCKBOLTS - 16mm/⅝in. ø at 2.5m/98in. centres. Mechanical anchor. Tbf = 0.11MN/25 000 lb.	0.02 MPa 2.6 psi	0.02 MPa 2.6 psi	0.02 MPa 2.6 psi	0.02 MPa 2.6 psi
I - LIGHT ROCKBOLTS - 19mm/¾" ø at 2.0m/79in. Mechanical anchor. Tbf = 0.18MN/40 000 lb.	0.045 MPa 6.5 psi	0.045 MPa 6.5 psi	0.045 MPa 6.5 psi	0.045 MPa 6.5 psi
J - MEDIUM ROCKBOLTS - 25mm/1" ø at 1.5m/59in centres. Mechanical anchor. Tbf = 0.267MN/60 000 lb.	0.12 MPa 17 psi	0.12 MPa 17 psi	0.12 MPa 17 psi	0.12 MPa 17 psi
K - HEAVY ROCKBOLTS - 34mm/1⅜" at 1m/39in centres. Resin anchored. Tbf = 345 MN/ 150 000 lb.	0.34 MPa 49 psi	0.34 MPa 49 psi	0.34 MPa 49 psi	0.34 MPa 49 psi

Examples of rock-support interaction analysis

Example 1

An 8m (26 ft) diameter mine haulage tunnel is excavated in very good quality quartzite at a depth of 1000m (3280 ft) below surface. Due to the proximity of mining operations, the stresses acting on this tunnel are increased locally by varying amounts. It is required to investigate the stability of the tunnel and to consider the support measures which should be used in different stress environments.

For the purposes of this analysis it is assumed that the shape of the tunnel can be approximated by a circular cross-section and that the vertical and horizontal stresses acting on the rock mass surrounding the tunnel are equal. The input data assumed for the determination of the required support lines are as follows:

Uniaxial compressive strength of
    intact rock σ_c = 300 MPa   (43,500 lb/in²)
Material constants for original rock mass m = 7.5, s = 0.1   (see table 12 on page 176)
Modulus of elasticity of rock mass E = 40,000 MPa   (5.8 x 10⁶ lb/in²)
Poisson's ratio of rock mass ν = 0.2
Material constants for broken rock mass m_r = 0.3, s_r = 0.001 (see table 12 on page 176)
Unit weight of broken rock γ_r = 0.02 MN/m³   (0.074 lb/in³)
In situ stress magnitudes
     A - p_o = 27 MPa   (3915 lb/in²)
     B - p_o = 54 MPa   (7830 lb/in²)
     C - p_o = 81 MPa   (11745 lb/in²)
     D - p_o = 108 MPa   (15660 lb/in²)
Tunnel radius r_i = 4m   (157.5 in)

Substitution of the values listed above into the equations given in table 14 on page 259 give the results plotted in figure 131. Note that, in order to compare the required support lines on the same graph, a dimensionless plot of tunnel deformation against support pressure has been used.

From equation 106 on page 251, the critical support pressure below which a zone of broken rock is formed around the tunnel is found to be

$$p_{icr}/p_o = (1 - M\sigma_c/p_o)$$

This relationship is plotted as a broken line in figure 131. For values of p_i/p_o greater than the critical support pressure, the behaviour of the rock surrounding the tunnel is elastic.

The results used in plotting figure 131 are summarised below:

Line	In situ stress p_o-MPa	Critical support pressure p_{icr}/p_o	Deformation u_i/r_i%	Remarks
A	27	<0	0.09	Elastic behaviour
B	54	0.0184	0.22	Minor failure
C	81	0.0741	0.98	Moderate failure
D	108	0.1139	4.40	Substantial failure

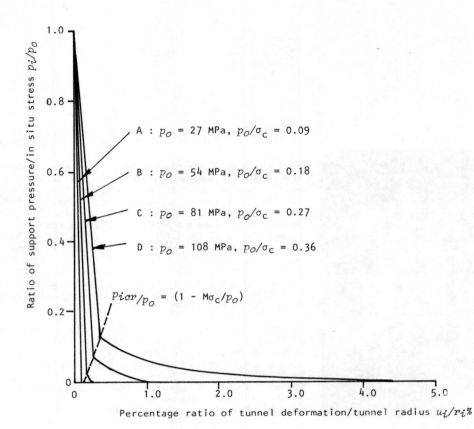

Figure 131 : Required support lines for a very good quality quartzitic rock mass surrounding an 8m (26 ft) diameter tunnel subjected to different in situ stresses.

In the case of the tunnel subjected to an in situ stress of p_o = 27 MPa (line A), the deformation of the surrounding rock mass is elastic and no support is required.

For an in situ stress level of p_o = 54 MPa, the critical support pressure is about 1 MPa (145 lb/in²) and hence, without support, some fracturing will occur around the tunnel. Figure 131 shows that the tunnel deformation is u_i/r_i = 0.22% (u_i = 8.8mm = 0.35 in). The authors consider it likely that the relatively minor amount of spalling which would occur under these circumstances could be tolerated in a mining application and that support would not be necessary. If small rockfalls and ravelling proved to be a problem, wire mesh pinned to the tunnel walls with short grouted bolts would probably be sufficient to control the problem.

It should be noted that any attempt to prevent non-elastic deformation of the tunnel under in situ stress conditions of p_o = 54 MPa would be totally uneconomic. Table 17 on page 269 shows that a support pressure of 1 MPa for a tunnel of 4m radius would require a concrete lining at about 300mm thickness or the use of heavy steel sets. As pointed out above, the non-elastic deformation which occurs without support is probably unimportant in this case and, hence, any attempt to prevent this deformation would serve no useful purpose.

When the in situ stress magnitude reaches p_o = 81 MPa (line C), the critical support pressure p_{icr} = 6 MPa (870 lb.in²) and the deformation which occurs without support is u_i = 39mm (1.5 inches). Under these conditions the authors consider

Cracking of concrete used for support of a heavily stressed draw-point in a mine.

In spite of the severe deformation indicated by the buckled posts, spalling of this mine excavation has been controlled by the application of a thin shotcrete layer and stability is adequate for temporary access.

it probable that the spalling and fracturing problems would be severe enough that some form of support would be required.

An obvious solution to this support problem appears to be the provision of a concrete lining or heavy steel sets which would apply a support pressure of 1 or 2 MPa (see table 17 on page 269) to limit the deformation to say 20mm. Dr P.K.Kaiser* points out that this would be the incorrect solution since, in this case, support should be used to control spalling rather than prevent deformation. In spite of the fact that the deformation is non-elastic and that its magnitude is relatively large, the tunnel does reach a state of equilibrium without support. Hence, provided that the fractured rock can be kept in place to prevent progressive ravelling, there is little danger of the fracture zone propagating to the point where the tunnel would collapse.

A good solution to this problem is to use light support which is installed close to the advancing tunnel face and which accommodates the tunnel deformation while preventing ravelling and spalling. Light steel sets with relatively soft blocking would be an acceptable solution but an even more economical solution would be the installation of untensioned grouted reinforcing bars or of "split sets" (see discussion on this type of support in Chapter 9).

An example of the problem discussed above is illustrated in figures 132 and 133. In this case, heavy steel sets were used in an attempt to control fracturing around a mine tunnel and, as shown in figure 132, these sets proved to be too stiff and were unable to accommodate the deformation to which they were subjected. An alternative approach to the same problem was to install untensioned grouted reinforcing bars in holes drilled into the roof and sidewalls of the tunnel. These bars were installed close to the advancing face and were tensioned by the deformation of the rock mass as the support provided by the face was removed and the rock allowed to relax. Figure 133 shows that this system has worked very well and there is very little spalling in the roof and sidewalls. The lower sidewalls, which were not supported, have suffered from fairly severe spalling.

As pointed out on page 256, the support provided by grouted reinforcing elements cannot easily be quantified and their action is believed to be related to the "knitting together" of the rock mass rather than to the control of deformation. In spite of this lack of adequate theoretical background, practical experience has demonstrated that untensioned reinforcing elements can be very effective when used in the correct way. This subject will be discussed further in a later section of this chapter.

When the in situ stress level reaches p_o = 108 MPa, the deformation which occurs without support is u_i = 176 mm (7 in) and this case represents a more severe example of the problem discussed above. While it may be possible to control spalling and ravelling with grouted reinforcing bars with the addition of mesh, it may have to be accepted that it is not practical to maintain an 8m diameter tunnel under these stress conditions.

---

* Personal communication with the authors. Dr Kaiser is a member of the Department of Civil Engineering at the University of Alberta, Edmonton, Canada.

Figure 132 : Use of heavy steel sets in an attempt to control fracturing around a mine tunnel subjected to high stresses. The steel sets are too stiff for this application and cannot accommodate the deformation to which they are subjected.

Figure 133 : Grouted untensioned reinforcing bars installed close to the advancing tunnel face have been used to control spalling in the roof and upper sidewalls. Note that severe spalling has occurred in the unreinforced lower sidewalls. The projections from the reinforcing holes are wooden wedges used to keep the bars in place during the setting of the grout.

In plotting the required support lines for the massive quartzite considered in example 1 (figure 131), the influence of the dead weight of the broken rock surrounding the tunnel has been ignored. This is because, under the high stress levels considered in this example, the weight of the broken rock makes a very small contribution to the stress levels around the tunnel. This can be checked by calculating a few points on the required support lines for the roof, sidewalls and floor of the tunnel by means of steps l, m and n listed in table 14 on page 259. It will be found that these points are practically coincident and hence, only one line has been plotted for each stress level in figure 131.

Example 2

A 35 ft (10.7m) diameter highway tunnel is driven in fair quality gneiss at a depth of 400 ft (122m) below surface. The following data are required to calculate the required support lines for the rock mass surrounding the tunnel :

Uniaxial compressive strength of rock	σ_c = 10,000 lb/in^2	(69 MPa)
Material constants for original rock mass	m = 0.5	
	s = 0.001	
Modulus of elasticity of rock mass	E = 2 x 10^5 lb/in^2	(1380 MPa)
Poisson's ratio of rock mass	ν = 0.2	
Material constants for broken rock	m_r = 0.1	
	s_r = 0	
Unit weight of broken rock	γ_r = 0.074 lb/in^3	(0.02 MN/m^3)
In situ stress magnitude	p_o = 480 lb/in^2	(3.31 MPa)
Tunnel radius	r_i = 210 in	(5.33m)

The required support lines for the roof, sidewalls and floor of the tunnel are plotted in figure 134. In this example the dead weight of the broken rock surrounding the tunnel plays an important role in determining the stability of the tunnel. It will be seen in figure 134 that the sidewalls stabilise at a deformation of about 5 inches but that the roof deformation continues to increase in an unstable manner for support pressures of less than about 12 lb/in^2.

A traditional approach to the choice of a support system for this tunnel would be to use steel sets designed on the basis of Terzaghi's Rock Load Factor (see table 1 on page 17). Assuming that the rock mass surrounding the tunnel can be described as "moderately blocky and seamy", table 1 shows that the rock load to be supported by the steel sets ranges from about 8 lb/in^2 (0.055 MPa) to about 25 lb/in^2 (0.172 MPa). This range compares well with that given by the analysis presented in figure 134 and, from table 17 on page 269, medium weight steel sets (category F) are chosen as the most appropriate support system.

The input data required for an analysis of the support available from these steel sets are :

Flange width of steel set	W = 4.16 in	(0.1059 m)
Depth of section of steel set	X = 8 in	(0.2023 m)
Cross-sectional area of steel	A_s = 6.71 in^2	(0.0043 m^2)
Moment of inertia of steel section	I_s = 64 in^4	(2.67 x 10^{-5} m^4)
Young's modulus of steel	E_s = 30 x 10^6 lb/in^2	(207000 MPa)
Yield strength of steel	σ_{ys} = 36000 lb/in^2	(245 MPa)
Tunnel radius	r_i = 210 in	(5.33 m)
Set spacing	S = 60 in	(1.52 m)
Half angle between blocking points	θ = 11.25°	

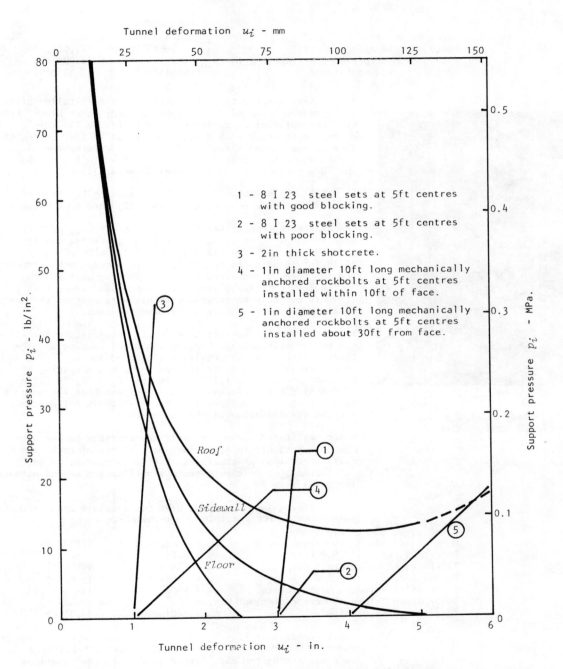

Figure 134 : Rock-support interaction analysis of a 35ft diameter tunnel in fair quality gneiss at a depth of 400 feet below surface.

Block thickness	t_B = 10in	(0.25 m)
Modulus of blocking material	E_B = 1.5 × 10⁶ lb/in²	(10,000 MPa)
Deformation before support installation	u_{io} = 3in	(0.075 m)
In situ stress magnitude	p_o = 480 lb/in²	(3.31 MPa)

Inadequately blocked steel sets in a large hard rock tunnel.

It has been assumed that the steel sets are installed after a deformation u_{io} of 3 inches has already taken place. From the load-deformation curves plotted in figure 134 this can be estimated to occur at a distance of about 25 feet behind the face of the tunnel since the full tunnel deformation of 5 inches would be expected to occur about 40 to 50 feet behind the face (between 1 and 1½ tunnel diameters).

It has also been assumed that stiff blocking is correctly installed at an even spacing of $2\theta = 22\frac{1}{2}°$ around the tunnel.

The available support for these 8 I 23 steel sets at 5 ft spacing is given as curve number 1 in figure 134 and it will be seen that the maximum support capacity of about 25 lb/in² is adequate to stabilise the broken rock surrounding the tunnel. However, since the spacing of the sets (5 feet) is relatively large for "blocky and seamy rock", it would probably be necessary to use wire mesh or shotcrete to support small rock pieces between the sets.

The importance of correct blocking in steel set installations can be demonstrated by changing the block spacing and the block stiffness in the analysis presented above. Curve 2 in figure 134 has been calculated for a block spacing of $2\theta = 40°$ and an elasticity modulus of E_B = 72,000 lb/in² (500 MPa) for the blocks. It will be seen that the support capacity has now dropped to a level which is not adequate for the stabilisation of the tunnel roof.

Since, as stated above, it would be necessary to use wire mesh or shotcrete to stabilise small pieces of rock, the obvious question which arises is - can shotcrete or mesh-reinforced shotcrete be used to stabilise the tunnel without the steel sets ?

The following input data are used to calculate the available support curve for a 2 inch thick shotcrete layer :

Modulus of elasticity of shotcrete	E_c = 3 × 10⁶ lb/in²	(20700 MPa)
Poisson's ratio of shotcrete	ν_c = 0.25	
Thickness of shotcrete layer	t_c = 2in	(0.05 m)
Tunnel radius	r_i = 210in	(5.33 m)
Compressive strength of shotcrete	σ_{cc} = 5000 lb/in²	(34.5 MPa)
Deformation before shotcrete placing	u_{io} = 1in	(0.025 m)
In situ stress magnitude	p_o = 480 lb/in²	(3.31 MPa)

Curve number 3 in figure 134 represents the available support characteristics of a 2 inch stotcrete layer, calculated by substituting the data listed above into the equations summarised in table 14 on pages 260 and 261. This curve shows that the shotcrete layer has adequate strength and stiffness to stabilise the broken rock surrounding the tunnel.

Many engineers find it difficult to believe that a thin layer of shotcrete can provide effective tunnel support and these engineers would generally prefer to use steel sets rather than shotcrete. However, the rising price of steel and of the labour required to manufacture and place steel sets will force these engineers to examine alternative support systems. Those who have had experience of using shotcrete will have little doubt that this is a viable alternative which will find more and more applications in underground excavation support.

The brittle behaviour of concrete is one of the problems associated with the use of shotcrete or concrete lining in tunnels. This problem is particularly troublesome in mining operations in which mining activities induce a constantly changing stress field in the rock surrounding underground excavations. In excavations such as stope access tunnels and draw-points, the use of shotcrete or concrete should be avoided because these excavations suffer from large stress changes during mining.

In order to overcome this problem, the use of wire mesh reinforcement is common and an increasing amount of attention is being given to steel fibre or glass fibre reinforcement of shotcrete. This subject will be dealt with in greater detail in Chapter 9.

A further problem with the use of shotcrete is associated with the irregular excavation profile which is usually achieved in a drill and blast tunnelling operation. Due to the jointed nature of the rock mass and to careless blasting practices, substantial overbreak is common in hard rock tunnelling and, while this situation can be improved by the use of correct blasting techniques, it is not possible to avoid it completely. A study of the derivation of the support reaction equations for concrete lining will soon convince the reader that the action of the lining depends upon a uniform stress distribution in a continuous ring in intimate contact with the rock mass. In the case of a thick concrete lining, the variation of lining thickness due to variations in the tunnel profile does not give rise to serious problems. However, in the case of a thin shotcrete lining, the abrupt changes in tunnel profile can induce high stress concentrations in the shotcrete lining, causing cracking and a serious reduction in the load bearing capacity of the lining. In general, the authors do not recommend the use of shotcrete as the sole means of excavation support in situations in which the tunnel profile deviates by more than a few percent from the design profile.

Rockbolts are obvious candidates for consideration as the support system for this 35 foot diameter tunnel. Table 17 on page 269 suggests that a pattern of 1 inch diameter bolts spaced at about 5 feet centres should provide adequate support capacity for this application and the following data are used in the calculation of the available support curves for such a system.

Rockbolt length	$l = 120$ in	(3 m)
Rockbolt diameter	$d_b = 1$ in	(0.025 m)
Modulus of elasticity of bolt steel	$E_b = 30 \times 10^6$ lb/in^2	(207,000 MPa)
Anchor/head deformation constant	$Q = 2.5 \times 10^{-5}$ in/lb	(0.143 m/MN)
Ultimate strength of bolt system	$T_{bf} = 65,000$ lb	(0.285 MN)
Tunnel radius	$r_i = 210$ in	(5.33 m)
Circumferential bolt spacing	$s_c = 60$ in	(1.52 m)
Longitudinal bolt spacing	$s_l = 60$ in	(1.52 m)
Deformation before bolt installation	$u_{io} = 1$ in	(0.025 m)
In situ stress magnitude	$p_o = 480$ lb/in^2	(3.31 MPa)

Curve number 4 in figure 134 represents the available support for this rockbolt system and it shows that adequate support is provided for the tunnel.

Curve number 5 has been calculated using identical input data to those listed on the previous page with the exception of the value of u_{io}, the deformation assumed to occur before the installation of the rockbolts. If the bolt installation is delayed until 4 inches of tunnel deformation have occurred, the support reaction curve of the rockbolt system will not intersect the characteristic curve for the rock mass in the roof until it is too late. Under these circumstances, the rock mass will continue to ravel around the bolts and the bolts will eventually end up sticking out of a pile of rock on the floor of the tunnel!

This hypothetical example demonstrates the extreme importance of timing when using rockbolts for underground excavation support. Because of the flexibility of the rockbolt system, late installation can allow excessive deformation to occur before the bolt reaction is fully mobilised. As a general rule, rockbolts should be installed as early as possible and, ideally, the rockbolt installation procedure should be integrated into the drill-blast-muck cycle. This subject will be examined in greater detail later in this chapter and in Chapter 9 when some of the practical aspects of underground support design are discussed.

Curve number 4 illustrates a further important aspect of rockbolting as a support system for underground excavations. Obviously, a higher support pressure is required to stabilise the broken rock in the roof of the tunnel than that in the sidewalls. This variation in support pressure can be achieved by varying the spacing between individual bolts and, in a typical application, the spacing could be increased from 1.5 x 1.5m in the roof to 2 x 2m in the sidewalls. This reduction in bolt density in the sidewalls can result in significant reductions in cost, due mainly to the reduced amount of drilling required. Alternatively, if it is necessary to maintain a constant pattern, smaller diameter and shorter bolts can be used in the sidewalls to reduce the overall cost of the bolting system.

Example 3

A large cavern is to be excavated to house a number of turbines for power generation in a hydro-electric project. The cavern is 250m (820 ft) below surface in a good quality quartzite. The span of the excavation is to be 25m (82 ft).

In situ stress measurements have been carried out in an exploration tunnel and finite element studies have been performed to investigate the stress distribution around the proposed excavation. These studies have shown that the stresses in the rock above the roof of the cavern are very similar to those which would be induced in the rock surrounding a circular tunnel of 25m diameter subjected to equal horizontal and vertical stresses of p_o = 10MPa (1450 lb/in^2). On the basis of these studies it has been decided to use the rock-support interaction analysis, presented earlier in this chapter, to study the characteristics of possible support systems.

The required support line for the rock mass above the cavern roof is calculated by means of the equations listed in table

14 on page 259, using the following input data :

Uniaxial compressive strength of rock	σ_c = 200 MPa	(29,000 lb/in^2)
Material constants for original rock mass	m = 1.5 s = 0.004	
Modulus of elasticity of rock mass	E = 12,000 MPa	(1.77 x 10^6 lb/in^2)
Poisson's ratio of rock mass	ν = 0.2	
Material constants for broken rock mass	m_r = 0.1 s_r = 0	
Unit weight of broken rock mass	γ_r = 0.02 MN/m^3	(0.074 lb/in^3)
Equivalent in situ stress	p_o = 10 MPa	(1450 lb/in^2)
Equivalent tunnel radius	r_i = 12.5m	(492 in)

The resulting required support line is marked A in figure 135.

The proposed method of excavation involves mining a pilot tunnel along the axis of the cavern at roof level, as shown in the margin sketch. The cavern arch will be opened out from this central pilot tunnel and, once the roof has been opened to full span, the remainder of the cavern will be excavated by benching down from this top heading. As the pilot tunnel is mined and as each successive stage of the final roof profile is excavated, rockbolts will be installed to provide both temporary and permanent support for the rock.

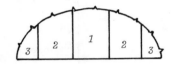

Excavation sequence for cavern roof

In order to investigate the adequacy of rockbolting as a means of supporting this cavern roof, the available support line for a 2 x 2m pattern of 5m long 25mm diameter mechanically anchored bolts is calculated from the data listed below.

Length of rockbolts	l = 5 m	(197 in)
Diameter of rockbolts	d_b = 0.025 m	(1 in)
Modulus of elasticity of bolt steel	E_b = 207000 MPa	(30 x 10^6 lb/in^2)
Anchor/head deformation constant	Q = 0.143 m/MN	(2.5 x 10^{-5} in/lb)
Ultimate strength of bolt system	T_{bf}= 0.285 MN	(65,000 lb)
Equivalent tunnel radius	r_i = 12.5 m	(492 in)
Circumferential bolt spacing	s_c = 2 m	(78.7 in)
Longitudinal bolt spacing	s_l = 2 m	(78.7 in)
Deformation before bolt installation	u_{io}= 0.005 m	(0.2 in)
Equivalent in situ stress	p_o = 10 MPa	(1450 lb/in^2)

The resulting available support line is marked 1 in figure 135 and it will be seen that this curve does not intersect the characteristic curve A for the rock mass in the cavern roof. Consequently, the proposed bolt pattern will not provide adequate support for the cavern roof.

An obvious solution to this lack of support capacity is to reduce the bolt spacing. Curve number 2 is the support reaction curve obtained by placing an additional bolt at the centre of each 2m x 2m grid, thereby reducing the effective bolt spacing to 1.41m x 1.41m. From equation 128 on page 256 it will be seen that the maximum support pressure p_{smax} provided by a bolt system installed in a square pattern is proportional to the square of the bolt spacing. Hence, a relatively small reduction in bolt spacing can be used to give a significant increase in maximum support pressure.

An alternative solution is to add a second support system, such as a layer of shotcrete, to the original 2 x 2m rockbolt pattern. This solution, provided that it gives an adequate increase in support capacity, would make a great deal of practical sense because it would probably be necessary to

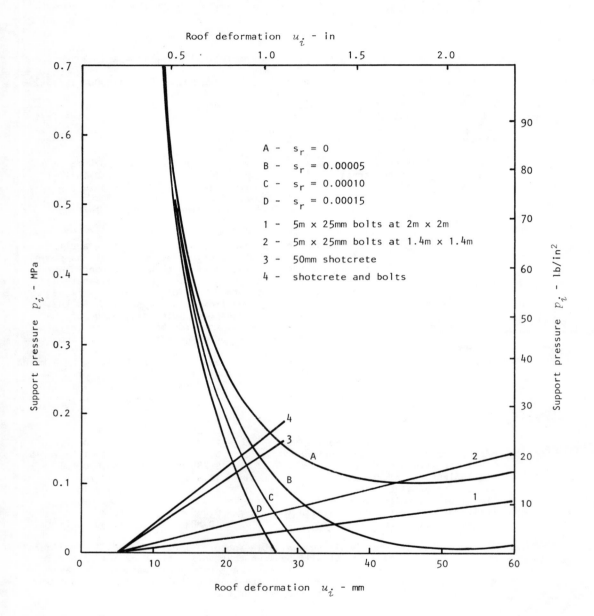

Figure 135 : Required and available support curves for a 25m span cavern roof arch in good quality quartzite.

stabilise the rock surface between the bolts in order to protect men and equipment in the cavern from small falling rocks. Consequently, a support system comprising rockbolts on a 2 x 2m grid with a layer of mesh reinforced shotcrete covering the rock surface may meet all the support requirements.

The input data needed to calculate the available support for a shotcrete layer are listed below :

Modulus of elasticity of shotcrete	E_c = 20700 MPa	(3×10^6 lb/in^2)
Poisson's ratio of shotcrete	ν = 0.25	
Thickness of shotcrete layer	t_c = 0.05 m	(2 in)
Tunnel radius	r_i = 12.5 m	(492 in)
Compressive strength of shotcrete	σ_{cc} = 40 MPa	(5800 lb/in^2)
Deformation before shotcrete placing	u_{io} = 0.005 m	(0.2 in)
Equivalent in situ stress	p_o = 10 MPa	(1450 lb/in^2)

The available support line for this shotcrete layer is marked 3 in figure 135 and it will be noted that the support capacity of this layer is only just equal to that required to stabilise the rock in the cavern roof. However, taking the combined support reaction of the 2 x 2m rockbolt pattern and the 50mm shotcrete layer (using the equations listed in table 14 on page 261) gives the available support line marked 4 in figure 135. Considering that, at this stage, the rockbolts have not been grouted, this combined support reaction is believed to be adequate to provide support during excavation of the cavern roof.

Since the rockbolts are intended to provide permanent support of the cavern roof as well as the support during excavation, it will be necessary to provide some form of long term corrosion protection for the bolts. The simplest and cheapest solution to this problem would normally be to inject grout into the annulus surrounding the bolt. The question which then arises is - what does grouting the bolt do to its support reaction curve ?

This question has already been discussed on pages 256 and 258 where it was pointed out that a direct calculation of the support reaction of grouted rockbolts cannot be made on the basis of the rock-support interaction analysis presented in this chapter. However, a very approximate analysis can be made by assuming that the effect of grouting is to knit the broken rock mass together and to increase its apparent "tensile" strength. This analysis can be carried out by increasing the values of m_r and/or s_r in the equations listed in table 14 on page 259.

In order to avoid confusion, only one of these material constants has been varied in the analysis presented in figure 135. The curves marked A, B, C and D have been calculated using the following values of the constant s_r: 0, 0.00005, 0.0001 and 0.00015. These relatively modest increases (compare the values with those listed in table 12 on page 176), result in a remarkable improvement in the stability of the broken rock in the roof of the cavern. The authors are uncertain of how realistic this analysis is but it does tend to confirm practical observations which suggest that grouted rockbolts perform very well in providing support in underground excavations.

A further question which arises from the discussion presented above is - if the bolts are to be grouted, is it necessary to tension these bolts before grouting ?

The answer to this question is directly related to the timing of both bolt installation and grouting. If the bolts are installed very close to the face of a tunnel or during each stage of a multi-stage excavation sequence, the deformation of the rock mass before installation of the bolts will be minimal. Subsequent excavation will induce further deformation which will mobilise the support reaction characteristics of the bolt. If the bolt is grouted immediately after installation then there is no need to pre-tension the bolt. In fact, there is no need to anchor the bolt and a dowel consisting of a length of reinforcing steel will do just as well.

Many large underground excavations in Scandinavia are supported by means of "Perfobolts" which are perforated hollow tubes which are filled with grout and inserted into drilled holes; the grout is then extruded when reinforcing rod is pushed down the centre of the grout-filled tube. A number of mines have simplified this system even further by pumping a thick grout into boreholes and then pushing the dowels into this grout before it sets. These systems will only work if the dowels are installed and grouted very soon after the rock to be supported has been exposed and before significant deformation has taken place.

When the grouting has to be delayed for any reason or if there is a danger that a contractor will fail to adhere to a very rigid bolt installation and grouting schedule, the authors recommend that bolts be anchored and tensioned . This removes the critical time constraint and it means that the bolts will provide useful support until the grouting can be carried out.

In the case of the 25m span cavern under consideration in this example, a suggested support system and schedule for the rock above the roof is as follows :

1. Excavate the pilot tunnel and install 5m mechanically anchored bolts on a 2m x 2m grid. Note that the pilot tunnel should be about 6m wide and 6 to 8m high to permit efficient drilling with a multi-boom jumbo and to provide enough room for the installation of 5m long bolts.

2. Tension the bolts to about 15 tonnes (half the bolt capacity), preferably with a hydraulic bolt tensioner. A gradual drop in bolt load will occur as the anchor works in the rock under the influence of adjacent blasting and general movement in the rock. When the bolts are finally tensioned before grouting, the measured load may be as low as 10 tonnes.

3. Excavate the remainder of the cavern roof stage by stage, installing and tensioning 5m long bolts on a 2m x 2m grid at each stage.

4. On completion of the cavern roof excavation and while good access is still available, ie before benching down commences, retension all the bolts using a hydraulic bolt tensioner. Inject grout into all the bolt holes, ensuring that grouting takes place within 24 hours of retensioning to avoid loss of load.

5. Install weldmesh over the entire cavern roof, attaching it

to the grouted rockbolts by means of a second washer and nut and fixing it at intermediate points with short grouted mesh fixing pins. Details of bolting, grouting, mesh and mesh attachment will be discussed in the next chapter.

6. Apply a 50mm layer of shotcrete, ensuring that it penetrates the mesh and adheres to the rock and that the mesh is completely covered. Thicker shotcrete layers may be required in areas where the excavation profile is irregular. It is important that the weld-mesh be completely covered in order to avoid later problems with corrosion of the mesh.

There are many alternative methods of carrying out the job described above. These will vary with the excavation schedule and with the equipment being used by the contractor. Provided that the basic principles outlined on the preceding pages are used, the exact method of execution will not have a significant influence on the final result.

Example 4

Cycloned mill tailings play an important role in the mining of massive base metal orebodies. The coarse sand fraction from these tailings is placed hydraulically in stopes and, when drained, this sand provides support for the surrounding rock. The mechanics of the support process are generally poorly understood and it is interesting to consider whether the rock-support interaction analysis presented in this chapter can be used to investigate this problem.

Suppose that a near vertical orebody stope is approximated by a vertical shaft which is subjected to equal stresses in a horizontal plane of p_o = 3000 lb/in^2 (20.7 MPa). The equivalent radius of the excavation is assumed to be r_i = 20 feet (6.1 m). The following input data are used in the calculation of the required support line for the rock mass surrounding the stope :

Uniaxial compressive strength of rock	σ_c = 10,000 lb/in^2	(69 MPa)
Material constants for original rock mass	m = 1.5 s = 0.004	(see table 12 on page 176)
Modulus of elasticity of rock mass	E = 4 x 10^6 lb/in^2	(27,600 MPa)
Poisson's ratio of rock mass	ν = 0.2	
Material constants for broken rock mass (see table 12 on page 176)	m_r = 0.30 (Fair) m_r = 0.08 (Poor) m_r = 0.015 (Very poor) s_r = 0	
Unit weight of broken rock mass	γ_r = 0.074 lb/in^3	(0.02 MN/m^3)
In situ stress magnitude	p_o = 3000 lb/in^2	(20.7 MPa)
Equivalent shaft radius	r_i = 240 in	(6.1 m)

Note that the material constant m_r has been varied to give the approximate required support lines for fair, poor and very poor rock surrounding the excavation. The condition of the broken rock surrounding the stope will depend not only upon the inherent quality of the original rock mass but also upon the excavation method used and the amount of damage inflicted on the rock by the stope blasting. Methods such as cut-and-fill mining or vertical crater retreat stoping will cause less damage to the rock than open stoping techniques involving massive blasts.

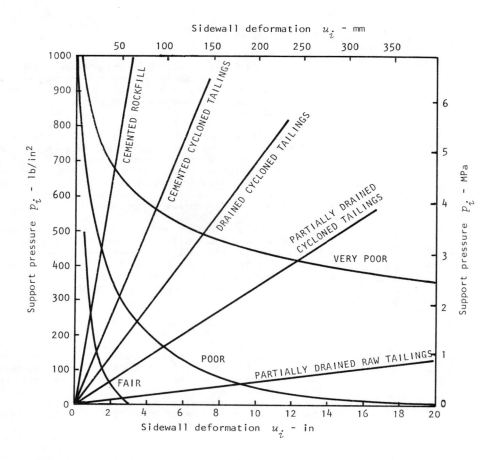

Figure 136 : Approximate analysis of support characteristics of fill in base metal mining.

In order to investigate the available support characteristics of different types of fill, the equations for concrete lining (table 14 on page 260) are used with $t_c = r_i$. Values for the fill moduli have been estimated from published papers dealing with mine filling[234-237]. The following input data have been used to calculate the available support lines plotted in figure 136.

Modulus of elasticity of fill :
    Partially drained raw tailings E_c= 1000 lb/in²   (6.9 MPa)
    Partially drained cycloned tailings E_c= 5000 lb/in²   (34.5 MPa)
    Drained cycloned tailings E_c= 10,000 lb/in²   (69 MPa)
    Cemented cycloned tailings E_c= 20,000 lb/in²   (138 MPa)
    Cemented waste rock fill E_c= 50,000 lb/in²   (345 MPa)
Poisson's ratio of fill $\nu = 0.25$
Thickness of fill t_c = 240 in   (6.1 m)
Equivalent shaft radius r_i = 240 in   (6.1 m)
Uniaxial strength of fill σ_{cc} (see note below)
Deformation before fill placing u_{io} = 0
Equivalent in situ stress p_o = 3000 lb/in²   (20.7 MPa)

Since the fill is almost completely confined by the stope walls, failure of the fill does not result in a significant loss of stiffness. Consequently, the compressive strength of the fill is arbitrarily chosen as some high value for the calculation of the available support curves.

The initial deformation before filling has been assumed at $u_{io} = 0$ to correspond to immediate filling of the stope. In order to construct the available support line for a particular fill for a given value of initial deformation, the plotted curve is shifted parallel to itself to pass through the specified vaule of u_{io}.

The following general conclusions can be drawn from the results presented in figure 136 :

When the rock mass is of inherently good quality and when sufficient care is taken to minimise damage to the stope walls by careful blasting, the stope will probably stabilise without the need for internal support. This situation is illustrated by the rock mass characteristic curve marked "fair" in figure 136.

As the rock mass quality deteriorates and/or as the damage to the rock surrounding the stope increases, the need for internal support becomes greater. This is illustrated by the deformation of more than 20 inches which occurs before the rock mass designated "poor" reaches stability. In a typical cut-and-fill operation in which the fill is placed close to the excavated face, the available support provided by cycloned tailings should be enough to stabilise the stope. If filling is delayed as would be the case in a post-filled open stoping operation, it may be necessary to use cemented cycloned tailings in order to achieve the required degree of support.

In poor rock masses or when the damage to the stope walls is very severe, the required support line marked "very poor" shows that unusually high support pressures would be required in order to achieve stability. Under such circumstances, the use of cemented cycloned tailings or of cemented waste rock fill may be required. In extreme cases, it may be necessary to introduce the rockfill into the top of the stope at the same time as the ore is drawn from the bottom to ensure that the stope walls are not left unsupported at any time during the mining operation.

Obviously, this analysis is extremely crude but it does illustrate many of the facts related to the use of fill in stope support. The authors do not believe that any useful purpose would be served by attempting to carry out a more refined analysis than that presented above and they emphasise, once again, that the purpose of this example is to illustrate some of the basic principles rather than to obtain quantitative information.

Discussion on rock-support interaction analysis

The simplified analysis and examples presented on the preceding pages are intended to provide the reader with a basic understanding of the mechanics of rock support. As stated earlier in this chapter, a number of assumptions (listed on page 249) have been made in deriving the equations used in the rock-support interaction analysis and these assumptions impose some serious limitations upon the accuracy and applicability of this analysis. The most serious limitation is that it is only strictly correct for a circular tunnel subjected to equal horizontal and vertical in situ stresses.

In spite of these limitations, the authors have found this analysis to be of great value in developing their own understanding of the mechanics of rock support and in teaching others some of the basic concepts of this relatively poorly understood field. It is hoped that the reader will find the discussion presented earlier in this chapter to be of value in the same way.

It will be clear to the more mathematically inclined reader that some of the concepts used in developing the rock-support interaction analysis are amenable to considerable refinement. As our knowledge of rock mass behaviour improves and as our ability to realistically model this behaviour by computer techniques increases, it is probable that techniques for analysing rock-support interaction will advance far beyond those currently available. Numerical studies of tunnel support using finite element techniques have been published by a number of authors[238-240] and Cundall and his colleauges at the University of Minnesota[241,242] have discussed the use of "distinct element" numerical models in which individual blocks of rock can be incorporated in rock-support interaction studies. The authors have no doubt that these numerical studies are forerunners of a vast array of numerical techniques which will be developed in the future.

The authors have no objection to these developments provided that both the numerical analysts and the users of these methods bear in mind the real world in which elegant theoretical models can be rendered invalid by the presence of a clay filled fault, the careless design of a blast resulting in a poor excavation profile, or the failure of a grout pump during a concrete lining operation resulting in voids behind the lining. The essential ingredient of any successful rock support programme is the ability of the underground excavation engineer to adapt to the actual conditions encountered in the field. It is important that this engineer should start out with a clear understanding of what he or she is trying to achieve and a sound idea, based upon theoretical studies or on precedent experience, of what support options are available. It is even more important that this engineer should be able to adapt these options to the constantly changing conditions encountered underground. The rigid adherence to a single support design, however elegant, does not produce an effective or economical underground excavation.

Use of rock mass classifications for estimating support

An alternative to the theoretical approach to rock support is to use precedent experience as a basis for estimating the support requirements for underground excavations. This approach tended to develop in a rather haphazard manner until the advent of rock mass classification systems provided a rational framework for relating previously isolated pieces of practical experience.

A full discussion of the development of rock support design on the basis of precedent experience is beyond the scope of this chapter and the interested reader is referred to chapter 2 and to the numerous references cited in that chapter. In the following pages, the discussion will be concentrated upon the use of the rock mass classification systems developed by Barton et al of the Norwegian Geotechnical Institute[1] and by Bieniawski of the South African Council for Scientific and Industrial Research[25,26].

Maximum unsupported excavation spans

Barton[29] has compiled a collection of about 30 cases of permanently *unsupported* excavations in a variety of rock masses. The spans of these excavations have been plotted against the rock mass quality (Q) in figure 137 and the authors consider that this plot provides an excellent basis for estimating the maximum unsupported span which could be excavated in a rock mass of known quality.

As pointed out by Barton, there is no way of knowing how close these excavations are to failure and hence it could be argued that figure 137 will always provide a conservative estimate of unsupported excavation span. While this is certainly true, the reader would do well to remember that it has taken very brave men to mine some of these excavations and that, before deciding upon a less conservative design it is worth spending a few moments contemplating the consequences of being too daring.

In figure 138, the maximum unsupported spans for different Excavation Support Ratios (ESR - see page 30) have been plotted against the rock mass quality Q. The equation which defines the lines plotted in figure 138 is :

$$\text{Span} = 2 \cdot \text{ESR} \cdot Q^{0.4} \qquad (131)$$

Alternatively, the critical value of Q for a given excavation span can be found by rearranging equation 131 as follows :

$$Q = (\text{Span}/2 \cdot \text{ESR})^{2.5} \qquad (132)$$

Barton suggests that, during mapping of a tunnel route or the evaluation of an underground excavation site from drill core, a knowledge of the critical value of Q (from equation 132) will enable the geologist or engineer to identify those sections which can probably be left unsupported and those which will require detailed consideration in terms of their support requirements.

A comparison between Scandinavian, South African and Austrian estimates of maximum unsupported span for different rock mass qualities has been made by Bieniawski[26] and a part of one of his graphs is reproduced in figure 139. Note that this graph has been plotted in terms of "stand up" time (see discussion on page 27) and, consequently, is not directly comparable with Barton's results presented in figure 137. However, figure 139 does show that the maximum unsupported span estimated from figure 138 will usually be less conservative than that estimated from Bieniawski's[25,26] or Lauffer's[8] classifications.

A final word of warning on estimating the maximum unsupported span of an excavation - none of the classification systems adequately covers the situation in which a few isolated structural features intersect to release a block or wedge from the roof or sidewalls of an excavation. Consequently, when excavating a tunnel or a cavern in which no *systematic* support is considered necessary on the basis of figure 138, the need for isolated support to stabilise blocks or wedges must always be kept in mind. The techniques discussed on pages 185 to 194 are applicable in these situations.

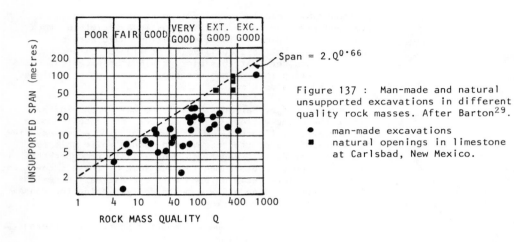

Figure 137 : Man-made and natural unsupported excavations in different quality rock masses. After Barton[29].
- man-made excavations
- natural openings in limestone at Carlsbad, New Mexico.

Span = $2.Q^{0.66}$

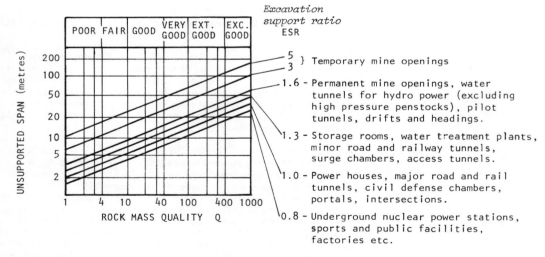

Excavation support ratio ESR

5
3 } Temporary mine openings

1.6 - Permanent mine openings, water tunnels for hydro power (excluding high pressure penstocks), pilot tunnels, drifts and headings.

1.3 - Storage rooms, water treatment plants, minor road and railway tunnels, surge chambers, access tunnels.

1.0 - Power houses, major road and rail tunnels, civil defense chambers, portals, intersections.

0.8 - Underground nuclear power stations, sports and public facilities, factories etc.

Figure 138 : Recommended maximum unsupported excavation spans for different quality rock masses. After Barton[29].

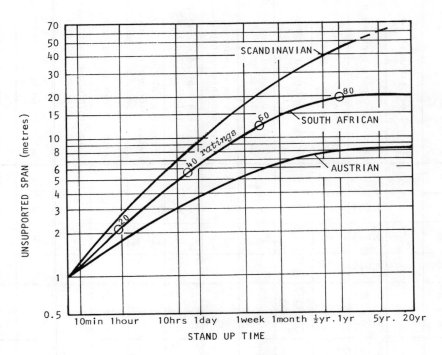

Figure 139 : Comparison between stand up times for unsupported excavation spans predicted by Scandinavian, South African and Austrian rock mass classification systems. Ratings are for CSIR Geomechanics Classification. After Bieniawski[26].

Norwegian Geotechnical Institute support predictions

When the span of an underground excavation exceeds that predicted by equation 132 or figure 138, it becomes necessary to install some form of support in order to maintain the rock mass surrounding the excavation in an acceptably stable condition. Barton, Lien and Lunde[1], in their original paper on rock mass classification, proposed 38 categories of support, depending upon the Tunnelling Quality Index Q and upon the Excavation Support Ratio ESR. The interested reader is urged to consult this original paper for details of these categories and for a very comprehensive discussion on the derivation of the support recommendations proposed.

Since there is little point in simply repeating the details published by Barton, Lien and Lunde, the authors have taken it upon themselves to present the NGI support recommendations in a different format. Care has been taken to adhere to the original recommendations as accurately as possible but it is hoped that this presentation will assist the reader in the practical application of these recommendations. The authors have added their own comments to those offered by Barton, Lien and Lunde when they consider that some point needs additional clarification or when it is felt that the Scandinavian bias of the original data should be kept in mind when applying these support recommendations to situations in which different conditions apply.

Rock mass quality Q	Equivalent dimension $\frac{\text{Span}}{\text{ESR}}$	Block size $\frac{RQD}{J_n}$	Inter-block strength $\frac{J_r}{J_a}$	Approx. support pressure p MPa	Spot reinforcement with untensioned grouted dowels	Untensioned grouted dowels on grid spacing indicated	Tensioned rockbolts on grid spacing indicated	Chainlink mesh anchored to bolts and intermediate points	Shotcrete applied directly to rock, thickness indicated	Shotcrete reinforced with weldmesh, thickness indicated	Unreinforced cast concrete arch, thickness indicated	Steel reinforced cast concrete arch, thickness indicated	Notes by Barton, Lien and Lunde	Notes by Hoek and Brown
1000-400	20-100			<0.001	✓								1	a
400-100	12-88			0.005	✓								1	a
100-40	8.5-19	≥20		0.025	✓								1	a
100-40	8.5-19	<20		0.025		2.5-3m								
100-40	14-30	≥30		0.025		2-3m								
100-40	14-30	<30		0.025		1.5-2m		✓						b
100-40	23-72	≥30		0.025		2-3m								
100-40	23-72	<30		0.025		1.5-2m		✓						b
40-10	5-14	≥10	≥1.5	0.05	✓								2	
40-10	5-14	≥10	<1.5	0.05		1.5-2m							2	
40-10	5-14	<10	≥1.5	0.05		1.5-2m							2	
40-10	5-14	<10	<1.5	0.05		1.5-2m				20-30mm			2	
40-10	15-23	≥10		0.05			1.5-2m	✓					2,3	b
40-10	15-23	<10		0.05			1.5-2m		50-100mm				2,3	c
40-10	9-15			0.05		1.5-2m		✓					2,4	b
40-10	15-40	>10		0.05			1.5-2m	✓					2,3,5	b
40-10	15-40	≤10		0.05			1.5-2m		50-100mm				2,3,5	c
40-10	30-65	>15		0.05			1.5-2m	✓					2,6,7,13	b
40-10	30-65	≤15		0.05			1.5-2m		100-150mm				2,6,7,13	c
10-4	3.5-9	>30		0.10	✓								2	a
10-4	3.5-9	≥10 ≤30		0.10		1-1.5m							2	

TABLE 18 - RECOMMENDED SUPPORT BASED UPON NGI TUNNELLING QUALITY INDEX Q

Rock mass quality Q	Equivalent dimension Span/ESR	Block size RQD/J_n	Inter-block strength J_r/J_a	Approx. support pressure p MPa	Spot reinforcement with untensioned grouted dowels	Untensioned grouted dowels on grid spacing indicated	Tensioned rockbolts on grid spacing indicated	Chainlink mesh anchored to bolts and intermediate points	Shotcrete applied directly to rock, thickness indicated	Shotcrete reinforced with weldmesh, thickness indicated	Unreinforced cast concrete arch, thickness indicated	Steel reinforced cast concrete arch, thickness indicated	Notes by Barton, Lien and Lunde	Notes by Hoek and Brown
10-4	6-9	<10		0.10		1-1.5m			20-30mm				2	
10-4	<6	<10		0.10					20-30mm				2	
10-4	10-15	>5		0.10			1-1.5m	✓					2,4	a
10-4	7-10	>5		0.10		1-1.5m		✓					2	a
10-4	10-15	≤5		0.10		1-1.5m			20-30mm				2,4	
10-4	7-10	≤5		0.10		1-1.5m			20-30mm				2	
10-4	20-29			0.10			1-2m			100-150mm			2,3,5	c
10-4	12-20			0.10			1-1.5m			50-100mm			2,3	c
10-4	35-52			0.10			1-2m			200-250mm			2,6,7,13	c
10-4	24-35			0.10			1-2m			100-200mm			2,3,5,13	c
4-1	2.1-6.5	≥12.5	≤0.75	0.15		1m			20-30mm				2	
4-1	2.1-6.5	<12.5	≤0.75	0.15					20-30mm				2	
4-1	2.1-6.5		>0.75	0.15		1m							2	
4-1	4.5-11.5	>10	<30 >1	0.15		1m		✓					2	a
4-1	4.5-11.5	≤10	>1	0.15					25-75mm				2	
4-1	4.5-11.5	<30	≤1	0.15		1m			25-50mm				2	c
4-1	4.5-11.5	≥30		0.15		1m							2	
4-1	15-24			0.15			1-1.5m			100-150mm			2,3,5,8	c
4-1	8-15			0.15			1-1.5m			50-100mm			2	c
4-1	30-46			0.15			1-1.5m			150-300mm			2,6,7,13	c

TABLE 18 - RECOMMENDED SUPPORT BASED UPON NGI TUNNELLING QUALITY INDEX Q

TABLE 18 - RECOMMENDED SUPPORT BASED UPON NGI TUNNELLING QUALITY INDEX Q

Rock mass quality Q	Equivalent dimension $\frac{Span}{ESR}$	Block size $\frac{RQD}{J_n}$	Inter-block strength $\frac{J_r}{J_a}$	Approx. support pressure p MPa	Spot reinforcement with untensioned grouted dowels	Untensioned grouted dowels on grid spacing indicated	Tensioned rockbolts on grid spacing indicated	Chainlink mesh anchored to bolts and intermediate points	Shotcrete applied directly to rock, thickness indicated	Shotcrete reinforced with weld-mesh, thickness indicated	Unreinforced cast concrete arch, thickness indicated	Steel reinforced cast concrete arch, thickness indicated	Notes by Barton, Lien and Lunde	Notes by Hoek and Brown
4-1	18-30			0.15			1-1.5m			100-150mm			2,3,5	c
1-0.4	1.5-4.2	>10	>0.5	0.225		1m		✓					2	d
1-0.4	1.5-4.2	≤10	>0.5	0.225		1m				50mm			2	c
1-0.4	1.5-4.2		≤0.5	0.225		1m				50mm			2	c
1-0.4	3.2-7.5			0.225			1m			50-75mm			14,11,12	c
1-0.4	3.2-7.5			0.225		1m			25-50mm				2,10	
1-0.4	12-18			0.225			1m			75-100mm			2,10	c
1-0.4	6-12			0.225		1m				50-75mm			2,10	c
1-0.4	12-18			0.225			1m				200-400mm		14,11,12	e
1-0.4	6-12			0.225			1m			100-200mm			14,11,12	c
1-0.4	30-38			0.225			1m			300-400mm			2,5,6,10,13	c,f
1-0.4	20-30			0.225			1m			200-300mm			2,3,5,10,13	c
1-0.4	15-20			0.225			1m			150-200mm			1,3,10,13	c
1-0.4	15-38			0.225			1m					300mm-1m	5,9,10,12,13	
0.4-0.1	1-3.1	>5	>0.25	0.3		1m			20-30mm					
0.4-0.1	1-3.1	≤5	>0.25	0.3		1m				50mm				c
0.4-0.1	1-3.1		≤0.25	0.3			1m			50mm				c
0.4-0.1	2.2-6	≥5		0.3			1m			25-50mm			10	c
0.4-0.1	2.2-6	<5		0.3						50-75mm			10	c
0.4-0.1	2.2-6			0.3			1m			50-75mm			9,11,12	c
0.4-0.1	4-14.5	>4		0.3			1m			50-125mm			10	c

Rock mass quality Q	Equivalent dimension $\frac{Span}{ESR}$	Block size $\frac{RQD}{J_n}$	Inter-block strength $\frac{J_r}{J_a}$	Approx. support pressure p MPa	Spot reinforcement with untensioned grouted dowels	Untensioned grouted dowels on grid spacing indicated	Tensioned rockbolts on grid spacing indicated	Chainlink mesh anchored to bolts and intermediate points	Shotcrete applied directly to rock, thickness indicated	Shotcrete reinforced with weldmesh, thickness indicated	Unreinforced cast concrete arch, thickness indicated	Steel reinforced cast concrete arch, thickness indicated	Notes by Barton, Lien and Lunde	Notes by Hoek and Brown
0.4–0.1	4–14.5	≤4 ≥1.5		0.3						75–250mm			10	c
0.4–0.1	4–14.5	<1.5		0.3			1m				200–400mm		10,12	e
0.4–0.1	4–14.5			0.3			1m					300–500mm	9,11,12	
0.4–0.1	20–34			0.3			1m			400–600mm			3,5,10 12,13	f
0.4–0.1	11–20			0.3			1m			200–400mm			4,5,10 12,13	c
0.4–0.1	11–34			0.3			1m					400mm –1.2m	5,9,11 12,13	
0.1–0.01	1–3.9	≥2		0.6			1m			25–50mm			10	c
0.1–0.01	1–3.9	<2		0.6						50–100mm			10	c
0.1–0.01	1–3.9			0.6						75–150mm			9,11	c
0.1–0.01	2–11	≥2	≥0.25	0.6			1m			50–75mm			10	c
0.1–0.01	2–11		<0.25	0.6						150–250mm			10	c
0.1–0.01	2–11			0.6			1m				200–600mm		9,11,12	
0.1–0.01	15–28			0.6			1m			300–1m			3,10 12,13	c,f
0.1–0.01	15–28			0.6			1m					600mm –2m	3,9,11 12,13	
0.1–0.01	6.5–15			0.6			1m			200–750mm			4,10 12,13	c,f
0.1–0.01	6.5–15			0.6			1m					400mm –1.5m	3,9,11 12,13	
0.01–0.001	1–2			1.2						100–200mm			10	c
0.01–0.001	1–2			1.2			0.5–1m			100–200mm			9,11 12	c
0.01–0.001	1–6.5			1.2						200–600mm			10	c,f
0.01–0.001	1–6.5			1.2			0.5–1m			200–600mm			9,11 12	c,f
0.01–0.001	10–20			1.2								1m–3m	10,14	

TABLE 18 - RECOMMENDED SUPPORT BASED UPON NGI TUNNELLING QUALITY INDEX Q

TABLE 18 - RECOMMENDED SUPPORT BASED UPON NGI TUNNELLING QUALITY INDEX Q

Rock mass quality Q	Equivalent dimension $\frac{Span}{ESR}$	Block size $\frac{RQD}{J_n}$	Inter-block strength $\frac{J_r}{J_a}$	Approx. support pressure p MPa	Spot reinforcement with untensioned grouted dowels	Untensioned grouted dowels on grid spacing indicated	Tensioned rockbolts on grid spacing indicated	Chainlink mesh anchored to bolts and intermediate points	Shotcrete applied directly to rock, thickness indicated	Shotcrete reinforced with weldmesh, thickness indicated	Unreinforced cast concrete arch, thickness indicated	Steel reinforced cast concrete arch, thickness indicated	Notes by Barton, Lien and Lunde	Notes by Hoek and Brown
0.01-0.001	10-20			1.2			1m					1-3m	3,9,11 12,14	
0.01-0.001	4-10			1.2						700mm -2m			10,14	c,f
0.01-0.001	4-10			1.2			1m			700mm -2m			4,9,10 11,14	c,f

Supplementary notes by Barton, Lien and Lunde

1. The type of support used in extremely good and exceptionally good rock will depend upon the blasting technique. Smooth wall blasting and thorough barring-down may remove the need for support. Rough wall blasting may result in the need for a single application of shotcrete, especially where the excavation height exceeds 25m.

2. For cases of heavy rock bursting or "popping", tensioned bolts with enlarged bearing plates often used, with spacing about 1m (occasionally 0.8m). Final support when "popping" activity ceases.

3. Several bolt lengths often used in same excavation, ie 3, 5 and 7m.

4. Several bolt lengths often used in same excavation, ie 2, 3 and 4m.

5. Tensioned cable anchors often used to supplement bolt support pressures. Typical spacing 2 to 4m.

6. Several bolt lengths often used in same excavation, ie 6, 8 and 10m.

7. Tensioned cable anchors often used to supplement bolt support pressures. Typical spacing 4 to 6m.

8. Several older generation power stations in this category employ systematic or spot bolting with areas of chain link mesh, and a free span concrete arch roof (250 - 400mm) as permanent support.

9. Cases involving swelling, for instance montmorillonite clay (with access of water). Room for expansion behind the support is used in cases of heavy swelling. Drainage measures are used where possible.

10. Cases not involving swelling clay or squeezing rock.

11. Cases involving squeezing rock. Heavy rigid support is generally used as permanant support.

12. According to author's experience (Barton et al), in cases of swelling or squeezing, the temporary support required before concrete (or shotcrete) arches are formed may consist of bolting (tensioned shell-expansion type) if the value of RQD/J_n is sufficiently high (ie > 1.5), possibly combined with shotcrete. If the rock mass is very heavily jointed or crushed (ie RQD/J_n <1.5, for example a "sugar cube" shear zone in quartzite), then the temporary support may consist of up to several applications of shotcrete. Systematic bolting (tensioned) may be added after casting the concrete (or shotcrete) arch to reduce the uneven loading on the concrete, but it may not be effective when RQD/J_n <1.5, or when a lot of clay is present, unless the bolts are grouted before tensioning. A sufficient length of anchored bolt might also be obtained using quick setting resin anchors in these extremely poor quality rock masses. Serious occurrences of swelling and/or squeezing rock may require that the concrete arches be taken right up to the face, possibly using a shield as temporary shuttering. Temporary support of the working face may also be required in these cases.

TABLE 18 - RECOMMENDED SUPPORT BASED UPON NGI TUNNELLING QUALITY INDEX Q

Supplementary notes by Barton, Lien and Lunde (Continued)

13. For reasons of safety the multiple drift method will often be needed during excavation and supporting of roof arch. For Span/ESR > 15 only.

14. Multiple drift method usually needed during excavation and support of arch, walls and floor in cases of heavy squeezing. For span/ESR > 10 in exceptionally poor rock only.

Supplementary notes by Hoek and Brown

a. In Scandinavia, the use of "Perfobolts" is common. These are perforated hollow tubes which are filled with grout and inserted into drillholes. The grout is extruded to fill the annular space around the tube when a piece of reinforcing rod is pushed into the grout filling the tube. Obviously, there is no way in which these devices can be tensioned although it is common to thread the end of the reinforcing rod and place a normal bearing plate or washer and nut on this end.(See figure 154 on page 328).
In north America the use of "Perfobolts" is rare. In mining applications a device known as a "Split set" or "Friction set" (developed by Scott[243]) has become popular. This is a split tube which is forced into a slightly smaller diameter hole than the outer diameter of the tube. The friction between the steel tube and the rock, particularly when the steel rusts, acts in much the same way as the grout around a reinforcing rod. For temporary support these devices are very effective. (See figure 153 on page 326).
In Australian mines, untensioned grouted reinforcing is installed by pumping thick grout into drillholes and then simply pushing a piece of threaded reinforcing rod into the grout. The grout is thick enough to remain in an up-hole during placing of the rod.

b. Chainlink mesh is sometimes used to catch small pieces of rock which can become loose with time. It should be attached to the rock at intervals of between 1 and 1.5m and short grouted pins can be used between bolts. Galvanised chainlink mesh should be used where it is intended to be permanant, eg in an underground powerhouse.

c. Weldmesh, consisting of steel wires set on a square pattern and welded at each intersection, should be used for the reinforcement of shotcrete since it allows easy access of the shotcrete to the rock. Chainlink mesh should never be used for this purpose since the shotcrete cannot penetrate all the spaces between the wires and air pockets are formed with consequent rusting of the wire. When choosing weldmesh, it is important that the mesh can be handled by one or two men working from the top of a high-lift vehicle and hence the mesh should not be too heavy. Typically, 4.2mm wires set at 100mm intervals (designated 100 x 100 x 4.2 weldmesh) are used for reinforcing shotcrete.

d. In poorer quality rock, the use of untensioned grouted dowels as recommended by Barton, Lien and Lunde depends upon immediate installation of these reinforcing elements behind the face. This depends upon integrating the support drilling and installation into the drill-blast-muck cycle and many non-Scandinavian contractors are not prepared to consider this system. When it is impossible to ensure that untensioned grouted dowels are going to be installed immediately behind the face, consideration should be given to using tensioned rockbolts which can be grouted at a later stage. This ensures that support is available during the critical excavation stage.

e. Many contractors would consider that a 200mm thick cast concrete arch is too difficult to construct because there is not enough room between the shutter and the surrounding rock to permit easy access for pouring concrete and placing vibrators. The US Army Corps of Engineers[244] suggests 10 inches (254 mm) as a normal minimum while some contractors prefer 300mm.

f. Barton, Lien and Lunde suggest shotcrete thicknesses of up to 2m. This would require many separate applications and many contractors would regard shotcrete thicknesses of this magnitude as both impractical and uneconomic, preferring to cast concrete arches instead. A strong argument in favour of shotcrete is that it can be placed very close to the face and hence can be used to provide early support in poor quality rock masses. Many contractors would argue that a 50 to 100mm layer is generally sufficient for this purpose, particularly when used in conjunction with tensioned rockbolts as indicated by Barton, Lien and Lunde, and that the placing of a cast concrete lining at a later stage would be a more effective way to tackle the problem. Obviously, the final choice will depend upon the unit rates for concreting and shotcreting offered by the contractor and, if shotcrete is cheaper, upon a practical demonstration by the contractor that he can actually place shotcrete to this thickness.
In north America, the use of concrete or shotcrete linings of up to 2m thick would be considered unusual and a combination of heavy steel sets and concrete would normally be used to achieve the high support pressures required in very poor ground.

South African Council for Scientific and Industrial Research support predictions

Based upon his Geomechanics Classification, Bieniawski of the South African Council for Scientific and Industrial Research proposed a guide for the choice of support for underground excavations. The most recent version of this guide[26] is reproduced in table 19 below.

TABLE 19 - GEOMECHANICS CLASSIFICATION GUIDE FOR EXCAVATION AND SUPPORT IN ROCK TUNNELS
SHAPE: HORSESHOE; WIDTH: 10 m ; VERTICAL STRESS: BELOW 25 MPa; CONSTRUCTION: DRILLING AND BLASTING.

Rock mass class	Excavation	Support		
		Rockbolts (20mm dia. fully bonded)	Shotcrete	Steel sets
Very good rock I RMR: 81-100	Full face. 3 m advance.	Generally no support required except for occasional spot bolting.		
Good rock II RMR: 61-80	Full face. 1.0-1.5 m advance. Complete support 20m from face.	Locally bolts in crown, 3m long spaced 2.5m with occasional mesh.	50mm in crown where required	None
Fair rock III RMR: 41-60	Top heading and bench, 1.5-3m advance in heading. Commence support after each blast. Complete support 10m from face.	Systematic bolts 4m long, spaced 1.5-2m in crown and walls with mesh in crown.	50-100mm in crown, 30mm in sidewalls.	None
Poor rock IV RMR: 21-40	Top heading and bench, 1-1.5m advance in heading. Install support concurrently with excavation - 10m from face.	Systematic bolts 4-5m long, spaced 1-1.5m in crown and walls with wire mesh.	100-150mm in crown, and 100mm in sides.	Light ribs spaced 1.5m where required.
Very poor rock V RMR: < 20	Multiple drifts. 0.5-1.5m advance in top heading. Install support concurrently with excavation. Shotcrete as soon as possible after blasting.	Systematic bolts 5-6m long, spaced 1-1.5m in crown and walls with wire mesh. Bolt invert.	150-200mm in crown, 150mm on sides and 50mm on face.	Medium to heavy ribs spaced 0.75m with steel lagging and forepoling if required. Close invert.

It should be noted that the support recommendations listed in table 19 are for civil engineering tunnels of approximately 10 metres span excavated by drill and blast methods at depths of less than 1000 metres (3300 feet) below surface.

Modification of CSIR support recommendations for mining applications

Laubscher and Taylor[245], on the basis of their experience in applying rock mass classification to over 50,000 metres of mining development and drill core, have proposed a number of modifications to Bieniawski's Geomechanics Classification and to the support recommendations presented in table 19. These modifications are designed to refine some of the rock mass characteristics incorporated in table 5 on page 26 and to allow for the differences in support practices between mining and civil engineering.

The adjustments proposed by Laubscher and Taylor are made by modifying the original value of a particular parameter, chosen for incorporation in table 5, by a percentage determined by the influence of weathering, stress change or blasting on that parameter. Hence, if an RQD value of 75 is established from an examination of diamond drill core and it is known that the blasting techniques to be used for mining in this rock are poor, the RQD is reduced by 20% to give an adjusted value of 60. The guidelines for making these adjustments, published by Laubscher and Taylor[245], are summarised below.

a. Weathering.
Certain rock types weather rapidly on exposure and this aspect must be taken into account in deciding upon permanent support measures. Weathering affects three of the parameters listed in table 5 on page 26:
Intact rock strength - decrease by up to 96% if weathering takes place along micro-structures in the rock.
Rock Quality Designation - decrease by up to 95% as the rock weathers resulting in an increase in fractures.
Condition of joints - Reduce the rating for the condition of joints by up to 82% if weathering is considered to cause deterioration of the joint wall rock or the joint filling.

b. Field and induced stresses.
Field and induced stresses can influence the condition of joints by keeping the joint surfaces in compression or by allowing joints to loosen and hence increase the possibility of shear movement.
Condition of joints - when the stress conditions are such that joints will be kept in compression, increase the rating by up to 120%. If the possibility of shear movement is increased, decrease the rating by up to 90%. If the joints are open and can be equated to joints with thin gouge filling, decrease the rating by up to 76%.

c. Changes in stress.
When large stress changes are induced by mining operations, for example during the extraction of crown pillars or the over-mining of draw-points, the condition of the joints will be changed in the same way as in b. above.
Condition of joints - when stress changes are such that joints will always be in compression, increase rating by up to 120%. When stress changes are likely to cause serious shear movement or joint opening, decrease rating by up to 60%.

d. Influence of strike and dip orientations.
The size and shape and direction of advance of an underground excavation, when considered in relation to the jointing in a rock mass, will have an influence upon the stability of the excavation. This is taken into account in table 6 and table 5b on page 26 in Bieniawski's original classification but Laubscher and Taylor consider that a further adjustment is necessary. They consider that the stability of an excavation in jointed rock depends upon the number of joints and excavation faces which are inclined away from vertical and they recommend the following

adjustments:
Spacing of joints rating adjustment -

Number of joints	Percentage adjustment depending on number of inclined excavation faces (shown below)				
	70%	75%	80%	85%	90%
3	3		2		
4	4	3		2	
5	5	4	3	2	1
6	6		4	3	2,1

The following adjustments to the joint spacing rating are proposed for shear zones encountered in developments:

0-15° = 76%   15-45° = 84%   45-75° = 92%

e. Blasting effects
Blasting creates new fractures and causes movement on existing joints. The following reductions are proposed for the ratings for
Rock Quality Designation and *Condition of Joints*

Boring	100%
Smooth wall blasting	97%
Good conventional blasting	94%
Poor conventional blasting	80%

f. Combined adjustments.
In some situations the Geomechanics Classification will be subjected to more than one adjustment. Laubscher and Taylor suggest that the total adjustment should not exceed 50%.

g. Support recommendations.
Based upon the adjusted classification ratings and taking into account typical mining support practices, Laubscher and Taylor have proposed the support guide reproduced in table 20 on page 299.

Comparison of underground excavation support predictions

The authors consider that the most effective way of comparing the relative merits of the various approaches to underground excavation support design, outlined in the first part of this chapter, is to apply them to actual cases. One such case for which sufficient detailed information has been published is the Kielder Experimental Tunnel. The predicted and actual support systems for this tunnel will be examined on the following pages.

Kielder experimental tunnel

Figure 140, reproduced from the paper by Houghton[246], shows some of the features of the Kielder experimental tunnel mined in the Weardale Valley, near Stanhope, in north-east England. This tunnel was driven as part of an initial scheme relating to the Kielder Water Project which, in turn, is part of a larger project to control river flow. Further details on the Kielder Project and on the experimental tunnel are given in papers by Berry and Brown[247], Nicholson[248], Freeman[249] and Ward, Coats and Tedd[250].

TABLE 20 - MODIFIED GEOMECHANICS CLASSIFICATION GUIDE FOR SUPPORT OF MINING EXCAVATIONS									
Adjusted ratings	Original Geomechanics ratings								
	90-100	80-90	70-80	60-70	50-60	40-50	30-40	20-30	10-20	0-10
70-100									
50-60		a	a	a	a				
40-50			b	b	b	b			
30-40				c,d	c,d	c,d,e	d,e		
20-30					g	f,g	f,g,j	f,h,j	
10-20						i	i	h,i,j	h,j
0-10							k	k	l	l

- a — Generally no support but locally joint intersections might require bolting.
- b — Patterned grouted bolts at 1m collar spacing.
- c — Patterned grouted bolts at 0.75m collar spacing.
- d — Patterned grouted bolts at 1m collar spacing and shotcrete 100mm thick.
- e — Patterned grouted bolts at 1m collar spacing and massive concrete 300mm thick and only used if stress changes not excessive.
- f — Patterned grouted bolts at 0.75m collar spacing and shotcrete 100mm thick.
- g — Patterned grouted bolts at 0.75m collar spacing with mesh reinforced shotcrete 100mm thick.
- h — Massive concrete 450mm thick with patterned grouted bolts at 1m spacing if stress changes are not excessive.
- i — Grouted bolts at 0.75m collar spacing if reinforcing potential is present, and 100mm reinforced shotcrete, and then yielding steel arches as a repair technique if stress changes are excessive.
- j — Stabilise with rope cover support and massive concrete 450mm thick if stress changes not excessive.
- k — Stabilise with rope cover support followed by shotcrete to and including face if necessary, and then closely spaced yielding arches as a repair technique where stress changes are excessive.
- l — Avoid development in this ground otherwise use support systems j or k.

Supplementary notes

1. The original Geomechanics Classification as well as the adjusted ratings must be taken into account in assessing the support requirements.
2. Bolts serve little purpose in highly jointed ground and should not be used as the sole support where the joint spacing rating is less than 6.
3. The recommendations contained in table 20 are applicable to mining operations with stress levels less than 30 MPa.
4. Large chambers should only be excavated in rock with adjusted total classification ratings of 50 or better.

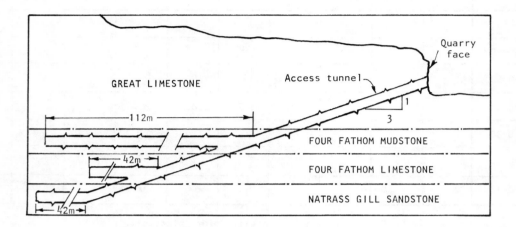

Figure 140 : Diagrammatic cross-section through Kielder Experimental Tunnel. After Houghton[246].

The four rock types through which the tunnel passes are described by Houghton as follows :

Great Limestone - A moderately thickly bedded, grey, fine to medium grained crystalline limestone with bituminous stylolitic partings. It is generally fresh to only slightly weathered. There is a well developed jointing reflecting the regional structure, with vertical to sub-vertical joints, locally lined with carbonaceous material.

Four Fathom Mudstone - A thinly laminated, dark grey mudstone, containing silty and sandy material in its upper sections and characterised by frequent micaceous wavy partings. When exposed the mudstone is prone to rapid deterioration and quickly forms a soil-like mass, although at depth the rock is comparatively unweathered.

Vertical joints are common, often ironstained reflecting the movement of groundwater. With depth there is an increasing fissility, as a result of short, non-continuous multi-directional joints.

Four Fathom Limestone - This is very similar in character to the Great Limestone, the major difference being stratigraphic rather than lithological. It forms a 10m band of moderately bedded, grey, fine grained crystalline limestone with occasional muddy horizons occurring towards the top. Joints are few and those that do occur are often tight and lined with calcite.

Natrass Gill Sandstone - Medium to coarse grained sandstone, thickly bedded with occasional shaly partings and well developed jointing. Many of the joints are clay filled at depth and show considerable alteration.

On the basis of the information contained in Houghton's paper, the rock masses have been classified using both the CSIR and the NGI classifications. In addition, the properties required to determine the required support lines for the rock mass have been estimated and are given in table 21 .

TABLE 21 - CLASSIFICATION OF ROCK MASSES ENCOUNTERED BY KIELDER EXPERIMENTAL TUNNEL				
	GREAT LIMESTONE	FOUR FATHOM LIMESTONE	FOUR FATHOM MUDSTONE	NATRASS GILL SANDSTONE
a) CSIR Geomechanics Classification (see table 5 on page 26 - ratings in italics)				
Intact rock strength - MPa	100-200 *(12)*	100-200 *(12)*	25-50 *(4)*	50-100 *(7)*
Rock Quality Designation RQD%	79-95 *(18)*	82-97 *(19)*	25-80 *(10)*	50-99 *(13)*
Joint spacing	0.3-1m *(20)*	1-3m *(25)*	50-100mm *(8)*	0.3-1m *(20)*
Condition of joints	Continuous, no gouge, opening 5mm. *(6)*	Continuous with gouge, opening <1mm. *(6)*	Continuous, no gouge, opening <5mm. *(6)*	Continuous, no gouge, opening 0.1mm. *(12)*
Groundwater	Moist *(7)*	Moist *(7)*	Moist *(7)*	Moist *(7)*
Total rating	63	69	35	59
b) NGI Tunnelling Quality Index Q (see table 7 on pages 31 to 33)				
Rock Quality Designation RQD%	79-95	82-97	25-80	50-100
Joint set number J_n	3-4	4-15	6-15	4-9
Joint roughness number J_r	1-3	1.5-2.0	1.0-2.0	0.5-1.5
Joint alteration number J_a	1-4	.25-1.0	1-3	.75-2.0
Joint water reduction factor J_w	1	1	1	1
Stress reduction factor SRF	2.5-5	1-5	5-7.5	2.5-10
Quality Index Q	7-16.5	1.1-16.3	.009-0.15	0.39-7.4
c) Input data for rock-support interaction characteristic curves (table 14, page 259)				
Intact rock strength σ_c MPa	150	150	37	75
Original rock constant m	0.7	1.0	0.1	0.3
Original rock constant s	0.004	0.01	0.00008	0.0001
Modulus of elasticity E	15,000 MPa	20,000 MPa	5000 MPa	10,000 MPa
Poisson's ratio ν	0.25	0.25	0.25	0.25
Broken rock constant m_r	0.14	0.14	0.05	0.08
Broken rock constant s_r	0.0001	0.0001	0.00001	0.00001
Unit weight γ_r	0.02 MN/m^3	0.02 MN/m^3	0.02 MN/m^3	0.02 MN/m^3
In situ stress magnitude p_o	1.02 MPa	2.82 MPa	2.56 MPa	3.07 MPa
Tunnel radius r_i	1.65 m	1.65 m	1.65 m	1.65 m

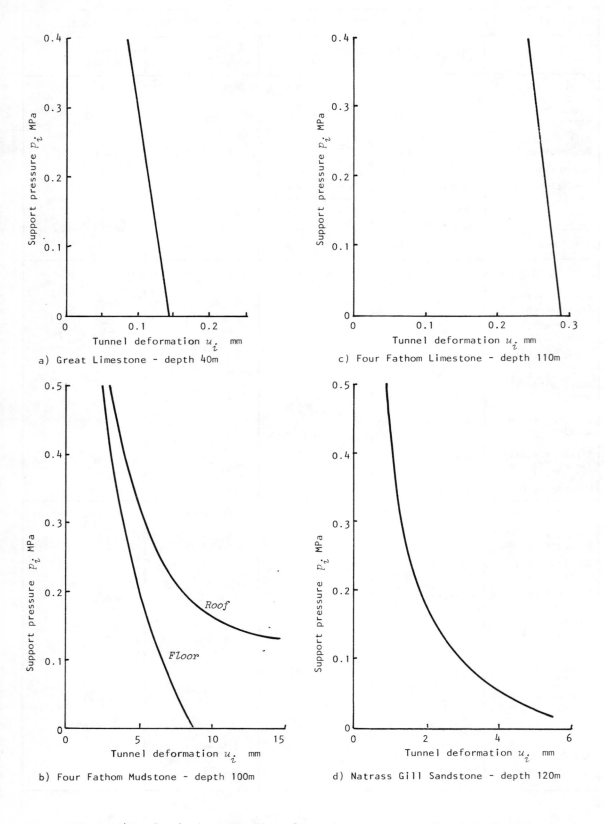

Figure 141 : Required support lines for rock masses surrounding 3.3m diameter tunnel.

The required support lines for the different rock masses are plotted in figure 141. These show that the tunnels in both the Great Limestone and the Four Fathom Limestone behave elastically. Assuming that this analysis is reasonably representative of the average rock conditions surrounding these two tunnels, they will remain stable without the assistance of support except that required to secure local blocks and wedges defined by intersecting structural features.

The rock surrounding the tunnel in the Four Fathom Mudstone is highly unstable with the deformation of the roof showing no sign of reaching stability for support pressures of less than 0.2 MPa. The recommended and actual support measures for this tunnel will be discussed later in this example.

The tunnel in Natrass Gill Sandstone is considered to be marginally stable. In a mining situation it is possible that the roof deformation of approximately 6 mm could be tolerated and that, for temporary use, the tunnel could be left unsupported. In most civil engineering applications and for permanent mining access, some support would be required in this tunnel.

On the basis of the rock mass classifications listed in table 21 and the rock mass characteristic curves given in figure 141, a listing of support recommendations has been compiled and is presented in table 22. Note that, in addition to the CSIR and NGI support recommendations, a series of support recommendations derived by Houghton[246] from a paper by Wickham et al[21] have been included in table 22. It should also be noted that the support recommendations for the CSIR and the NGI systems have been derived from tables 19 and 18 and do not correspond precisely with the recommendations derived by Houghton. Such variations in interpretation between different users are not uncommon and emphasise the fact that these classifications are intended to provide general guidance rather than precise quantitative values.

In compiling table 22, the CSIR recommendations on bolt lengths given in table 19 on page 296 have been reduced to take into account the fact that the tunnel diameter under consideration is 3.3m as compared with the 10m tunnel for which table 19 was prepared.

The relative merits of the different support recommendations contained in table 22 will be discussed at the end of this example after consideration of the actual support systems used in the Kielder Experimental Tunnel.

Figure 142, reproduced from the paper by Freeman[249], shows the excavation methods and the different support systems used in the 112m long tunnel section in Four Fathom Mudstone. Detailed description of the support systems are listed hereunder :
1) 5 in x 4½ in x 20 lb/ft H section ribs at 1 m centres. ($W = 127$mm, $X = 114$mm, $A_s = 0.0038$ m^2, $I_s = 1.042 \times 10^{-5}$ m^4 for use in equations 122 and 123). Steel bank bars and light galvanised sheeting for backing with waste rock packing as required.
2) Rows of seven 25mm diameter x 1.8m long untensioned grouted bolts (dowels) at about 0.9m centres with 50 x 50 x 3.2mm weldmesh.

TABLE 22 - SUPPORT RECOMMENDATIONS DERIVED FROM ROCK MASS CLASSIFICATIONS AND FROM ROCK-SUPPORT INTERACTION ANALYSES FOR KIELDER EXPERIMENTAL TUNNEL.

Basis for support recommendation	Great Limestone	Four Fathom Limestone	Four Fathom Mudstone	Natrass Gill Sandstone
CSIR Geomechanics Classification. (Table 19, page 296)	20mm diameter 2m long bolts at 1.5m centres with occasional mesh and 50mm shotcrete where required.	20mm diameter 2m long bolts at 1.5m centres with occasional mesh and 50mm shotcrete where required.	20mm diameter 3-4m grouted bolts at 1-1.5m centres with mesh and 100-150mm shotcrete. Ribs at 1.5m centres if required.	20mm diameter 3m grouted bolts at 1.5-2m centres with mesh and 50-100mm shotcrete.
NGI Tunnelling Quality Index (Table 18, pages 290-295)	No support or untensioned grouted dowels at 1-1.5m centres where required.	No support or untensioned grouted dowels at 1-1.5m centres where required.	75-100mm shotcrete or tensioned bolts at 1m centres with 25-50mm shotcrete. Approx. support pressure 0.6 MPa	Untensioned grouted dowels at 1.5m centres. Approximate support pressure 0.1 MPa.
Rock Structure Rating from Wickham et al[21]	No support	No support	Rockbolts at 2m centres with 250-350mm shotcrete.	250mm shotcrete
Rock-support interaction analysis. (Table 17, page 269)	No support	No support	Assume support pressure 0.3-0.5 MPa : 50-100mm shotcrete with 25mm, 5m grouted bolts at 1m centres or medium steel sets at 1 m centres.	Assumed support pressure 0.15 MPa: 50mm shotcrete or 25mm, 2.5m long grouted bolts at 1.5m centres.

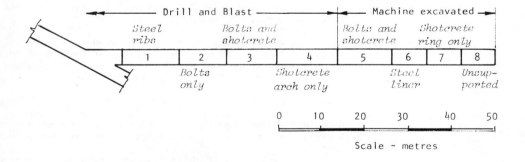

Figure 142 : Actual support systems used in the Kielder Experimental Tunnel in Four Fathom Mudstone. After Freeman[249].

3) One layer of shotcrete in roof plus rockbolting as in 2), followed by 200 x 200 x 6.4mm weldmesh with a further layer of shotcrete to form an arch covering roof and walls (but not floor). Average shotcrete thickness approximately 140mm.

4) Identical to 3) but without rockbolts.

5) The same as 3) but with rows of 5 (as compared with 7) rockbolts.

6) A special circular liner of 12.7mm thick mild steel plates consisting of eleven 0.7m long rings, each ring comprising four equal segments, butt-welded together in groups of two, three and four rings. The space between the liner and the rock was filled with a weak, fast-setting grout consisting of 4 parts of pulverized fuel ash to 1 part of Portland cement with 0.2 times the cement weight of sodium metasilicate added.

7) A complete circular ring of shotcrete applied in two layers with 200 x 200 x 6.4mm weldmesh between. The assumed total shotcrete thickness is 140mm.

8) Completely unsupported but with steel arches, mesh and timber providing a protective screen for safe access.

The first 72m of the tunnel was excavated by normal drill and blast methods. The remaining 40m was mined by means of a Dosco road header with the invert being mined by hand to complete the circular profile.

Extensometers were installed at various locations along the tunnel and these instruments were used to monitor the movement of the rock mass surrounding the tunnel. One such extensometer array was installed in the unsupported tunnel section (8) and the locations and responses of these extensometers are shown in the margin drawing. In this case the extensometer was installed in a borehole drilled from surface before excavation of the tunnel and it provides an excellent record of the complete movement pattern in the rock surrounding the tunnel.

According to Ward, Coats and Tedd[250], less than 1mm of movement of extensometer head A (in the roof of the tunnel) had occurred before the face reached the extensometer position. When the face had advanced 2m beyond this position, the point A had moved downwards about 6mm. The next day the face was advanced to its final position 5.6m beyond the extensometer position and point A moved a further 2mm. After about 200 days of observation, this point had moved down about 30mm. Note that, as would be expected from figure 141b, the floor of the tunnel has risen by a significant amount.

In the rock-support interaction analysis presented earlier in this chapter, the influence of time-dependent movement of the rock mass was not included. Hence, only the short term movement which occurred within a few days of excavation can be compared with the calculated displacements shown in figure 141b.

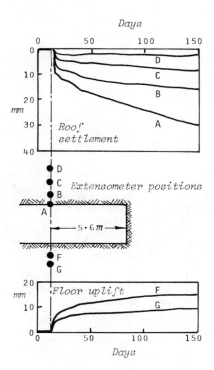

Movement pattern in the unsupported tunnel section. After Ward et al[250].

Displacement versus time plots for each of the 8 support sections in the tunnel are given in figure 143 and these will be discussed in detail on the following pages.

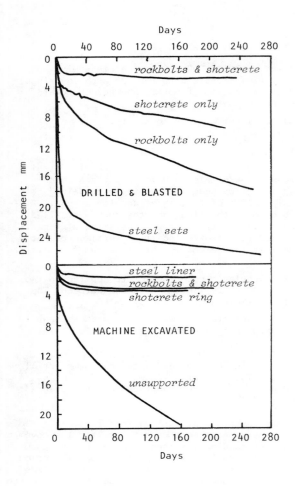

Figure 143 : Typical displacement curves for extensometer points 0.3m above roof of excavation in different support sections. After Ward, Coats and Tedd[250].

The required support curves for the rock mass surrounding the experimental tunnel in Four Fathom Mudstone, given in figure 141b, are plotted in greater detail in figure 144. Available support curves for a number of support systems are also plotted in this figure, assuming that all support systems are installed when the 3.3m tunnel has deformed 5mm. Note that the thickness of the broken rock zone surrounding the tunnel ($r_e - r_i$ from equation 104) has been plotted for different support pressures in figure 144.

It is instructive to discuss the results presented in figures 143 and 144 together and this is done in the following notes:

1) Steel ribs or sets - from figure 143 it is evident that these steel supports provide the least effective method of supporting this particular tunnel. The reason for this poor performance is suggested by figure 144 which shows that the support reaction curve is very sensitive to the effectiveness of the blocking. The two support reaction curves in figure 144 were calculated for block spacings of $22\frac{1}{2}°$ and $40°$, keeping all other parameters constant. Ward et al[250] state that "the steel ribs are loaded irregularly according to the position of the bank bars, the irregularity of the positions of rock

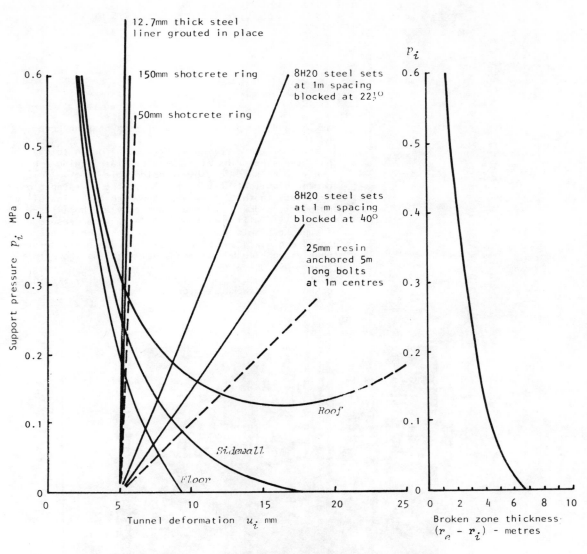

Figure 144: Required and available support curves and variation of broken zone thickness for 3.3m diameter tunnel in Four Fathom Mudstone.

INPUT DATA FOR SUPPORT REACTION ANALYSIS (See Table 14 on page 260)

1. Steel liner grouted in place
   (Use equation 120)

   E_s = 207000 MPa
   ν_s = 0.3
   t_s = 0.0127 m
   r_i = 1.65 m

2. Shotcrete rings 50mm and 150mm thick (equations 120 and 121)

   E_c = 28000 MPa
   ν_c = 0.25
   t_c = 0.15 and 0.05m
   r_i = 1.65 m
   σ_{cc} = 35 MPa

3. 8H20 steel sets at 1 m spacing
   (equations 122 and 123)

   W = 0.127 m
   X = 0.114 m
   A_s = 0.0038 m^2

   Steel sets (continued)
   I_s = 1.042 × 10^{-5} m^4
   E_s = 207000 MPa
   σ_{ys} = 245 MPa
   r_i = 1.65 m
   S = 1 m
   θ = 11.25 and 20°
   t_B = 0.2 m
   E_B = 500 MPa

4. Resin anchored rockbolts
   (equations 127 and 128)

   l = 5m (use two 2½m bolts coupled)
   d_b = 0.025 m
   E_b = 207000 MPa
   Q = 0 (resin anchored)
   T_{bf} = 0.3 MN
   r_i = 1.65 m
   s_c = 1 m
   s_l = 1 m

packing and the protuberances of rock bearing here and there."

In spite of these comments, Ward et al go on to say that "the steel ribs with their partial lagging are likely to be safe for some considerable time". This comment highlights one of the important advantages of steel sets in that, even when they are badly installed and poorly blocked and even when they are badly deformed, sets will continue to provide some measure of support. This support is *visible* and most underground engineers and miners will enter a badly deformed tunnel supported by twisted steel sets, as shown in figure 145, while they would feel extremely nervous about entering a similar tunnel supported by straining rockbolts or cracked shotcrete.

Figure 145 : Deformed steel sets providing support for a critical access tunnel in very poor quality rock under high stress.

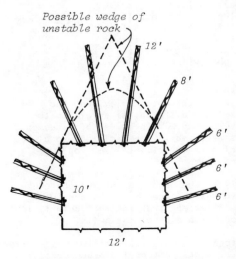

Rockbolt pattern used in drives and cross-cuts in 500 copper orebody at Mount Isa, Australia. Bolt pattern is 2½ ft. circumferential and 4 ft. longitudinal, lengths as indicated. After Mathews and Edwards[251].

2) Rockbolts only - the behaviour of these polyester resin grouted untensioned bars is discussed in the papers by Freeman[249] and Ward et al[250] and it is clear that the poor performance of this support system is due to the failure of the bond between the resin and the rock.

Figure 144 shows that the thickness of the zone of broken rock surrounding the tunnel varies from about 2 to about 7 metres, depending upon the support pressure acting on the rock. Even if this analysis is wrong by a factor of 2, the conclusion which must be drawn is that the 1.8m dowels used in this experiment were too short. This means that the entire length of the grouted dowel would have been achored in poor quality rock in the process of breaking up and that the untensioned dowels never really had a fair chance of working.

In general, the length of a rockbolt should be chosen so that it is anchored beyond the boundary of the broken rock zone in the original rock mass. It is appreciated that this depth is difficult to determine but some indication can be obtained by rock-support interaction analyses or by numerical model studies such as those described on pages 219 and 225. In this case, figure 144 suggests that, for a support pressure of about 0.2 MPa, the bolts should be about 5m long. Obviously a single 5m bolt length cannot be installed in a 3.3m diameter tunnel and this would necessitate coupling two 2½m bolts together.

In some cases it may not be possible to anchor the bolts or dowels in good rock and Mathews and Edwards[251] have described a very interesting case at Mount Isa in which bolts were used to support very friable leached ground at depths of 1200 feet (366 m). The bolting layout is illustrated in the margin, utilising 3/4 inch (19mm) high tensile bolts of 6 to 12 feet length (2 to 4m). The ends of these bolts were deformed into "pigtails" to improve the anchorage and a mixture of sand, cement and water was pumped into the holes through two plastic tubes. This "grout" anchor extended about one half the length of the bolt, leaving a free bolt length for tensioning. Weldmesh was used to support the small pieces of rock between the washer plates which were spaced at 30 inches (0.76m) circumferentially and 48 inches (1.22m) longitudinally. In most cases this bolt pattern was effective in stabilizing the rock surrounding these drives and crosscuts and in limiting the deformation. In very poor ground at depths of 1200 feet the deformation was very large - up to 5 feet (1.5m) - but the rock mass above the roof was maintained in a reasonably stable condition as illustrated in figure 146.

3) Rockbolts and shotcrete - this support section provided the most satisfactory support in the blasted tunnel length, stabilising the deformation at about 2.5mm as shown in figure 143. Ward et al report that strains, which were measured by means of gauges attached to the shotcrete surface, and observed crack development, indicated that the lining is influenced by the joint pattern in the rock mass and the fact that the shotcrete was applied in the form of an arch rather than a closed ring. This means that the lateral stiffness of the open end of the arch is very low and the sidewalls of

Figure 146 : Roof deformation and sidewall failure in a drive in the 500 copper orebody in Mount Isa mine, Australia. The height of the drive has decreased from 10 feet to about 5 feet as shown in this photograph but the bolts and weldmesh have maintained reasonably stable roof conditions. After Mathews and Edwards[251]. Mount Isa Mines Limited photograph.

the tunnel can move inwards more easily than the roof can move downwards. The differential strain in the shotcrete induced cracking within a week of placement. These observations resulted in the decision to place a full shotcrete ring in experimental section 7 of the tunnel.

4) Shotcrete arch only - the behaviour of the shotcrete arch was similar to that of the shotcrete arch with bolts discussed in 3) except that the roof deformation was greater due to the absence of the bolts.

5) Rockbolts and shotcrete - this test section was similar to section 3) except that the tunnel had been excavated by machine rather than by drill and blast methods. No significant difference between the behaviour of this section and section 3) was observed.

6) Steel liner - as shown in figure 143, this liner provided the stiffest support system of all those used. A crude analysis of the support reaction curve for this steel ring was carried out using equation 120 and the results are plotted in figure 144. This analysis confirms that the ring is very stiff but the authors suggest that a critical factor in this behaviour is the effectiveness of the grouting of the gap between the rock and the steel. Any deficiencies in this grouting would result in eccentric loading and a serious drop in the support capacity of the ring. For this reason it is suggested that this form of support should only be used in very special circumstances in which the expense of fabrication and grouting would be justified.

7) Shotcrete ring - the behaviour of this support system was very satisfactory and conforms closely to that which would be expected from an idealised analysis such as that used in deriving equation 120. As shown in figure 143, the tunnel has been stabilised by this shotcrete ring and Ward et al report that it has continued to behave in a satisfactory manner.
A question which does arise is whether the shotcrete lining is too thick at 140mm in a 3.3m tunnel. As shown in figure 144, a shotcrete thickness of 50mm should do an adequate job in this small tunnel. A factor which would influence the choice of shotcrete thickness would be the accuracy of the tunnel profile. As pointed out earlier in this chapter, thin shotcrete linings should not be used when the tunnel walls are very irregular since the stress distribution in the shotcrete layer will be significantly different from the ideal and cracking will be induced in corners and at changes of section.

On the basis of the detailed analysis presented on the preceding pages and with the benefit of a great deal of hindsight, the authors offer the following conculsions on the Kielder Experimental Tunnel and on the comparison between the support predictions summarised in table 22 and the support systems used in the tunnel.

a) On the basis of Houghton's description[246] of the Four Fathom Mudstone as "prone to rapid deterioration" upon exposure, the authors feel that the application of a thin layer of shotcrete as soon as possible after excavation should be the first step in the support programme. This layer, about 25mm thick, would serve to seal the surface of the rock and would prevent the rapid moisture content changes which induce slaking in shales and mudstones.

b) Because of the bedded nature of the mudstone and the presence of a number of near vertical joint sets, it is anticipated that it would be very difficult to achieve a good excavation profile by drill and blast methods and that immediate wedge failures may occur even in the case of machine excavation. This suggests that support systems such as steel sets, which require uniform load transfer from the rock to the steel, would be difficult to use. Blocking steel sets, if it were to be done correctly, would be very time consuming and hence expensive.

c) In jointed rock masses of the type under consideration here, the authors consider rockbolting to be the most convenient and economical support system. Consequently, the second stage in their support programme would be to attempt to establish the correct bolt length and bolt pattern to ensure that the money invested in rockbolts served a useful purpose. It is felt that the rock-support interaction analysis or some similar form of numerical model study should form the basis for the choice of the correct bolt length which, in this case should be about 5 metres. None of the rock mass classification systems appear to provide an adequate prediction of rockbolt length.
The support pressure and the associated rockbolt size

and spacing seems to be subject to a wide range of possible interpretations. The bolt spacing derived from the CSIR and the NGI classifications compare reasonably well with those suggested by the rock-support interaction analysis (table 22). The spacing suggested by Wickham et al[21] is clearly too wide for this small tunnel and would not induce a high enough support pressure.

d) Because of the poor quality of the rock mass and the low strength of the intact mudstone, mechanical anchors are considered to be inappropriate for the bolts to be used in this tunnel. It is recommended that a two-stage grout system, either resin or cement, would be the most appropriate anchoring and grouting method for this application. This would involve placing a grout anchor, tensioning the rockbolt and then grouting the remainder of the bolt length. Techniques for carrying out this operation will be discussed in the next chapter.

e) Because of the irregular tunnel profile and the danger of the initial shotcrete layer cracking and releasing small pieces of rock which could fall out between bolts, it is recommended that a layer of weldmesh be attached to the bolts and anchored at intermediate points to the shotcrete. A final layer of shotcrete, 25 to 50mm thick, placed over the weldmesh would complete the support system. This would give a final layer of 50 to 75mm of mesh reinforced shotcrete which, together with the tensioned and grouted rockbolts, is considered to be the most economical support system which could be used to stabilise this tunnel.

Pre-reinforcement of rock masses

In the discussion presented up to this point, it has been assumed that support is installed *after* an excavation has been created. While this is certainly the normal approach to support design, there is no reason why support should not be installed *before* excavation, provided that access is available to place the reinforcement.

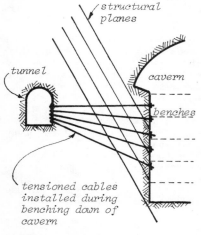

Pre-reinforcement of cavern sidewall from a parallel access tunnel.

A good example of pre-reinforcement is illustrated in the margin sketch which shows part of a cavern for an underground hydroelectric project. Strongly developed structural features were known to dip out of the proposed cavern sidewall as shown and it was feared that sliding could occur on one or more of these features when they were exposed in the excavation. In order to overcome this potential problem, a gallery was mined parallel to the cavern and cables were installed from this gallery as the cavern was benched down. Hence, before the dipping structural features were exposed in the sidewall, the blocks or wedges resting on these surfaces had already been reinforced.

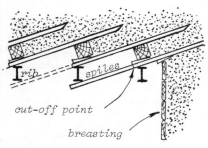

Forepoling through running ground. After Proctor and White[6].

One of the most common pre-reinforcement techniques used in tunnelling is *spiling* or *forepoling* which involves installing reinforcing rods in angled holes drilled from the advancing face. In very soft poor ground, the spiles can simply be driven into the face as illustrated in figures 147 and 148. In more competent rock of better quality, the spiles or forepoles are usually forced into holes drilled ahead of the face. The lower margin sketch, taken from Proctor and White[6], illustrates a forepoling technique which can be used for tunnelling through cohesionless sand or gravel or "running ground". Breasting to prevent ravelling of the face is not

Figure 147 : Spiling through extremely poor fault material in a tunnel in India.

Figure 148 : Detail of figure 147 showing spiles driven between steel sets to provide an umbrella under which excavation of the face and the installation of the next set could be carried out. In this case, the fault material was soft enough for the spiles to be driven in by means of a heavy hand-held hammer.

usually required if a slight amount of cohesion is present in the rock mass and if a vertical face will stand long enough for the rib installation and spiling to proceed.

Brekke and Korbin[252] describe some interesting tunnelling applications of spiling or forepoling. They discuss the action of spiles in terms of the rock-support interaction concept and suggest that, because the support from these spiles is available at the very start of tunnel deformation, they provide very effective support.

A support method which is increasingly being used in the mining industry is the pre-placement of grouted dowels or cables before the mining of part of an excavation or before the excavation is subjected to stress.

A good example of this technique has already been discussed on page 272 and illustrated in figure 133. In this case reinforcing bars were grouted into holes drilled into the roof and sidewalls during the excavation of a mine tunnel. This tunnel was subjected to a high stress level at a later stage when adjacent ore was being mined and the grouted bars provided very effective support during the period in which the tunnel was subjected to high stress. Note that these bars were grouted in place *in anticipation* of the high stress conditions encountered *later* in the life of the tunnel. One of the difficulties in using this approach is convincing the miners that the installation of support during excavation, when the stability of the tunnels is usually excellent, is not a complete waste of money. It is very difficult to believe that untensioned grouted reinforcing bars will work as well as they do unless one sees practical evidence such as that illustrated in figure 133.

Another example of the use of grouted dowels is illustrated in figure 149 which shows the system which can be used to provide support for a drawpoint in a large mechanised metal mine. Stability of drawpoints is critical in mining operations in which broken ore is drawn from the bottom of a mined opening. In particular, failure of the brow of the drawpoint can result in complete loss of control of the drawing operation.

The drawpoint is excavated before the ore is broken and, at this stage, it is normally subjected to low stresses and its stability is generally adequate. During blasting and drawing of the ore, the drawpoint is subjected to high stresses and considerable wear and, unless the rock is correctly supported, failure of the drawpoint can occur. Clearly, once the drawpoint is in use, support is very difficult to install and it is extremely important that the required support be installed *before* the ore is broken and drawn.

A used drawpoint photographed from the trough after it had been drawn empty. This drawpoint is unreinforced and has lasted very well. Note the signs of wear above the brow.

A typical support pattern is shown in figure 149 and consists of untensioned grouted reinforcing bars of 2 to 3m length placed on a gird of about $1\frac{1}{2}$ x $1\frac{1}{2}$m in the roof and upper sidewalls of the drawpoint and scram excavations. The brow area, shown shaded in figure 149 is blasted last and only when it has been pre-supported as shown. Angled grouted reinforcing bars are installed from the drawpoint and from the trough drive. These grouted bars reinforce the rock mass which forms the brow of the drawpoint and they have been found to provide excellent stability and resistance to wear.

Practical experience suggests that grouted reinforcing bars provide the best form of support in the drawpoint brow area. Since the scram and the intersection of the scram and the drawpoint are not subjected to the same stress and abrasion as the brow area, lighter reinforcement can be used to support the rock in these parts of the excavation. Split sets, described in Chapter 9, have been found to be a convenient and economical form of support for these areas. Mechanically anchored bolts are also suitable for this support duty but they are generally more expensive and take longer to install than split sets.

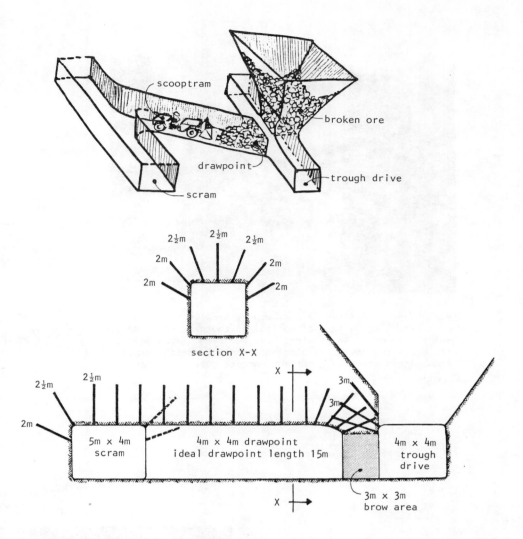

Figure 149 : Use of grouted reinforcing bars to pre-support a drawpoint in a large mechanised mine. The brow area, shown shaded, is blasted last after reinforcement has been installed from the drawpoint and from the trough drive.

Figure 150 illustrates an example of drawpoint brow failure in a large mechanised metal mine. In this case the drawpoint was unreinforced and progressive failure of the rock surrounding the opening occurred when the draw commenced. Attempts to reinforce the drawpoint brow with a heavy steel beam set in concrete were not successful in this case.

In the same mine, the pre-support system detailed in figure 149 was used. Grouted reinforcing bars were placed during development of the drawpoint and before the brow area or the ore had been blasted. Figure 151 shows the conditions in this pre-reinforced drawpoint during operation and it will be seen that, as compared with the situation illustrated in figure 150, excellent stability has been achieved. The total cost of this pre-reinforcement was very much less that that caused by the production delay and loss of ore in the example shown in figure 150.

Figure 150 : Failure of a drawpoint brow in a mechanised mine resulting in loss of control of the process of drawing the broken ore. No reinforcement had been placed before the ore was blasted and attempts to reinforce the drawpoint brow with a steel beam set in concrete were not successful.

Figure 151 : Pre-reinforced drawpoint in a mechanised mine. Reinforcing bars were grouted in place, as illustrated in figure 149, during development of the drawpoint and before the ore was blasted. Excellent stability was achieved in this example.

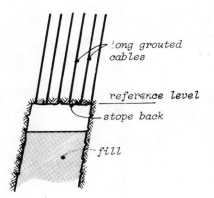
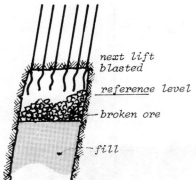
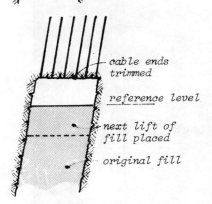
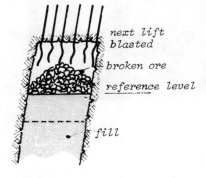

Since the early 1970s, the Australian and Canadian metal mining industries have used long grouted cables and continuously threaded reinforcing bars grouted into long holes to pre-reinforce mine openings. In particular, this pre-reinforcement has been used to support the backs of cut and fill stopes as illustrated diagrammatically in the margin sketches.

In cut and fill mining, ore is removed in a series or horizontal slices with fill being placed after each slice has been removed in order to provide a working platform for the removal of the next slice. Because the method involves having men and equipment working in the stope under the freshly blasted back, it is essential that the stope back be stabilised. Conventional rockbolts and timber support were used before the introduction of long grouted cables, but these proved to be both expensive and, in the case of the rockbolts, unreliable.

Fully grouted tensioned cables were introduced for cut and fill stope back support at the New Broken Hill Consolidated Ltd. mine in 1973[253]. Twenty metre long cables were placed in 64mm diameter holes drilled on a 1.8m x 1.8m pattern in the back of the stope. The cables, illustrated in figure 152, consisted of seven 7mm diameter high tensile steel wires with an expansion shell anchor at one end. The cables were tensioned to 180 kN and then grouted in place with a low viscosity cement based grout. The grouting procedure is illustrated in figure 153. Clifford[253] showed that, in trial areas in which this type of pre-reinforcement was used, a dramatic improvement in stope back conditions was achieved. Similar trials carried out at Mount Isa Mine, using cables of the same configuration, were equally successful[254].

At the CSA mine at Cobar in Australia, Palmer et al[255] used similar seven strand cable to that used at Broken Hill and Mount Isa but they left it *untensioned*. Similar improvements in back stability were achieved and the elimination of the tensioning stage greatly simplified the installation procedure. Very simple anchors were found to be sufficient to support the weight of the cable during grouting and 18m long cables were used.

As shown in the margin sketch, the ends of the cables are trimmed (using a friction cutter or a welding torch) after each lift has been blasted. An 18 to 20m long set of cables will generally pre-reinforce the rock for three or four lifts and then another set of cables is installed. An overlap of at least 2m has been found to provide sufficient continuity of support between one set of cables and the next. Because of the difficulty experienced with hole deviation for drillhole lengths of 20m, some mines are tending to reduce the length of the cables and hence the number of lifts which can be supported with one installation. No deterioration of the effectiveness of the support has been observed with these shorter cables.

While it is now accepted that pre-placed untensioned grouted cables give a spectacular improvement in the overall stability of the rock mass above the back of a stope, there is still a problem with small pieces of loose rock which can fall from the surface of the back. Some mines use conventional rockbolts to stabilise this near surface rock but the installation of these bolts is a time-consuming and expensive operation. In order to overcome this problem, Cobar Mines Pty. Ltd.

Figure 152 : Seven strand high tensile steel cable used for pre-reinforcement of cut and fill stope backs at the New Broken Hill Consolidated Ltd. mine in Australia. The expansion shell anchor is used to anchor the cable at the top of a long hole to permit tensioning before the cable is fully grouted.

Figure 153 : Grouting a stressed 20m long high tensile steel cable in a stope back at the New Broken Hill Consolidated Ltd. mine.

Stressed end of a 20m long grouted high tensile steel cable used for cut and fill stope back support at the New Broken Hill Consolidated Ltd. mine in Australia.

Trimmed end of a grouted seven strand high tensile steel cable after the removal of a lift at the New Broken Hill Consolidated Ltd. mine.

have used 16mm diameter threadbar in place of multi-strand cables. While this threadbar is significantly more expensive than the cable, it does have the advantage that a nut and washer can be attached to the end projecting from the back after a blast and this generally eliminates the need to place additional rockbolts[255].

Since 1974, the Commonwealth Scientific and Industrial Research Organisation have been sponsored by the Australian metal mining industry to carry out a number of research studies into the mechanics of pre-placed support. The results of this research are summarised in a series of papers and reports[254,256,257,258] and the interested reader is strongly recommended to consult these publications for further details.

Fuller and Cox[257] found that the effectiveness of pre-placed support is critically dependent upon the grout to steel bond to transfer shear stress to the rock mass. Lightly rusted steel with no loose surface material is found to give the best bond. A cement grout with a water-cement ratio of less than 0.5 and with an additive to minimise "bleeding" is recommended for use with lightly rusted steel.

Fuller and Cox[258] have also proposed a theoretical model based upon the role of pre-placed support in limiting displacement on rock joints to explain the effectiveness of this type of reinforcement. This theoretical model differs in detail but not in overall concept from the mechanisms discussed earlier in this chapter in which the retention of the interlocking of a rock mass is proposed as one of the most critical factors in the behaviour of rock masses.

The opening sentence in this chapter reads - the principal objective in the design of underground excavation support is to help the rock mass to support itself. Pre-placed grouted reinforcing elements are probably the most effective means of achieving this objective and the authors have no doubt that the future will see a great increase in the use of this support technique.

Suggestions for estimating support requirements

From the discussion presented in this chapter it will be obvious to the reader that it is not possible to list a simple set of rules to govern the choice of underground support. The optimum support system for a particular underground excavation will depend upon the mechanical characteristics of the rock mass, the in situ stress field, the loading history to which the rock mass will be subjected and also upon the availability and cost of different types of support.

In order to provide some guidance to the reader, the following steps are suggested for a preliminary evaluation of support requirements:

1. During the site investigation or exploration adit stage of an underground excavation project, classify the rock mass by means of either the CSIR or the NGI rock mass classification systems (preferably both) as set out in chapter 2.

2. Make a preliminary evaluation of support systems on the

basis of the recommendations summarised in tables 18, 19 and 20 in this chapter. Note that these recommendations are based upon precedent experience and tend to emphasise support systems used in particular countries (Scandinavia and South Africa). They also tend to be conservative.

3. Estimate the in situ stress conditions from measurements of stress at the site or at adjacent sites in similar rock. If no measurements are available, use figures 40 and 41 on pages 99 and 100 to obtain a crude first estimate of in situ stresses.

4. From figure 118 on page 222, estimate the maximum boundary stresses in the rock surrounding the proposed excavation. Note that very high stress concentrations at sharp corners are not included in this figure.

5. If figure 118 shows that tensile stresses can occur on the excavation boundary, turn to Appendix 3 and find the excavation shape and in situ stress conditions closest to those under consideration. Consider the extent of the tensile stress zone (from the minor principal stress contours) and explore the extent to which this zone could be eliminated or minimised by small changes in the excavation shape. If no such change is possible, consider the length of rockbolts or cables which would be necessary to stabilise this zone and estimate the capacity of this support from the volume of material subjected to tensile stress.

6. If figure 118 shows that only compressive stresses are present around the opening, calculate the maximum boundary stress (excluding sharp corners) from the stress concentration value given by figure 118 and the in situ stress estimated in step 3. Compare this maximum stress value with the unconfined compressive strength of the rock mass given by $\sigma_{cm} = \sqrt{s\sigma_c^2}$, where the value of s is estimated from table 12 on page 176. If the boundary stress exceeds the unconfined compressive strength of the rock mass, carry out a more detailed analysis of the extent of the potential failure zone as shown in figures 119 and 120. This last step is not a trivial one and the inexperienced reader may wish to call for help. The experienced geotechnical engineer would probably turn to a numerical technique such as the boundary element method presented in Appendix 4.

7. Whatever the outcome of the analyses described in steps 5 and 6 above, an examination must be made of the potential for structurally controlled instability. Using whatever geological information is available, examine the possibility of wedges or blocks forming by means of the stereographic techniques described at the beginning of chapter 7. If the potential for this type of failure exists, consider the length and capacity of rockbolts or cables required to stabilise the excavation.

8. Taking all of these analyses into account, consider the consequences of excavation sequence and possible variations in the timing of support installation.

9. If rockbolts are to be used for support, check that the various requirements do not conflict with one another. Check the rockbolt design against precedent experience.

The following empirical rules, originally devised during the Snowy Mountains project in Australia by Lang[259], provide a useful check for proposed bolt lengths and spacings :

Minimum bolt length
Greatest of :
a. Twice the bolt spacing.
b. Three times the width of critical and potentially unstable rock blocks defined by average joint spacing in the rock mass.
c. For spans of less than 6 metres (20 feet), bolt length of one half the span.
   For spans of 18 to 30 metres (60 to 100 feet), bolt length of one quarter of span in roof.
   For excavations higher than 18 metres (60 feet), sidewall bolts one fifth of wall height.

Maximum bolt spacing
Least of
a. One half the bolt length.
b. One and one half times the width of critical and potentially unstable rock blocks defined by the average joint spacing in the rock mass.
c. When weldmesh or chain-link mesh is to be used, bolt spacing of more than 2 metres (6 feet) makes attachment of the mesh difficult (but not impossible).

10. Once these estimates have been made, sit down at a drawing board and draw a typical cross-section of the excavation. Superimpose an approximate pattern of structural features and try to imagine what the final excavation profile will look like and the size and shape of the potential failure zone around the excavation. Draw the support system to scale and check that it looks right and that there is enough room available to drill holes, install bolts, place concrete.
    This final step may be the most important since it will often highlight deficiencies and anomalies which are not obvious when examining the results of a numerical model study, the predictions of a rock-support interaction analysis or the recommendations derived from a rock mass classification system.
    A typical support drawing is presented in figure 154.

Additional reading

Within the scope of this book it is obviously impossible to cover all the theoretical and practical details of rock support. An attempt has been made to give the reader a basic understanding of the behaviour of rock masses surrounding underground excavations and of the different approaches which can be taken to designing support systems to stabilize these rock masses. The reader is strongly urged to consolidate this basic understanding with additional reading in fields of particluar interest and a selection of reference material is listed hereunder to assist the reader in finding relevant information.

Timber support :
   Because of increasing concern with environmental factors and an increasing cost of both timber and labour, the use of timber for underground excavation support is not nearly as common as in the earlier part of this century. Many of the techniques used in timbering, particularly in mining, have been passed down from generation to generation as part of the "art" of mining.

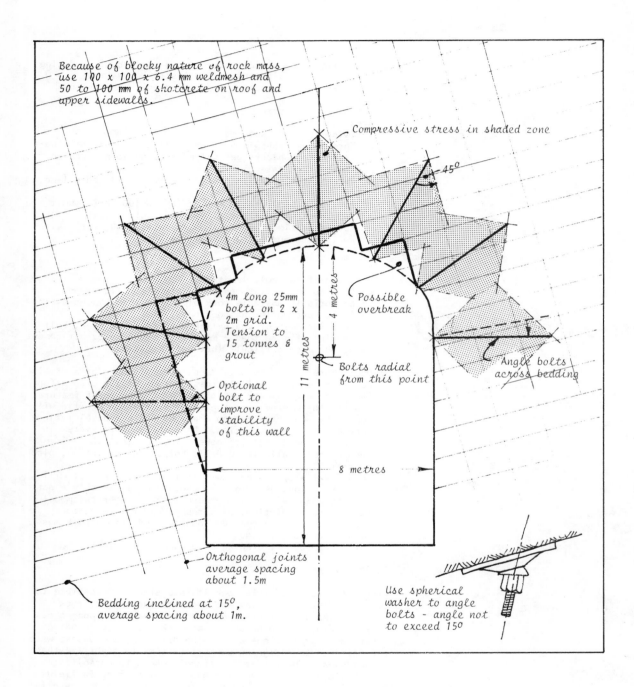

Figure 154 : Typical working sketch used during preliminary layout of rockbolting pattern for a tunnel in jointed rock.

One of the best references of which the authors are aware is the three volume mining textbook by Woodruff, timbering being dealt with in Chapter 5 of Volume 2^{260}.

Steel sets :
The most widely used reference on steel sets in America is "Rock Tunnelling with Steel Supports" by Proctor and White [6] which contains Terzaghi's paper on rock loads (see pages 14 to 17) plus a wealth of detail on steel supports. Woodruff[260], Volume 2, chapters 4 and 6, discusses both yielding and conventional steel sets. Szechy[192] gives a number of design details and the US Army Corps of Engineers[244] give some detailed examples on the design of steel set supports.

Concrete lining :
Szechy's book, "The Art of Tunnelling"[192] contains a great deal of detail on methods of calculating the support loads provided by tunnel lining, particularly concrete. This book is interesting because it makes extensive reference to work carried out in the USSR and in Europe and which is not readily available in the English language.
The US Army Corps of Engineers manual on Shafts and Tunnels in Rock[244] contains a substantial amount of information on the design of concrete tunnel linings as well as a listing of 104 tunnelling projects with details of geology, construction methods, support systems and cost.
Woodruff[260], Volume 2, Chapter 8, gives a very practical discussion on concrete lining methods and equipment with particular emphasis on mining applications.

Shotcrete :
The most comprehensive reference of shotcrete of which the authors are aware is a report entitled "Shotcrete Practice in Underground Construction" prepared for the US Department of Transportation by Mahar, Parker and Wuellner of the University of Illinois[261]. This large report, available from the National Technical Information Service, Springfield, Virginia 22151, USA, is essential reading for anyone seriously interested in the use of shotcrete for underground support.
The American Society of Civil Engineers published an excellent volume in 1974 containing the proceedings of a conference entitled "Use of Shotcrete for Underground Structural Support"[262]. This volume contains 35 papers and summaries of discussions held at the conference.

Rockbolts :
In spite of the extensive use of rockbolts for the support of underground excavations, surprisingly few comprehensive publications on this subject are readily available.
A comprehensive report entitled "Rock Reinforcement" (Engineering and Design) has recently been published by the US Army Corps of Engineers[263] and this book will be very useful for civil engineering applications. Books such as those by Szechy[192], Woodruff[260] and Obert and Duvall[68] contain rather sketchy discussions on rockbolting and the authors consider that the most useful information currently available is contained in individual papers.
The most important of these papers are those published

by Lang[259 264], Panek[265,266], Benson et al[267] and Cording et al[268]. Because of the current lack of readily available information on the use of rockbolts for underground excavation support, an additional chapter containing a summary of some of the practical aspects of rockbolting has been added to this book.

Cable anchors :

Grouted cables are sometimes used for the reinforcement of large structures and critical underground excavations. In a book entitled "Anchoring in Rock", Hobst and Zajic[269] have given a very extensive discussion on the design and many of the practical details of reinforcement by means of cables. An excellent series of papers on this subject has also been published by Littlejohn and Bruce[270].

Chapter 8 references

224. DAEMEN, J.J.K. Problems in tunnel support mechanics. *Underground Space*, Vol. 1, 1977, pages 163-172.

225. DAEMEN, J.J.K. Tunnel support loading caused by rock failure. *Tech. Report* MRD-3-75, Missouri River Division, Corps of Engineers, Omaha, 1975.

226. FRANKLIN, J.A. and WOODFIELD, P.F. Comparison of a polyester resin and a mechanical rockbolt anchor. *Trans. Inst. Min. Metall.*, London, Vol.80, Section A, Bulletin No 776, 1971, pages A91-100.

227. RABCEWICZ, L.V. The new Austrian tunnelling method. *Water Power*, Vol. 16, 1964, pages 453-457 (part 1) and Vol. 17, 1965, pages 19-24 (part 3).

228. LADANYI, B. Use of the long-term strength concept in the determination of ground pressure on tunnel linings. *Proc. 3rd Congr., Intnl. Soc. Rock Mech.*, Denver, 1974, Vol. 2B, pages 1150-1156.

229. LOMBARDI, G. The influence of rock characteristics on the stability of rock cavities. *Tunnels and Tunnelling*, Vol.2, 1970, pages 104-109.

230. LOMBARDI, G. Dimensioning of tunnel linings with regard to constructional procedure. *Tunnels and Tunnelling*, Vol. 5, 1973, pages 340-351.

231. EGGER, P. Gebirgsdruck im Tunnelbau und Stützwrikung der Ortsburst bei Uberschreiten der Gebirgsfestigkeit. *Proc. 3rd Congr., Intnl. Soc. Rock Mech.*, Denver, 1974, Vol. 2B, pages 1007-1011.

232. PANET, M. Analyse de la stabilité d'un tunnel creusé dans un massif rocheux en tenant compte du comportement apres la rupture. *Rock Mechanics*, Vol. 8, 1976, pages 209-223.

233. LAMA, R.D. and VUTUKURI, V.S. *Handbook on Mechanical Properties of Rocks.* Volume III, Appendix III. Trans Tech Publications, Aedermannsdorf, Switzerland, 1978.

234. Mc NAY, L.M. and CORSON, D.R. Hydraulic sandfill in deep metal mines. *U.S. Bureau of Mines Information Circular* 8663, 1975.

235. SINGH, K.H. Cemented hydraulic fill for ground support. *Canadian Inst. Min. Bull.*, Vol.69, No.765, 1976, pages 69-74.

236. THOMAS, E.G. Cemented fill practice and research at Mount Isa. *Proc. Australian Inst. Min. Metall.*, No. 240, 1971, pages 33-51.

237. MATHEWS, K.E. and KAESEHAGEN, F.E. The development and design of a cemented rock filling system at Mount Isa. *Jubilee Symposium on Mine Filling*, Mount Isa, Australia, 1973, pages 13-23.

238. EWOLDSEN, H.M. and Mc NIVEN, H.D. Rockbolting of tunnels for structural support. Part 11 - Design of rockbolt systems. *Intnl. J. Rock Mechanics and Mining Sciences*, Vol. 6, 1969, pages 483-497.

239. SAKURAI, S. and YAMAMOTO, Y. A numerical analysis of the maximum earth pressure acting on a tunnel lining. *Proc. 2nd Conf. on Numerical Methods in Geomechanics*, published by ASCE, 1976, Vol. 2, pages 821-833.

240. MANFREDINI, G., MARTINETTI, S., RIBACCHI, R, and RICCIONI, R. Design criteria for anchor cables and bolting in underground openings. *Proc. 2nd Conf. on Numerical Methods in Geomechanics*, published by ASCE, 1976, Vol. 2, pages 859-871.

241. CUNDALL, P.A. Rational design of tunnel supports ; a computer model for rock mass behaviour using interactive graphics for the input and output of geometrical data. *U.S. Army Corps of Engineers Technical Report* MRD-2-74, 1974.

242. VOEGELE, M., FAIRHURST, C. and CUNDALL, P.A. Analysis of tunnel support loads using a large displacement, distinct block model. *Proc. 1st Intnl. Symp. Storage in Excavated Rock Caverns*, Stockholm, 1977, Vol. 2, pages 247-252.

243. SCOTT, J.J. Friction rock stabilizers - a new rock reinforcement method. *Proc. 17th Symposium on Rock Machanics*, Snowbird, Utah, 1976, pages 242-249.

244. U.S. ARMY CORPS OF ENGINEERS. *Shafts and Tunnels in Rock.* Engineer Manual EM 1110-2-2901, 1978.

245. LAUBSCHER, D.H. and TAYLOR, H.W. The importance of geomechanics classification of jointed rock masses in mining operations. *Proc. Symp. Exploration for Rock Engineering*, Johannesburg, 1976, pages 119-128.

246. HOUGHTON, D.A. The role of rock quality indices in the assessment of rock masses. *Proc. Symp. Exploration for Rock Engineering*, Johannesburg, 1976, pages 129-135.

247. BERRY, N.S.M. and BROWN, J.G.W. Performance of full facers on Kielder tunnels. *Tunnels and Tunnelling*, Vol. 9, No. 4, 1977, pages 35-39.

248. NICHOLSON, K. Coping with difficult ground on the fullface mechanised tunnel drive at Kielder. *Tunnels and Tunnelling*, Vol. 11, No. 5, 1979, pages 55-57.

249. FREEMAN, T.J. The behaviour of fully-bonded rock bolts in the Kielder Experimental tunnel. *Tunnels and Tunnelling*, Vol. 10, No. 5, 1978, pages 37-40.

250. WARD, W.H., COATS, D.J. and TEDD, P. Performance of tunnel support systems in the Four Fathom Mudstone. *Proc. "Tunnelling '76"*, Published by Inst. Min. Metall., London, 1976, pages 329-340.

251. MATHEWS, K.E. and EDWARDS, D.B. Rock mechanics practice at Mount Isa Mines Limited, Australia. *Proc. 9th Commonwealth Mining and Metallurgical Congress*, London, 1969, paper 32.

252. BREKKE, T.L. and KORBIN, G. Some comments on the use of spiling in underground openings. *Proc. 2nd Intnl. Cong. Intnl. Assn. Engg. Geol.*, Sao Paulo, Brazil, Vol. 2, 1974, pages 119-124.

253. CLIFFORD, R.L. Long rockbolt support at New Broken Hill Consolidated Limited. *Proc. Australian Inst. Min. Metall.*, Vol. 251, 1974, pages 21-26.

254. FULLER, P.G. Pre-reinforcement of cut and fill stopes. *Proc. Conf. Application of Rock Mechanics to Cut and Fill Mining*, Lulea, Sweden, June 1980, *in press*.

255. PALMER, W.T., BAILEY, S.G. and FULLER, P.G. Experience with preplaced supports in timber cut and fill stopes. *Proc. Australian Mineral Industries Research Association Technical Meeting*, Woolongong, Australia, 1976, pages 45-75.

256. WILLOUGHBY, D.R. Rock mechanics applied to cut and fill mining in Australia. *Proc. Conf. Application of Rock Mechanics to Cut and Fill Mining*, Lulea, Sweden, June 1980, *in press*.

257. FULLER, P.G. and COX, R.H.T. Mechanics of load transfer from steel tendons to cement based grout. *Proc. 5th Australian Conf. Mech. of Struct. and Mat.*, Melbourne, 1975, pages 189-203.

258. FULLER, P.G. and COX, R.H.T. Rock reinforcement design based on control of joint displacement - a new concept. *Proc. 3rd Australian Tunnelling Conf.*, Sydney, 1978, pages 28-35.

259. LANG, T.A. Theory and practice of rockbolting. *Trans. Amer. Inst. Mining Engineers*, Vol. 220, 1961, pages 333-348.

260. WOODRUFF, S.D. *Methods of Working Coal and Metal Mines*, Vol. 2 - Ground Support Methods, Pregamon Press, New York, 1966, 430 pages.

261. MAHAR, J.W., PARKER, H.W and WUELLNER, W.W. *Shotcrete Practice in Underground Construction*. US Department of Transportation Report FRA-OR&D 75-90, 1975. Available from the National Technical Information Service, Springfield, Virginia 22151.

262. A.S.C.E.. *Use of Shotcrete for Underground Structural Support*. Proc. Engineering Foundation Conference, Maine 1973, published by American Society of Civil Engineers, New York, 1975.

263. US ARMY CORPS OF ENGINEERS. *Rock Reinforcement - Engineering and Design*. US Army Corps of Engineers, Engineer Manual EM 1110-1-2907, 1980.

264. LANG, T.A. Rock reinforcement. *Bull. Association Engineering Geologists*, Vol. IX, No. 3, 1972, pages 215-239.

265. PANEK, L.A. The combined effects of friction and suspension in bolting bedded mine roofs. *U.S. Bureau of Mines Report of Investigations* 6139, 1962.

266. PANEK, L.A. Design for bolting stratified roof. *Transactions American Institute of Mining Engineers*, Vol. 229, 1964, pages 113-119.

267. BENSON, R.P., CONLON, J.,MERRITT, A.H., JOLI-COEUR, P. and DEERE, D.U. Rock mechanics at Churchill Falls. *Proc. Symp. Underground Rock Chambers*, Pheonix, 1971, Published by ASCE, 1971, pages 407-486.

268. CORDING, E.J., HENDRON, A.J. and DEERE, D.U. Rock engineering for underground caverns. *Proc. Symp.Underground Rock Chambers*, Pheonix, 1971, Published by ASCE, 1971, pages 567-600.

269. HOBST, L. and ZAJIC, J. *Anchoring in Rock*. Elsevier, New York, 1977, 390 pages.

270. LITTLEJOHN, G.S. and BRUCE, D.A. Rock anchors : state of the art. *Ground Engineering,* Vol. 8, No. 3, 1975, pages 25-32; Vol. 8, No. 4, 1975, pages 41-48; Vol. 8, No. 5, 1975, pages 34-45; Vol. 8, No. 6, 1975, pages 36-45; Vol. 9, No. 2, 1976, pages 20-29; Vol. 9, No. 3, 1976, pages 55-60; Vol. 9, No. 4, 1976, pages 33-44.

Chapter 9: Rockbolts, shotcrete and mesh

Introduction

The use of steel sets and concrete linings for the support of tunnels and other underground excavations has been common in civil engineering for several decades. Consequently, the practical details associated with the use of these support systems have been dealt with in a number of readily available text books and papers.

The same is not true for support systems incorporating rockbolts as the principal support element. In spite of the widespread use of rockbolts and dowels, particularly by the mining industry, there are surprisingly few comprehensive books or even papers which deal with this support method. Because the authors are convinced that rockbolts, together with mesh reinforced shotcrete, will become the dominant support system in the future, they have decided to include this chapter which summarises some of the most important practical aspects of the use of rockbolts, shotcrete and mesh.

Organisation of a rockbolting programme

Figure 155 illustrates a traditional excavation and support sequence commonly used by the civil engineering industry in the construction of large underground projects. Note that two distinct cycles are involved - one for excavation and the installation of "temporary" support and the second for "permanent" support installation.

If one reflects on the historical development of the underground civil engineering construction industry, it is easy to see why these two cycles have developed. When using steel sets or concrete lining it is very difficult to organise an efficient working cycle at the face if the support has to be installed after each mucking cycle. The cumbersome operation required to place steel sets or the bulky formwork required to cast a concrete lining do not mix well with the relatively mobile equipment used in the drill-blast-muck cycle. Consequently, it has become common to use rockbolts for "temporary" support and to install these as part of the excavation cycle. Once the face has advanced a convenient distance, the "permanent" support system is installed by a different crew working on a completely independent cycle.

A civil engineering tradition which has grown up in parallel with the use of independent excavation and support cycles is that the Contractor is responsible for the choice and installation of the "temporary" support while the Engineer is responsible for the design and supervision of construction of the "permanent" support system. The authors regard this as one of the least helpful traditions in civil engineering since, in their view, there is no logical distinction between support systems used in underground construction. It will be clear to anyone who has read the previous chapter that each support system plays a vital part in controlling the ultimate behaviour of the excavation, even if two or three support systems are used at widely spaced time intervals. The only exception to this comment is the pinning of small loose blocks in an excavation which is inherently stable and where a concrete or steel lining is installed at a later stage for hydraulic or other reasons.

When rockbolts are used as the principal element in a support

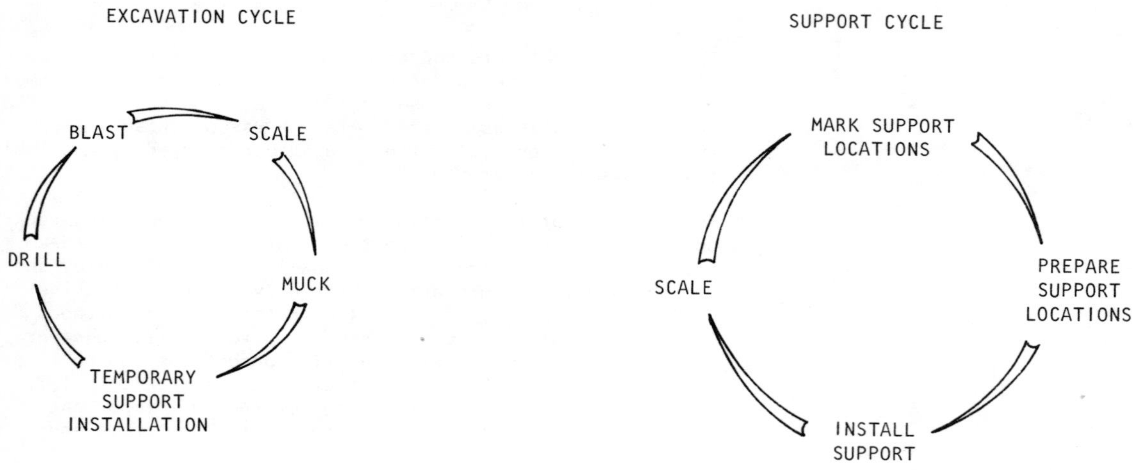

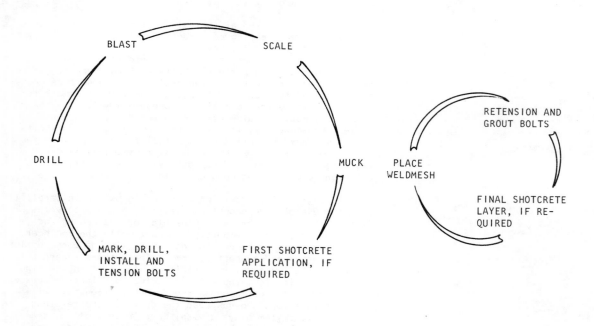

Figure 155 : Traditional excavation and support cycles used for steel set and concrete lining support in the civil engineering industry.

Figure 156 : Integrated excavation and support cycle used when rockbolts are used as the principal support element.

system, there is no logic at all in referring to one rockbolt system, installed during the excavation cycle, as "temporary" while a second system installed later is called "permanent". Any rockbolt which does not fulfil a "permanent" support function is simply a waste of money (again, pinning of small loose blocks excepted). Consequently, it is much more logical to consider the use of an integrated excavation and support cycle such as that illustrated in figure 156.

The equipment used to drill for and to install rockbolts is very similar to that used in the drill-blast-muck cycle and there is no major problem in integrating these two activities. In addition to the organisation of the physical activities, it is essential that the division of responsibilities between the Contractor and the Engineer be fully understood and agreed before commencement of the work. The traditional arrangement in which the Contractor is responsible for part of the support work while the Engineer is responsible for another part will not work when an integrated excavation/support system is used. In the authors' experience, it is preferable for the Engineer to assume full responsibility for the design and construction supervision of all support and to work closely with the Contractor in sorting out practical details and day to day construction problems.

Readers who are not familiar with civil engineering practice may regard some of the preceding comments as obvious and even redundant. Unfortunately, civil engineering traditions are very deep rooted and change very slowly and anyone who has experienced the difficulties which can arise from an inappropriate division of responsibilities between Contractor and Engineer on the question of support will appreciate that this is not a trivial problem.

A further difficulty which must be resolved, once the integrated system illustrated in figure 156 has been adopted, is the organisation of the actual support installation activity. In the authors' experience, this activity requires a different approach to that required for the normal drill-blast-muck cycle in which the emphasis must be on speed and efficiency rather than on detail. The emphasis on "producing rock" which characterises a good section engineer or foreman responsible for advancing a heading is inappropriate for someone who must ensure that each bolt is correctly installed, tensioned and grouted.

A good solution, adopted on many jobs where there has been conflict between production and support interests, is to create a specialist support crew. This crew should be led by an engineer with an interest in engineering detail and quality rather than in the grander scale achievement in which details are blurred. Depending upon the size of the job, this engineer should have one or more crews of about six men who carry out the bolt installation, tensioning and grouting.

After each blast, the support engineer should examine the rock conditions, determine the support requirements and mark out the rockbolt drillhole positions, inclinations and lengths (a can of spray paint is an invaluable aid in this exercise). The holes are drilled by the normal jumbo crew, immediately before or after drilling the blastholes, and the specialist support crew then moves in and installs and tensions the rockbolts. At a later stage, as shown in figure

156, the support crew can return to retension the bolts, place weldmesh and shotcrete and grout the bolts. All of these activities should be carried out under the close supervision of the support engineer who should carry sufficient rank that he can stop the job if, in his opinion, conditions are unsafe.

It has been found that this system can work very well if sufficient attention is paid to choice of the correct individual to lead the support team and the clear definition of his responsibilities in relation to the production engineers with whom he will have to work. Experience shows that there is minimal disruption of the overall cycle if the system is working correctly and that rapid and safe advance can be achieved.

Review of typical rockbolt systems

There are hundreds of different types of rockbolts and dowels used throughout the world and it is obviously impossible to cover all of these in the space of this chapter. A selection of representative types has been made and details are presented in a series of drawings in figures 157 to 165. The comments on these drawings are largely self-explanatory and the following comments are intended to provide a general summary of typical applications.

Wooden dowels - developed to minimise damage to coal cutting equipment and conveyor belts and used during war-time steel shortages, these untensioned dowels are only suitable for very light support duty. They are very rarely used today and are included mainly for historical interest.

Untensioned anchored or grouted steel dowels - including "Perfobolts", Worley bolts, Split Sets and grouted rebar, are used in situations where very rapid installation of support is possible. These dowels can only accept load as they are strained by deformation in the surrounding rock and, if the dowels are installed too far behind the advancing face, most of the short-term rock deformation will have taken place and the dowels will be ineffective. In the hands of a skilled contractor using a fully integrated excavation/support cycle, these devices can be very effective. When improperly used they can be a waste of money at best and a disaster at worst.

Tensioned mechanically anchored bolts - most commonly used in the mining industry and, with subsequent grouting, in underground civil engineering construction. Slotted bolt and wedge anchors are only effective in very good rocks and have largely been replaced by expansion shell anchors of which there are many varieties. Very high bolt tension, approaching the yield strength of the bolt shank, can be achieved in good quality rock masses but local crushing of the rock at the anchor points will result in anchor slip in poorer quality rock.

Grouted tensioned rockbolts - developed to give improved anchorage in poor ground and corrosion protection for steel bolt shanks, this type may well become the dominant rockbolt system in the future. The most sophisticated system is the two stage resin cartridge system (figure 165) giving a fully tensioned high capacity bolt in one simple and rapid operation. In spite of high resin cost, installed cost of the complete system compares well with other rockbolts due to reduced labour costs.

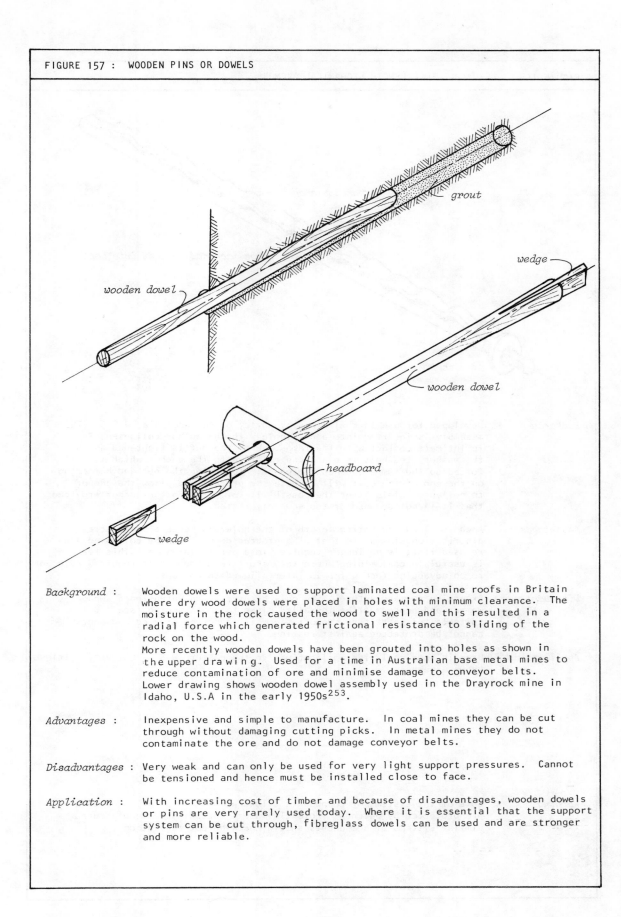

FIGURE 157 : WOODEN PINS OR DOWELS

Background : Wooden dowels were used to support laminated coal mine roofs in Britain where dry wood dowels were placed in holes with minimum clearance. The moisture in the rock caused the wood to swell and this resulted in a radial force which generated frictional resistance to sliding of the rock on the wood.
More recently wooden dowels have been grouted into holes as shown in the upper drawing. Used for a time in Australian base metal mines to reduce contamination of ore and minimise damage to conveyor belts.
Lower drawing shows wooden dowel assembly used in the Drayrock mine in Idaho, U.S.A in the early 1950s[253].

Advantages : Inexpensive and simple to manufacture. In coal mines they can be cut through without damaging cutting picks. In metal mines they do not contaminate the ore and do not damage conveyor belts.

Disadvantages : Very weak and can only be used for very light support pressures. Cannot be tensioned and hence must be installed close to face.

Application : With increasing cost of timber and because of disadvantages, wooden dowels or pins are very rarely used today. Where it is essential that the support system can be cut through, fibreglass dowels can be used and are stronger and more reliable.

FIGURE 158 : RE-USABLE FULL LENGTH MECHANICAL ANCHOR

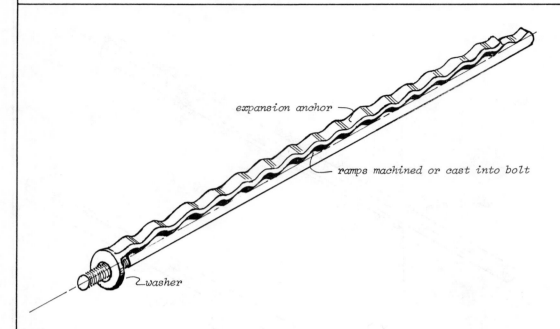

Background : Developed for use in coal mines by Worley of Philadelphia, U.S.A.. The assembly is inserted into a drillhole with the anchor collapsed, ie in intimate contact with the ramps. When the nut is tightened against the washer, the bolt is displaced relative to the anchor which is forced up the ramps and hence expands. Loosening the nut and hammering on the end of the bolt will reverse the process and allow the anchor to collapse. This gives the possibility of re-using the anchor provided that it is not damaged or too severely rusted.

Advantages : Anchors along the entire length of the hole and hence gives a stress distribution similar to that of a grouted dowel. Can be collapsed and re-used if it is no longer required in a particular area. This feature is useful in coal mining where very short term support is required adjacent to an advancing face which is later allowed to collapse.

Disadvantages : Expensive to manufacture. Can only be used as an untensioned support member and hence must be installed close to an advancing face. Only suitable for short term support since it cannot be grouted and hence cannot be protected against rusting.

Applications : Used in some coal mine applications in the eastern U.S.A. but very little known elsewhere.

Warning : Some of the systems or components illustrated in this series of drawings are protected by patents. Potential users intending to manufacture similar systems or components should check patent regulations.

FIGURE 159 : FRICTION ANCHOR OR SPLIT SET

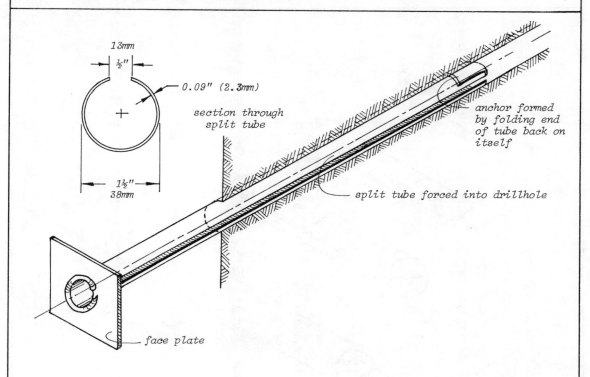

Background : Developed by Scott[243] in conjunction with the Ingersoll-Rand Company in the U.S.A.. This device has gained considerable popularity in the mining industry.
The 1½ inch (38mm) diameter split tube is forced into a 1⅜ inch (35mm) diameter drillhole. The spring action of the compressed tube applies a radial force against the rock and generates a frictional resistance to sliding of the rock on the steel. This frictional resistance increases as the outer surface of the tube rusts.

Advantages : Simple and quick to install and claimed to be cheaper than a grouted dowel of similar capacity.

Disadvantages : Cannot be tensioned and hence is activated by movement in the rock in the same way as a grouted dowel. Its support action is similar to that of an untensioned dowel and hence it must be installed very close to a face. The drillhole diameter is critical and most failures during installation occur because the hole is either too small or too large.
In some applications, rusting has occurred very rapidly and has proved to be a problem where long term support is required. The device cannot be grouted.

Applications : Increasingly used for relatively light support duties in the mining industry, particularly where short term support is required. Very little application in civil engineering at present.

Warning : Some of the systems or components illustrated in this series of drawings are protected by patents. Potential users intending to manufacture similar systems or components should check patent regulations.

FIGURE 160 : "PERFOBOLT" SYSTEM FOR ANCHORING AND GROUTING DOWELS

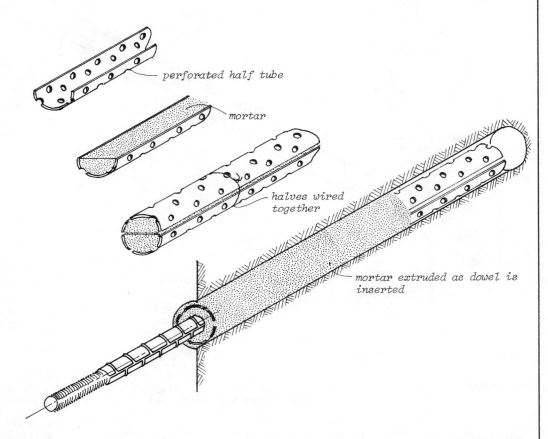

Background : Developed in Scandinavia for grouting and anchoring dowels in boreholes. Perforated half tubes are packed with mortar as shown and the halves are then wired together and inserted in the drillhole. The mortar is extruded when the dowel is pushed down the centre of the tube as shown.
Recommended sizes are as follows :

Reinforcing bar		Drillhole diameter		Sleeve diameter	
3/4 in	19mm	1 1/4 in	32mm	1 1/16 in	27mm
1 in	25mm	1 1/2 in	38mm	1 1/4 in	32mm
1 1/8 in	29mm	1 3/4 in	44mm	1 1/2 in	38mm
1 1/4 in	32mm	2 in	51mm	1 3/4 in	44mm
1 3/8 in	35mm	2 1/4 in	57mm	2 in	51mm

Advantages : Simple and effective if recommended sizes are strictly adhered to. Short length can be used to form anchor for tensioned bolt.

Disadvantages : Relatively expensive compared with grouted dowels.

Applications : Widely used in civil engineering in Scandinavia. Application elsewhere relatively limited.

Warning : Some of the systems or components illustrated in this series of drawings are protected by patents. Potential users intending to manufacture similar systems or components should check patent regulations.

FIGURE 161 : UNTENSIONED GROUTED DOWEL

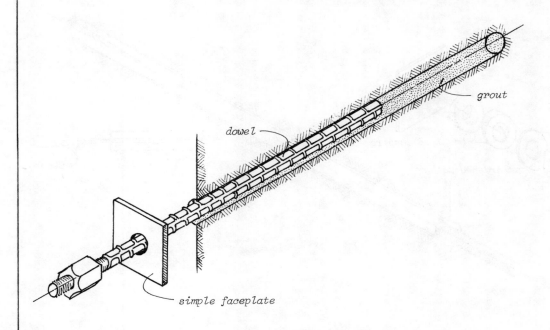

Background : Developed as an inexpensive alternative to the Scandinavian "Perfobolt" system where use of untensioned dowels is appropriate. A thick grout is pumped into the drillhole by means of a simple hand pump or a mono-pump. The dowel is pushed into the grout as shown in the illustration. For up-holes, the dowel is sometimes held in place by a small wooden or steel wedge inserted into the collar of the hole. A faceplate and nut can be added if required although, for very light support, a plain dowel is sometimes used.

Advantages : Simple and inexpensive.

Disadvantages : Cannot be tensioned and hence must be installed before significant deformation of the rock mass has taken place.

Applications : Widely used in the mining industry for light support duties and in civil engineering for mesh fixing and for supporting ventilation tubing, pipework and similar services.

FIGURE 162 : SLOTTED BOLT AND WEDGE

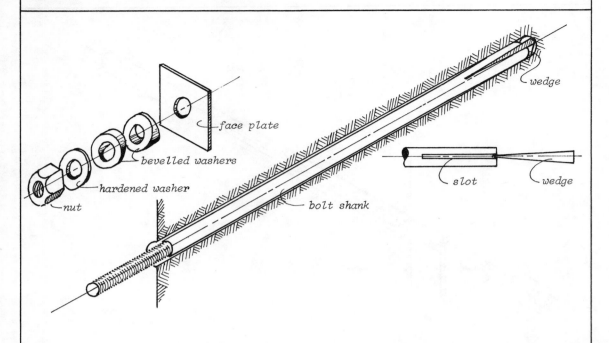

Background :   Probably the earliest type of mechanically anchored rockbolt. Very simple and inexpensive to manufacture and widely used throughout the world. The end of the bolt shank is slotted as illustrated and the wedge is driven home by pushing the assembly against the end of the drillhole. The wedge expands the end of the bolt shank and anchors it in the rock.
Also illustrated are two bevelled washers which are used to accommodate an inclined rock face. The hardened washer is used when the bolt is tensioned by applying a measured torque to the nut.

Advantages :   Simple and inexpensive. In hard rock it provides an excellent anchorage and permits immediate tensioning of the bolt.

Disadvantages : Due to the small contact area between the expanded anchor and the rock, local crushing of the rock with consequent slip of the anchor can occur when the intact rock strength is less than about 10 MPa (1500 lb/in^2).

Applications : Because of the unreliability of the anchor in poor quality rock, the slotted bolt and wedge has given way to the more versatile expansion shell anchor. Relatively rarely used today.

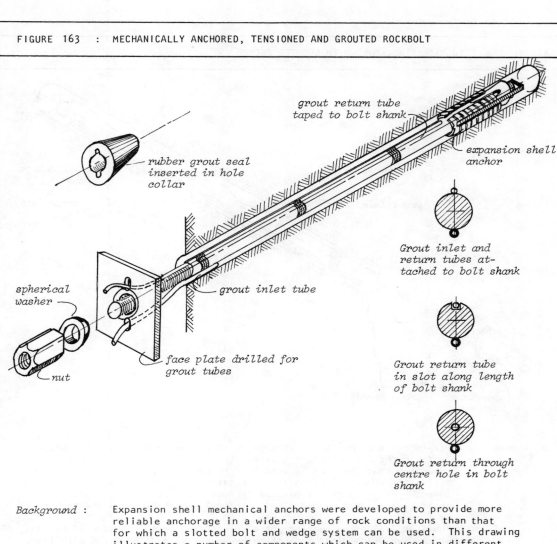

FIGURE 163 : MECHANICALLY ANCHORED, TENSIONED AND GROUTED ROCKBOLT

Background : Expansion shell mechanical anchors were developed to provide more reliable anchorage in a wider range of rock conditions than that for which a slotted bolt and wedge system can be used. This drawing illustrates a number of components which can be used in different combinations. The expansion shell anchor is one of a large number of different types, all of which operate in basically the same way. A wedge, attached to the bolt shank, is pulled into a conical anchor shell forcing it to expand against the drillhole walls.
The rubber grout seal is used to centre the bolt in the hole and to seal the collar of the hole against grout leakage. An alternative system is to use a quick setting plaster to seal the hole collar.
Different grout tube arrangements are illustrated. In all cases the grout is injected into the collar end of the hole (except in down-holes) and the return pipe is extended for the length of the hole. Grout injection is stopped when the air has been displaced and when grout flows from the return tube.

Advantages : Bolt can be tensioned immediately after installation and grouted at a later stage when short term movements have ceased. Very reliable anchorage in good rock and high bolt loads can be achieved.

Disadvantages : Relatively expensive. Correct installation requires skilled workmen and close supervision. Grout tubes are frequently damaged during installation and check by pumping clean water before grouting is essential.

Applications : Very widely used for permanent support applications in civil engineering. Mechanically anchored bolts without grout are widely used in mining.

Warning : Some of the components illustrated are protected by patents.

FIGURE 164 :   TENSIONED ROCKBOLT WITH GROUTED ANCHOR

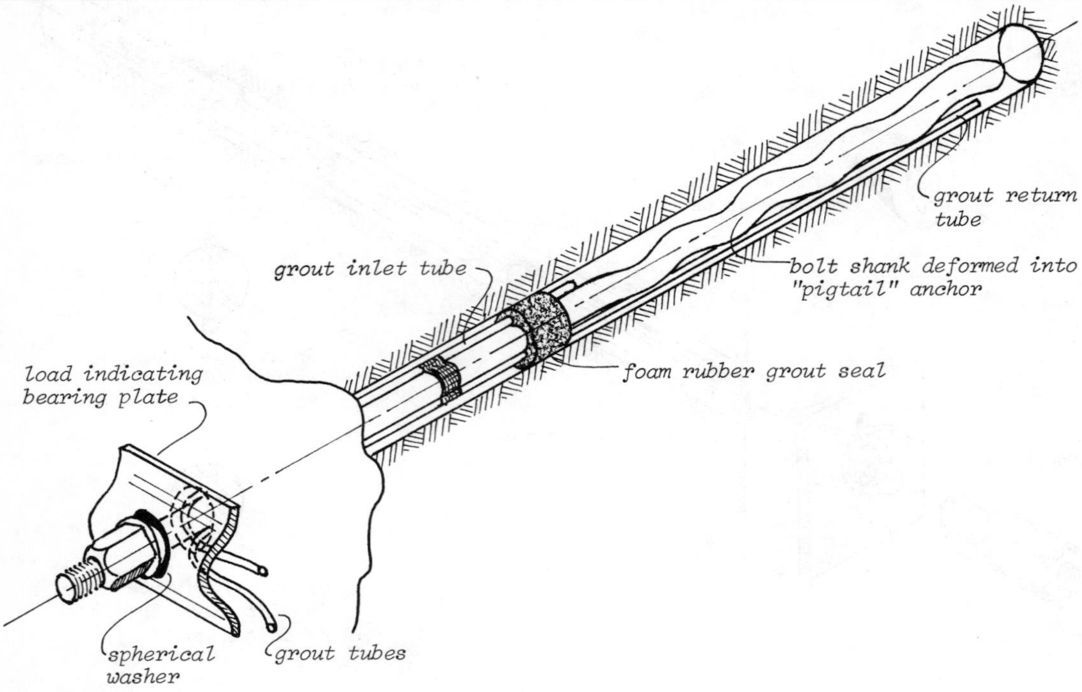

Background :   Drawing is a composite of various systems used in rockbolting, particularly in the Australian mining industry. Grouted anchors have the advantage that they can be used in very poor quality rock masses. One system for grout injection is illustrated. An alternative system is to inject a dry sand/cement mixture through one pipe and a measured water quantity through a second pipe, withdrawing both pipes as the anchor is formed.
The load indicating bearing plate illustrated is one of several designs which give visual load indication by progressive deformation with load.

Advantages :   Inexpensive system with good anchorage characteristics in a wide range of rock conditions. Load bearing plate gives good visual indication of bolt load and adds "spring" to bolt for certain applications.

Disadvantages :   Care required to form good anchor. Bolt cannot be tensioned until grout has set. Stiffness of bolt and bearing plate may be too low for some applications.

Applications :   Principally used in the mining industry where relatively short term support requirements do not require complete grouting of bolt shank for corrosion protection. Ungrouted bolt length acts as a spring in cases where large stress changes are anticipated during the life of the bolt.

Warning :   Some of the systems or components illustrated in this series of drawings are protected by patents. Potential users intending to manufacture similar systems or components should check patent regulations.

FIGURE 165 : RESIN GROUTED, TENSIONED THREADED BAR

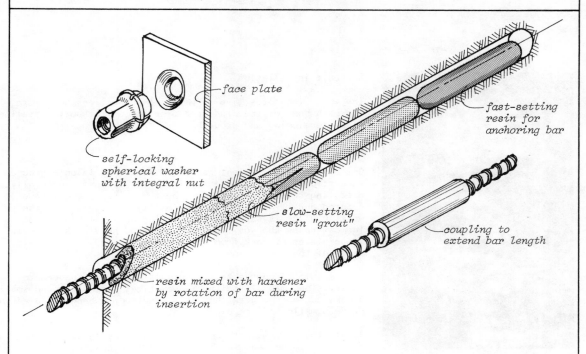

Background : The most sophisticated rockbolt system currently in use, combines most of the advantages of other bolt systems. Resin and a catalyst are contained in plastic "sausages", the catalyst being separated in a glass or plastic container in the resin. These capsules are pushed into the hole with a loading stick and the bar is then inserted. Rotation of the bar during insertion breaks the plastic containers and mixes the resin and catalyst.
In the application illustrated, a fast-setting resin capsule is inserted first and forms a strong anchor which permits tensioning of the bolt a few minutes after mixing. Slow-setting resin then "grouts" the remainder of the bar.
The bar illustrated has a very coarse rolled thread which gives good bonding and allows the length of the installation to be adjusted very easily.

Advantages : Very convenient and simple to use. Very high strength anchors can be formed in rock of poor quality and, by choosing appropriate setting times, a "one shot" installation produces a fully grouted tensioned rockbolt system.

Disadvantages : Resins are expensive and many suffer from a limited shelf-life, particularly in hot climates.

Applications : Increasingly used in critical applications in which cost is less important than speed and reliability.

Warning : Some of the systems or components illustrated in this series of drawings are protected by patents. Potential users intending to manufacture similar systems or components should check patent regulations.

Rockbolt installation

The following notes give a number of suggestions on how to approach some of the practical problems of rockbolt installation, tensioning and grouting.

Scaling

One of the most frequent causes of accidents in underground excavations is inadequate scaling after a blast. Generally, a scaling crew moves in after the fumes from a blast have cleared and it is their responsibility to ensure that the working place is safe for the mucking and drilling crews. In most cases the scaling crew uses hand-held bars to lever loose pieces of rock from the freshly blasted roof and sidewalls of the excavation.

As a result of poor access, poor visibility or inadequate supervision, manual scaling is sometimes incomplete and small blocks can become detached as a result of subsequent blasting or deformation of the excavation.

A solution to this problem, increasingly being adopted in mechanised mines and in large underground construction projects, is to use a heavy pneumatic or hydraulic hammer mounted on an articulated arm carried on a rubber tyred vehicle. This mobile unit can be moved in after the blast and used to scale the roof and sidewalls mechanically. The reach of the articulated arm should be such that easy access to all freshly blasted areas is available and that the muckpile need not be moved before scaling commences. The weight and the power of the hammer should be sufficient to ensure that all loose rock is brought down.

Mechanical scaling is not only safer for the scaling crew but it also tends to reduce the need for spot bolting to secure loose blocks. This spot bolting is always expensive since it is generally carried out as an extra activity and may cause delays in the construction schedule.

Installation

Problems of rockbolt installation are generally problems of access. With larger and larger excavations being used in mechanised mines and civil engineering projects, access must generally be provided by some form of lift vehicle. A good example is illustrated in figure 166 which shows a twin boom jumbo fitted with a rockbolting cage. Highly mobile equipment of this type, used by skilled operators, can reduce the time and hence the expense of rockbolt installation significantly.

In laying out rockbolt patterns, the underground excavation designer should always keep this problem of access in mind. Rockbolt length is just as important as location and, from the rockbolter's point of view, the installation of a very long rockbolt may be a difficult and dangerous operation when it has to be carried out from a small cage on a highlift vehicle. If long rockbolts are required for support, the use of coupled lengths of rod may be safer for the operator and just as effective for the designer.

In the case of large caverns in which cranes are to be installed, consideration should be given to anchoring the crane

Scaling with a hand held pinchbar from a high lift vehicle in a large cavern.

A heavy hydraulic hammer fitted to the articulated arm of a mobile vehicle being used to break boulders in a grizzly. This type of equipment is ideal for mechanical scaling in hard rock excavations.

Figure 166 : Twin boom jumbo for drilling holes and installing rockbolts. Photograph reproduced with permission from Atlas Copco.

beams to the sidewalls of the cavern so that they are available during an early stage of the construction for the installation of a construction or permanent crane.

An example is illustrated in the three photographs reproduced in figures 167, 168 and 169 showing a large cavern for a hydroelectric project in which the crane beams were anchored to the cavern walls. Figure 167 shows the crane beams in place before the completion of the cavern excavation. A temporary construction crane is illustrated in figure 168. This crane was used to gain access to the roof and upper sidewalls of the cavern to permit final tensioning and grouting of rockbolts which had been installed during the sequential excavation of the upper part of the cavern. The availability of this access meant that the completion of the roof support programme could be carried out as an independent operation, allowing the benching down of the lower part of the cavern to proceed without interference.

Figure 169 shows the permanent crane in operation, assisting in the excavation of the turbine pits and available for all the subsequent equipment installation.

Figure 167 : Crane beams anchored to the walls of a large cavern in an underground hydroelectric project. These beams were installed during an early stage of the cavern excavation, before the main benching down illustrated in this photograph commenced.

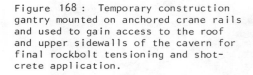

Figure 168 : Temporary construction gantry mounted on anchored crane rails and used to gain access to the roof and upper sidewalls of the cavern for final rockbolt tensioning and shotcrete application.

Figure 169 : Machine hall crane supported on anchored crane beams. Because of the early installation of the crane, it was available to assist in the excavation of the turbine pits and the construction of the base of the cavern.

Anchoring

When rockbolts are to be tensioned, some form of anchor must be used to secure the end of the bolt in the hole. The three most common anchor types are mechanical, cement grout and chemical (synthetic resin).

Mechanical anchors, such as the expansion shell anchor illustrated in figure 163, are very commonly used in both mining and civil engineering applications. A large number of different expansion shell designs are manufactured commercially and there is little point in attempting to list all of these in this text. Basically, all expansion shell anchors operate in much the same way and the choice of anchor type for a particular job will usually depend more on price and availability than upon anchoring efficiency.

In good quality hard rock, mechanical anchors are very efficient and are both fast and convenient to install. In weaker or softer rocks the effectiveness of the anchor is reduced by local crushing of the rock by the ribbed sleeves. In very weak shales, mudstones and weakly cemented sandstones, the use of mechanical anchors is not recommended.

Cement grout or cement mortar anchors are less convenient than either mechanical anchors or resin cartridge anchors but they are probably the cheapest form of anchor. Placing the grout at the end of the hole is the most difficult problem and several methods have been used, with varying degrees of success.

One method which has been found to be convenient and effective is to use a short length of "Perfobolt" sleeve such as that illustrated in figure 160. This sleeve is packed with a stiff mortar and pushed up to the end of the hole by the bolt. Driving the bolt into the mortar will cause the mortar to be extruded through the perforations and to fill the space between the bolt and the hole. Because the mortar is stiff it will remain in place at the end of the hole.

A typical mix for a stiff mortar is as follows :

Type III Portland cement	100 parts by weight
Sand , clean angular with maximum grain size of about 2mm	100 parts by weight
Fluidifier and expanding agent such as "Interplast-C" or equivalent	1.4 parts by weight
Water : Approximately 0.3 water-cement ratio by weight. A good mix is one that will "pack like a snowball without exuding free water"[271].	

The US Bureau of Mines have developed a cement and water cartridge for anchoring and grouting rockbolts[272]*. A loose mixture of dry cement powder and pinhead-size water droplets encased in waxy globules is packed into a sausage-shaped cartridge. The cartridge is inserted in the drillhole and a bolt is thrust in and rotated. Crushing of the water cap-

---

* Details are available from the Technology Transfer Office, Spokane Mining Research Centre, US Department of the Interior, US Bureau of Mines, Spokane, Washington, 99207, USA.

sules releases the water uniformly and rotation of the bolt mixes the water and cement. The grout hardens very quickly and anchoring strengths of 3500 lb/ft of hole (51 MN/m) at 2½ minutes and 8000 lb/ft (117 MN/m) at 5 minutes have been demonstrated. The shelf life of these water-cement cartridges is reported to be in excess of six months.

A variety of techniques have been used to pump thick grouts into holes or to form grout anchors in place by simultaneously injecting a dry mixture of sand and cement and water[261]. These techniques tend to vary from site to site, depending upon the equipment available and the ingenuity of the operators.

Resin cartridges, such as those illustrated in figures 165 and 170 are increasingly being used in applications in which high anchor strength and rapid installation are required. The high cost of these cartridges is usually justified by the speed and convenience of installation and this is particularly important when it is necessary to integrate the excavation and support cycles, as discussed on pages 329 to 331.

A short shelf life was a problem with some of the early resin cartridges and, while this problem has been partly overcome, the intending user should check the manufacturer's specifications very carefully. In critical applications which are a long way from sources of supply, air freight is sometimes a justifiable means of transporting resin cartridges in order to minimise this shelf life problem. Currently, the most widespread use of resin cartridges is in countries in which they are locally manufactured and therefore readily available.

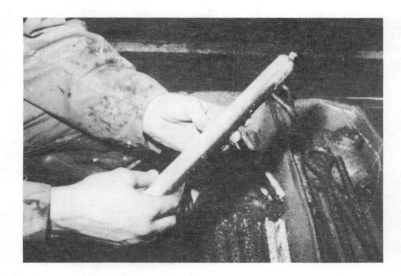

Figure 170 : Typical resin cartridge used for anchoring and grouting rockbolts as illustrated in figure 165. The resin and catalyst are contained in separate compartments in the same cartridge and are mixed when the rockbolt is inserted and rotated in the hole.

Bolt tensioning

Rockbolts can be tensioned by applying a calibrated torque to the nut or by applying a direct tension to the bolt.

In the case of rockbolt applications in which tensions of less than about 10 tons are required, the use of a torque wrench or an impact wrench set to stall at a calibrated torque value is usually adequate. At higher loads, the uncertainty in the relationship between the torque applied to the nut and the tension induced in the bolt can give rise to very large variations in bolt tension. Apart from possible inaccuracies in the torque measurement, the tension in the bolt is influenced by such factors as the rusting of the bolt threads and the gouging of the washer by the sharp corners of the nut. The substitution of a hardened washer for a mild steel washer can change the tension in the bolt by a factor of two.

The US Bureau of Mines carried out a number of tests on the relationship between applied torque and induced tension in $3/4$ and $5/8$ inch diameter rockbolts [273,274,275] and the results of these tests have been widely used. Rockbolt manufacturers will sometimes supply torque-tension calibrations for their products and these can be used provided that the field conditions are similar to those under which the calibrations were carried out.

The authors recommend that, whenever a torque wrench or impact wrench is to be used for tensioning rockbolts, the calibration be checked by the direct measurement of bolt tension (as described below) in a random sample of installed bolts.

Use of a torque wrench to tension rockbolts.

When high bolt loads are required or when it is considered necessary to determine the actual bolt tension with a reasonable degree of accuracy, direct hydraulic tensioning of the bolt assembly is the most practical approach.

A typical hydraulic bolt tensioner is illustrated in figures 171 and 172. The important factors to take into account when buying or building such a device are:
  a. A direct tension must be applied to the bolt in such a way that it can be measured without interfering with the functioning of the bolt as a support element.
  b. It should be possible to adjust the load in the bolt and to lock the nut at a pre-determined load value.
  c. The load should be applied in such a way that separation can only occur between the nut and the surface of the washer with which it is in direct contact.
  d. The capacity of the jack should be adequate to permit pull-out testing of the entire rockbolt assembly.

The tensioner illustrated in figures 171 and 172 satisfies all of these requirements but there are many other designs, some of which are commercially available, which do the job equally well. Note that, as shown in figure 171, there is sufficient room between the spacer posts to allow a spanner to be inserted to lock the nut and, also, the reaction plate bears directly onto the spherical washer surface in the application illustrated.

Hydraulic tensioning of rockbolts.

When checking the existing tension in a rockbolt, the hydraulic pressure is increased slowly until the nut is finger-tight

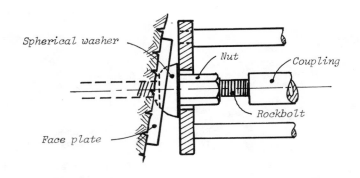

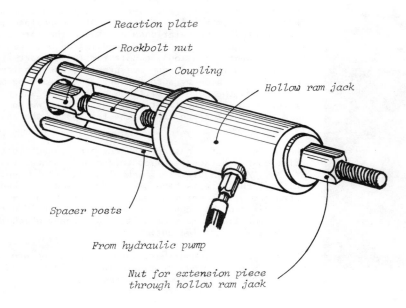

Figure 171 : Typical hydraulic rockbolt tensioner.

Figure 172 : Hydraulic tensioner and pump. The unit illustrated was fabricated on a construction site but there are several commercially available hydraulic rockbolt tensioners.

Installation of a rockbolt washer across the "sawtooth" profile of a blocky rock face makes subsequent sealing of the hole difficult.

Application of rapid setting plaster to a rockbolt washer assembly to seal it for grouting.

at which stage the load is measured. In the case of pull-out testing of bolts, it is advisable to attach a safety chain to the hydraulic tensioner to prevent accidents when the bolt fails.

If it is required to keep a constant check on bolt tension as part of a monitoring programme, an electronic or hydraulic load cell can be installed between the nut and the washer. Note that rockbolt load measurement becomes meaningless when the bolt shank is grouted and hence, bolts which are to be used for load monitoring should not be grouted or should be fitted with a bond-breaker. A plastic sleeve fitted over a greased bolt will act as an adequate bond-breaker when the hole is to be filled with grout.

Grouting

Grouting serves two purposes in rockbolt applications :
1. It bonds the bolt shank to the rock making the bolt an integral part of the rock mass. As discussed in the previous chapter, this improves the interlocking of the individual elements in the rock mass and results in a significant improvement in the properties of the rock mass.
2. It protects the bolt installation against corrosion.

It is strongly recommended that all rockbolts intended for long term applications should be grouted.

A typical pumpable liquid grout mixture is as follows[263]:

Type III Portland cement	100 parts by weight
Flyash (optional)	40 parts by weight
Fluidifier and expanding agent such as "Interplast-C" or equivalent	1.4 parts by weight
Water approximately 0.4 water-cement ratio by weight.	

When added, the flyash improves the fluidity and plasticity of the mix without influencing its strength. This grout will set relatively slowly and the bolt should not be tensioned until the anchor has set for about 48 hours. If more rapid tensioning is required, accelerators can be added to the grout mix in accordance with manufacturer's specifications.

One of the most critical steps in rockbolt grouting is the sealing of the collar of the hole. The first important step in this sealing is the care taken in preparing the rock face and placing the washer. As illustrated in the upper margin photograph, careless location of the bolt and lack of any face preparation makes it almost impossible to seal the borehole collar.

When it is impossible to place a bolt so that the rock face is reasonably flat, a pad of mortar can be placed under the washer to improve the contact and to improve the subsequent sealing of the hole. This sealing can be achieved by means of a rubber seal ring such as that illustrated in figure 163 or by applying a fast setting mortar or plaster to the bolt head assembly as illustrated in the lower margin photograph.

Flushing the bolt hole with clean water before commencement of grouting serves to clean the hole and also to check the

Grouted hollow rockbolt.

effectiveness of the collar seal.

Figure 163 on page 339 illustrates a typical grout tube assembly for up-hole grouting. A short plastic tube (about 8mm inside diameter and 11mm outside diameter) is used to inject the grout. A 6mm inside diameter and 8mm outside diameter plastic tube is taped to the rockbolt for its full length to act as a grout return tube. Note that these two tubes should be of different dimensions to prevent injection into the grout return tube. The diameter of the grout injection fitting should be such that it will only fit into the larger grout inlet tube. In the case of down-hole grouting, the grout should be injected at the bottom of the hole and a short grout return tube fitted to the top of the hole.

A grout return tube taped to the outside of the rockbolt shank in easily damaged during handling and installation of the rockbolt and special care must be taken to avoid such damage. When the tube is damaged, the bolt should be removed and the tube replaced. Alternatively, if the bolt cannot be removed, a new grout tube can be pushed into the hole on a stiff wire and the wire then withdrawn to leave the tube in place.

As shown in figure 163, alternative grout tube arrangements which are less susceptible to damage are to place the grout return tube in a slot machined into the bolt shank or to use a hollow bolt*. While more expensive than the system described above, these alternative arrangements for the grout return tube do provide a much greater degree of reliability.

Grout injection pressures should generally be kept below 172 kPa (25 lb/in^2) in order to avoid "jacking" of the rock mass. Grouting should be continued until there is a full flow of grout through the return tube.

Figure 173 : Grouting rockbolts in a large underground excavation.

---

* Patented by the Williams Form Engineering Corporation.

Wire mesh

Wire mesh is used to support small pieces of loose rock or as reinforcement for shotcrete. Two types of wire mesh are commonly used in underground excavations: chainlink mesh and weldmesh.

Chainlink mesh

This is the type of wire mesh commonly used for fencing and it consists of a woven fabric of wire. The wire can be galvanised for corrosion protection and, because of the construction of the mesh, it tends to be flexible and strong. A typical mining application is illustrated in figure 174 below which shows chainlink mesh attached to the roof of a haulage by means of rockbolts. Small pieces of rock which become detached from the roof are supported by the mesh which, depending upon the spacing of the support points, can carry a considerable load of broken rock.

As shown in the margin photograph, chainlink mesh is not ideal for reinforcing shotcrete because of the difficulty of getting the shotcrete to penetrate the woven fabric of the mesh. The authors do not recommend chainlink mesh for this application and prefer to use weldmesh as discussed on the next page.

Chainlink mesh is not recommended for the reinforcement of shotcrete.

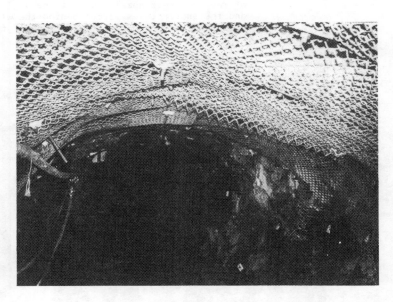

Figure 174: Chainlink mesh used to prevent falls of small pieces of broken rock from the roof of a mine haulage.

Weldmesh

Weldmesh is commonly used for reinforcing shotcrete and it consists of a square grid of steel wires, welded at their intersection points. A typical weldmesh for underground use has 4.2mm wires at 100mm centres (designated 100x100x4.2 mesh) and it is supplied in panel sizes which are convenient for installation by one or two men.

Generally, weldmesh is attached to the rock by means of a second washer plate and nut placed on each existing rock-bolt. Intermediate anchorage is provided by means of short grouted bolts or expansion shell anchors. Sufficient intermediate anchors should be placed to ensure that the mesh is drawn close to the rock face. While a good shotcrete operator can work with mesh as far as 200mm (8 inches) from the rock, this tends to be very wasteful of shotcrete as it is essential that the mesh be fully covered by shotcrete.

Mesh is easily damaged by flyrock from nearby blasts and its installation should be delayed until the blasting is far enough away or it should be protected from flyrock by means of blasting mats. Damaged mesh should be replaced by cutting out the damaged section and providing a good overlap to ensure continuity of the reinforcement. Weldmesh has the advantage of not unravelling when damaged as is the case with chainlink mesh.

Galvanised weldmesh is difficult to obtain and hence the steel will suffer from serious corrosion if it is not fully encased in shotcrete. Care should be taken to ensure that air pockets are not formed behind the wires or the intersection points and this can be achieved by constant movement of the shotcrete nozzle so that the angle of impact is varied and the shotcrete is forced behind the wires. This will be discussed further in a later section of this chapter.

Figure 175 : Weldmesh is usually attached to the rock by placing a second washer and nut on each existing rockbolt. Short pins or bolts can be used for intermediate mesh fixing.

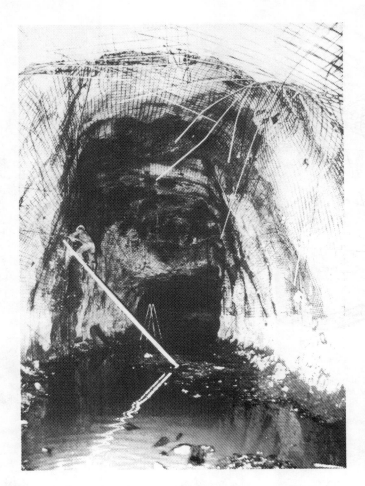

Figure 176 : Weldmesh being placed in a permanent mine excavation in preparation for shotcreting.

Shotcrete

Pneumatically applied mortar and concrete (generally known as "Gunite" or "Shotcrete") are increasingly being used for the support of underground excavations. Essential reading for any serious intending user of this type of reinforcement are the US Department of Transportation report *Shotcrete Practice in Underground Construction*[261] and the proceedings of an ASCE conference on *Use of Shotcrete for Underground Structural Support*[262]. These excellent publications give a wealth of practical detail on the use of pneumatically applied mortar or concrete and only a short summary of some of the most important points will be presented in this chapter.

There are two basic types of shotcrete, a term which will be used in a generic sense in this chapter. Dry-mix shotcrete, as the name implies, is mixed dry and the water is added at the nozzle. Wet-mix shotcrete is mixed as a low slump concrete which is then pumped to the nozzle. In the case of the dry-mix, accelerator can be added to the mix but, in the case of the wet-mix process, it must be added at the nozzle. Typical equipment layouts for the two processes are illustrated diagrammatically in figures 177 and 178 and the processes are compared in table 23 on page 355. The choice of which process is to be used on a particular project will depend upon a number of considerations which are not directly related to the quality of the final product.

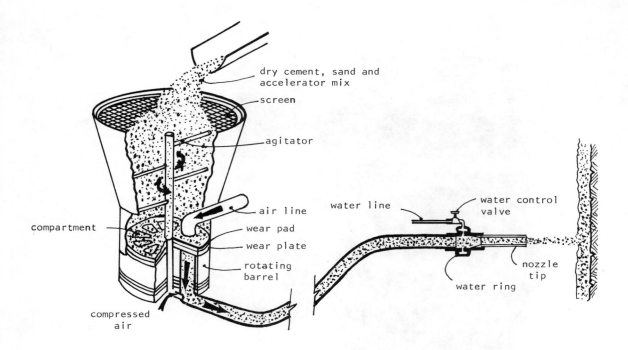

Figure 177 : Typical dry-mix shotcrete operation. Drawing compiled from *Shotcrete Practice in Underground Construction*[261].

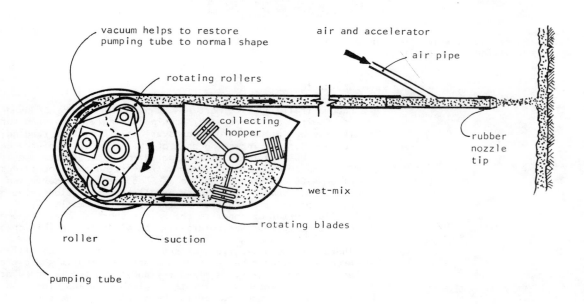

Figure 178 : Typical wet-mix shotcrete operation using squeeze type machine. Drawing compiled from *Shotcrete Practice in Underground Construction*[261].

TABLE 23 - COMPARISON BETWEEN WET- AND DRY-MIX SHOTCRETE PROCESSES*	
Wet-mix	*Dry-mix*
• Lower rebound when spraying. • Lower dusting. • Control of water/cement ratio. • Quality control in the preparation of the materials is easier because the manufacture of materials is nearly identical to concrete. • Quality of in-place shotcrete is not so sensitive to the performance of the nozzleman since he does not adjust water flow. • Nozzleman directly controls the impact velocity of the particles and thus compaction by regulating air flow at the nozzle. • Easier to clean. • Lower maintenance costs. • Higher production rates.	• More adaptable to varying ground conditions, particularly where water is involved. • Dry-mix equipment is typically less expensive and a larger inventory of used equipment is available. • Dry-mix machines are typically smaller and are thus more adaptable to tunnels with limited space.

Mix design

The following paragraph is quoted directly from *Shotcrete Practice in Underground Construction*[261]:

" The process of mix design is a long, complicated process which is an integral part of setting up the operation, selecting the materials and equipment and training the crew. Any change in this process results in a change in the final product. A mix shot at a later stage of pre-construction testing will probably result in a significantly better product than the same mix shot at an early stage. However certain arbitrary assumptions in the mix design must be made at the beginning to get trial mixes started."

The overall approach to mix design is similar for both wet- and dry-mix processes but there are some important differences in detail depending upon which process is used. In either process, the mix design must satisfy the following criteria:
1. *Shootability* - must be able to be placed overhead with minimum rebound.
2. *Early strength* - must be strong enough to provide support to the ground at ages less than 4 to 8 hours.
3. *Long-term strength* - must achieve a specified 28 day strength with the dosage of accelerator needed to achieve shootability and early strength.
4. *Durability* - long-term resistance to environment.
5. *Economy* - low cost of materials and minimum losses due to rebound.

* From reference 261.

Selection of materials

The following comments apply to the selection of materials typically used in shotcrete:

Portland Cement - Type I Portland cement is most widely used in shotcrete applications since it is the most readily available type and it satisfies most of the normal shotcrete requirements.

Type II (moderate sulfate resistance) and Type V (high sulfate resistance) may be required when the rock, groundwater or mixing water contain sulfates. The rate of strength gain of these cements is relatively slow.

Type III cement, because of its composition and fineness, provides high early strength.

Regulated-set cement is a new commercially available cement which contains calcium fluoroaluminate. This results in a very high rate of strength gain for the first few hours, without the use of accelerators[261]. After about one day, the rate of strength increase and the physical properties of the shotcrete are similar to those of a Type I cement shotcrete.

Generally, Type I Portland cement is used in conjunction with accelerators because of the availability of these materials and because of the flexibility which can be achieved by small variations in mix design. Note that it is essential to check the compatibility of the cements and accelerators used in shotcrete applications since both early and ultimate behaviour may be influenced by mixing components which are not compatible.

Aggregates - Natural gravels are preferred over crushed stone because of the better pumping characteristics of the rounded natural aggregate particles. Otherwise, the quality of aggregate required for shotcrete is the same as that for good quality concrete.

Aggregates should be clean, hard, tough, strong and durable. No more than 2% of the aggregate should pass a No. 200 sieve (0.075mm). The aggregate should be free from an excess of silt, soft or coated grains, mica, harmful alkali and organic matter. Alkali-reactive aggregates should be avoided.

Generally, the maximum size of aggregate should not exceed one third of the smallest constriction in the hose line. Shotcrete machines which will accept aggregate sizes of up to 32mm (1¼ inches) are available but the normal practice is to use maximum aggregate sizes of 19mm (0.75 inch) or less.

Aggregate gradation is critical in mix design, pumpability, flow through hoses, hydration at the nozzle, adherence to the area sprayed and the density and economy of the final product. Typical gradation recommendations for concrete aggregates are listed in table 24 and these recommendations can be used as a basis for shotcrete aggregate selection. A recommended range of gradation for combined coarse and fine aggregates for use in shotcrete is presented in figure 179.

An increase in the percentage of coarse aggregate will give better compaction, increased density, lower water and cement requirements, less shrinkage and higher bond

TABLE 24 - AGGREGATE GRADATION LIMITS FOR SHOTCRETE [261]

Fine aggregate

Sieve	Percentage passing, by weight
3/8 in. (9.5mm)	100
No. 4 (4.75mm)	95 to 100
No. 8 (2.36mm)	80 to 100
No. 16 (1.18mm)	50 to 85
No. 30 (0.60mm)	25 to 60
No. 50 (0.30mm)	10 to 30
No. 100 (0.15mm)	2 to 10

Coarse aggregate

Sieve	No.8 to 3/8 in.	No.4 to 1/2 in.	No.4 to 3/4 in.
1 in. (25.0mm)	-	-	-
3/4 in. (19.0mm)	-	100	90-100
1/2 in. (12.5mm)	100	90-100	-
3/8 in. (9.5mm)	85-100	40-70	20-55
No. 4 (4.75mm)	10-30	0-15	0-10
No. 8 (2.36mm)	0-10	0-5	0-5
No. 16 (1.18mm)	0-5	-	-

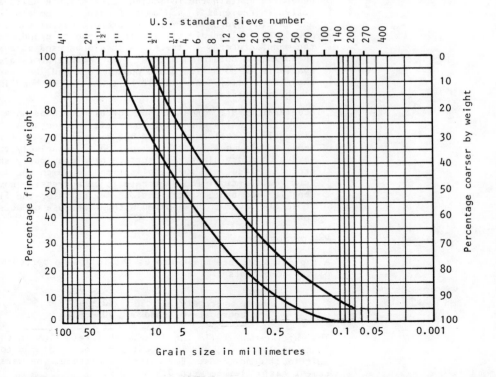

Figure 179 : Recommended gradation range for combined fine and coarse aggregate [261].

and flexural strength. An increase in the coarse aggregate fraction will make the shotcrete more difficult to pump and will give more rebound during shooting. Hence, a compromise must be reached and it is recommended that figure 179 be used as a starting point in establishing the optimum aggregate gradation.

Water - Water used in shotcrete should meet the same standards as that used in concrete. It should be clean and free from injurious amounts of oil, grease, salts, alkali and organic matter. Generally, water which is suitable for drinking and having no pronounced taste or odour is suitable for use in shotcrete.

Accelerators - When a rapid gain in the early strength of shotcrete is required in order to provide immediate support to the rock, accelerating admixtures or accelerators are added to the mix. The addition of accelerators can also be used to improve shooting conditions and to reduce rebound, particularly when working overhead.

Calcium chloride, a common accelerator used for concrete, is sometimes used in shotcrete but is not sufficiently fast-acting for most underground applications. Proportions of the order of 5% do provide a rapid set but at the cost of a decrease in ultimate strength and durability. Hence, the use of calcium chloride is not recommended.

A number of special accelerators are marketed for use in shotcrete and these are much faster acting than those used for conventional concrete. These accelerators generally contain the following water soluble salts as active ingredients : sodium carbonate, sodium aluminate and calcium hydroxide[261]. The proportions of these and other active ingredients vary from brand to brand and the manufacturer's instructions should be used as a starting point in obtaining a trial mix. Accelerators are available in both liquid and powder form and, because of their causticity, care should be taken in the handling of these materials.

Accelerators are normally used for all overhead work and on vertical walls where the shotcrete thickness is considerable. Accelerators need not be used in shotcrete applied to tunnel inverts or applied in relatively thin layers on a dry clean rock surface or a previously shotcreted surface. In some applications in which the long term resistance of the shotcrete to an aggressive environment is important, specifications may call for the final shotcrete layer to be applied without the use of accelerators.

Mix proportions

A typical shotcrete mix contains the following percentages of *dry* components :

cement	15 - 20%
coarse aggregate	30 - 40%
fine aggregate or sand	40 - 50%

The water-cement ratio for dry-mix shotcrete lies in the range of 0.3 to 0.5 and is adjusted by the nozzleman to suit local conditions. For wet-mix shotcrete, the water-cement ratio lies between 0.4 and 0.6.

TABLE 25 - TYPICAL SHOTCRETE MIXES AND ENGINEERING PROPERTIES

PROJECT	MATERIALS			MIX DESIGN Percent of total batch weight					COMPRESSIVE STRENGTH MPa**				ELASTIC MODULUS GPa***				REFERENCE
	wet- or dry-mix shotcrete	cement type	maximum aggregate size - mm	cement	coarse aggregate	fine aggregate	sand	water*	1-3 hours	3-8 hours	1 day	28 days	6-7 hours	1 day	3-8 days	28 days	
Washington DC metro	dry	I	13	18.7	40.65	40.65											Bawa[276]
New Melones Dam	dry	II	19	18.8	38.2		43.0			3.7-4.1	6.9-18.6	16.3-32.3			17.2-23.9	19.7-27.5	US Corps of Engrs.[277]
Vancouver tunnel	dry	I	19	16.6	23.0	21.7	38.7		4.5			35.9					Mason[278]
Illinois Institute of Technology Research Institute project on shotcrete	dry	III	13	13.5	31.5		55.0		0.4-1.3	0.4-5.2	14.9	27.8					Bortz et al[279]
	dry	III	13	17.9	29.9		52.2		0.76-3.45	3.45-10.7	20.3	29.6	4.0-7.1	13.6-23.4	18.8-21.3	17.8-23.1	Singh and Bortz[280]
	dry	III	13	21.8	28.5		49.7		0.34-3.58	3.58-9.03	20.2	30.5					
Waterways Experiment Station research	dry	II	19	17.7	32.7		49.6					36.0					Tynes and McCleese[281]
University of Illinois research project	dry	I	9	18.4	41.4		40.2			4.96-6.37		27.6-41.4	6.2-15.9		21.4-50.3		Fernandez et al[282]
Lakeshore project, Hecla Mining Co.	wet	II	13	15.8	34.5	41.3		8.4			6.9	27.6					Chitunda[283]
Waterways Experiment Station research	wet	II	19	16.3	30.8		44.8	8.1				57.7					Tynes and McCleese[281]
Illinois Institute of Technology Research Institute project on shotcrete	wet	III	13	12.8	29.9		52.2	5.1	0-0.55	1.17	6.14-10.4	25.9-34.3					Bortz et al[279]
	wet	III	13	16.7	27.9		48.7	6.7	1.17	1.17-5.59	18.9-20.3	33.3-39.4		12.3-28.0	22.3-27.0	23.8-35.9	Singh and Bortz[280]
	wet	III	13	20.1	26.2		45.7	8.0	1.51	1.51-4.97	24.1	41.9					
St. John's Abbey, St. Paul, Minnesota	wet	I	16	16.4	33.2		50.4										Hoffmeyer[284]
Henderson mine	wet	I	9	17.8	27.4		54.8										Jones[285]

* If water quantity is not stated, quantities are expressed as a percentage of the *dry* components.
** 1 MPa = 10.2 kgf/cm² = 145 lbf/in².   *** 1 GPa = 10.2 × 10³ kgf/cm² = 0.145 × 10⁶ lbf/in².

Engineering properties of shotcrete

A number of engineering properties have an influence on the behaviour of shotcrete in underground excavation support applications. These include the compressive strength, bond strength, flexural strength, tensile strength and modulus of elasticity. A full discussion on the significance of all of these properties will be found in *Shotcrete Practice in Underground Construction*[261]. For the purposes of this discussion, only the compressive strength and modulus of elasticity will be considered.

A rapid gain in compressive strength with age is essential if shotcrete is to fulfil a role as an effective support member in poor ground. Typical compressive strength values for shotcrete are as follows :

	1 - 3 hours	3 - 8 hours	1 day	28 days
Shotcrete with no accelerator	0	0.2 MPa 30 lb/in^2	5.2 MPa 750 lb/in^2	41.4 MPa 6000 lb/in^2
Shotcrete with 3% accelerator	0.69 MPa 100 lb/in^2	5.2 MPa 750 lb/in^2	10.3 MPa 1500 lb/in^2	34.5 MPa 5000 lb/in^2
Regulated set shotcrete (estimated[279])	8.27 MPa 1200 lb/in^2	10.3 MPa 1500 lb/in^2	13.8 MPa 2000 lb/in^2	34.5 MPa 5000 lb/in^2

These values are presented for general guidance only and should not be used as a substitute for actual field tests. The authors regard the early strength values for regulated set shotcrete as rather optimistic but they have been included because they indicate the extent to which research into cements and cement-accelerator combinations may improve the early strength characteristics of shotcrete.

The modulus of elasticity of shotcrete is very closely related to the compressive strength and, as would be expected, shows very similar gains with time to those for compressive strength.

In order to provide the reader with a starting point in choosing a shotcrete mix, a list of typical shotcrete mixes with their compressive strengths and elastic moduli has been compiled and is presented in table 25. More detailed information can be obtained from *Shotcrete Practice in Underground Construction*[261] from which table 25 was compiled.

Placement of shotcrete

The quality of placed shotcrete depends upon the materials used and upon the mix design, as discussed above, but it is also heavily dependent upon the method of placement. In particular, the skill of the nozzleman in preparing the surface, controlling the delivery rate and thickness and, in the dry-mix process, the water-cement ratio, has a significant influence on the final product.

The success of a shotcrete operation depends very heavily upon the skill of the nozzleman.

Preparation of the surface to be shotcreted is an essential part of the shotcreting operation. Effective scaling, as discussed on page 342, is important for the safety of the operators and also to reduce the chances of "drummy" shotcrete caused by spraying onto loose rock. Obviously, if the rock is very poor, scaling may not be possible and the shotcrete may have to be applied as quickly as possible after

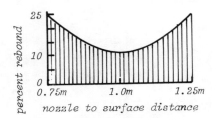

Effect of nozzle distance on rebound. After Kobler[286].

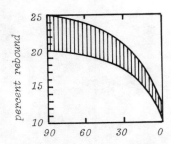

Effect of nozzle angle on rebound. After Kobler[286].

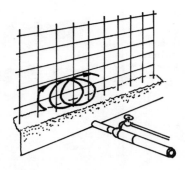

Recommended motion for hand-held shotcrete nozzle. Loops are elliptical 500mm long, 200mm high and advance between loops is 100mm. After Ryan[287].

exposure of the rock face in order to provide support. In such a case, a second shotcrete layer, reinforced by means of weldmesh, may be required to complete the treatment of the surface.

Surfaces to be shotcreted should be free of all loose or foreign matter if a proper bond is to be obtained. Dust from the blasting operation and gouge from the rock joints should be washed off the surface and this is most easily achieved by jetting the surface with an air-water mixture. This jetting can be carried out with the shotcrete machine operating at normal shotcreting pressure (0.3 to 0.4 MPa, 45 to 60 lb/in^2) with water added in sufficient quanity to dislodge all the loose materials on the rock surface. The nozzle should be held approximately 1 to 2m (3 to 6 ft) from the surface being cleaned.

Slickensided surfaces are not easily removed by water jetting and sandblasting is sometimes used to improve such surfaces for shotcreting. A normal dry-mix shotcrete machine can be used for sandblasting if the regular nozzle is replaced by a sandblasting nozzle. Since this operation is expensive in terms of time and materials, it should only be used for critical applications and only after air-water jetting has been tried.

Once the surface to be shotcreted has been properly cleaned, the shotcreting can commence. The nozzleman selects the air pressure and the gunman adjusts the material feed rate to match this pressure. Too low a feed rate will result in slugs of material rather than a steady stream. Too fast a feed rate will cause the machine to plug. Obviously, it is very difficult to provide general guidance on this part of the shotcreting process since it depends upon the characteristics of the actual machine being used and upon the skill of and the cooperation between the nozzleman and the gunman.

The optimum distance between the nozzle and the surface being shotcreted is approximately 1 metre (3.3 feet). The amount of rebound is significantly influenced by this distance as shown in the upper margin drawing. Rebound is also influenced by the angle of the nozzle to the horizontal as shown in the centre margin drawing.

Whether the nozzle is hand held or operated on a robot machine, it must be kept moving in a controlled path while delivering shotcrete. Failure to keep the nozzle moving results in a shotcrete of non-uniform compaction and thickness. When using a hand-held nozzle, it is recommended that the shotcrete be applied in a continuous series or overlapping circular or elliptical loops.

When spraying shotcrete onto weldmesh reinforcement, care should be taken to ensure that voids are not formed behind individual wires. Bringing the nozzle closer to the face is one method which is used to overcome this problem[261]. Another technique which is particularly successful when using a robot machine is to vary the angle of the nozzle with respect to the face. This prevents void formation by forcing the shotcrete under individual wires.

The thickness of a shotcrete layer is generally estimated from the volume of material placed with an appropriate allowance for losses due to rebound. When spraying very

Figure 180 : Robot machine set up to spray shotcrete in a large cavern in an underground hydroelectric project.

Figure 181 : Shotcrete spraying using a robot machine. The machine can be controlled from the cab mounted on the truck or from a remote control box as is being done in this application.

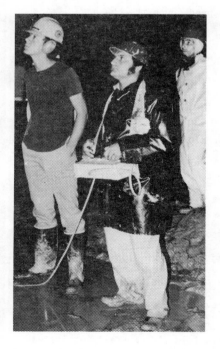

Figure 182 : Remote control box for shotcrete placement using a robot machine. The nozzle position is controlled by means of the two small "joysticks" on the control box.

irregular surfaces it is almost impossible to achieve a uniform thickness and it may be necessary to apply more shotcrete than originally planned in order to ensure that all the rock is covered. When weldmesh reinforcement is attached to the rock face, as illustrated in figure 175, this mesh can be used as a gauge for shotcrete thickness. Similarly, if grouted reinforcing bars are used for rock support, the ends of these bars can be left sticking out of the hole to provide a gauge for shotcrete thickness.

Some designers specify that short steel pins of the required length should be attached to the rock at frequent intervals in order to provide a gauge for shotcrete thickness. This specification is seldom met in practice because of the expense and the danger of installing the pins before shotcreting takes place.

Fibre reinforced shotcrete

One of the disadvantages of normal shotcrete is its low tensile strength and it is not uncommon to see shotcrete which has been severely cracked by movements in the rock mass after the shotcrete has set. The placing of weldmesh reinforcement, as described on page 352, can be used to overcome this problem but the installation of the mesh is a time consuming and therefore expensive operation. The idea of mixing steel wire reinforcement directly with the shotcrete during application has attracted a great deal of attention and a considerable amount of research on this possibility has been carried out during the past decade[261].

Much of the early experimental work involved mixing 25mm long 0.25mm steel wires with the cement and aggregate in proportions of 3 to 6% by weight. Fibre contents of greater than this were found to be difficult to mix and to shoot. Difficulties with balling of the fibres and rebound losses of up to 60% have been reported[261]. More recent work by Sandell[288] involving mixing the shotcrete and the fibres at the nozzle appears to reduce fibre losses to about 15% and to reduce the problem of fibre balling. Sandell reports a 28 day tensile strength of 8 MPa (1160 lb/in^2) for steel fibre reinforced shotcrete as compared with 2 to 5 MPa (290 to 725 lb/in^2) for unreinforced shotcrete.

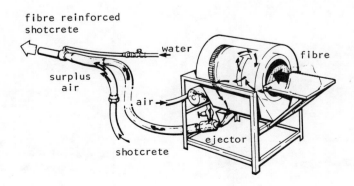

Figure 183: Equipment for mixing steel fibre reinforcement into shotcrete. After Sandell[288].

Not only is the tensile strength of fibre reinforced shotcrete higher than that of conventional shotcrete but, even after the maximum strength has been exceeded, the fibres have a considerable capacity for keeping the shotcrete together. This means that the shotcrete will stay in place and continue to play a useful although diminished support role.

A number of projects have been successfully completed using fibre reinforced shotcrete[288,289,290] and the authors have little doubt that the use of this material will increase in the future.

Chapter 9 references

271. US ARMY CORPS OF ENGINEERS. Rock reinforcement systems. *Engineer Technical Letter* No. 1110-1-39, 1970, 18 pages.

272. HOPPE, R. Winning the battle against bad ground. *Engineering and Mining Journal*, Feb. 1979, pages 66-73.

273. BARRY, A.J., PANEK, L.A. and McCORMICK, J.A. Use of torque wrench to determine load in roof bolts. Part 1, Slotted type bolts. *US Bureau of Mines Report of Investigations*, No. 4967, 1953.

274. BARRY, A.J., PANEK, L.A. and McCORMICK, J.A. Use of torque wrench to determine load in roof bolts. Part 2, Expansion-type, $3/4$-inch bolts. *US Bureau of Mines Report of Investigations*, No. 5080, 1954.

275. BARRY, A.J., PANEK, L.A. and McCORMICK, J.A. Use of torque wrench to determine load in roof bolts. Part 3, Expansion-type, $5/8$-inch bolts. *US Bureau of Mines Report of Investigations*, No. 5228, 1956.

276. BAWA, K.S. Development of shotcrete for metro construction in Washington. In *Use of Shotcrete for Underground Structural Support*, published by American Society of Civil Engineers, New York, 1975, pages 33-49.

277. US ARMY CORPS OF ENGINEERS. Shotcrete for multi-purpose tunnel, New Melones Lake, Stanislaus River, California. *US Army Corps of Engineers*, Sacramento District, California, 1974, 87 pages.

278. MASON, E.E. The function of shotcrete in the support and lining of the Vancouver railway tunnel. In *Rapid Excavation-Problems and Progress*, published by the Society of Mining Engineers of AIME, 1970, pages 334-346.

279. BORTZ, S.A., ALESHIN, E., WADE, T.B. and CHUG, Y.P. Evaluation of present shotcrete technology for improved coal mine ground control. *IIT Research Institute, Bureau Mines Report* ORF-54-73, 1973, 219 pages.

280. SINGH, M.M. and BORTZ, S.A. Use of special cements in shotcrete. In *Use of Shotcrete for Underground Structural Support*, published by American Society of Civil Engineers, New York, 1975, pages 200-231.

281. TYNES, W.D. and McCLEESE, W.F. Investigation of shotcrete. *US Army Waterways Experiment Station Technical Report* C-74-5, Vicksburg, Miss., 1974, 38 pages.

282. FERNANDEZ-DELGADO, G., MAHAR, J. and CORDING, E.J. Shotcrete, structural testing of thin layers. *Report for Department of Transportation, Federal Railroad Administration*, Washington, DC., 1975, 219 pages.

283. CHITUNDA, J.K. Shotcrete methods at Lakeshore Mine aid overall ground support program. *Mining Engineering*, Dec. 1974, pages 35-40.

284. HOFFMEYER, T.A. Wet-mix shotcrete practice. In *Shotcreting*, American Concrete Inst. Special Publication No. 14, 1966, pages 59-74.

285. JONES, R.C. Shotcrete practices at the Henderson Mine. In *Use of Shotcrete for Underground Structural Support*, published by American Society of Civil Engineers, New York, 1975, pages 50-57.

286. KOBLER, H.G. Dry-mix coarse-aggregate shotcrete as underground support. In *Shotcreting*, American Concrete Inst. Special Publication No. 14, 1966, pages 33-58.

287. RYAN, T.F. *Gunite - A Handbook for Engineers*, Cement and Concrete Association, London, England. 1973, 62 pages.

288. SANDELL, B. Sprayed concrete with wire reinforcement. *Tunnels and Tunnelling*, Vol. 10, No. 3, 1978, pages 29-30.

289. VAN RYSWYK, R. Steel fibrous shotcrete used to repair railway tunnel in Fraser Canyon. *Construction West*, Sept. 1979, pages 10-11.

290. RYAN, T.F. Steel fibres in gunite - an appraisal. *Tunnels and Tunnelling*, Vol. 7, No. 4, 1975, pages 74-75.

Chapter 10: Blasting in underground excavations

Introduction

The following quotation is taken from a paper by Holmberg and Persson[291]:

" The innocent rock mass is often blamed for insufficient stability that is actually the result of rough and careless blasting. Where no precautions have been taken to avoid blasting damage no knowledge of the real stability of the undisturbed rock can be gained from looking at the remaining rock wall. What one sees are the sad remains of what could have been a perfectly safe and stable rock face."

In underground excavation engineering, good blasting is just as important as the choice of the correct excavation shape to conform to the in situ stress field or the design of the correct support system to help the rock mass support itself.

Two of the most important factors to be considered in relation to blasting in underground excavations are :

1. The blast should break the rock efficiently and economically and should produce a well fragmented muckpile (or orepile) which is easy to remove, transport, store and process.
2. The rock mass left behind should be damaged as little as possible in order to reduce the need for scaling and support to a minimum.

Obviously, these two conflicting requirements can only be satisfied by making a number of carefully planned compromises in the design of the blast. Such compromises can only be made on the basis of an understanding of the mechanics of explosive rock breaking.

Basic mechanics of explosive rock breaking

When an explosive contained in a borehole is detonated, the high pressure gasses generated by the explosion impact the walls of the borehole and generate an intense pressure wave which travels outwards into the rock. In the immediate vicinity of the borehole wall, the stresses can exceed the strength of the rock and shattering and crushing of the rock can occur. Since the intensity of the stresses generated by the explosion falls off rapidly with distance from the borehole, the rock behaviour will range from plastic deformation to brittle elastic fracturing and the particle size will increase rapidly with distance from the borehole wall.

Outside this zone in which the compressive strength of the rock is exceeded, a zone of radial cracks will be formed by the tangential tensile stress (hoop stress) component of the stress field induced by the explosion. These radial cracks will continue to propagate radially as long as the tangential tensile stress at the crack tips exceeds the tensile strength of the rock. Kutter and Fairhurst[292] have reported theoretical and experimental studies on the length of these cracks in homogeneous rock. In actual rock masses, this cracking will be influenced by anisotropy in the rock, pre-existing fissures in the rock and the in situ state of stress.

When the borehole is close to a free face created by a previous blast or by an uncharged relief hole, the fracture

The sad remains of what could have been a perfectly safe and stable rock face.

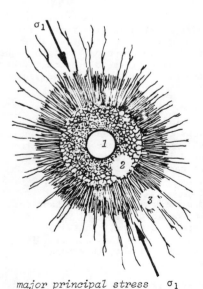

Idealised picture of fracturing induced by detonation of an explosive in a borehole.

1. *borehole,*
2. *pulverized zone,*
3. *radial cracks with preferential growth parallel to σ_1*

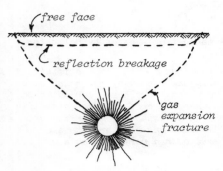

Blasting close to a free face. After Kutter and Fairhurst[292].

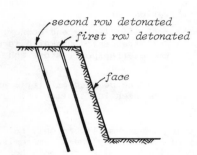

Typical bench blast layout

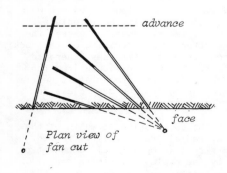

Plan view of fan cut

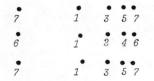

Face view showing typical initiation sequence

pattern around the hole is influenced to a very significant extent by the presence of this free face. This is because the radial compressive stress wave moving outwards from the borehole is converted to a reflected *tensile* stress wave when it encounters a free surface. This tensile stress wave moves back from the free surface towards the borehole and, in addition to causing spalling at the free surface, can alter the stress field surrounding the borehole and alter the resulting crack pattern. As will be seen later, blasting **must** always take place towards a free face in order to allow for the "swell" of the rock and to prevent "freezing" or "choking" of the blast.

In addition to the dynamic stress effects discussed above, the gas pressure generated by the explosion also plays an important part in wedging the cracks open and causing further propagation of the cracks. This gas pressure plays a critical role in techniques such as presplitting and smooth blasting in which the cracks are encouraged to grow in certain directions. In addition, the gas pressure is responsible for "heaving" the broken rock away from the hole and providing an adequate expansion volume for subsequent blasts.

Creation of a free face

In bench blasting operations, a free face is normally available since the blast is initiated from the holes closest to the bench face created by the previous blast. In a carefully designed bench blast, each row of holes will heave the rock away and create a new free face for the next row of holes to be blasted [2].

In a tunnel blasting operation, no free face is available since the boreholes are drilled parallel to the tunnel axis and there are no free surfaces parallel to the boreholes. Hence an essential first step in a tunnel blast is to produce a free cut. The *cut*, which is fired first to produce a free face, can be achieved in a number of ways which are discussed below.

Fan cut

When the tunnel advance is less than the span of the tunnel, as shown in the margin sketch opposite, there is enough room to allow inclined holes to be drilled into the face. Under these circumstances a fan cut can be used to create the expansion volume required for the subsequent blasting.

A typical fan cut layout, adapted from the book by Langefors and Kihlstrom[293], is shown in the margin sketch. Note that only the fan cut holes and not the other blastholes are shown. The detonation sequence shown in the face view is designed to break the rock sequentially so that each hole is working towards a free face.

Obviously there are a number of alternative ways in which a fan cut can be drilled and detonated and the reader is referred to the specific examples given by Langefors and Kihlstrom[293]. The charging of the holes is also critical for the success of the fan cut but a full discussion on charging exceeds the scope of this text. Once again, specific details can be obtained from Langefors and Kihlstrom or from the excellent handbooks issued by a number of companies concerned with the manufacture of explosives.

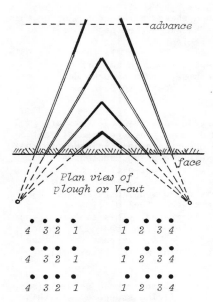

Plan view of plough or V-cut

Face view and initiation sequence

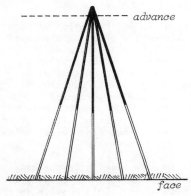

Plan view of instantaneous cut

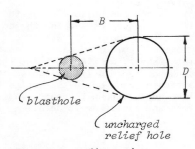

Burn cut configuration

Plough or V-cut

The layout of a plough or V-cut is illustrated in the upper margin sketch. A typical advance for this type of cut is 45 to 50% of the tunnel width. In wide tunnels, this advance is limited by drillhole deviation which is typically of the order of ±5%. Hence, in a 5m long hole, the end of the hole can deviate by ±0.25m and this can cause problems with flashover from one hole to another during ignition of the charges. Langefors and Kihlstrom[293] report a case in which an advance of 5.4m was achieved in a 9m span tunnel (60% of the tunnel width) by accurately drilling slightly larger than normal holes (44mm diameter) and by using detonators of high delay accuracy.

Instantaneous cut

A variation of the V-cut involves drilling the holes more nearly parallel, as shown in the centre margin sketch, and igniting all the charges simultaneously. Advances of up to 83% of the tunnel width can be achieved with this cut. Langefors and Kihlstrom[293] report a case in which an average advance of 5.5m was obtained and where the width of the base of the cut was only 3.3m in a tunnel of about 6.6m span.

A disadvantage of the instantaneous cut is the large throw which results in the muckpile being distributed over a considerable distance from the tunnel face. A modification of the layout shown in the centre margin sketch which involves truncating the end of the pyramid, is claimed to reduce the throw[293].

Parallel hole cuts

Because of the space required to accommodate a drilling machine and the rods for drilling angled holes in a face, the cuts described above can only be used in relatively large tunnels. In many civil and mining engineering applications, the tunnels are too small to allow these cuts to be used and hence parallel hole cuts have to be used to obtain the required advance.

When the blastholes are drilled perpendicular to the tunnel face and parallel to one another, the only effective way in which to provide a free face for the initial detonation is to drill a relief hole which is left uncharged.

The relationship between the burden B and the diameter D of the relief hole (see lower margin sketch) is critical for the success of the detonation of the first charge in a parallel hole cut. Langefors and Kihlstrom[293] suggest that for good fragmentation and clean ejection of the broken material the burden B should be less than 1.5 times the relief hole diameter D. Hagan[294] suggests that, even for 200mm diameter relief holes, the burden should not exceed 200mm.

The amount of explosive energy per metre of blasthole is critical in the detonation of the first hole in blasting a parallel hole cut. If the amount of energy is too low, the interlocking network of fracture surfaces will not be unmeshed and the segment of broken rock will remain in place. In this case the charge will blow out or "rifle" rather than break the burden. On the other hand, if the charge is excessive, a large number of radial cracks (see page 367) will be

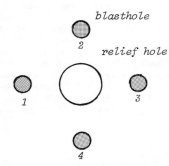

Typical parallel hole cut

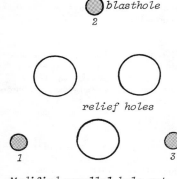

Modified parallel hole cut. After Hagan[294]

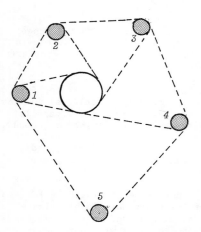

Spiral cut configuration

formed. The increase in volume or "swell" of the broken rock tends to be greater when these radial cracks occur than when spalling or reflection breakage (see page 368) takes place. Consequently, insufficient expansion volume is available and the broken rock tends to "freeze" in the relief hole. According to Hagan[294], freezing tends to be more common in fine grained rock than in coarser grained rocks because of the greater tendency for radial cracks to develop in fine grained rocks.

Additional problems in blasting parallel hole cuts are sympathetic detonation and dynamic pressure desensitization. Sympathetic detonation can occur in a hole adjacent to that being detonated when the explosive being used has a high level of initiation sensitivity. Most nitroglycerine based explosives exhibit such sensitivity. Dynamic pressure desensitization occurs in many explosives, particularly ANFO-type explosives (Ammonium Nitrate / Fuel Oil), when the density of the explosive is increased. This can occur when the pressure wave from an earlier-firing charge in an adjacent blasthole acts on a charge. In extreme cases, the density can be increased to a point where the explosive will not detonate.

Dynamic pressure desensitization problems can be minimised by using adequate delays in the firing sequence. In this way, the firing of each successive blasthole is delayed long enough for the pressure wave from the previous blast to have passed and for the explosive to have recovered to its normal density and sensitivity level.

Hagan[294] gives a detailed discussion on the design of parallel hole cuts and suggests that many of the problems discussed above are minimised by using the layout illustrated in the centre margin sketch. The three relief holes provide a larger expansion volume for the blasthole and they also shield the blastholes from each other which reduces the problems of sympathetic detonation and dynamic pressure desensitization. Hagan reports that of 31 parallel hole cut configurations tested in a dolomitic limestone with a strong tendency to freeze, the pattern with the three relief holes (centre margin sketch) was the most successful.

Because each successive blast in a parallel hole cut enlarges the expansion volume available for later-firing blasts, the burden can be increased for each blast. This results in the spiral cut configuration illustrated in the lower margin sketch.

Langefors and Kihlstrom[293] give a number of examples of parallel hole cut configurations and they also discuss the charge concentrations, initiation sequences and the choice of delays. They point out that the problems of drilling deviation can give rise to severe problems in parallel hole cuts and suggest that templates should be used to guide the drill, particularly when using hand-held drilling equipment. Any reader who is likely to become involved in the design or supervision of tunnel blasting is strongly recommended to study the text by Langefors and Kihlstrom in detail.

Rock damage

Good blasting starts with the correct design of the cut and of the remaining blastholes required to break the rock around the cut. In order to minimise the damage to the rock which is to remain, special consideration must be given to the layout and charging of the final line of blastholes. The design of these perimeter holes must be based upon an understanding of the factors which control rock damage adjacent to a blasthole.

Holmberg and Persson[291] suggest that rock damage is related to the peak particle velocity induced by the blast. This peak particle velocity may be estimated by means of the following empirical equation :

$$v = \frac{k \cdot W^\alpha}{R^\beta} \qquad (133)$$

where v is the peak particle velocity in mm/sec.,
W is the charge weight in kg.,
R is the radial distance from the point of detonation in metres,
k, α and β are constants which depend upon the structural and elastic properties of the rock mass and which vary from site to site.

The constants k, α and β used in this equation depend upon the type of blast and the condition of the rock mass in which the blast is carried out. Holmberg and Persson[291] suggest values of $k = 700$, $\alpha = 0.7$ and $\beta = 1.5$ for tunnel blasting conditions in competent Swedish bedrock and they have used these values in calculating the results plotted in figure 184.

As will be shown later in this chapter, the values of these constants can vary by large amounts, depending upon the conditions and the assumptions made in interpreting the data. Ideally, these constants should be determined for each site by conducting a series of trial blasts and monitoring the induced particle velocity at different distances from the points of detonation.

Note that equation 133 is based upon the assumption that the detonation occurs at a single point and hence it is only valid when the distance R is large compared with the length of the charge. When the point under consideration is close to a long charge, as is the case in tunnel blasting, the peak particle velocity v must be obtained by integration over the charge length. Holmberg and Persson[291] have discussed this problem and have derived the curves presented in figure 184 for typical tunnel blasting conditions.

Rock fracture is associated with a peak particle velocity of 700 to 1000 mm/sec and the zone of damage surrounding a 45mm diameter blasthole charged with ANFO, with a linear charge density of 1.5 kg/m, is approximately 1.5m in radius. This amount of damage is unacceptable when the blasthole is close to the final tunnel wall and steps must be taken to reduce the damage by reducing the charge. Figure 184 shows that the radius of the zone of damaged rock will be reduced to about 0.3m when the charge density is reduced to 0.2 kg/m. The use of charge densities of this magnitude in closely spaced holes is the basis of the technique of *smooth* or *perimeter* blasting which is used to control blast damage in underground excavations. This technique is discussed in the next section.

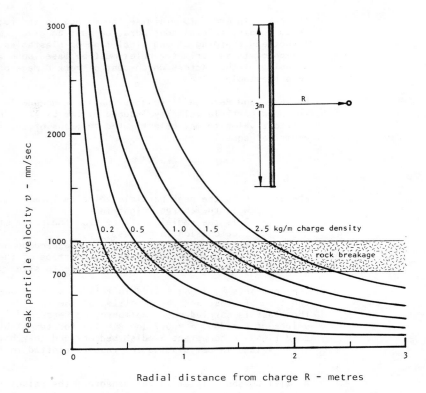

Figure 184 : Peak particle velocities at various distances from a borehole charged to different charge density levels. After Holmberg and Persson[291].

Smooth blasting and presplitting

The techniques most commonly used to control damage in the final walls of rock excavations are smooth blasting and presplitting. The basic mechanics of failure for these two techniques are almost identical and will be examined before the differences are discussed.

When two adjacent boreholes are detonated simultaneously, the circumferential tensile stresses induced by the explosion reinforce one another and cause an increase in the tensile stress acting perpendicular to a line drawn between the two holes. This tensile stress, which is higher that that acting across any other radial line drawn from either borehole, tends to cause preferential crack growth along the line between the two boreholes. By carefully choosing the correct borehole spacings and charge densities, a clean fracture can be caused to run from borehole to borehole around the perimeter of an excavation.

In addition to the need to detonate a line of smooth blast or presplit holes simultaneously in order to cause the stress reinforcement effect discussed on the previous page, it is also necessary to *decouple* the charge in each hole in order to mimimise the extent of the pulverized zone and encourage the growth of radial cracks (see margin sketch on page 367). This can be achieved by making the diameter of the charge smaller than the diameter of the borehole so that an annular air space is provided around the charge. This air space absorbs some of the initial explosive energy and reduces the magnitude of the initial high pressure impact which is responsible for the crushing of the rock immediately surrounding the borehole.

Once the crack running between the simultaneously detonated boreholes has been initiated by the interacting stress fields, the gas pressure in the boreholes plays an important role in wedging the cracks open and causing them to propagate cleanly between the holes.

When these conditions of simultaneous detonation of closely spaced blastholes with decoupled charges are satisfied, the required crack can be induced to propagate at a low charge density which, as shown in figure 184, reduces the amount of damage to the rock surrounding the borehole. Consequently, not only do the techniques of smooth blasting and presplitting give a cleanly fractured final surface but they also reduce the amount of damage inflicted upon the rock mass behind this final surface.

Smooth blasting

Smooth blasting involves drilling a number of closely spaced parallel boreholes along the final excavation surface, placing low charge density decoupled charges in these boreholes and detonating all of these charges together after the detonation of the remainder of the blastholes in the face. This means that the initial cut and the normal tunnel blast are carried out as discussed earlier in this chapter and a final skin of rock is peeled off the undamaged final excavation surface by the smooth blast.

In most cases, the smooth blast holes are drilled, charged and blasted in the same tunnelling cycle as the cut and main blastholes. Consequently, apart for the requirement for additional drilling, the use of smooth blasting does not introduce any delay into a normal tunnelling production cycle. In some cases the cut and main blast are carried out during one cycle and the smooth blast during the following cycle. This has the effect of advancing a pilot tunnel one round ahead of the final trimming blast and some operators consider that this produces better results than a single cycle blast. The authors believe that the main difference between these two methods is operational and the reader is left to choose the most convenient method.

In smooth blasting, the spacing between the boreholes is usually 15 to 16 times the hole diameter and the burden (the distance between the boreholes and the free face created by the previously detonated blastholes) is 1¼ times the spacing. The minimim linear charge concentration for both smooth blasting is given by[295]

$$w = 90.d^2 \qquad (134)$$

where w is the linear charge density of ANFO-equivalent

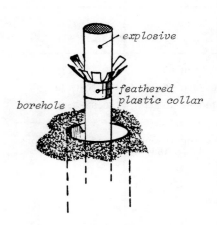

Decoupled charge used for smooth blasting and presplitting

explosive in kg/m and
d is the borehole diameter in mm.

Svanholm et al[295] give the following recommendations for smooth blasting in underground excavations:

Drill hole diameter mm	Charge diameter mm	Charge concentration kg ANFO/m	Burden m	Spacing m
25-32	11	0.08	0.30-0.45	0.25-0.35
25-48	17	0.20	0.70-0.90	0.50-0.70
51-64	22	0.44	1.00-1.10	0.08-0.90

Some examples of smooth blasting in underground excavations are illustrated in figures 185 to 188.

Figure 185 : Smooth blasting results in massive quartzite in a deep level gold mine in South Africa. Photograph reproduced with permission of the South African Chamber of Mines.

Figure 186 : Smooth blasting results in a tunnel for a hydroelectric project in South Africa. The rocks are weak horizontally bedded sandstones, mudstones and shales.

Figure 187 : Results achieved in a hard rock mining tunnel using crude smooth blasting techniques. In spite of the wide spacing of the boreholes (marked with white paint), the results are better than those obtained with normal bulk blasting.

Figure 188 : Controlled excavation using smooth blasting in the Swedish State Power Board's Stornorrfors Power Station. Photograph reproduced with permission of Atlas Copco.

Presplitting

The difference between smooth blasting and presplitting is that the presplit holes are generally more closely spaced than those used for smooth blasting and the charges are detonated simultaneously *before* the main blast. This means that the presplit crack exists before the main blast is detonated and this crack tends to limit the propagation of cracks from the main blastholes by providing a path to allow venting of the expanding gasses.

Because of the need to detonate the presplit charges in advance of the main blast, the use of pre-splitting in underground excavations may involve a separate drilling and charging cycle ahead of the main blast[295]. The inconvenience and the delay caused by this additional operation tends to limit the use of presplitting in underground excavations. In addition, the cost of the additional drilling and the tendency of the presplit crack to be deflected by high in situ stresses (since there is no adjacent free face to relieve these stresses) makes presplitting less attractive than smooth blasting for underground excavations.

Presplitting is generally used in benching operations where horizontal stresses tend to have been relieved and where there is generally more room in which to carry out different steps in the overall blasting process. Figure 189 shows the overall improvement in the stability of a bench face which can be achieved by presplitting (on the left) as compared with uncontrolled bulk blasting (on the right).

For presplitting, the borehole spacing is normally 8 to 12 times the borehole diameter[295] and the burden can be considered infinite. The charge diameters and charge concentrations listed on page 374 can be used for presplitting.

Figure 189 : Difference in the appearance and stability of a bench face in gneiss achieved by presplitting (left) and by uncontrolled bulk blasting (right).

Design of blasting patterns

When blasting damage occurs during the excavation of underground openings, there is a tendency to attempt to remedy the problem by the introduction of smooth blasting of all final walls. If this is done without consideration of the main blast design, the results can be less than satisfactory because the rock beyond the smooth blast line may already have suffered excessive damage from the main blast.

Good blasting design can never be done on a piecemeal basis; the entire blasting pattern including the cut, the main part of the blast and the perimeter holes must be considered if a satisfactory result is to be achieved.

A typical tunnel face blasting pattern is illustrated in figure 190. The reader can follow the logic of this pattern layout by carefully considering the ignition sequence shown in the figure. If a triangle is drawn with its apex at each blasthole and its base formed by the free face created by earlier firing holes (see lower margin sketch on page 370), the sequence of rock breakage can be followed.

Langefors and Kihlstrom[293] give a number of examples of blasthole patterns and ignition sequences and the reader is strongly advised to examine these patterns in order to gain a better understanding of the logic of blasting pattern design than it is possible to convey in this brief chapter.

Holmberg[296] has described the use of computers for designing blasting patterns for both surface and underground blasting and the authors of this book are convinced that the future will see more and more applications of the computer in the optimisation of blasting patterns.

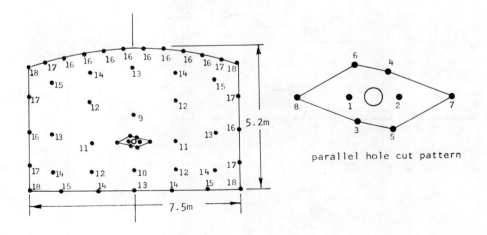

Figure 190 : Blasthole pattern and initiation sequence for a typical tunnel blast using a parallel hole cut and smooth blasting for the final walls. The relief hole diameter in the cut is 125mm and the blasthole diameter is 50mm. The charge density ranges from 0.4 to 2.5 kg/m and the advance is 4.3m. After Langefors and Kihlstrom[293].

Damage to adjacent underground excavations

Diehl and Sariola[297] report an experiment carried out in the Kiruna iron ore mine in Sweden where a 20kg blast was detonated and the induced particle velocities were monitored in a series of parallel tunnels adjacent to the blast site. The results of this study are summarised in figure 191 which gives a brief description of the damage associated with different ranges of peak particle velocity.

A reasonable approximation of the relationship between the particle velocity v and the distance R for this blast can be obtained by substituting $k = 200$, $\alpha = 0.7$ and $\beta = 1.5$ in equation 133 on page 371.

This equation has been used to construct the graph presented in figure 192 which can be used to estimate the particle velocity which would be induced at any distance R from a blast of W kg of explosives. This type of graph can be of considerable assistance in designing blasts on a particular site but it must be emphasised that figure 192 applies to Kiruna only and that it is essential to determine the constants k, α and β for each site. This can be done by carrying out a series of trial blasts and monitoring the induced particle velocities at different distances from these blasts.

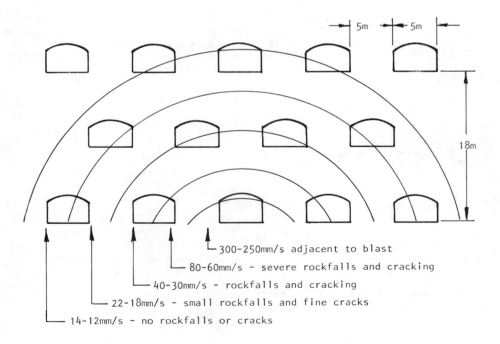

Figure 191 : Monitored particle velocities and observed damage in a series of parallel tunnels adjacent to a 20kg detonation at a depth of approximately 400m below surface in the Kiruna iron ore mine in Sweden. After Diehl and Sariola[297].

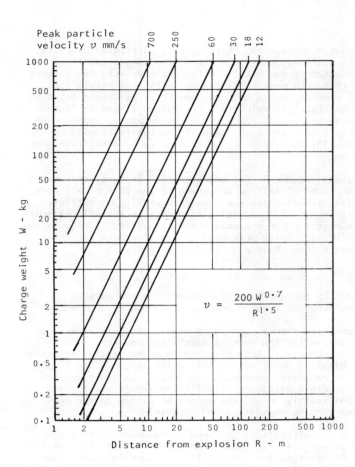

Figure 192 : Graph for predicting induced particle velocity at a distance R from a blast of W kg of explosives. Note that this graph applies to Kiruna only and that the constants k, α and β may be different at different underground blasting sites.

$$v = \frac{200 W^{0.7}}{R^{1.5}}$$

Typical published values for the constants k, α and β are listed below.

k*	α	β	conditions	reference
730	0.66	1.54		Lundborg et al[298]
2 083	0.53	1.60	$R/W^{1/3} > 3.97$	Ambraseys and Hendron[299]
11 455	0.93	2.80	$R/W^{1/3} < 3.97$	Ambraseys and Hendron[299]
1 686	0.71	1.78		Holmberg[300]
707	0.68	1.56		Vorob'ev et al[301]
700	0.70	1.50	Average Swedish bedrock	Holmberg and Persson[291]
193-1930	0.80	1.60	Down hole bench blasting	Oriard[302]
37-148	0.55	1.10	Coyote blasting**	Oriard[302]
5 958	0.80	1.60	Presplit blasting	Oriard[302]

* To calculate the particle velocity v in in/sec for a distance R in feet from a charge of W lb of explosive, divide k by 7.42 and use the same values for α and β.

** Coyote blasting involves placing a large concentration of explosives in a tunnel or cavity and it is used to loosen large volumes of rock close to surface.

The amount of damage associated with a particular particle velocity will obviously depend upon the condition of the rock mass before the blast. Unfortunately, very few authors report this rock mass condition in sufficient detail for general conclusions to be drawn on the relationship between damage and particle velocity. The values listed in figure 191 may be taken as a rough guide for damage to underground excavations but, as in the case of the constants used in equation 133, the damage criteria need to be established for each site.

The authors recommend that whenever this is done, the rock mass should be classified according to the CSIR or NGI rock mass classification systems (see chapter 2). If sufficient information is published relating damage to particle velocity in rock masses which are adequately classified, it should be possible to derive more general relationships for rock mass damage than those available at present.

Conclusion

> *" Blasting for underground construction purposes is a cutting tool, not a bombing operation."*

This quotation from a paper by Svanholm et al[295] emphasises the message which the authors have attempted to convey in this brief chapter. Good blasting involves careful attention to the blasting pattern, the initiation sequence and to the amount of explosive detonated per delay. In addition, the use of smooth blasting or presplitting (where appropriate) can significantly reduce the amount of damage to the rock mass surrounding an underground excavation.

There is ample practical evidence to demonatrate that the simple principles which have been summarised in this chapter work. Considering the availability of excellent books such as that by Langefors and Kihlstrom[293] in which these principles are clearly explained, the authors feel that there is no excuse for poor quality blasting in underground construction.

Chapter 10 references

291. HOLMBERG, R. and PERSSON, P.-A. Design of tunnel perimeter blasthole patterns to prevent rock damage. *Trans. Inst.Min. Metall.*, London, Vol. 89, 1980, pages A37-40.

292. KUTTER, H.K. and FAIRHURST, C. On the fracture process in blasting. *Intnl. J. Rock Mech. Mining Sci.*, Vol. 8, 1971, pages 181-202.

293. LANGEFORS, U. and KIHLSTRÖM, B. *The Modern Technique of Rock Blasting*. John Wiley and Sons, New York, 1973, 405 pages.

294. HAGAN, T.N. Understanding the burn cut - a key to greater advance rates. *Trans. Inst. Min. Metall.*, London, Vol. 89, 1980, pages A30-36.

295. SVANHOLM, B.-O., PERSSON, P.-A. and LARSSON, B. Smooth blasting for reliable underground openings. *Proc. 1st Intnl. Symp. on Storage in Excavated Rock Caverns*, Stockholm, Vol. 3, 1977, pages 37-43.

296. HOLMBERG, R. Computer calculations of drilling patterns for surface and underground blasting. *Proc. 16th Symp. Rock Mech.*, Minn., 1975, S.L. Crouch & C. Fairhurst, eds., pages 357-364.

297. DIEHL, G.W. and SARIOLA, P.J. The small-hole drilling method in rock store excavations. *Proc. 1st Intnl. Symp. on Storage in Excavated Rock Caverns*, Stockholm, Vol. 3, 1977, pages 23-27.

298. LUNDBORG, N., HOLMBERG, R. and PERSSON, P.-A. Relation between vibration, distance and charge weight. *Report Swedish Committee for Building Research*, R11, 1978.

299. AMBRASEYS, N.N. and HENDRON, A.J. Dynamic behaviour of rock masses. In *Rock Mechanics in Engineering Practice*, edited by K.G. Stagg and O.C. Zienkiewicz, published by John Wiley and Sons, London, 1968, pages 203-236.

300. HOLMBERG, R. Results form single shot ground vibration measurements. *Report Swedish Detonic Research Foundation,* DS 1979:9, 1979.

301. VOROB'EV, I.T. et al. Features of the development and propagation of the Rayleigh surface wave in the Dzhezkazgan deposit. *Soviet Mining Science*, Vol. 8, 1972, pages 634-639.

302. ORIARD, L.L. Blasting effects and their control in open pit mining. *Proc. 2nd Intnl. Conf. on Stability in Open Pit Mining*, Vancouver, Canada, 1971, published by AIME, New York, 1972, pages 197-222.

Chapter 11: Instrumentation

Introduction

In the early days of rock mechanics, a fairly common approach to underground problems seemed to be - " If you cannot think of anything else to do, go and measure something ". While this was good for the companies manufacturing instruments, it did not result in the solution of too many practical problems and it gave rise to severe scepticism on the part of owners and clients who had to pay for all of the gadgets used.

As the subject has matured, the approach to the use of instrumentation in underground construction projects has become more responsible and there is now a tendency to use instrumentation as part of an overall design and construction control package.

A vast array of instruments is available for use underground and no attempt will be made to deal with all of these instruments in this chapter. A few of the most important instruments and measuring techniques will be discussed and the reader will be left to follow up details in the ever growing number of catalogues available from instrument manufacturers.

Objectives of underground instrumentation

Instrumentation is used for the following purposes before, during and after the construction of underground excavations:

Before construction - to determine information required for the design of the excavations. Such information includes the modulus of deformation of the rock mass, the strength of the in situ rock and the in situ state of stress.

During construction - to confirm the validity of the design and to provide a basis for changes to the design. In addition, monitoring of displacements plays an important role in providing information which can be used to improve the safety of the underground construction sites.

After construction - to check the overall behaviour of the excavation during operation (in civil engineering applications) or to monitor the response of an excavation to the mining of adjacent excavations (in mining applications).

The objective of an underground instrumentation programme should be to satisfy these requirements as efficiently and as economically as possible. The requirements for simplicity, ruggedness and reliability cannot be over-stated since underground instrumentation is required to operate under severe conditions of temperature, humidity and rough handling. A typical underground instrumentation site is illustrated in figure 193.

Common inadequacies in instrumentation programmes

Lane[303] has listed the most common inadequacies in instrumentation programmes used in the construction of tunnels for civil engineering purposes. Lane's list is equally applicable to other types of underground construction and the following list has been adapted from that published by Lane.

1. Little systematic pre-planning - what information needed, what variables should be regulated for a controlled experiment; lack of concern for geologic conditions and for the need to control certain construction methods.

Figure 193 : Typical stress measurement site in an underground mine. Instruments for use underground have to be extremely rugged and reliable to withstand the severe conditions of temperature, humidity and rough handling to which they are subjected.

2. Initial behaviour patterns not well established (from which significant subsequent movements measured). Instruments installed too late, or installed and not read due to difficulty of access.

3. Inexperience of crew installing and observing, often failing to detect instrument misbehaviour or to recognise warnings of instability.

4. Initial evaluation and interpretation not made immediately in the field. In some cases, the delay caused by sending the results to a distant office for interpretation can result in the opportunity for immediate remedial action to be missed. Worse still, the office staff may not be able to interpret the results correctly without direct knowledge of the conditions in the field.

5. Instruments damaged by construction operations. Lack of back-up redundancy in measurements, particularly desirable for more sophisticated devices.

6. Contractor's responsibilities (assistance and payment therefor) not well defined, usually resulting in lack of cooperation.

7. Instruments inadequate. Not designed to withstand severe exposure in underground environment. Sophisticated devices not fully de-bugged before use.

8. Loss of results when cost cutting by management results in

curtailment of observations, analysis and reporting of results.

Instrumentation for the collection of design data

Instruments and equipment for the collection of structural geology data and rock strength and deformation characteristics have been mentioned in previous chapters and no further discussion will be included in this chapter. A large amount of published information, summarised in books such as that by Lama and Vutukuri[233], is available to the interested reader seeking further details on this type of equipment.

From the discussion presented in chapter 7 it will be evident to the reader that the in situ state of stress is one of the most important items of design data required by the underground excavation engineer. The magnitudes and directions of the principal stresses which exist in the rock before the creation of an excavation play an important role in controlling the stability of the excavation.

Many methods for measuring in situ stresses in rock have been proposed and these may be grouped under the following headings :

1. *Hydraulic fracture techniques* - the only method available for the measurement of stresses at distances of more than about 50m from the point of access. Fractures are induced in the rock by the application of a hydraulic pressure to the internal walls of a borehole and, from a knowledge of the pressure at which fracture occurs and the directions of the fractures, the in situ stresses can be estimated. Unfortunately, certain assumptions regarding the principal stress directions and the magnitude of one of the three principal stresses are required in order to interpret the data and this tends to limit the usefulness of the method. Haimson and his co-workers have described the hydraulic fracturing technique for stress measurement and have given a number of examples of practical application of the method [88-92].

2. *Direct stress measurement using flat jacks* - one of the oldest methods of stress measurement involving the measurement of the pressure required to restore a set of measuring pins on either side of a slot to the positions which they occupied before the slot was cut. Obert and Duvall[68] have summarised the work carried out by the US Bureau of Mines on the interpretation of the results of this type of measurement. Londe[304] has described the use of flat jacks inserted into slots cut by means of a diamond saw and this technique is illustrated in figures 194, 195 and 196. Obviously this method can only be used where access is available in an exploration adit or pilot tunnel and the technique is not suitable for heavily jointed rock or rock which has been severely damaged by blasting.

3. *Borehole methods of stress measurement* - currently the most popular type of stress measurement since the complete state of stress can be determined from a single borehole. Leeman[107], Rocha and Silverio[305], Worotnicki and Walton[306] and Blackwood[307] have all described instruments incorporating strain gauges which can be used for a complete stress determination in a single operation. Hast[75], Obert[93] and others have described other types of borehole instruments which can be used to determine the in situ stress field by combining measurements from several boreholes.

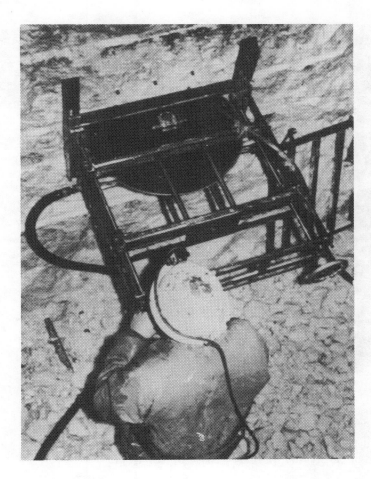

Figure 194 : Cutting a semi-circular slot in a rock face by means of a diamond blade driven by a hydraulic motor. Photograph reproduced with permission of Pierre Londe of Coyne et Bellier, Paris.

Figure 195 : Specially constructed semi-circular flat jack designed to fit into slot cut by saw shown in figure 194. No grouting is required and the jack can be reused. Photograph reproduced with permission of Pierre Londe of Coyne et Bellier, Paris.

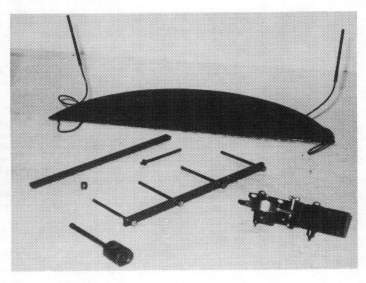

Figure 196 : Measurement of the hydraulic pressure required to restore the pins on either side of the flat jack slot to their original positions. Note that the pins must be installed before the slot is cut in order to establish the zero position. Photograph reproduced with permission of Pierre Londe of Coyne et Bellier, Paris.

A full discussion on the advantages and disadvantages of all of these borehole stress measuring instruments would exceed the scope of this chapter and the following comments are confined to one instrument. This is the hollow inclusion stress cell described by Worotnicki and Walton[306]. This instrument is considered by the authors of this book to be the most practical and reliable of the currently available instruments. The installation procedure is almost identical to that used for most of the other borehole instruments mentioned earlier in this discussion.

Figures 197 to 202 illustrate the principles of the stress measuring operation and give some details of some of the steps required in order to carry out the complete operation. Note that the strain gauges are fully encapsulated in an epoxy resin tube and that the read-out cable is permanently attached to the instrument (see figure 198). This construction ensures that the cell is extremely rugged and fully waterproof. Monitoring of the strain gauges is carried out during the over-coring operation and this permits identification of any malfunction in any of the gauges. Since only six strain readings are required for a complete determination of the stress field, the availability of nine gauges gives an adequate degree of redundancy to allow for malfunction in one or two of the gauges.

The authors recommend that extreme care be taken in the recording and analysis of the strain readings since it is very easy to mix readings from different gauges or to note down an incorrect sign for one of the readings.

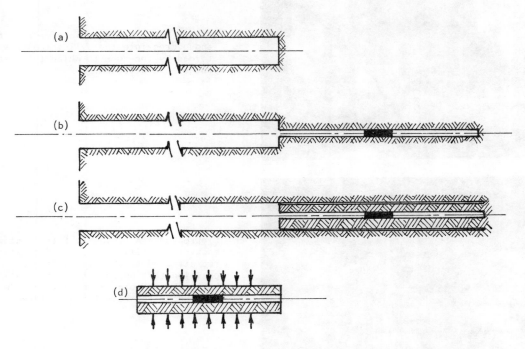

Figure 197 : Steps involved in the complete determination of the in situ state of stress in a single operation .
  a. Drill a large diameter (150mm or 6 in) borehole to the depth at which the stress determination is to be carried out.
  b. Drill a small diameter borehole (38mm or 1.5 in) a distance of about 500mm or 18in beyond the end of the large borehole. Install the stress measuring instrument in this small borehole.
  c. Overcore the stress measuring instrument using a thin wall diamond core barrel of the same outer diameter as the large diameter borehole.
  d. Subject the core containing the stress measuring instrument to radial pressure in a hydraulic cell in order to determine the modulus of elasticity at each strain gauge location.

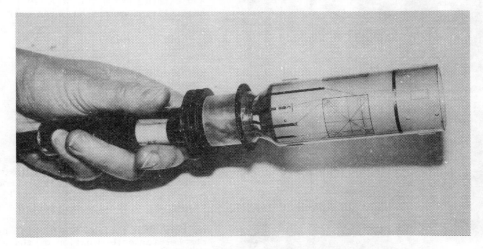

Figure 198 : Hollow inclusion stress cell with nine strain gauges fully encapsulated in epoxy resin and a permanently attached cable to permit readout of all strain gauges during over-coring. The instrument illustrated was manufactured by Rock Instruments, P.O.Box 245, East Caulfield, Melbourne, Victoria 3145, Australia from the original design by Worotnicki and Walton[306].

Figure 199 : Epoxy resin cement is mixed and poured into the cylindrical cavity formed by the tube to which the strain gauges are attached.

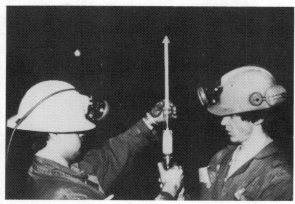

Figure 200 : A piston is fitted into the epoxy resin filled cavity. The extension rod attached to the piston is used to position the instrument in the small borehole and to push the piston in when the end of the hole is encountered.

Figure 201 : The instrument, filled with activated epoxy resin, is carried to the end of the hole by means of a special tool fitted with guide wheels to centre the instrument. A mercury switch in the tool gives the orientation of the gauges and a trip wire within the cell indicates that all the cement has been extruded.

Figure 202 : Once the cement has set (usually overnight), the cell is over-cored and the strain gauges are monitored during the over-coring process. If all gauges are operating correctly, a final set of strain readings is taken and the cell is then placed in a biaxial pressure chamber for determination of the elastic constants of the core.

A full discussion on the analysis of the strain gauge readings obtained from over-coring a hollow inculsion stress cell has recently been published by Duncan Fama and Pender[308]. Similar discussions, more limited in scope than that mentioned above, have been published by Leeman[107], Rocha and Silverio[305] and Worotnicki and Walton[306].

In some underground excavations, particularly those associated with water distribution or hydroelectric projects, the movement of groundwater through the rock mass is of considerable importance. Rock instability induced by reduction of the effective stress acting on joints (see page 153), leakage from the tunnel and erosion of soft discontinuity fillings are all important practical problems which have to be anticipated by the excavation designer. Many of these problems are difficult to quantify with any degree of precision but it is possible to obtain reasonable estimates of the rock mass permeability and the groundwater pressure distribution in the rock mass. These estimates can be used in analytical or numerical models to study the sensitivity of the design to groundwater pressure.

A comprehensive discussion on the permeability of jointed rock masses has been published by Rissler[309] and the interested reader is also referred to papers by Louis and Maini[310], Snow[311], Sharp and Maini[312] and Wilson and Witherspoon[313] for further details.

In the past, most of the theoretical work and field studies on the influence of groundwater has been restricted to slopes and foundations[2,5,304]. This is because water pressures in these structures may be of the same order as the stresses acting across discontinuities and this can give rise to serious instability. In the case of underground structures the stresses in the rock mass are generally very much higher than the groundwater pressures and the dangers of instability induced by a reduction in effective stress are not as high. Consequently, most underground designers tend to treat groundwater as a nuisance, to be dealt with when encountered, rather than a threat.

As pointed out, an exception to this general rule is the case of tunnels designed to conduct water under pressure. More recently, work on the underground disposal of radioactive waste[314,315] and on the storage of oil and gas[316-321] has emphasised the fact that the movement of even small quantities of groundwater may be important in certain applications. This work has also highlighted deficiencies in instrumentation and techniques for the measurement of low groundwater flows and pressures. The authors anticipate that many of these deficiencies will be remedied in the next few years as the results of current research come to fruition.

Monitoring of underground excavations during construction

Measurements carried out during construction should be designed to provide information which can be used to check the validity of the design or to permit the completion of on-going design work. In addition, these measurements should provide warnings of potential problems in order that remedial measures can be implemented before the problems have developed to a stage where the remedial measures are either very expensive or impossible to execute.

Londe[322], in discussing the use of instrumentation during tunnel construction, has emphasised the need for very simple and rugged instruments installed and monitored in such a way that interference with construction activities is minimal. Londe[322] and also Bieniawski and Maschek[323] consider that the measurement of displacement is the most effective means of monitoring rock mass behaviour during the construction of an underground excavation.

Several methods for monitoring rock mass displacements are available and these are summarised in the following notes.

1. *Optical surveying* - where suitable access is available and where the measurements can be related to a remote stable base, normal high quality surveying techniques such as levelling and triangulation can be used to determine the absolute displacements of targets fixed to the surfaces of underground excavations. The advantage of this approach is that surveying equipment of the required quality is normally available on site and most competent surveyors can carry out the measurements required. The disadvantages are that the measurements and the computations are time consuming and tend to interfere with the normal duties of the surveyor and that, in long tunnels, the results may not be precise enough to detect movements in a hard competent rock mass.

The advent of electro-optical distance measurement devices has made the task of surveyors a little easier and, for large underground caverns in which access is difficult, measurement from a fixed instrument position to a number of reflecting targets attached to the cavern roof and walls can give useful information.

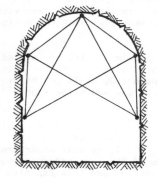

Typical five point convergence array in a tunnel.

2. *Convergence measurements* - normally carried out by means of a tape or rod extensometer between targets attached to the walls and roof of an excavation as illustrated in the margin sketch. A number of convergence measuring instruments are available and figure 203 shows one of these - a tape extensometer manufactured by the Slope Indicator Company.

Figure 203: Tape extensometer attached to a convergence measurement point fixed to the wall of an underground exavation. The instrument illustrated is a model 51855 tape extensometer manufactured by the Slope Indicator Company of 3668 Albion Place N., Seattle, Washington 98103, USA. The extensometer has an accuracy of ± 0.003 inch (0.08mm) over a distance of 100 feet (30m).

3. *Borehole extensometers* - are used to measure displacements in the rock mass surrounding an underground excavation. These extensometers consist of either sliding rods or tensioned wires, anchored at selected points within boreholes. In the case of the rod extensometers, the rods are usually sheathed in plastic tubing or run through nylon bearings to ensure that the friction between the different components is kept to a minimum. In the case of tensioned wire extensometers, the wires are generally manufactured from creep resistant alloy steels with low coefficients of thermal expansion. These wires are usually maintained under constant tension by a spring loaded tension head.

A typical rod extensometer with two anchor positions is illustrated in figure 204. Up to six anchor points are available in the case of rod extensometers and eight for wire extensometers.

Figure 205 shows an instrument station in a large underground cavern in a hydroelectric project. A three point rod extensometer head is shown in the upper right portion of the photograph and several pneumatic piezometer connections can be seen in the lower left part of the picture. Note that the entire station is housed in a drilled recess in the rock and that a circular steel cover plate (not shown) can be bolted onto the surface to protect the instrument connections.

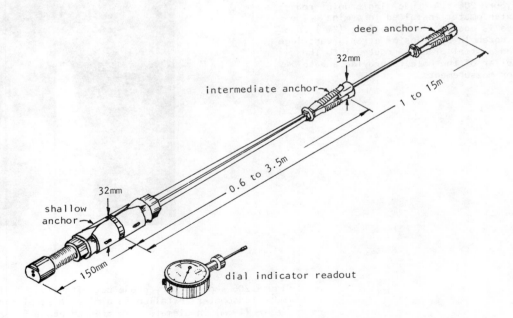

Figure 204 : Typical rod extensometer for the measurement of displacements in rock masses surrounding underground excavations. The instrument illustrated is a model E-2 double-point rod extensometer manufactured by Irad Gage, 14 Parkhurst Street, Lebanon, New Hampshire 03766, USA.

Figure 205 : Instrument station in a large cavern in an underground hydro-electric project. The station consists of a three point rod extensometer head and four pneumatic piezometer connections.

Figure 206 : A crude single point rod extensometer installed in a mine. The rod is anchored about 6 feet (2m) into a borehole by means of a resin cartridge and a steel pipe is grouted into the collar of the hole to provide a base for measurements.

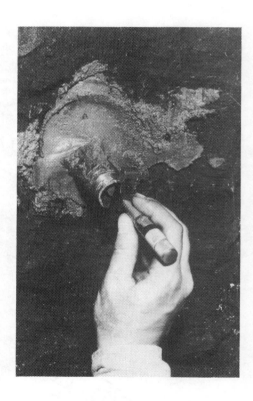

Figure 206 shows a very crude but effective "do-it-yourself" rod extensometer installed in a mine. A steel rod, about $^3/_4$ in (19mm) in diameter, is grouted about 6 feet (2m) into a borehole and a short length of steel pipe is grouted into the collar of the hole. The changes in distance between the end of the rod and the front face of the pipe are monitored by means of a depth micrometer or dial gauge. Note that the free length of the rod must be measured after it has been grouted in place in order to determine the extensometer length.

The monitoring of stress changes during construction is sometimes attempted but it is not quite as simple a task as it would appear. Most of the in situ stress measuring tools discussed earlier in this chapter are manufactured from plastics or are bonded to the rock by means of resin cements. These synthetic materials have a tendency to creep with time and hence the instruments are more suitable for short term measurements than long term monitoring.

In some cases these problems are overcome by repeating the absolute stress measurement by over-coring at different time intervals and, in other cases, a flat jack is left in place and the cancellation pressure measured whenever required.

Monitoring of underground excavations after construction

Once construction of an underground excavation has been completed, it is sometimes necessary to monitor its behaviour. Typical examples of such monitoring include displacement measurements in permanent hydroelectric caverns or mine crusher chambers to ensure that no long term instability is developing, monitoring of leakage from high pressure penstocks and monitoring of surface subsidence above shallow tunnels or underground mines.

The monitoring techniques used are identical to those discussed in the previous section with the emphasis on convergence measurements and extensometers because of the excellent long term stability offered by these instruments.

In underground mining projects in which caving techniques are employed, monitoring the development of the cave can be an important requirement, particularly when critical surface installations are located close to the potential cave boundary. Precise surface survey techniques are most commonly used for this problem and these generally give a good indication of the time-dependent development of the cave. In other cases, a simple indicator instrument can be constructed by anchoring a series of wires at different depths in a borehole drilled towards the cave. As illustrated in figure 207, the wires are brought out from the borehole, passed over a pulley system and tensioned by means of dead weights. When a weight drops to the floor, the indications are that the cave has progressed beyond the corresponding anchor point.

Monitoring of trial excavations

The monitoring of trial underground excavations is a special application of instrumentation that is being increasingly used to gather design data[303,324]. In some projects, there are major practical difficulties involved in obtaining and interpreting monitoring information and implementing design changes during construction. If adequate time is allowed before the main excavation work is undertaken, the results obtained from instrumented trial excavations, which should preferably be of full prototype width, can be most helpful in staging excavation and finalising support designs for the full-scale excavation. In one case known to the authors, the results obtained from an instrumented trial enlargement showed that the rockbolt and rock anchor reinforcement being planned for the excavation could be substantially reduced. The consequent savings in support costs exceeded the cost of the trial excavation.

Figure 207 : Simple borehole instrument to monitor the development of a cave in a large metal mine. A number of wires are anchored at different distances in the borehole and the wires are tensioned by means of dead weights. Movement of the weights indicates movement of the corresponding anchors in the rock mass.

The ability to adequately monitor the behaviour of the rock around the trial excavation and to measure loads in support elements is essential to the success of such an approach. In an invaluable paper on this subject, Sharp, Richards and Byrne[324] have presented a case-history describing the successful use of an instrumented trial excavation and, perhaps more importantly, have given a full discussion of the instrumentation requirements for trial enlargements in general.

Conclusion

The brief discussion presented in this chapter is intended to cover some of the most important aspects of instrumentation for underground construction or mining projects. The authors make no claim that the entire subject has been covered nor do they suggest that the instruments mentioned are the best available. The purpose has been to give the reader a basic understanding of the concepts and some of the practical approaches used in underground excavation engineering.

Instrumentation should never be used as an end in itself but always as a means towards an end. The end product of most underground construction projects is an opening which has been constructed safely and economically and which satisfies the need for which it was designed. If instrumentation can be used to improve the economy or safety of this construction process, its use is justified.

Chapter 11 references

303. LANE, K.S. Field test sections save cost in tunnel support. *Report from Underground Construction Research Council*, published by ASCE, New York, 1975, 59 pages.

304. LONDE, P. The role of rock mechanics in the reconnaissance of rock foundations, water seepage in rock slopes and the stability of rock slopes. *Quarterly Journal of Engineering Geology*, Vol. 5, 1973, pages 57-127.

305. ROCHA, M. and SILVERIO, A.A. A new method for the complete determination of the state of stress in rock masses. *Geotechnique*, Vol. 19, 1969, pages 116-132.

306. WOROTNICKI, G. and WALTON, R.J. Triaxial "Hollow Inclusion" gauges for the determination of rock stress in situ. *Proc. ISRM Symp. on Investigation of Stress in Rock and Advances in Stress Measurement*, Sydney, 1976, pages 1-8.

307. BLACKWOOD, R.L. An instrument to measure the complete stress field in soft rock and coal in a single operation. *Proc. Intl. Symp. on Field Measurements in Rock Mechanics*, K. Kovari, ed., A.A.Balkema, Rotterdam, 1977, Vol. 1, pages 137-150.

308. DUNCAN FAMA, M.E. and PENDER, M.J. Analysis of the hollow inclusion technique for measuring in situ rock stress. *Intnl. J. Rock Mech. Min. Sci.*, Vol.17, No. 3, 1980, pages 137-146.

309. RISSLER, P. Determination of water permeability of jointed rock. *Publication Inst. Found. Engg., Soil Mech., Rock Mech. and Water Ways Constr.*, RWTH (University) Aachen, Germany, Vol. 5, 1978, 150 pages.

310. LOUIS, C. and MAINI, Y.N.T. Determination of in situ hydraulic parameters in jointed rock. *Proc. 2nd Congr. Intnl. Soc. Rock Mech.*, Belgrade, Vol. 1, paper 1-32, 1970.

311. SNOW, D.T. Rock fracture, spacings, openings and porosities. *J. Soil Mech. Found. Div., Proc. ASCE*, Vol. 94, No. SM 1, 1968, pages 73-91.

312. SHARP, J.C. and MAINI, Y.N.T. Fundamental considerations on the hydraulic characteristics of joints in rock. *Proc. ISRM Symp. on Percolation through Fissured Rock*, Stuttgart, 1972, pages 1-15.

313. WILSON, C.R. and WITHERSPOON, P.A. Steady state flow in rigid networks of fractures. *Water Resources Research*, Vol. 10, No. 2, 1974, pages 328-335.

314. CHARLWOOD, R.G. and GNIRK, P.F. Conceptual design studies for a high-level waste repository in igneous rock. *Proc. 1st Intnl. Symp. on Storage in Excavated Rock Caverns*, Stockholm, Vol. 2, 1977, pages 339-346.

315. WITHERSPOON, P.A., GALE, J.E. and COOK, N.G.W. Radioactive waste storage in argillaceous and crystalline rock masses. *Proc. 1st Intnl. Symp. on Storage in Excavated Rock Caverns*, Stockholm, Vol. 2, 1977, pages 363-368.

316. BERGMAN, S.M. Groundwater leakage into tunnels and storage caverns. A documentation of factual conditions at 73 caverns and tunnels in Sweden. *Proc. 1st Intnl. Symp. on Storage in Excavated Rock Caverns*, Stockholm, Vol. 2, 1977, pages 51-58.

317. REINIUS, E. Groundwater flow to rock caverns. *Proc. 1st Intnl. Symp. on Storage in Excavated Rock Caverns*, Stockholm, Vol. 2, 1977, pages 119-124.

318. WESSLEN, A., GUSTOFSON, G. and MARIPUU, P. Groundwater and storage in rock caverns, pumping tests as an investigation method. *Proc. 1st Intnl. Symp. on Storage in Excavated Rock Caverns*, Stockholm, Vol. 2, 1977, pages 137-144.

319. ABERG, B. Prevention of gas leakage from unlined reservoirs in rock. *Proc. 1st Intnl. Symp. on Storage in Excavated Rock Caverns*, Stockholm, Vol. 2, 1977, pages 175-190.

320. LINDBLOM, U.E., JANELID, I. and FORSELLES, T.A. Tightness test for underground cavern for LPG. *Proc. 1st Intnl. Symp. on Storage in Excavated Rock Caverns*, Stockholm, Vol. 2, 1977, pages 191-198.

321. MILNE, I.A., GIRAMONTI, A.J. and LESSARD, R.D. Compressed air storage in hard rock for use in power applications. *Proc. 1st Intnl. Symp. on Storage in Excavated Rock Caverns*, Stockholm, Vol. 2, 1977, pages 199-206.

322. LONDE, P. Field measurements in tunnels. *Proc. Intl. Symp. on Field Measurements in Rock Mechanics*, K. Kovari, ed., A.A. Balkema, Rotterdam, 1977, Vol. 2, pages 619-638.

323. BIENIAWSKI, Z.T. and MASCHEK, R.K. Monitoring the behaviour of rock tunnels during construction. *The Civil Engineer in South Africa*, Vol. 17, No. 10, 1975, pages 255-264.

324. SHARP, J.C., RICHARDS, L.R. and BYRNE, R.J. Instrumentation considerations for large underground trial openings in civil engineering. *Proc. Intl. Symp. on Field Measurements in Rock Mechanics*, K. Kovari, ed., A.A. Balkema, Rotterdam, 1977, Vol. 2, pages 587-609.

Appendix 1: Bibliography on large underground excavations

Introduction

This appendix presents bibliographic details and, in most cases, short summaries of approximately 350 published items dealing with geotechnical aspects of the design and construction of large permanent underground excavations in hard rock. The emphasis is on case histories rather than more general or theoretical material. The bibliography is based on one previously published by the senior author (item 258 below) which, in turn, draws heavily on an excellent bibliography on underground hydroelectric power plants covering the years to 1957 prepared by Cooke and Strassburger (item 78). In the preparation of the present bibliography, many of the items published before 1957 and included in these earlier bibliographies have been omitted, only those dealing with geotechnical problems in some detail having been retained.

The choice of material for the present bibliography is largely subjective, and depends to a great extent on publications known to the authors and which they have found to be informative and useful. Generally, though not invariably, the publications referenced are written in the English language, and only excavations of a permanent nature in which the smallest span is at least 10m have been included. This means that some papers dealing with the larger transportation tunnels excavated in rock are included, but that many papers dealing with underground mining operations are excluded. Unfortunately, there is a paucity of published data on the larger underground mining excavations, many of which are of the same order of size as the hydro-electric power plant excavations with which a majority of the publications listed here are concerned. Because of their temporary nature, significantly less support is provided for many of these mining excavations than for permanent excavations of equivalent size, and so they could be regarded as representing a practical lower bound for support requirements.

In addition to the bibliography in which publications are listed in chronological order, this appendix contains a tabulation of major underground excavations in rock, listed by country. This tabulation does not seek to provide a catalogue of existing underground excavations for the obvious reasons that the task of compiling such a catalogue would be a most daunting one, and the results would be of limited value. Rather, an attempt has been made to give details of the sizes, rock types and conditions, and support systems used for a selected number of well-documented cases in the hope that this data will serve as a useful point of reference for the engineer faced with the task of designing an excavation in similar conditions.

PART 1 - BIBLIOGRAPHY

1. JAEGER, C. Underground hydro-electric power stations. *Civil Engineering and Public Works Review*, Vol 43, Dec 1948, pages 620-623; Vol 44, Jan-Feb 1949, pages 38-41 and 85-86. Review of underground power stations in Sweden, Switzerland, Scotland and Australia; diagrams of four different types of underground stations; technical drawings of plant at Innertkirchen, Pfaffenspring and Erstfeld, all in Switzerland; bibliography.

2. ROUILLARD, R.J. The excavation and support of an underground engine chamber at Durban Roodepoort Deep Ltd. *Papers and Discussions, Association of Mine Managers of South Africa*, 1948-49, pages 1-18. An engine chamber 19m x 16.5m x 8.9m high was excavated in hard, unjointed quartzite at a depth of 1500m. Roof support by concrete reinforced with steel arches.

3. WRIGHT, L.G.C. and KNIGHT, A. The cutting and support of large deep level excavations. *Papers and Discussions, Association of Mine Managers of South Africa*, 1948-49, pages 19-37. Engine and fan chambers excavated at depths of 2378 and 2073m in hard quartzite. Reinforced concrete roofs and concrete wall support used. A 5m wide pump chamber in a friable dyke also described.

4. EBERSBERGER, M. Lavey hydro-electric station of Lausanne city electricity supply. *Brown Boveri Review*, Vol 36, Oct-Nov 1949, pages 330-347. Geological factors decided adoption of underground scheme.

5. ADLER, G.F.W. Model tests on Clachan underground power station. *English Electric Journal*, Vol 11, Jun 1950, pages 119-127.

6. SEMENZA, C. An underground station. *Water Power*, Vol 2, Jul-Aug 1950, pages 144-151. Principal features of Lumiei hydroelectric project in Italy; most of article devoted to high concrete dam. Brief description of underground power plant; transformers and switchgear underground. Due to poor rock conditions, walls and concrete pillars were constructed with reinforced buttresses and intermediate relieving arches.

7. WESTERBERG, D. and HELLSTROM, B. Swedish practice in water power development. *Transactions, 4th World Power Conference*, London, 1950, Vol 4, pages 2071-2080.

8. HEGGSTAD, R. Norwegian hydroelectric power stations built into rock. *Transactions, 4th World Power Conference*, London, 1950, Vol 4, pages 2250-2266.

9. WITTROCK, K.J.P. and PIRA, K.G.G. Designing and dimensioning of the hydroelectric power plants of the Swedish State Power Board. *Swedish State Power Board*, Publication no 8, 1950.

10. GALLIOLI, L. Underground power stations. *Water Power*, Vol 2, Nov-Dec 1950, pages 227-236 and 246. Factors to be considered in underground schemes. Stazzona project, Milan is typical underground scheme; plant placed underground for security reasons. Transformers, switchboards and auxiliary equipment in two lateral tunnels; tunnel completely concrete lined, power station housed in an inner structure which is independent of outer structure. Cranes carried on reinforced pillars located in very poor formation.

11. ILLINGWORTH, F. Harspranget ambitious Swedish undertaking. *Mine and Quarry Engineering*, Vol 17, Jan 1951, pages 11-13. Excavation of Harspranget power station 225 feet down in granite.

12. JAEGER, C. A study tour in the Alps. *Water Power*, Vol 3, Jun 1951, pages 213-218. Discussion on Santa Massenza II and Santa Guistina projects. Latter has heavy reinforced struts in turbine floor to resist plastic movement of rock.

13. ANON. First underground power plant? *Engineering News-Record*, Vol 147, Dec 6, 1951, page 38. Snoqualmie Falls, Washington, plant completed 1899, placed underground to avoid freezing spray from falls. Another plant built 1904 at Fairfax Falls, Vermont, destroyed by 1927 flood which broke through shallow rock roof.

14. STEPHENS, F.H. Kemano. *Western Miner*, Vol 25, Jun 1952, pages 55-62. Mining work necessary for construction of hydroelectric power plant at Kemano; construction of powerhouse 1400 feet underground; driving of raises and tunnels.

15. RICE, H.R. Alcan's Nechako-Kemano-Kitimat project. *Canadian Mining Journal*, Vol 73, Jun 1952, pages 79-87; Jul, pages 63-70. Rock elasticity investigation, powerhouse construction, general outline of complete project. Part II gives descriptive account of rock and tunnel excavation work and drilling and blasting procedures used in power chamber.

16. ANON. Santa Giustina. *Water Power*, Vol 4, Aug 1952, pages 289-298; Sep, pages 324-333. Outstanding features of hydroelectric scheme on River Noce in northern Italy. First part deals with Santa Giustina dam and pressure tunnel, second part with underground power house. Well illustrated. Surge chamber in two parts - vertical shaft and helicoidal shaft which ascends with increasing inclination and variable horizontal area. Power chamber fully lined with false ceiling; tailrace tunnel lined.

17. BÄCHTOLD, J. Experience gained during the construction of the Handeck II power plant. (In German). *Schweitz Bauzeitung*, Vol 70, Oct 4, 1952, pages 573-577; Oct 11, pages 587-590; Oct 25, pages 612-614.

18. ANON. Trollhättan. *Water Power*, Vol 4, Oct 1952, pages 364-70 and 387. Hydroelectric installations at Trollhättan on Göta River in Sweden. One of the new plants is Hojum underground station, briefly described. Arched roof concreted with crane supported on columns, walls exposed rock.

19. ANON. 350 Mw hydroelectric generating station at Harspranget, Sweden. *Engineering,* Vol 174, Nov 7, 1952, pages 585-587; Nov 14 1952, pages 617-19. Article in two parts giving complete and detailed description of scheme; second installment devoted to power station. Upper portion of four vertical penstocks reinforced concrete designed to take external pressure; rock carries internal pressure; lower portion steel lined. Separate generator room and transformer cubicles; roof sound granite reinforced by 25mm diameter bars on 1m x 1m pattern, grouted to depth of 3 to 5m and covered with reinforced gunite; drainage provided; provision made for false ceiling over generator room if necessary. Notes that Swedish powerplants are placed underground for economic reasons, with safety being an additional but not decisive factor.

20. WISE, L.L. World's largest underground power station. *Engineering News-Record,* Vol 149, Nov 13, 1952, pages 31-36. Illustrated discussion on powerhouse, penstock and power tunnel excavation at Kitimat - Kemano scheme. Lists reasons for going underground.

21. HUBER, W.G. Alcan - British Columbia power project under construction. *Civil Engineering,* ASCE, Vol 22, Nov 1952, pages 938-943. Reinforced concrete roof arch, concrete block curtain walls, transformers in vaults in main chamber.

22. WEISSEL, W. Underground power stations. *Institution of Engineers (India) Journal,* Vol 33, Dec 1952, pages 195-208. Advantages and disadvantages and factors in design of underground power stations; data on plants in Sweden, Norway, Germany, Russia, France, Spain, Austria, Switzerland, Italy and Australia.

23. TALOBRE, J. The present state of underground penstock technique. *La Houille Blanche,* Vol 7, 1952, pages 513-531. (Translation by Jan C. van Tienhoven). Comprehensive article on the design of pressure shafts. Important to underground plants because of economics involved in pressure shaft design.

24. HUBER, W.G. Tunnels and underground penstocks require a million cubic yards of excavation (Alcan - British Columbia hydro project). *Civil Engineering,* ASCE, Vol 23, Feb 1953, pages 102-107. Two 3.35m diameter penstocks in concrete backfilled shaft; penstock slope 48° to horizontal; 1361.2m long; steel plate up to 49.2mm thick; 1 penstock to 4 units.

25. JOHANSON, E.A. Underground hydro plant boosts Rio power. *Electrical World,* Vol 139, Mar 23, 1953, pages 130 - 133. Forcacava (now Nilo Pecanha) underground power plant - engineering details, plans.

26. BUENAVENTURA, A.P. Engineering considerations of Ambuklao hydroelectric project in Agno River, Luzon. *Phillipine Engineering Record,* Vol 14, Apr 1953, pages 6-11.

27. ANON. Snowy Mountains. *Water Power,* Vol 5, Apr 1953, pages 131-39; May, pages 164-72; Jun, pages 204-12. Description of Snowy Mountains project in Australia. Part I: history of project. Part II - technical and economic background. Part III: illustrates proposed underground power station Tumut 1.

28. BACHTOLD, J. Construction of Oberaar power plant. (In German). *Schweitz Bauzeitung,* Vol. 71, May 1953, pages 271-277.

29. HUBER, W.G. Complex excavation pattern cuts out underground powerhouse. *Civil Engineering,* ASCE, Vol 23, Jun 1953, pages 396-401. Description of Kemano excavations, concrete arch machine hall roof, columns and girts anchored to walls.

30. JAEGER, C. Isère-Arc development. *Water Power,* Vol 5, Jul 1953, pages 256-262; Aug, pages 301-304. One of the projects of Electricité de France. Special reference to design of dam, surge chamber and pressure shaft; power plant data given.

31. ANON. The Paraiba-Parai Diversion. *Water Power,* Vol 5, Aug 1953, pages 287-293. Diversion scheme to obtain additional power for Sao Paulo, Brazil; Forcacava underground power station described.

32. SEETHARAMIAH, K. Underground hydroelectric power stations. *Indian Journal of Power and River Valley Development*, Vol 111, Sept 1953, pages 5-13 and 30. Discussion of various types of underground power stations - advantages and disadvantages, history of underground station development, plants operating under heads from 56 to 680 metres, claims first plant built in Switzerland in 1897, modern tunneling techniques chiefly responsible for development, various lining schemes, trend to place transformers underground, brief discussions on Kagginfon, Norway; Brommat, France; Innertkirchen, Switzerland and Santa Massenza, Italy.

33. RABCEWICZ, L.v. The Forcacava hydroelectric scheme. *Water Power*, Vol 5, Sep 1953, pages 333-337; Oct, pages 370-377; Nov, pages 429-435. Describes excavations for underground power station at Forcacava, Brazil - working methods adopted in unstable rock; power chamber fully concrete lined - broken rock caused overbreak; inclined slip planes encountered. Concrete columns erected during excavation heavily anchored to rock and concrete struts required to hold some columns. See also items 25, 31, 39 and 43.

34. MATTIAS, F.T. and ABRAHAMSON, C.W. Tunnel and powerhouse excavations at Kemano, B.C. for Alcan hydro power. *Canadian Mining and Metallurgical Bulletin*, Vol 46, Oct 1953, pages 603-621. Describes geology of region and excavation and driving methods used.

35. KENDRICK, J.S. Civil engineering features of the Kitimat project. *Boston Society of Civil Engineers*, Vol 41, Jan 1954, pages 88-112. General outline of project, brief statistics on underground power plant, diagrams showing method of excavation of main block of powerhouse. See also items 14, 15, 20, 29, 34, 38 and 45.

36. ESKILSSON, E. Tåsan power station. *ASEA Journal*, Vol 27, Mar 1954, pages 39-43. Swedish power station Tåsan, 740 feet below ground level; penstock from inlet tunnel slopes 1:1, is concrete upper 2/3 and sheet metal pressure tube grouted in concrete lower 1/3; control room, transformers and switchgear underground.

37. AUROY, F. et al. Les travaux de performation. (In French). *Travaux*, Vol 38, Mar 1954, pages 143-155. Work carried out in connection with the construction of the Montpezat power scheme in France.

38. ANON. Kitimat. *Water Power*, Vol 6, Mar 1954, pages 89-99; Apr, pages 124-135. Discusses in detail the excavation of the Kemano tunnel, shafts and powerhouse. Well illustrated.

39. RABCEWICZ, L.v. Bolted support for tunnels. *Water Power*, Vol 6, Apr 1954, pages 150-155; May, pages 171-175. Technique of roof bolting - its advantages, particularly when driving through unstable rock, theoretical principles of bolting. Uses the Forcacava tunnel and underground power station in Brazil to illustrate practical application.

40. ANON. Some recent Swiss hydro-electric schemes. *The Engineer*, Vol 197, Part 1, Apr 9, 1954, pages 518-520; Part 11, Apr 16, pages 554-555; Part IV, May 14, pages 698-701; Part IX, Jun 18, pages 878-882; Part X, Jun 25, pages 914-918. Review of seven major hydroelectric projects in ten articles - Grande Dixence, Mauvoisin, Innertkirchen, Handeck 11, Grimsel, future Grimsel 11, Peccia, Cavergno and Verbano. Illustrations.

41. MERRILL, R.H. Design of underground mine openings, oil-shale mine, Rifle, Colo. *U.S. Bureau of Mines Report of Investigation* 5089, 1954. A 60m long room with a horizontal roof in oil shale was progressively widened until it failed at a width of 24.4m. Roof sag measurements were made over a period of two years. Failure took place in a 50 cm. thick layer of shale. See also item 123.

42. HAGRUP, J.F. Swedish underground hydro-electric power stations. *Proceedings, Institution of Civil Engineers*, Vol 3, Aug 1954, pages 321-344. Sweden's water power resources; design of underground water power stations; blasting tunnels and underground stations; costs for power stations with underground machine halls.

43. RABCEWICZ, L.v. and FOX, P.P. Forcacava hydro-electric scheme. *Water Power*, Vol 6, Sep 1954, pages 353-354. Letter to editor of Water Power discussing article listed as item 39 above. Discusses excavation and construction methods used. Name of plant changed to Nilo Pecanha. See also items 25, 31, 33, 39 and 43.

44. EBERSBERGER, M. Salamonde hydro-electric station in Portugal. *Brown Boveri Review*, Vol 41, Oct 1954, pages 359-370. Description of power plant; emphasis on electrical equipment. Separate valve chamber with bypass tunnel around power plant, transformers on surface power chamber concrete lined; crane rail beams supported on cavern sides and tied to arch roof.

45. MATTIAS, F.T. The Nechako-Kemano-Kitimat Development (Kemano Underground). *Engineering Journal*, Vol 37, Nov 1954, pages 1398 - 1412. Part of a symposium on the complete project by various authors. Detailed review of construction details - plant, methods, equipment etc. used in excavation and concreting of Kemano power chamber, valve chamber, access tunnels, pressure conduits etc.

46. FERNANDES, L.H.G. The Salamonde hydro-electric scheme. *Water Power*, Vol 6, Nov 1954, pages 408-418; Dec, pages 449-456. Part 1 - entire Salamonde scheme; Part 11 - underground power plant. Well illustrated. See also item 44.

47. WESTERBERG, G. Building underground pays in Sweden. *Engineering News-Record*, Vol 153, Dec 9, 1954, pages 33-39. Reasons why underground power plants, industrial plants etc. are so economically constructed in Sweden; development of techniques and mass production basis of tunnelling.

48. NILSSON, T. Over 7 million yards of rock excavated for two power projects. *Engineering News-Record*, Vol 153, Dec 16, 1954, pages 41-43. Excavation of Kilforsen and proposed Stornorrfors plant. Rock walls at Kilforsen exposed with roof secured by rock bolts and gunited.

49. ANON. Oberhasli power schemes. *Engineering*, Vol 179, Jan 7, 1955, pages 13-16; Jan 14, pages 50-53. Features of Swiss development including Innertkirchen and Handeck 11 underground power stations, plans and profiles of schemes, powerhouse pictures and sections. Innertkirchen was first Swiss underground station in 1942; features of site and good rock made underground economic. Chamber unsupported but fully lined with false ceiling for seepage control. Penstock design discussed. Handeck 11 underground because no suitable surface site and rock is hard and compact; separate valve chamber in case of valve rupture; machine hall unsupported but fully lined, 20 inch arch thickness.

50. JAEGER, C. Present trends in the design of pressure tunnels and shafts for underground hydro-electric power stations. *Proceedings, Institution of Civil Engineers*, Part 1, Vol 4, Mar 1955, pages 116-200.

51. JONSSON, S. The Sog development. *Water Power*, Vol 7, Mar 1955, pages 84-92. Irofoss plant of Iceland described. Tailrace surge chamber; machine hall fully lined with false roof.

52. LORDET, J., OUQUENNOIS, H. and GUILHAMAN, J. Hydro-electric development of the Qued Agrioun. *Travaux (Edition of Science et Industrie)*. Special supplement, Fifth International Congress on Large Dams, May 1955, pages 203-206. Darguinah power station on the Qued Ahrzerousftis in Algeria - photographs and drawings; powerhouse 60 metres underground in rocky tributary canyon wall with tailrace tunnel to develop full head; poor ground heavily supported; fully concrete lined; valves in machine hall; transformers above ground.

53. JAEGER, C. The new technique of underground hydro-electric power stations. *The English Electric Journal*, Vol 14, Jun 1955, pages 3-29.

54. ANON. Developments in the Ångerman catchment. *Water Power*, Vol 7, Jun 1955, pages 202-213; July, pages 247-253; Aug, pages 292-300. Part 1 - Ångerman River catchment in central Sweden; Kilforsen power station; penstocks vertical steel lined shafts; sides of chamber exposed; cranes on concrete columns; transformers and switchgear in separate hall; machine hall roof bolted, wire netted and gunited. Part 11 - mechanical and electrical equipment. Part 111 - Lasele underground plant.

55. BOWMAN, W.G. Swedes make rock tunnel history. *Engineering News-Record*, Vol 155, Sep 1, 1955, pages 34-37 and 40-44. Latest projects in continuing underground programme. Two hydro plants and an air-raid shelter make new advances in size of tunnels and speed of driving. Hydro plants are Harrsele and Stornorrfors on the Ume River.

56. RANKIN, R.I. Excavation of no. 4 power station, Kiewa. *Chemical Engineering and Mining Review*, Vol 47, Sep 10, 1955, pages 489-493. Design and construction consideration for underground plants; geology and site excavation methods; poor rock resulted in concreting arch and walls at Kiewa; some construction schedule problems peculiar to underground plants.

57. CHANDRASEHARAN, A.S. Underground hydro electric power stations. *Indian Journal of Power and River Valley Development*, Vol 5, No 9, Sept 1955, pages 15-19, 28.

58. MOYE, D.G. Engineering geology for the Snowy Mountains Scheme. *Journal of the Institution of Engineers, Australia*, Vol 27, No 10-11, Oct-Nov 1955, pages 287-298.

59. FOX, A.J. Hydro-Quebec is developing more than horsepower at Bersimis. *Engineering News-Record*, Vol 155, Nov 17, 1955, pages 34-39. Bersimis project in eastern Quebec. Concrete lined roof arch, suspended ceiling and bare walls in underground powerhouse.

60. ANON. High speed tunneling techniques. *Water Power*, Vol 8, No 1, Jan 1956, pages 8-16. Breadalbane scheme in Scotland, St. Fillans powerhouse.

61. CAMPBELL, D.E. et al. The upper Tumut works. *Journal of the Institution of Engineers, Australia*, Vol 28, Jan-Feb 1956, pages 1-27. A symposium, half devoted to T-1 underground powerhouse; design criteria for access and plant dimensions.

62. CORDELLE, F. and DURAND, M. The upper Oued Djen Djen project. *Water Power*, Vol 8, Feb 1956, pages 46-54. Algerian development includes one dam and two powerplants - Mansouria underground - valve chamber, machine hall and substation are separate chambers.

63. ANON. Vinstra. *Water Power*, Vol 8, Mar 1956, pages 86-93. Underground plant of Vinstra development in Norway. Three separate tunnels for transportation, cables and tailwater-penstock tunnel isolated from station by self-closing steel bulkhead; connecting tunnel bypasses machine hall to tailrace; transformer hall parallel to machine hall.

64. VAUGHAN, E.W. Steel linings for pressure shafts in solid rock. *Journal of the Power Division, ASCE*, Vol 82, Apr 1956. Factors in the design of steel linings for pressure shafts for underground hydro-electric power plants; methods used in the design of two such shafts in Brazil. Nilo Pecanha and Cubatao discussed in detail and general features compared with those of other shafts constructed elsewhere. General approach to problem together with pertinent details of design and construction.

65. ROUSSEAU, F. Bersimis - Lac Casse. *The Engineering Journal*, Vol 39, Apr 1956, page 373. Cavern has concrete roof arch, transformers outside. Discusses steel lined penstocks and powerhouse excavation; photographs. Underground for economic reasons.

66. ANON. North of Scotland hydro-electric schemes - the Glen Shira scheme. *The Engineer*, Vol 201, Apr 20, 1956, page 364. Clachan underground plant is cut and cover excavation with concrete lined walls and arch roof. Review of pressure shaft design.

67. NILSSON, T. Recent development of Swedish water power design and construction. *Fifth World Power Conference*, Vienna, 1956, Vol 12, Paper 60H/12, pages 4167-4182. General characteristics of Swedish plants. Where head exceeds 35 metres, invariably cheaper to build underground. Table of statistics of 29 underground plants existing or under construction. New practices which save cost such as draft tube above scroll case; improved crane design for reduced clearances.

68. HEGGSTAD, R. Trends in Norwegian practice in water power development. *Fifth World Power Conference*, Vienna, 1956, Vol 12, Paper 109 H/21, pages 4285-4298. Intensive building of hydroelectric power stations from 1956 -56 new plants of which 24 are underground. 26 underground plants existing in Norway with 12 under construction. Advantages mostly economic. Good results with smooth blasting. In all new plants, transformers and in many cases high voltage circuit breakers underground; prevention of ice in air intakes important.

69. LAWTON, F.L. Kemano - advances in design and construction. *Fifth World Power Conference*, Vienna, 1956, Vol 13, Paper 154 H/25, pages 4365-4385.

70. HAYATH, M. and VIJ, K.L. Hydro-electric developments in India. *Fifth World Power Conference*, Vienna, 1956, Vol 13, Paper 179 H/29, pages 4411-4424. Progress and plans - Damodar Valley project includes Maithon and Konar underground. Projected Koyna project calls for three stage development underground.

71. ANON. Kariba hydro-electric scheme on the Zambesi River. *The Engineer*, Vol 202, Aug 3, 1956, pages 154-156. Kariba dam and underground power station with future second power station on Zambesi river, Rhodesia. Power plants to be under left and right abutments of dam. Principal elements of powerplants and dam described. See also items 97 and 108.

72. VERJOLA, V. Development of the water power resources in the north of Finland. *Water Power*, Vol 8, Aug 1956, pages 295-302. Jumisko, first underground powerplant in Finland - Exposed rock walls. Remotely controlled from 110 miles away.

73. ANON. The Aura development. *Water Power*, Vol 8, Sep 1956, pages 327-334; Oct, pages 389-396. Aura power station is Norway's largest. Part 1 describes development in general. Part 11 - powerplant, 2 machine halls 50 metres apart, tunnel branches at surge chamber, false ceilings, south hall has separate valve chamber.

74. TRYGGVASON, T. The rock series at Irafoss. *Water Power*, Vol 9, No 1, Jan 1957, pages 13-19. Discusses the geological investigations undertaken and the geological problems encountered in the construction of the Irafoss underground power station on the river Sog in Iceland. The rocks are Quaternary basalts, tuffs, sandstones, siltstones, agglomerates and volcanic ash and moraine.

75. ZERNICHOW, C.D. and MOYNER, H. Underground warehouse excavated in granite. *Civil Engineering*, ASCE, Vol 27, No 1, Jan 1957, pages 50-53. Underground warehouse in Oslo, Norway, consists of six parallel chambers 193 m. long by 14 m. wide connected by two access tunnels 160 m. long lined with precast concrete units fabricated underground.

76. TALOBRE, J. *La mécanique des roches, appliquée aux travaux publics*, Dunod, Paris, 1957. See in particular tables on pages 356-358 and 404-405 giving data on underground power stations and pressure tunnels in France and elsewhere.

77. NARAYANSWAMI, B.S.S. Geologic conditions affecting design and construction of pressure conduits and power house underground. *Indian Journal of Power and River Valley Development*, Vol 7, No 8, Aug 1957, pages 1-10, 22.

78. COOKE, J.B. and STRASSBURGER, A.G. Bibliography: underground hydroelectric power plants. *Journal of the Power Division*, ASCE, Vol. 83; No P04, 1957, 36 pages. Bibliography of 213 items covering the years 1912-1956. Gives list of plants with tabulated data but does not include geotechnical data.

79. PATTERSON, F.W., CLINCH, R.L. and McCAIG, I.W. Design of large pressure conduits in rock. *Journal of the Power Division*, ASCE, Vol 83, No P04, 1957, 30 pages. Discusses design of pressure conduits in rock - role of available cover and rock strength, proportioning of internal pressure in steel lining between steel and rock and design of steel lining against external pressure. Reviews theoretical principles and design assumptions. Describes the design, fabrication and construction of pressure conduits for Bersimis 1, Bersimis 11 and Chute-des-Passes schemes in Quebec.

80. NILSSON, T. Sweden excavates 2,100,000 cu yd for tailrace tunnel of underground power plant. *Civil Engineering*, ASCE, Vol 28, 1958, pages 19-21. Describes excavation methods for the 14.4m to 16m wide and 26.5m high Stornorrfors tailrace tunnel in granite and gneiss. See also items 55, 89, 92 amd 114.

81. ROBERTS, C.M., WILSON, E.B., THORNTON, J.H. and HEADLAND, H. The Garry and Moriston hydro-electric schemes. *Proceedings, Institution of Civil Engineers*, Vol 11, 1958, pages 41-68. The Ceannacroc and Glenmoriston power stations are underground in schists with injections of igneous rocks. The rock at Glenmoriston was more heavily jointed and therefore weaker than that at Ceannacroc. See item 88 for further details.

82. LANG, T.A. Rock bolting speeds Snowy Mountains project. *Civil Engineering*, ASCE, Vol 28, No 2, Feb 1958, pages 40-42.

83. MARCELLO, C. Underground power houses in Italy and other countries. *Journal of the Power Division*, ASCE, Vol 84, No PO1, 1958, 43 pages. Gives drawings and photographs of several Italian underground power stations but little geotechnical data.

84. MIZUKOSHI, T. The Sudagai underground power plant, Japan. *Journal of the Power Division*, ASCE, Vol 84, No PO1, 1958, 17 pages. Good description of excavation and concreting methods. Main cavern is 35m long, 16.6m wide and 31m high in a jointed and weathered course-grained granite.

85. COOKE, J.B. The Hass hydroelectric power project. *Journal of the Power Division*, ASCE, Vol 84, No PO1, 1958, 40 pages. Gives very complete details of this high head project on the North Fork of the King's River, California.

86. EBERHARDT, A. Ambuklao underground power station. *Journal of the Power Division*, ASCE, Vol 84, No PO2, 1958, 30 pages. Describes the design of the power features of this project on the Agno River in Luzon in the Philippines. Principal rocks are diorite and metamorphosed rocks that are highly fractured and weathered to great depth. A cave-in occurred during excavation of the main chamber.

87. MCQUEEN, A.W.F., SIMPSON, C.N. and MCCAIG, I.W. Underground power plants in Canada. *Journal of the Power Division*, ASCE, Vol 84, No PO3, 1958, 22 pages. Presents a review of factors affecting design practice in Canada and fully describes the Bersimis No 1 and Chute-des-Passes stations in North-central Quebec.

88. ROBERTS, C.M. Underground power plants in Scotland. *Journal of the Power Division*, ASCE, Vol 84, No PO3, 1958, 29 pages. Describes the development, layout and construction of the Ceannacroc and Glenmoriston underground power stations in Inverness-shire. Grouted rock bolts and concrete roof supports were used. Heavy water inflows were experienced during construction of the main chamber at Glenmoriston.

89. ANON. Tunnel blasting at Stornorrfors. *Water Power*, Vol 10, No 12, Dec 1958, pages 465-469. Describes techniques used in drilling and blasting the 360m^2 tailrace tunnel for the Stornorrfors power station on the Ume river in northern Sweden.

90. LAWTON, F.L. Underground hydro-electric power stations. *Engineering Journal*, Vol 42, No 1, Jan 1959, pages 33-51, 67.

91. PICHLER, E. and de CAMPOS, F.B. Rock characteristics at the Paulo Afonso power plant. *Journal of the Soil Mechanics and Foundations Division*, ASCE, Vol 85, No SM4, 1959, pages 95-113. The Paulo Afonso project on the Rio Sao Francisco was the first underground power station in Brazil. This paper describes the geology (rock mainly migmatite) and the in situ tests. The power station is 60m long, 15m wide and 31m high.

92. LANGEFORS, U. Smooth blasting. *Water Power*, Vol 11, No 5, May 1959, pages 189-195. Gives the basis of smooth blast design and illustrates its application to the Stornorrfors underground power station. See also items 55, 80, 89 and 114.

93. AHLSTROM, R. and JORGENSEN, J. The Harrsele tailrace tunnel. *Water Power*, Vol 11, No 7, July 1959, pages 267-274. This tunnel is 15m wide, 18.4m high and 3400m long. Excavation was by a full face top heading 9.7m high and a bench 8.7m high blasted as a unit.

Concrete lining cast in top heading before bench excavated.

94. SHEPHERD, E.M., SHARMAN, A.E., BOYLE, E.F. and CARD, G.M. The Tully Falls Hydro-electric Power Project. *Journal of the Institution of Engineers, Australia*, Vol 31, No 9, Sep 1959, pages 197-225. Describes all engineering features of this relatively small project including a 2m dia inclined pressure tunnel and the Kareeya Power Station excavated in massive rhyolite.

95. MOYE, D.G. Rock mechanics in the investigation and construction of the Tumut 1 underground power station, Snowy Mountains, Australia. *Engineering Geology Case Histories*, Geological Society of America, No 3, 1959, pages 13-44. See also items 58, 61, 99, 103 and 120.

96. FINZI, D., MAINARDIS, M. and SEMENZA, C. Underground power stations in Italy. *Journal of the Power Division*, ASCE, Vol 85, No P06, Dec 1959, pages 63-99, Vol 87, No P01, Jan 1961, pages 56-61 (Bibliography). Statistical data on 65 Italian power stations, sections and photographs of many, and a bibliography of 63 items are presented. Contains little geotechnical data.

97. ANDERSON, D., PATON, T.A.L. and BLACKBURN, C.L. Zambezi hydro-electric development at Kariba, first stage. *Proceedings, Institution of Civil Engineers*, Vol 17, 1960, pages 39-60. Machine hall is 142m long, 23m wide and 40m high with a 100m rock cover beneath the reservoir. Rock above the underground excavations (gneiss) is heavily grouted. Machine hall is fully concrete lined. Surge chamber 19.2 m dia. and 50.6 m high located underneath the dam. See also items 71 and 108.

98. SCISSON, S.E. Planning for mined underground LPG storage. *Oil and Gas Journal*, Vol 58, No 18, May 2, 1960, pages 141-142, 144. Rock in which LPG is to be stored must be impervious, massive enough to allow excavation of cavity, structurally sound and inert so that it will not react with stored material.

99. JAGGAR, B.K. The Tumut 1 project. *Water Power*, Vol 12, No 5, May 1960, pages 169-175 (Part 1); No 6, June 1960, pages 231-236 (Part 2). Excavation work for the underground power station described in Part 2. See items 95 and 103 for fuller geotechnical details.

100. SERATA, S. and GLOYNA, E.F. Design for underground salt cavities. *Journal of the Sanitary Engineering Division*, ASCE, Vol 86, No SA3, 1960, pages 1-21.

101. FRITZ, R.W. Titan construction for Titan missile. *Civil Engineering*, ASCE, Vol 31, No 4, Apr 1961, pages 50-53. Construction of base for 9 missiles involved excavation of 2 million cu.yd. of earth and rock, and installation of 500 tons of rock bolts.

102. PAINE, R.S., HOLMES, D.K. and CLARK, H.E. Presplit blasting at Niagara power project. *Explosives Engineer*, Vol 39, No 3, May-June 1961, pages 71-78 and 82-93. Twin conduits 14 m wide and 20 m high with arched tops constructed by cut and cover. Drilling and blasting methods including presplitting techniques described.

103. PINKERTON, I.L., ANDREWS, K.E., BRAY, A.N.G. and FROST, A.C.H. The design, construction and commissioning of Tumut 1 Power Station. *Journal of the Institution of Engineers, Australia*, Vol 33, No 7-8, Jul-Aug, 1961, pages 235-252. Describes the principal design and constructional features of this 320 MW underground power station completed by the Snowy Mountains Hydro-Electric Authority in 1959. The machine hall, 93m long, 18m wide and 34m high is located 330m underground in granitic gneiss intruded by biotite granite. Support is by a ribbed concrete arch roof and rock bolts.

104. RUFENACHT, A. Kiewa No 1 hydro-electric development. *Journal of the Institution of Engineers, Australia*, Vol 33, No 9, Sept 1961, pages 313-325. Gives engineering details of this small station on the Kiewa River in north-eastern Victoria, Australia. The power station is 60m underground, 13.7m wide, 21.4m high and 74.4m long. The rock is granodiorite containing a major set of discontinuities dipping at 15°. A major vertical fault-dyke complex some 15m wide running parallel and close to the power station on its west side largely governed the final positioning of the excavation.

105. JAEGER, C. Rock mechanics and hydro-power engineering. *Water Power*, Vol 13, Nos 9-10, 1961, pages 349-360 and 391-396.

106. YEVDJEVICH, V.M. Underground power plants in Yugoslavia. *Journal of the Power Division*, ASCE, Vol 87, No P03, Nov 1961, pages 81-92. In 1961, Yugoslavia had 12 underground hydroelectric power plants in operation and three under construction. The majority are constructed in limestone which has many practical rock engineering advantages but requires heavy pumping of seepage water.

107. CHAPMAN, E.J.K. Pressure tests on rock galleries for the Ffestiniog Pumped Storage Plant. *Transactions, 7th International Congress on Large Dams*, Rome, 1961, Vol 2, pages 237-260.

108. LANE, R.G.T. and ROFF, J.W. Kariba underground works. Design and construction methods. *Transactions, 7th International Congress on Large Dams*, Rome, 1961, Vol 2, pages 215-236. Main chamber is 143m long x 23m wide x 40m high. Problems caused by the jointed nature of the quartzite in the upper one-third of the works and a fault zone crossing part of the power station, gate shafts and surge chambers. Special drainage measures taken; concrete lining used throughout except on vertical powerhouse walls and tailrace tunnels. See also items 71 and 97.

109. LAURILA, L. The Pirttikoski tailrace tunnel. *Transactions, 7th International Congress on Large Dams*, Rome, 1961, Vol 2, pages 317-330. Give considerable detail of the drill and blast techniques used in excavating this 16m wide and 2500m long tunnel in granite and gneiss. See item 116 for details of the surge chamber on this project.

110. SCHULZ, W.G., THAYER, D.P. and DOODY, J.J. Oroville underground power plant. *Transactions, 7th International Congress on Large Dams*, Rome, 1961, Vol 2, pages 425-437. General description of this project on the Feather River, California.

111. OLIVEIRA NUNES, J.M. Underground works in Picote and Miranda hydro-electric developments (in French). *Transactions, 7th International Congress on Large Dams*, Rome, 1961, Vol 2, pages 607-636.

112. BOROVOY, A.A. and MAMASAKHLISOV, M.I. Some constructional features of underground structures in the U.S.S.R. *Transactions, 7th International Congress on Large Dams*, Rome, 1961, Vol 2, pages 833-850. Gives some details of the Ingouri, Nourek and Tcherkey hydro-electric stations.

113. BOYUM, B.H. Subsidence case histories in Michigan mines. *Proceedings, 4th Symposium on Rock Mechanics, Bulletin, Mineral Industries Experiment Station*, Penn. State University, No 76, 1961, pages 19-57. Detailed case histories of subsidence occurrences. Includes the Cliffs Shaft mine, Oshpeming, Michigan where hematite is mined by room and pillar methods. The maximum unsupported span area is 20 x 23m, or 27.4 x 70m when rock bolted.

114. ANON. The development of the Ume River - Part 3. *Water Power*, Vol 14, No 1, Jan 1962, pp. 25-35. Describes the construction of Stornorrfors, then Sweden's largest underground power station. The machine hall excavation is 124m long, 18.5m wide and 29m high. Underground complex also includes transformer hall and large tailrace tunnel. Smooth blasting techniques successfully used; roof supported with rock bolts, anchored in mortar, wire mesh and gunite. See also items 55, 80, 89 and 92.

115. KUDROFF, M.J. Titan ICBM hardened facilities. *Journal of the Construction Division*, ASCE, Vol 88, No C01, 1962, pages 41-57. Missiles are housed in concrete silos 13.4m dia and 49.m deep; control centre and power station are reinforced concrete lined domed structures spanning 30 and 37.5m respectively. Anti-shock mountings à major feature of the design. See also item 101.

116. PONNI, K., SISTONEN, H. and VOIPIO, E. Surge chamber of the low-head power plant at Pirttikoski. *Sixth World Power Conference*, Melbourne, 1962, Vol 6, pages 2277-2293. Describes engineering aspects of this 16m wide, 500m long and from 29 to 43m high unlined surge chamber excavated in granite.

117. MCLEOD, J.A.S. Choosing between surface and underground power stations for the Snowy Mountains Scheme. *Journal of the Institution of Engineers, Australia*, Vol 34, No 9, Sept 1962, pages 233-248. Includes a useful bibliography of 81 items on underground power plants.

118. ENDERSBEE, L.A. and HOFTO, E.O. Civil engineering design and studies in rock mechanics for Poatina underground power station, Tasmania. *Journal of the Institution of Engineers, Australia*, Vol 35, 1963, pages 187-207. Poatina powerhouse 150 metres underground in sedimentary rock; machine hall 91.4m long, 13.7m wide and 26m high. High horizontal stress in rock, compared with rock strength, gave rise to rock failure in exploratory openings. Design features developed to suit these conditions included stress-relief slots and special roof shape. Stresses in roof area were measured during excavation and compared with photoelastic predictions. Permanent support achieved by use of grouted, tensioned rockbolts and a relatively thin reinforced gunite lining.

119. DICKINSON, J.C. and GERRARD, R.T. Cameron Highlands hydro-electric scheme. *Proceedings, Institution of Civil Engineers*, Vol 26, 1963, pages 387-424. Gives full engineering details of this scheme in Malaya, including construction of the Jor underground power station. See also item 128.

120. ALEXANDER, L.G., WOROTNICKI, G. and AUBREY, K. Stress and deformation in rock and rock support, Tumut 1 and 2 underground power stations. *Proceedings, 4th Australia-New Zealand Conference on Soil Mechanics and Foundation Engineering*, Adelaide, 1963, pages 165-178. See also items 124 and 147.

121. COATES, D.F. Rock mechanics applied to the design of underground installations to resist ground shock from nuclear blasts. *Proceedings, Fifth Symposium on Rock Mechanics*, Minneapolis, 1963, pages 535-562.

122. ROSEVEARE, J.C.A. Ffestiniog pumped-storage scheme. *Proceedings, Institution of Civil Engineers*, Vol 28, 1964, pages 1-30. Full general engineering details but little discussion of geotechnical aspects. See also items 107 and 171.

123. EAST, J.H. and GARDNER, E.D. Oil-shale mining, Rifle, Colo., 1944-56. *United States Bureau of Mines Bulletin* 611, 1964, 163 pages. Gives a very detailed, well illustrated account of the design, development and operation of the U.S.B.M. demonstration room and pillar oil shale mine. See also item 41.

124. PINKERTON, I.L. and GIBSON, E.J. Tumut 2 underground power plant. *Journal of the Power Division*, ASCE, Vol 90, No PO1, 1964, pages 33-58. Complete description of geology, design, support, excavation procedures, and monitoring of the 97.5m long, 15.5m wide and 33.5 m high machine hall and ancillary works. The main rock types were a granitic gneiss and a biotite granite. A number of porphyry dykes intersected one end of the machine hall; the rock was generally sheared and jointed. Support of the machine hall was by a concrete arch roof and grouted rock bolts. See also items 120 and 147.

125. BLASCHKE, T.O. Underground command centre - problems in geology, shock mounting, shielding. *Civil Engineering*, ASCE, Vol 34, No 5, May 1964, pages 36-39. Several 3 storey steel frame buildings were built in excavated chambers; geological conditions required re-alignment of chambers from original design. In one area concrete lining was required to reinforce rock. See also items 130 and 138.

126. KNIGHT, G.B. Subway tunnel construction in New York City. *Journal of the Construction Division*, ASCE, Vol 90, No CO2, 1964, pages 15-36. Tunnel sections in the Manhattan schist have variable size, shape and proximity to existing structures. A combination of rock bolt, timber and structural steel is used for support. Concrete lining was used on the section of tunnel described.

127. THAYER, D.P., STROPPINI, E.W. and KRUSE, G.H. Properties of rock at underground power-house, Oroville Dam. *Transactions, 8th International Congress on Large Dams*, Edinburgh, 1964, Vol 1, pages 49-72. Gives complete details of in-situ stress measurements by flat jacks and borehole deformation gauge. An approximately hydro-static stress field of 3.5 MPa at a depth of 100 m was obtained.

128. KLUTH, D.J. Rock stress measurements in the Jor underground power station of the Cameron Highlands Hydro-electric Scheme, Malaya. *Transactions, 8th International Congress on Large Dams*, Edinburgh, Vol 1, 1964, pages 103-119. In-situ stresses measured in tunnels; lateral stresses of 1.8 and 2.6 times the vertical stress of 6.9 MPa which was close to the overburden stress at a depth of 275 - 290 m . Stress concentrations around the main excavation are determined and support dimensioned accordingly.

129. VLATSEAS, S. Rock excavation and consolidation of underground power station in Scottish highlands. *Water and Water Engineering*, Vol 69, No 830, Apr 1965, pages 146-149. Describes construction and support of the Dearie underground power station in the Strathfarrar and Kilmorack hydro-electric scheme.

130. SAMUELSON, W.J. Engineering geology of the NORAD combat operations center, Colorado Springs, Colorado. *Bull. Association of Engineering Geologists*, Vol 2, No 2, July 1965, pages 20-30. Approximately 360,000 m^3 of granite were mined to construct an array of openings varying in size from small 3.7m x 3.7m exhaust tunnels to 14m x 18m, facility chambers with a minimum of 240m of solid rock cover. The proposed chamber layout was reoriented to take account of joint orientations. Except in occasional major shear zones only rock bolting support was required. See also items 125 and 138.

131. DUTRO, H.B. Rock mechanics study determines design of tunnel supports and lining. *Civil Engineering*, ASCE, Vol 36, Feb 1966, pages 60-62. Twin 9.8 m wide highway tunnels with a centreline separation of 23.5m were excavated through marl, silstone, silty sandstone and shale near Green River, Wyoming. Displacements and support loads measured at four stations indicated the need for light but continuous initial support. Steel sets were used for temporary support.

132. STRANDBERG, H.V. Design and construction features: Boundary Project. *Journal of the Power Division*, ASCE, Vol 92, No PO2, 1966, pages 157-180. The Boundary Project is on the Pend Oreille River in Washington state near the Canadian border. The underground works including the machine hall are generally in a massive limestone. See item 139 for rock engineering details.

133. BREKKE, T.L. and SELMER-OLSEN, R. A survey of the main factors influencing the stability of underground constructions in Norway. *Proceedings, 1st Congress, International Society for Rock Mechanics*, Lisbon, 1966, Vol 2, pages 257-260. Gives a brief account of problems arising from discontinuity orientation, low strength of coated and filled discontinuities, solution of calcite, rock pressure, groundwater flow and swelling clays.

134. ROUSE, G.C. and WALLACE, G.B. Rock stability measurements for underground openings. *Proceedings, 1st Congress, International Society for Rock Mechanics*, Lisbon, 1966, Vol 2, pages 335-340. Describes the instrumentation program used at Morrow Point Underground Powerplant to give early warning of the potential instability of an extremely large wedge of rock isolated by major discontinuities. See also items 140, 167 amd 190.

135. MUIR, W.G. and COCHRANE, T.S. Rock mechanics investigations in a Canadian salt mine. *Proceedings, 1st Congress, International Society for Rock Mechanics*, Lisbon, 1966, Vol 2, pages 411-416. Rooms 18m wide and 12.8m high with 64m square pillars giving an average pillar stress of 23.4 MPa. Vertical and horizontal closure rates measured in rooms initially decrease with time finally reaching a constant rate. Non-uniform convergence causes beam flexure in the openings and local roof spalling.

136. YOUNG, W. and FALKINER, R.H. Some design and construction features of the Cruachan Pumped Storage Project. *Proceedings, Institution of Civil Engineers*, Vol 35, Nov 1966, pages 407-450. Underground construction involved the excavation of over 80,000m^3 of diorite and about 20,000m^3 in the associated caverns, tunnels and shafts. The rock was generally sound and jointed. Support was by 4.6m expansion shell rock bolts on 2.3m centres. A crushed zone about 6m wide ran transversely across the station requiring longer high tensile steel rock bolts and anchors on 3.5m centres prestressed to 60 tonnes each for support.

137. TANTON, J.H. The establishment of underground hoisting and crushing facilities. *Papers and Discussions, Association of Mine Managers of South Africa*, 1966-67, pages 139-162. Crusher and hoist chambers in quartzite with reinforced concrete and occasional rock bolts as roof support and mass concrete supporting the sidewalls.

138. UNDERWOOD, L.B. and DISTEFANO, C.J. Development of a rock bolt system for permanent support at NORAD. *Transactions Society of Mining Engineers*, AIME, Vol 238, No 1, Mar 1967, pages 30-55. The NORAD Command Center consists of three parallel chambers 14m wide, 18m high and 183m long separated by rock pillars 30m wide and connected by three intersecting chambers 10m wide, 17m high and 40m apart. Excavation was in granite. Paper includes discussions on excavation and ground support problems, rock strengthening following excavation, orienting bolts with respect to jointing systems, pull-out tests, intersection problems, grouting, and loss of tension and anchorage. See also items 125 and 130.

139. SCHILLING, A.A. Rock mechanics engineering for Boundary Project. *Journal of the Construction Division*, ASCE, Vol 93, No CO1, 1967, pages 27-46. The machine hall is an excavation 23m wide, 58m high and 146m long in massive limestone, and located parallel to but 36.6 m away from a near vertical natural rock surface. The paper describes rock tests, evaluation of design criteria, stress measurement, stability analyses and construction monitoring. As a result of jointing revealed during excavation, supplementary rock bolting was installed. Typical support was by grouted rock bolts on 1.8m centres with wire mesh and gunite as necessary. See also item 132.

140. SEERY, J.D. Construction of Morrow Point power plant and dam. *Journal of the Construction Division*, ASCE, Vol 93, No CO1, 1967, pages 47-58. Gives construction details of the underground power station described in items 134, 167 and 190.

141. BLIND, H. Excavating Söckingen cavern. *Water Power*, Vol 20, No 6, June 1968, pages 219-226 (Part 1); Vol 20, No 7, July 1968, pages 284-287 (Part 2). Full account of the excavation of the power house cavern for the Söckingen pumped storage scheme in the Black Forest area of Southern Germany. Excavation in generally good quality paragneiss. Support by rock-bolts (mainly perfo-anchors) and shotcrete.

142. IMRIE, A.S. and JORY, L.T. Behaviour of the underground powerhouse arch at W.A.C. Bennett dam during construction. *Proceedings, 5th Canadian Rock Mechanics Symposium*, Toronto, Dec 1968, pages 19-37. General description of Portage Mountain underground power house excavation on the Peace River in Canada. Measurements of displacements in arch during excavation showed considerable deviation from elastic theory. Authors suggest that excavation period most critical for support system and that dynamic loading from blasting should be controlled; suggest adaptation of tunnelling machines for powerhouse excavation. See also item 146.

143. HEDLEY, D.G.F., ZAHARY, G., SODERLUND, H.W. and COATES, D.F. Underground measurements in a steeply dipping orebody. *Proceedings, 5th Canadian Rock Mechanics Symposium*, Toronto, 1968, pages 105-125. Describes measurements made in an open stope and pillar iron ore operation at the MacLeod Mine, Wawa, Canada. Stopes are 18 to 23 m along strike and usually 70 m high. Pillars are 23-25 m along strike.

144. ACKHURST, A.W. Rock mechanics applications in design and excavation of No 2 crusher station at New Broken Hill Consolidated Limited. *Broken Hill Mines 1968*, Australasian Institute of Mining and Metallurgy Monograph Series No 3, 1968, pages 31-40.

145. DETZLHOFER, H. Rockfalls in pressure galleries. Translation from German original in *Felsmechanik und Ingenieurgeologie*, 1968, Suppl. 4, pages 158-180. *U.S. Army Corps Engrs. Cold Regions Research N.H. Tech. Report* AD 874 929, 1970, 23 pages. Paper deals with rockfalls which have occurred during the operation of pressure galleries at various power plants, resulting from decomposition of fissure filling and disintegration of rock structure under influence of varying pressure, particularly when gallery water gains access to the rock structure. Indications of danger of such rockfalls are difficult to detect during excavation. Describes the case of Kauner Valley pressure gallery in schist gneiss in which large scale pressure test was carried out in unlined section; reasons for large scale tests and method of execution discussed.

146. LAUGA, H. The underground powerplant. *Engineering Journal* (Engineering Institute of Canada), Vol 52, No 10, 1969, pages 35-42. Describes general design of the 20.3m wide, 46.7m high and 271m long power station excavation for the Peace River-Portage Mountain development in British Columbia, Canada. Excavation made approx. 150m below surface in layered sandstones and shales between two massive sandstone layers. Support by grouted and ungrouted rock bolts up to 6m long on a nominally 1.5 square grid. See also item 142.

147. WOROTNICKI, G. Effect of topography on ground stresses. *Rock Mechanics Symposium, University of Sydney, Australia,* Feb 1969, pages 71-86. Photoelastic and electrical analog study of stresses in rock surrounding Tumut 1 and Tumut 2 underground power plant excavations located in the steep eastern bank of the Tumut River valley, a 600 m deep V notch valley in the Snowy Mountains. Results suggest concentration of both vertical and horizontal stresses and rotation of principal stresses in location of excavations; site measurements tend to confirm these predictions. See also items 120 and 124.

148. MAMEN, C. Rock work at Churchill Falls powerplant. *Canadian Mining Journal,* Vol 90, No 3, 1969, pages 41-48. Describes tunnelling and construction work at Churchill Falls hydro-electric plant in Labrador. Tunnel drilling techniques and roof control by rock bolting discussed. See also items 164, 165, 180, 201 and 219.

149. ENDERSBEE, L.A. Applications of rock mechanics in hydro electric development in Tasmania. *Hydroelectric Commission of Tasmania Report*, May 1969, 46 pages. Comprehensive review of techniques used in Tasmania in designing underground powerhouse excavations and dams. Poatina used as an example to illustrate many of these techniques. See item 118.

150. GIANELLI, W.R. Oroville dam and Edward Hyatt powerplant. *Civil Engineering,* ASCE, Vol 39, No 6, June 1969, pages 68-72. The underground power station was one of the first to be analysed by the finite element method. The arched roof has 6.1m long rock bolts on 1.2m centres, steel chain link fabric and a 10 cm thick gunite coating.

151. BREKKE, T.L. A survey of large permanent underground openings in Norway. *Proceedings International Symposium on Large Permanent Underground Openings,* Oslo, 1969, pages 15-28.

152. BAWA, K.S. Design and instrumentation of an underground station for Washington Metro system. *Proceedings, International Symposium on Large Permanent Underground Openings,* Oslo, 1969, pages 31-42. Describes the du Pont Circle Station design and instrumentation. The station is a horseshoe shaped opening 236 m long, 23.5m wide and 13.4m high. Rock cover over the station arch varies from 7.5 to 9.0m and is overlain by 11 m of overburden. The rock is a schistose gneiss grading to a quartz hornblende or biotite gneiss in spots. Support is by a composite system of steel ribs and shotcrete supplemented by rock bolts.

153. ZAJIC, J. and HEJDA, R. Geotechnical survey applied to underground hydroelectric plants in Czechoslovakia. *Proceedings, International Symposium on Large Permanent Underground Openings,* Oslo, 1969, pages 43-55. Review of geotechnical surveys carried out in Czechoslovakia during past 15 years presented. Rock mainly granitic and surveys largely qualitative. In latter part of period more emphasis was placed on quantitative tests in situ and in the laboratory and the results were used in designing pumped storage schemes.

154. RAKIC, R. Engineering geological conditions during construction of Vrla-3 underground storage. *Proceedings, International Symposium on Large Permanent Underground Openings,* Oslo, 1969, pages 57-63. Collection and intepretation of geological data for Vrla-3 underground storage excavation. Geological mapping enabled designer to place excavation in good quality rock with significant saving in cost as a result of elimination of concrete lining.

155. OBRADOVIC, J. Investigation and structural analysis concerning the Mratinje underground power plant. *Proceedings, International Symposium on Large Permanent Underground Openings,* Oslo, 1969, pages 65-70. Reports the geological investigation of this site on the Piva River in Yugoslavia.

156. WEBER, H. Method of retaining or improving the properties of rock surrounding underground constructions. (In German). *Proceedings, International Symposium on Large Permanent*

Underground Openings, Oslo, 1969, pages 131-137. Changes in volume and structure of plastic and water-sensitive rocks and joint fillings can be reduced by use of "sandwich" lining using a compressible cushion between rock and lining.

157. SEELMEIER, H. Engineering geological considerations of underground power stations in the Austrian Alps. (In German). *Proceedings, International Symposium on Large Permanent Underground Openings*, Oslo, 1969, pages 207-213. Three underground power houses in the Austrian Alps are described, each situated in a different geological zone.

158. WEST, L.J. Rock mechanics application at projects involving underground excavations and a high cut slope. *Proceedings, 7th Engineering Geology and Soils Engineering Symposium*, Moscow, Idaho, Apr 1969, Published by Idaho Department Highways, 1970, page 2. Boundary hydroelectric project consists of 110 m high concrete arch dam with various underground excavations; machine hall in cavernous rock with a wide range of physical properties. Peach Bottom nuclear generating station on the west bank of the Susquehanna River in southern Pennsylvania is constructed in area created by excavation of one million cubic yards of rock resulting in a cut with maximum heights of 60 metres. Rock testing techniques and methods of evaluating stability are discussed.

159. CECIL, O.S. Shotcrete support in rock tunnels in Scandinavia. *Civil Engineering*, ASCE, Vol 40, No 1, Jan 1970, pages 74-79. Gives brief case histories of a number of underground rock construction projects in Sweden and Norway. Fuller details given in item 260.

160. FINE, J., TINCELIN, E., and VOUILLE, G. Determination of the stability of underground openings. (In French). *Practical Applications of Rock Mechanics Conference, French Committee on Rock Mechanics*, Paris, May 1970. Problem of determining underground excavation stability discussed in three sections: 1. Measurement of mechanical properties of rocks: 2. Mathematical analysis of stress distributions: 2. Field tests and measurements to verify theoretical predictions.

161. BURO, M.R. Prestressed rock anchors and shotcrete for large underground powerhouse. *Civil Engineering*, ASCE, Vol 40, No 5, May, 1970, pages 60-64. The Hongrin underground power station, a pumped-storage plant, is located on Lake Geneva in Switzerland. The cavern is 137m long, 30m wide and 27.4m high to the crown of its semi-circular arch roof. The limestone and limestone schist rock mass contains several groups of vertical fractures and a considerable amount of clay. Excavation began in small longitudinal galleries following the contour of the roof. Rock anchors and shotcreting were installed soon after excavation. A total of 650 anchors 11 - 13m long with working loads of 1110 - 1380 kN were installed.

162. LE FRANCOIS, P. In situ measurement of rock stresses for the Idikki hydroelectric project. *Proceedings, 6th Canadian Rock Mechanics Symposium*, Montreal, May 1970. Published by Department of Energy Mines and Resources, Ottawa, 1971, pages 65-90. Project located near the southern tip of India consists of three large dams and an underground powerhouse. Stress was measured to give quantitative information to supplement geological studies. Design of penstock steel lining and main powerhouse cavern dependent upon the stress field in the rock mass. Results obtained agree with recent data obtained in similar pre-Cambrian rocks.

163. LE COMPTE, P. Use of rock berms in underground power plants. *Proceedings, 6th Canadian Rock Mechanics Symposium*, Montreal, May 1970. Published by Department of Energy, Mines and Resources, Ottawa, 1971, pages 207-210. Rock berms excavated in the walls of an underground power house, just below the roof, are proposed as supports for crane rails. Experience of Hydro-Quebec with presplitting and flame cutting shows that adequate berms can be excavated, even when adverse jointing is present. If plant is favourably located with respect to major joint sets and rockbolting is used in critical areas, berms can be kept stable.

164. BENSON, R.P. *Rock mechanics aspects in the design of the Churchill Falls underground powerhouse, Labrador*. Ph.D. thesis, University of Illinois, 1970, 365 pages. Gives a full description of the rock engineering work on this scheme and an extensive bibliography. The main cavity is up to 24.4m wide, 21.3m high and 295m long. See also items 148, 165, 180, 201 and 219.

165. BENSON, R.P., MURPHY, D.K. and MCCREATH, D.R. Modulus testing of rock at the Churchill Falls underground powerhouse, Labrador. *American Society for Testing and Materials, Special Technical Publication* No 477, Jun 1970, pages 89-116. Rock comprises high quality gneissic assemblage. Modulus tests carried out on intact rock specimens and on in situ rock at depth of 30 metres. Plate jacking tests strongly influenced by blast damaged rock around excavations - low modulus inelastic rock behaviour near surface of excavation, high modulus elastic rock at depth; particularly important for pressure conduit design. See also items 148, 164, 180, 201 and 219.

166. ANON. Last of the big Snowy stations - massive rock excavation programme well ahead of schedule. *Mining and Minerals Engineering*, Vol 6, No 6, 1970, pages 13-14. Tumut 3 project on the Tumut River in the Snowy Mountains of Australia described briefly.

167. HAVERLAND, M.L. Installation, pressurisation and grouting of hydraulic flat jacks in Morrow Point power plant. *U.S. Bureau Reclamation Report.* No. REC-OCE-70-19, Jun 1970. Reaction against moving rock in a shear zone was provided by the installation of hydraulic flat jacks in the underground excavation for Morrow Point power plant, Colorado. Monitoring had indicated that one wall of powerhouse, located in a shear zone, was slowly moving towards river. 20 flat jacks 2.44m x 0.41m were installed and detailed description of installation procedure given. Rock movement was successfully stopped although no conclusion can be drawn as to whether this was due to jacks or to other remedial measures taken at the same time. See also items 134, 140 and 190.

168. BACKSTROM, A. and STROM, C.O. Underground storage structures. *Civil Engineering and Public Works Review*, Vol 65, No 766, 1970, pages 505-509. Underground storage structures have become increasingly competitive with surface storage. Description of projects carried out by Sentab, Sweden, in this field - storage of petroleum products and an underground freezing plant; advantages and economic benefits discussed.

169. THOMPSON, B.N. Geological investigation of the Aratiatia Rapids powerhouse area, Waikato River, New Zealand. *Proceedings, 1st International Congress, International Association of Engineering Geology*, Paris, Sep 1970, pages 1149-1158. Aratiatia Rapids power scheme was constructed on the crown of a partly exposed Pleistocene rhyolite dome consisting of hard, flow bonded rhyolite separated by superficial breccia zones. Surface mapping and drilling confirmed that the ends of the powerhouse would rest on hard rhyolite and that the central part would be on a breccia zone.

170. NEWBERY, J. Engineering geology in the investigation and construction of the Batang Padang hydroelectric scheme, Malaysia. *Quarterly Journal of Engineering Geology*, Vol 3, No 3, 1970, pages 151-181. The Batang Padang scheme in west Malaysia is located in an area of tropically weathered granite. Major factors in the siting, design and construction of the three dams, twelve miles of tunnel and powerhouse 270 metres underground were the nature of the residual soil mantle and the structure of the underlying rock. Site investigation involved 3000 metres of core drilling. Tunnels in competent granite with tight joints and hence unlined design chosen. Exploration of powerhouse site deferred until access available through tailrace tunnel. Site investigation included measurement of elastic modulus, stress measurement and discontinuity survey. Results confirmed that cavern would be excavated in intact rock suitable for permanent support by bolts, gunite and mesh. Intact rock in powerhouse proved to be "defective" in that it resulted in exfoliation of the cavern walls which necessitated more extensive guniting than expected. A major geological problem proved to be tunnelling through residual soil below water table which led to flow slides and necessitated tunnel diversions.

171. ANDERSON, J.G. Geological factors in the design and construction of the Ffestiniog pumped storage scheme, Merioneth, Wales. *Quarterly Journal of Engineering Geology*, Vol 2, No 3, 1970, pages 184-194. Scheme includes two dams, vertical shafts, pressure tunnels and penstocks. Influence of geology on choice of site and on subsequent design and construction discussed. See also items 107 and 122.

172. COMES, G. and BERNEDE, J. Effects of the shape and means of excavation used on the values of stresses measured at the walls of exploratory structures. *Proceedings, 2nd Congress, International Society for Rock Mechanics*, Belgrade, 1970, Vol 2, paper 3-16, 5 pages. Main access gallery to underground hydroelectric power station excavated by tunnelling machine while side galleries were excavated by blasting. Stresses measured in these two types of excavation are compared.

173. MANTOVANI, E., BERTACCHI, P. and SAMPAOLA, A. Geomechanical survey for the construction of a large underground powerhouse. *Proceedings, 2nd Congress, International Society for Rock Mechanics*, Belgrade, 1970, Vol 2, paper 4-24, 11 pages. Investigations carried out in connection with hydroelectric project at Lake Delio; geophysical tests and rock deformability tests described; measurement of rock deformation during excavation. See also items 181, 197 and 221.

174. HAYASHI, M. and HIBINO, S. Visco-plastic analysis on progressive relaxation of underground excavation works. *Proceedings, 2nd Congress, International Society for Rock Mechanics*, Belgrade, 1970, Vol 2, paper 4-25, 11 pages. Attempts to account for non-elastic, progressive deformations around underground excavations. The method developed is successfully applied to the construction of the Kisenyama underground power plant where damage due to blasting occurred. See items 175 and 185 for further details of this project.

175. YOSHIDA, M. and YOSHIMURA, K. Deformation of rock mass and stresses in concrete lining around the machine hall of Kisenyama underground power plant. *Proceedings, 2nd Congress, International Society for Rock Mechanics*, Belgrade, 1970, Vol 2, paper 4-29, 15 pages. Kisenyama pumped storage project has rockfill dam and machine hall 250 m below ground surface. The excavated cavity is 26.5m wide, 51m high and 60.4m long. Tests carried out in connection with machine hall design include geophysical survey, deformation modulus, shear tests, stress measurement, blasting tests. All tests were carried out in exploration tunnel. After excavation, further studies carried out to check validity of design, including photoelastic and finite element studies of stresses around machine hall, measurement of deformation in rock mass and stresses in concrete lining. See also item 185.

176. HEUZE, F.E. and GOODMAN, R.E. Design of room and pillar structures in competent jointed rock. Example: The Crestmore Mine, California. *Proceedings, 2nd Congress, International Society for Rock Mechanics*, Belgrade, 1970, Vol 2, paper 4-41. In situ stress and deformability measurements, deformation measurements, drill logging, laboratory tests on rock material and joints, and finite element analysis of this room and pillar limestone mining operation are described.

177. DIERNAT, F., COMES, G. and RIVOIRARD, R. Underground study of hygroscopic deformations of Sisteron calcareous marls. (In French). *Proceedings, 2nd Congress, International Society for Rock Mechanics*, Belgrade, 1970, Vol 2, paper 4-55, 6 pages. Strain gauges, microseismic measurements and core samples used in six month tests on deformations of hygroscopic origin in rock mass involved in construction of an underground hydroelectric plant.

178. BAUDENDISTEL, M., MALINA, H. and MULLER, L. The effect of geologic structure on the stability of an underground powerhouse (In German). *Proceedings, 2nd Congress, International Society for Rock Mechanics*, Belgrade, 1970, Vol 2, paper 4-56, 9 pages. Finite element study which allowed for effect of two sets of joints and a fault used to evaluate underground powerhouse stability. Stress and deformation patterns are illustrated. Allowance was made for rockbolt and concrete lining support.

179. DAGNAUX, J.P., LAKSHMANAN, J. and GARNIER, J.C. Seismic vibration testing of chalk subjected to laboratory and in situ stresses. (In French). *Proceedings, 2nd Congress, International Society for Rock Mechanics*, Belgrade, 1970, Vol 2, paper 4-57, 7 pages. Measurement of velocity and damping of seismic waves under variable stresses in an underground test opening in chalk in the Paris Basin in connection with investigation of an underground power plant site. Measurements compared with flat jack results and with stress distribution determined by finite element study.

180. BENSON, R.P., SIGVALDASON, O.T. and KIERANS, T.W. In situ and induced stresses at Churchill Falls underground power house, Labrador. *Proceedings, 2nd Congress, International Society for Rock Mechanics*, Belgrade, 1970, Vol 2, paper 4-60, 12 pages. In situ stress in rock mass at elevation of proposed machine hall measured by over-coring technique. Measured stresses used in finite element study of stress distribution around openings. Additional stress measurements carried out in rock pillar created by excavation. Test results indicated that in situ stresses varied in uniform and predictable manner throughout rock mass and this justified simple boundary conditions

used in finite element study. Measured stresses and deformations showed reasonable agreement with theoretical predictions. See also items 148, 164, 165, 201 and 219.

181. MANTOVANI, E. Method for supporting very high rock walls in underground power stations. *Proceedings, 2nd Congress, International Society for Rock Mechanics*, Belgrade, 1970, Vol 3, paper 6-5, 9 pages. Three underground power stations with vertical axis generating/pumping sets required cavities with 60m high walls to be excavated. In one excavation no special precautions were required but in other two rock walls had to be supported by anchored ropes during excavation. Ropes were 30m long with capacity of 100 tons each; rock locally supported by bolts and shotcrete. Lake Delio power station is the largest of the three; excavation stages are described and principles used in designing rope support given. See also items 173, 197 and 221.

182. MASUR, A. Efficiency of rock consolidation grouting in mountains around intake pressure galleries of hydroelectric power plants. (In German). *Proceedings, 2nd Congress, International Society for Rock Mechanics*, Belgrade, 1970, Vol 3, paper 6-17, 6 pages. Development of rock consolidation grouting techniques important in the construction of pressure tunnels for hydroelectric projects; reduction in time required for construction and minimisation of leakage are important economic considerations.

183. LANE, R.G. An investigation into the deformation of a combined dam and powerhouse structure. *Proceedings, 2nd Congress, International Society for Rock Mechanics*, Belgrade, 1970, Vol 3, paper 8-2, 4 pages. Finite element study of ground structure movements in case of powerhouse closely associated with intake dam; effect of varying elastic modulus and of a deep crack are also studied. Results show that, even for a very deformable rock, movements are small and should not give rise to serious operational problems.

184. HÖFER, K.H. Underground works - general report. (In English, French and German). *Proceedings, 2nd Congress, International Society for Rock Mechanics*, Belgrade, 1970, Vol 4, pages 346-372. General report based on 67 papers presented by authors from 20 countries.

185. YOSHIMURA, K. Measurement of rock deformation around cavity. *Rock Mechanics in Japan*, Vol 1, 1970, pages 103-105. Description of rock deformation measurements carried out in rock surrounding Kisenyama underground power station machine hall. See also item 175.

186. MATHUR, S.K. and SANGANERIA, J.S. Geotechnical considerations in selecting underground powerhouse for Mahi Hydel project, Banswara Dist., Rajasthan. *Journal of Engineering Geology* (India), Vol 5, No 1, Oct 1970, pages 191-196. An underground powerhouse is proposed as part of the hydropower development of the Mahi valley. Discusses the influence of geology in planning and design of cavity 70m below ground measuring 45m x 15m x 30m. Rock is Deccan basalt, overlying Pre-Cambrian granites and amphibolites. A 12 to 22m thick weathered rock zone has been encountered under the basaltic lava flow. Geological conditions favour siting machine hall in steeply dipping amphibolite band. In situ rock tests and photoelastic studies will establish the degree of anisotropy and stress distribution in the rock mass.

187. JALOTE, S.P. Geomechanical considerations in relation to the stability of the underground powerhouse cavity at Chibro, Yamuna, Hydel scheme, Stage 2. *Journal of Engineering Geology* (India), Vol 5, No 1, Oct 1970, pages 197-210. Machine hall is being excavated in thinly bedded limestones and slates of the Mandhali series, dipping at 45 to 50° with well developed bedding partings and numerous joints and slip planes. Statistical and stereographic analysis of structures indicated that 1. cavity has to be supported immediately after excavation, 2. power house arch may have rock load equivalent to a triangular mass of rock 20m wide at base and 15m high; 3. long walls of powerhouse contain potential triangular wedges which could slip out on shear zones if not adequately supported. Special steel arches have been designed and erected as permanent support for cavity roof and walls. These arches are supported by prestressed high tensile steel cables of 60 tonnes capacity spaced at 2 to 5m.

188. TANDON, G.N. Stress fields and the design of the powerhouse cavity at Chibro. *Journal of Engineering Geology* (India), Vol 5, No 1, Oct 1970, pages 255-266. Author claims that type of support required for roof and sidewalls of machine hall excavation had no known precedent and hence, in spite of careful choice of location and alignment of powerhouse, design had to be modified as excavation proceeded and as more information became available. The essential support requirement was that minimum delay should be allowed between excavation and installation of support. Paper describes considerations which governed location, layout and shape of cavity and details of the design and construction of the more important features of the cavity.

189. TANDON, G.N. Design of prestressed anchors for the support of the walls of the Chibro power house. *Journal of Engineering Geology* (India), Vol 5, No 1, Oct 1970, pages 267-274. Prestressed high tensile steel anchors were used to resist sliding along existing planes of weakness in the walls of the powerhouse. Such anchors must be installed with a few hours of blasting; provision made for adjustment of tension depending upon rock dilation or creep.

190. BROWN, G.L., MORGAN, E.D. and DODD, J.S. Rock stabilisation at Morrow Point power plant. *Journal of the Soil Mechanics and Foundations Division*, ASCE, Vol 97, No SM1, 1971, pages 119-139. During excavation of the underground powerhouse at Morrow Point, monitoring revealed inward movement along a longitudinal face of the chamber - movement associated with large wedge defined by two intersecting faults. Anchor bars, long rock bolts, poststressed tendons and flat jack were selected to restrain wedge. No further movement has taken place since installation of restraints. See also items 134, 140 and 167.

191. GOLZÉ, A.R. Edward Hyatt (Oroville) underground power plant. *Journal of the Power Division*, ASCE, Vol 97, No PO2, Mar 1971, pages 419-434. See also items 127, 150, 199 and 259.

192. GUNWALDSEN, R.W. and FERREIRA, A. Northfield Mountain pumped storage project. *Civil Engineering*, ASCE, Vol 41, No 5, May 1971, pages 53-57. General description of the then largest pure pumped storage project in the world and the first in U.S.A. to have an underground power house. The unlined main cavern in massive granite is 100m long, 21.3m wide and 47.2 m high. Support is by 35 mm dia. rock bolts on 1.5m centres and up to 10.7m long, and gunite reinforced with wire mesh fabric. See also item 198.

193. BAMFORD, W.E. Stresses induced by mining operations at Mount Charlotte. *Proceedings, 1st Australia-New Zealand Conference on Geomechanics*, Melbourne, 1971, Vol 1, pages 61-66. Low grade gold ore is mined by sub-level open stoping with open stopes 300 ft high, 180 ft wide and 80 to 180 ft long separated by rib pillars 80 - 90 ft thick. Following a "bump" or partial brittle fracture of a rib pillar in quartz dolerite a series of elastic stress calculations showed that failure was predictable.

194. DYSON, L.A. A rock mechanics survey and its use in an underground stability analysis at Kambalda, W.A. *Proceedings, 1st Australia-New Zealand Conference on Geomechanics*, Melbourne, 1971, pages 67-72. Structural geological survey, in situ stress measurement and mechanical property tests carried out at Kambalda nickel mines in Western Australia. Results combined with finite element model to study ground movements resulting from different mining methods.

195. JACOBS, J.D. Better specifications for underground work. *Civil Engineering*, ASCE, Vol 41, No 6, 1971, pages 47-49. Conflicts between owner and contractor on a construction project are frequently caused by badly written specification documents. Author suggests that contract should make provision for payment to contractor for over-excavation since this is unavoidable on most jobs. Contractor should be given a voice in determining the specifications for excavation support. An advisory committee may be helpful in this regard. To produce more attractive bid packages, unit price pre-stipulation should be avoided.

196. FARIS, C.O. Dworshak Dam underground crushing chamber. *Proceedings, Symposium on Underground Rock Chambers*, ASCE, New York, 1971, pages 147-165. This underground chamber houses the primary feeder, primary gyratory crusher, scalping screens and secondary core crushers for 13 million tons of granite gneiss needed for concrete aggregates for Dworshak dam. The chamber is 10 m wide, 25 m long and 31 m high, and is accessed by a

6 m wide by 5 m high horseshoe tunnel. Few geotechnical details are given, but construction details are described.

197. DOLCETTA, M. Problems with large underground stations in Italy - Lake Delio power plant. *Proceedings, Symposium on Underground Rock Chambers*, ASCE, New York, 1971, pages 243-286. Paper considers design approach used for three underground powerhouses under construction. Lake Delio station described in detail. See also items 173 and 221.

198. WILD, P.A. and MCKITTRICK, D.P. Northfield mountain underground power station. *Proceedings, Symposium on Underground Rock Chambers*, ASCE, New York, 1971, pages 287-331. Project located in mantled gneiss dome in the northern Appalachians. Although plant layout was dictated primarily by hydraulic and operating requirements, geotechnical considerations were of major significance in the design of underground excavations. Rock conditions exposed by excavation were in excellent agreement with test borings and geological studies; support system consisting primarily of rock bolts is described and long term stability is discussed. See also item 192.

199. KRUSE, G.H. Power plant chamber under Oroville dam. *Proceedings, Symposium on Underground Rock Chambers*, ASCE, New York, 1971, pages 333-379. Underground powerhouse under left abutment of dam gives least cost and best operating conditions. Feasibility of supporting underground excavation by rock reinforcement evaluated in three year study, including rock and bolt anchor testing and by finite element analysis. Observations during construction indicated that movement occurred during or immediately after blasting and that rock quickly stabilised; deformations agreed closely with finite element predictions. See also items 127, 150, 191 and 259.

200. RISING, R.R. and ERICKSON, G.A. Design of underground pumping chamber near Lake Mead. *Proceedings, Symposium on Underground Rock Chambers*, ASCE, New York, 1971. pages 381-405. Rock mechanics measurements and design of underground chambers described Hoover dam and powerplant constructed in a complex of metamorphic Pre-Cambrian biotite gneiss and schist, highly folded and fractured at the surface. To establish compressive strength and elastic properties, twelve exploratory boreholes were drilled, percolation tests and joint surveys were also performed. State of stress examined and finite element study carried out; borehole extensometers were installed to monitor movements.

201. BENSON, R.P., CONLON, R.J. and MERRITT, A.H. Rock mechanics at Churchill Falls. *Proceedings Symposium on Underground Rock Chambers*, ASCE, New York, 1971, pages 407-486. In powerhouse area, rocks comprise gneiss assemblage intruded by gabbro, diorite and lesser amounts of syenites and pegmatites. Geological exploration for the siting of the powerhouse included NX drilling, logging and photography of all core, detailed geological surface mapping and a petrographic study of rock specimens. Faults, shear zones, joints, groundwater and quality of rock mass are described, orientation, shape and size of excavations for maximum stability are discussed. Rockbolts major means of support for arch and walls. Finite element study and monitoring behaviour of underground excavations during construction described. See also items 148, 164, 165, 180 and 219.

202. HEUZE, F.E. and GOODMAN, R.E. Room and pillar structure in competent rock. *Proceedings, Symposium on Underground Rock Chambers*, ASCE, New York, 1971, pages 531-565. The state-of-the-art in room and pillar mining is considered under the headings theories of strata movement, stress distributions around openings, pillar loads and stresses, determination of pillar strength, roof stresses in the ground overlying pillars, and determination of roof strength.

203. CORDING, E.J., HENDRON, A.J. and DEERE, D.U. Rock engineering for underground caverns. *Proceedings, Symposium on Underground Rock Chambers*, ASCE, New York, 1971, pages 567-600. General review of rock mechanics problems in underground design. Dimensions and support details of 13 major underground projects given.

204. LANGBEIN, J.A. The Manapouri power project, New Zealand. *Proceedings, Institution of Civil Engineers*, Vol 50, 1971, pages 311-351. Describes the investigations, design and construction of this project which took place in the period 1960-1971. The machine hall is 111 m long, 18 m wide and 34 m high. Rock reinforcement is by grouted rock bolts with sprayed mortar and wire reinforcement added in the arch roof. Rocks are gneisses and intrusive granites.

205. KNILL, J.L. Engineering geology of the Turlough Hills pumped storage scheme. *Quarterly Journal of Engineering Geology*, Vol 4, No 4, 1971, pages 373-376. Scheme has been constructed in Leinster granite, 40 km south of Dublin in Ireland. Geological conditions are outlined with particular reference to the rock properties and the location of the underground works. For construction details see item 237.

206. EGGER, P., SCHETELIG, K. and STUBER, H. Anchoring for the power house shaft at Vianden, Luxemburg. (In German). *Rock Fracture, Proceedings, International Symposium on Rock Mechanics*, Nancy, France, Oct 1971, paper 3-4, 12 pages. 25m diameter, 50m deep shaft excavated in Devonian schists for extension of the Vianden pumped storage scheme. Bedding planes, containing mylonite zones up to 20cm thick, dip at 57° to the shaft. Rock mechanics investigations are described. To prevent wedge failure and to consolidate rock around shaft, anchors prestressed to 1400 kN were installed in shaft walls. Details of installation and results of control measurements described.

207. LOTTES, G. The development of European pumped-storage plants. *Water Power*, Vol 24, No 1, Jan 1972, pages 22-33. A study of the advances made in the design and construction of pumped storage plants in Europe in the 10 years since 1962 with particular reference to reservoirs and intakes, powerhouses, tunnel and cavern excavation, machines and costs. Gives some details of the design studies for Waldeck II for which see also items 228, 233, 235 and 244.

208. KIMMONS, G.H. Pumped storage plant at Raccoon Mountain in U.S.A. *Tunnels and Tunnelling*, Vol 4, No 2, Mar-Apr 1972, pages 108-113. Underground facilities include waterways, surge chamber, transformer vault, tunnels, switching and transmission facilities and power plant chamber 21.9m wide, 149m long and 50.3m high excavated in good quality horizontally bedded limestone. Details of excavation and grouted rock bolting techniques given. All excavation done using smoothwall blasting, cushion blasting or line drilling. At exterior portals tunnel peripheries were line drilled and a row of preset grouted rockbolts was 46cm. outside line prior to blasting.

209. ANON. Craig Royston under investigation, site test for pumped storage scheme. *Ground Engineering*, Vol 5, No 3, 1972, pages 27-28. Site on eastern bank of Loch Lomond in Stirlingshire. Scheme divided into three parts: 1.upper site for main dam and intake works, 2. lower site including various tunnels and underground power house, and 3. Loch Lomond which will act as lower reservoir. Description of geological and geophysical tests carried out to date.

210. CORDELL, R. Presplitting in drivages. (In French). *Revue L'Industrie Minerale*, Vol 54, No 3, 1972, pages 107-122. Three falls on the River Arc in the Savoy (French Alps) are used for the development of hydroelectric projects in three underground power stations requiring 5.3 million cu.ft. excavation. Strata varies from carboniferous sandstone to schist flysch and gneiss. Presplitting is used for all drivages resulting in considerable savings because of reduction of overbreak; rock bolts and guniting used for support.

211. ROSENSTROM, S. Kafue Gorge hydroelectric power project. *Water Power*, Vol 24, No 6, June 1972, pages 223-226 (Part 1); Vol 24, No 7, Aug 1972, pages 237-242 (Part 2). Fully describes the rock engineering associated with this station on the Zambian-Rhodesian border. The machine hall and transformer caverns are 32 x 15 x 130m and 21 x 17 x 125m (height x width x length), respectively, and are located 500m underground. The rocks are a complex of granites and gneisses mixed with highly metamorphosed mica-schists, amphibolites and quartzites.

212. MCCREATH, D.R., MERRITT, A.H. and MATERON, B.N. Rock excavations at Alto Anchicaya Project, Columbia. *Proceedings, 8th Canadian Rock Mechanics Symposium*, Toronto, 1972, pages 31-45. Brief description of geology, powerhouse, power tunnel and dam abutment excavations. A highly variable weathering profile and shear zones presented the major geotechnical problems.

213. MATHEWS, K.E. Excavation design in hard and fractured rock at the Mount Isa Mine, Australia. *Proceedings, 8th Canadian Rock Mechanics Symposium*, Toronto, 1972, pages 211-230.

214. SINGH, K.H. Reinforcement of an ore pass system - a case history. *Proceedings, 8th Canadian Rock Mechanics Symposium,* Toronto, 1972, pages 231-249. An existing ore pass system at Falconbridge Nickel Mines was supported by grouted cable bolts and pressure grouting to stabilize observed slip of the walls along fault planes. Extensometer measurements show a decrease in ground movements in the reinforced area.

215. BAWA, K.S. and BUMANIS, A. Design considerations for underground structures in rock. *Proceedings, First North American Rapid Excavation and Tunnelling Conference,* AIME, 1972, Vol 1, pages 393-417. About one fifth of the Washington Metro system will be designed as underground structures (tunnels and stations) in rock. Gives details of rock properties and shows typical tunnel and station cross-sections and support designs. Support typically by 3.7 - 5.5m grouted 25mm dia. rock bolts and 10 - 15cm shotcrete in double track tunnels. Station excavations supported by steel sets, rock bolts, shotcrete and precast concrete arch liners; span in the example given is 20.9m.

216. HOPPER, R.C., LANG, T.A., and MATHEWS, A.A. Construction of Straight Creek tunnel, Colorado. *Proceedings, First North American Rapid Excavation and Tunnelling Conference,* AIME, 1972, Vol 1, pages 501-538. This project consists of dual two-lane vehicular tunnels each approximately 15 m wide by 17m high as excavated and with maximum cover of of 440m. Excavation and support procedures are fully described. Extreme difficulty was caused by the squeezing behaviour of a fault zone and by the wide-spread occurrence of shear zones, faults and joints in the granite, gneiss and schist through which the tunnel was driven.

217. SMART, J.D. and SAGER, J.W. Instrumentation on recent Corps tunnelling projects. *Proceedings, First North American Rapid Excavation and Tunnelling Conference,* AIME, 1972, Vol 1, pages 623-657. Describes instrumentation used on the NORAD Cheyenne Mountain Complex Expansion and other projects. The NCMC project is constructed in a fine grained granite intersected by a number of shear zones. The largest excavations in the expansion are the power plant chamber - 52m long, 20m wide, and 15m high - and the cooling tower chamber 56m long, 12m wide and 14m high. Extensometer measurements used to check stability of the excavations are given. See also items 218, 337 and 342.

218. SMART, J.D. and FRIESTAD, R.L. Excavation quality for the NCMC expansion. *Stability of Rock Slopes, Proceedings, Thirteenth Symposium on Rock Mechanics,* ASCE, 1972, pages 643-664. This expansion of the NORAD Cheyenne Mountain Complex requires the excavation of 50,000 cu.yd. of granite for a new diesel power plant and associated cooling towers and air handling plant. Smooth-wall blasting and rock reinforcement techniques are described. Reinforcement is by grouted pre-bolting (an untensioned perfo sleeve type rock anchor) and tensioned hollow core deformed rock bolts. See also item 217.

219. GAGNE, L.L. Controlled blasting techniques for the Churchill Falls underground complex. *Proceedings, First North American Rapid Excavation and Tunnelling Conference,* AIME, 1972, Vol 1, pages 739-764. Details of this project are given in items 148, 164, 165 180 and 201.

220. BECKER, H. Yielding rock and its consequences in construction practice. *Proceedings, International Symposium on Underground Openings,* Lucerne, Sep 1972, Theme 1. General discussion on stress problems around underground excavations. Problems encountered in construction of the Ait Aadel dam in Morocco, Tarbela dam in Pakistan and underground powerhouses in Malaysia and Ireland are briefly described; rock anchor-gunite method was successfully applied in these constructions.

221. DOLCETTA, M. Rock load on the support structure of two large underground hydroelectric plants. *Proceedings, International Symposium on Underground Openings,* Lucerne, Sep 1972, Theme 3. Special construction solutions were required at Lake Delio and San Fiorano, two recent Italian projects. Based upon preliminary stability studies which included the geomechanical properties of the rock, it was decided to build solid supporting reinforced concrete structures which could be incorporated into the machine foundations. Measurements and finite element studies described. Wall deformations in very high openings were successfully controlled by rigid supporting structures. See also items 173 and 197.

222. JAEGER, C. *Rock Mechanics and Engineering*, Cambridge University Press, 1972, 417 pages. The first part of this text book deals with general rock mechanics principles. The second part covers practical applications including such topics as minimum overburden above a pressure tunnel and rock support systems. Brief details and discussion on several underground powerplants including Innertkirchen (Switzerland), Santa Massenza (Italy), Santa Giustina (Italy), Isere-Arc (France), Chute des Passes (Canada), Bersimis 2 (Canada) and Kemano (Canada). An extensive bibliography includes many papers published in Europe which are seldom included in English language bibliographies. Text contains many typographical errors and care should be taken in applying equations without first checking accuracy. Second edition published in 1979.

223. GILMOUR, L.W. New Zealand, tunnels for Tongariro power development. *Tunnels and Tunnelling*, Vol 4, No 6, 1972, pages 521-524. Description of powerplant and tunnels located in greywacke and argillites; concrete lining and rock support described.

224. BANG-ROLFSEN, P. Skjomen water power plant - excavation of tunnels in granitic gneiss in Norway. Rock bursting restrained by bolting. *Tunnels and Tunnelling*, Vol 5, No 1, 1973, pages 68-70.

225. NUSSBAUM, H. Recent development of the New Austrian Tunnelling Method. *Journal of the Construction Division*, ASCE, Vol 99, No CO1, 1973, pages 115-132. Describes the theoretical basis and practical application of the NATM. Gives some details of individual projects including the difficult Tauern tunnel.

226. RABCEWICZ, L.V. and GOLSER, J. Principles of dimensioning the supporting system for the New Austrian tunnelling method. *Water Power*, Vol 25, No 3, Mar 1973, pages 88-93. Explains the empirical approach to dimensioning recommended for use when applying the NATM and gives a brief mathematical analysis of the underlying theory. Gives some results for the Tauern tunnel, Austria for which see also items 264 and 295.

227. RABCEWICZ, L.V. Theory and practice at the underground works of a large dam project. (In German). *Rock Mechanics*, Supplement 2, 1973, pages 193-224. Full geotechnical details of the design and construction of the four Tarbela Dam tunnels, Pakistan. See also items 241 and 255.

228. RESCHER, O.J., ABRAHAM, K.H., BRAUTIGAM, F. and PAHL, A. The construction of an underground chamber with geomechanical conditions taken into consideration. (In German). *Rock Mechanics*, Supplement 2, 1973, pages 313-354. Full description of the geotechnical aspects of the design and construction of the Waldeck II underground power station in the Eder valley in northern Germany. The excavation in banded sands and dark shales alternating with greywacke sandstone, is approximately elliptical in cross-section with a height of 54m and a width of 33.5m. Support is by a systematic arrangement of rock bolts and prestressed anchors, and a two-layer shotcrete shell. The rock mass contains a number of open discontinuities and fault zones which control the rock mass behaviour. See also items 207, 235, 244 and 248.

229. EDLUND, S. and SANDSTROM, G.E. Stockholm puts sewage plants underground. *Civil Engineering*, ASCE, Vol 43, No 9, Sep 1973, pages 78-83. Describes lay-out and excavation techniques for underground plants consisting of a number of parallel chambers 10-11m wide and up to 135m long. Excavations are in sound rock and are lined with reinforced concrete.

230. MORFELDT, C.O. Storage of oil in unlined caverns in different types of rocks. *New Horizons in Rock Mechanics - Proceedings, Fourteenth Symposium on Rock Mechanics*, ASCE, 1973, pages 409-420.

231. SEDDON, B.T. Rock investigations for Camlough underground power station. *Proceedings, Symposium on Field Instrumentation in Geotechnical Engineering*, London, Butterworths, 1973, pages 370-381. The underground power station is located vertically below the upper reservoir in this pumped storage scheme in Northern Ireland. It has a minimum rock cover of 185m and is 100m long, 22m wide and 35m high. In situ stress measurements using flat jacks and stress calculations for the power station excavation are described.

232. TAKAHASHI, M. The largest hydro-power plant completed - Shintoyone pumped storage project. *Civil Engineering in Japan*, Vol 12, 1973, pages 59-74.

233. ABRAHAM, K.H. The surge chamber and tailrace tunnel for the Waldeck II plant. *Water Power*, Vol 25, No 10, Oct 1973, pages 385-392. Conditions necessitating provision of a surge chamber together with factors affecting its location, dimensions and construction are discussed. The chamber is 23.8m in diameter and 37.4m high and is heavily reinforced with prestressed anchors. Rocks are schist-shales and greywacke sandstones. A major fault zone and a number of minor faults cross the chamber.

234. STARFIELD, A.M. and MCCLAIN, W.C. Project Salt Vault: a case study in rock mechanics. *International Journal of Rock Mechanics and Mining Sciences*, Vol 10, No 6, Nov 1973, pages 641-657. Project Sault Vault is a feasibility study of radioactive waste disposal in an underground salt formation. Full scale studies of the performance of such a facility were carried out in part of a disused salt mine where pillar stresses and ground movements were monitored. The effects of creep and temperature on the load-deformation behaviour of the excavations was successfully modelled.

235. ABRAHAM, K.H. Construction progress at the Waldeck II plant. *Water Power*, Vol 25, No 12, 1973, pages 464-466. Support for the 1390m^2 underground machine cavern consists of shotcrete, 996 prestressed anchors, and 3800 4 or 6m rock bolts. See also items 207, 228, 233, 244 and 248.

236. DOSS WOOD, V. and VAN RYSWYK, R. Rock bolts at Churchill Falls. *Tunnels and Tunnelling*, Vol 6, No 1, Jan 1974, pages 19-23. For full descriptions of other aspects of this scheme see items 148, 164, 165, 180, 201 and 219.

237. O'DONOGHUE, J.D. and O'FLAHERTY, R.M. The underground works at Turlough Hill. *Water Power*, Vol 26, No 1, Jan 1974, pages 5-12 (Part 1); Vol 26, No 2, Feb 1974, pages 51-56 (Part 2); Vol 26, No 3, Mar 1974, pages 88-91 (Part 3). Complete description of construction of this 23m wide, 32m high and 82m long cavern in a uniform coarse-grained granite containing thin veins of pegmatite, aplite and quartz. Support is by systematic rock-bolting, pre-stressed anchors and gunite. Engineering geology described in item 205.

238. YANG, K-H. and NICHOLS, D.E. Bear Swamp pumped-storage plant will start up this summer. *Water Power*, Vol 26, No 5, May 1974, pages 157-163. Gives some details of the rock mechanics investigations and design of the powerhouse chamber. Excavation techniques used described in item 242.

239. PIESOLD, D.D.A., WALKER, B.C. and MURDOCH, G.B. Hydro-electric power development at the Victoria Falls on the Zambezi River. *Proceedings, Institution of Civil Engineers*, Vol 56, Part 1, 1974, pages 275-301. The underground machine hall is excavated in blocky and variable basalt as a horseshoe shaped chamber 13 m wide. The concrete lining is designed as a restraining arch. Other reinforcing is generally light except at recesses etc.

240. BENSON, R.P., MACDONALD, D.H. and MCCREATH, D.R. The effect of foliation shear zones on underground construction in the Canadian shield. *Proceedings, Second North American Rapid Excavation and Tunnelling Conference*, AIME, 1974, Vol 1, pages 615-641. Underground stability problems in the Bersimis I and II, Chute-des-Passes and Churchill Falls hydro-electric projects caused by foliation shear zones are described. A flexible design approach involving rapid installation of rigid support, rock bolts, pressure relief holes and instrumentation is advocated.

241. HILLIS, S.F., SZALAY, K.A., O'ROURKE, J.E. and SMITH, D. Instrumentation of Tarbela Dam tunnels. *Proceedings, Second North American Rapid Excavation and Tunnelling Conference*, AIME, 1974, Vol 2, pages 1275-1303. Four diversion tunnels with excavated diameters of up to 19m and 800m long were excavated in variable limestone, schists, phyllite and altered basic igneous rocks on the right bank of the Tarbela Dam, Pakistan. The tunnels were concrete and steel lined and instrumented to monitor deformations and lining loads. The average load was 75% of that predicted by applying Terzaghi's classification method. See also items 227 and 255.

242. BRADY, J.J. Excavation of Bear Swamp underground powerhouse. *Proceedings, Second North American Rapid Excavation and Tunnelling Conference*, AIME, 1974, Vol 2, pages 1351-1369. Excavation techniques used for the machine chamber - 69m long, 24m wide and 46m high -

and associated tunnels in this pumped storage scheme are given. All excavation was in two types of schist, but no geotechnical details are given. See also item 238.

243. BOCK, C.G. Rosslyn Station, Virginia: Geology, excavation and support of a large, near surface, hard rock chamber. *Proceedings, Second North American Rapid Excavation and Tunnelling Conference*, AIME, 1974, Vol 2, pages 1373 - 1391. Rosslyn Station, part of the Washington subway system, was excavated totally in rock with 16 - 21m of cover. The rock is a highly weathered, highly fractured gneiss at the top improving with depth. Fully grouted 5m and 7m long rock bolts were used for sidewall support. The roof was supported by steel ribs with a minimum of 15 cm of shotcrete being applied throughout. Overall station dimensions are 25m x 17m x 220m.

244. ABRAHAM, K.H. et al. Comparison of results from stress analysis, photoelastic models and in-situ measurements during excavation of the Waldeck II cavern. (In German) *Rock Mechanics*, Supplement 3, 1974, pages 143-166. Measured deformations were less than those anticipated from finite element analyses. See items 207, 228, 233 and 235 for further descriptions of this project.

245. JOHN, K.W. and GALLICO, A. Engineering geology of the site of the Upper Tachien Project. *Proceedings, Second International Congress, International Association of Engineering Geology*, Sao Paulo, Brazil, 1974, Vol 2, Paper VI-9, 11 pages. Describes the geology and geotechnical parameters of the site of the project discussed in item 246.

246. JOHN, K.W. and GALLICO, A. Design studies of underground powerhouse situated in jointed rock. *Proceedings, Second International Congress, International Association of Engineering Geology*, Sao Paulo, Brazil, 1974, Vol 2, Paper VII-9, 11 pages. The underground powerhouse is located in highly jointed slate and quartzite adjacent to the deep gorge of the Tachien River, Taiwan. The cavern is 24-35m high, 17-21m wide and 54-77m long. Roof support is by a concrete roof arch 0.8 - 1.5 m thick supplemented by rock bolts. Finite element and three-dimensional sliding wedge analyses were carried out in the design stage. In-situ measurements gave deformations similar to those predicted in the design studies.

247. HEITFELD, K.H. and HESSE, K.H. Engineering geological aspects on the lining of caverns in sedimentary rock. *Proceedings, Second International Congress, International Association of Engineering Geology*, Sao Paulo, Brazil, 1974, Paper VII-10, 12 pages. Outlines a design method for anchors for cavern walls in jointed rock.

248. PAHL, A. The cavern of the Waldeck II pump storage station - geomechanical investigations and critical analysis of control measurements. *Proceedings, Second International Congress, International Association of Engineering Geology*, Sao Paulo, Brazil, 1974, Vol 2, Paper VII-16, 8 pages. Deformation measurements used to monitor construction are fully described. For fuller details see items 207, 235 and 244.

249. ARAMBURU, J.A. La Angostura dam underground powerhouse: prediction and measurement of displacements during excavation. *Proceedings, Third Congress, International Society for Rock Mechanics*, Denver, 1974, Vol 2B, pages 1231-1241. Describes stress and deformability measurements in the limestone in which a test gallery for this underground powerhouse was excavated. Displacements in the prototype were then estimated using a linear elastic analysis. Actual displacements measured during construction agree well with finite element predictions if the dynamic field modulus is used. Modulus values derived from plate loading tests overestimated the measured values by a factor of two.

250. DI MONACO, A., FANELLI, M. and RICCIONI, R. Analysis of large underground openings in rock with finite element linear and non-linear mathematical models. *Proceedings, Third Congress, International Society for Rock Mechanics*, Denver, 1974, Vol 2B, pages 1256-1261. The finite element method is applied to the analysis of the Lagio Delio, Pelos and Taloro underground power stations. No tension analysis and the influence of reinforcement included.

251. MAURY, V. Stability, stress and operations of the May sur Orne mine as underground petroleum product storage. (In French). *Proceedings, Third Congress, International Society for Rock Mechanics*, Denver, 1974, Vol 2B, pages 1294-1301. Describes the analysis of the stability of a worked out iron mine in Normandy involved in the assessment of its suitability for the underground storage of 5 million m^3 of petroleum products. Large

rooms shown to be possible because pillars support only part of the immediate roof; rib pillars carry most of the load transferred by arching at higher levels.

252. MIZUKOSHI, T. and MIMAKI, Y. Measurements of in situ stresses and the design of a cavity for the construction of the Shintakasegawa underground power station. *Proceedings, Third Congress, International Society for Rock Mechanics*, Denver, 1974, Vol 2B, pages 1302-1307. The behaviour of the rock around the underground excavations is analysed by the elasto-plastic finite element method. The machine hall is 163m long, 27m wide and 54.5m high; the adjoining transformer hall is 107.8m long, 20m wide and 31.3m high and is separated from the machine hall by 41.5m. The ratio of horizontal to vertical in situ stress is 1.8. As a result of the stresss analysis predictions, the excavations were re-oriented. See also item 349.

253. PFISTERER, E., WITTKE, W. and RISSLER, P. Investigations, calculations and measurements for the underground power house Wehr. (In German). *Proceedings, Third Congress, International Society for Rock Mechanics*, Denver, 1974, Vol 2B, pages 1308-1317. Based on measured strength and deformability values, a number of finite element analyses were carried out. Reinforced shotcrete and systematic rock bolting designed using these results. 82 prestressed anchors of 1660 kN capacity installed to prevent failure of a 30m high rock wedge formed by a throughgoing master joint and the upstream wall of the cavern. Deformations and stresses monitored during construction agree well with predictions. Machine hall is 219m long, 19m wide and 33m high.

254. LLOVERA, L.O. and BECEDONIZ, J.F. Excavating works for the enlargement of Villarino power plant. *Proceedings, Third Congress, International Society for Rock Mechanics*, Denver, 1974, Vol 2B, pages 1365-1370. The Villarino underground power plant on the Tormes river, Spain, is being expanded to take two additional 135 MW reversible pump-turbine units. The excavations in strong granite, including enlargement of the main cavern, must be carried out without disturbing normal operation of the plant. The determination of admissible vibration levels and the control of blasting are described. Vibration velocities kept below 15mm/sec in machine foundations.

255. RABCEWICZ, L.V. and GOLSER, J. Application of the NATM to the underground works at Tarbela. *Water Power*, Vol 26, No 9, 1974, pages 314-321 (Part 1); Vol 26, No 10, 1974, pages 330-335 (Part 2). Describes construction of four large tunnels to be used to divert the Indus River during construction of the Tarbela Dam and subsequently for power generation and irrigation purposes. This is a good detailed description of the practical application of the NATM. See also items 227 and 241.

256. ANON. Ritsem: the last of the Lule River developments. *Water Power*, Vol 26, No 11, Nov 1974, pages 365-371. This is the fifteenth and final power station to be constructed on the Lule River in northern Sweden. Excavation of the underground machine hall (17m wide and 36m long) and the long headrace tunnel are described. Rocks at the site are hard schists and mylonites. Few geotechnical problems encountered.

257. BARTON, N., LIEN, R. and LUNDE, J. Engineering classification of rock masses for the design of tunnel support. *Rock Mechanics*, Vol 6, No 4, Dec 1974, pages 189-236. Presents details of a rock mass classification system using six parameters that can be estimated from mapping and judgement, and shows how a numerical estimate of rock mass quality can be used to predict support requirements for excavations of given dimensions. Several case histories are used to check the validity of the approach and illustrative worked examples are given.

258. HOEK, E. A bibliography on the geotechnical problems associated with the construction of large permanent underground excavations with particular emphasis on underground hydro-electric power plants. *International Journal of Rock Mechanics and Mining Sciences and Geomechanics Abstracts*, Vol 12, No 2, 1975, 31 pages.

259. KULHAWY, F.H. and FLANAGAN, R.F. Analysis of the behaviour of Edward Hyatt power plant. *Journal of the Geotechnical Engineering Division*, ASCE, Vol 101, No GT3, Mar 1975, pages 243-257. Incremental finite element analyses were conducted to determine whether measured displacements of the chamber excavation (21m wide x 43m high x 170m long and 91m below ground surface) could be predicted. A non linear model using stress-dependent rock and joint properties gave the best overall method of evaluating the performance of the opening. See also items 191 and 199.

260. CECIL, O.S. Correlations of rock bolt-shotcrete support and rock quality parameters in Scandinavian tunnels. *Swedish Geotechnical Institute Proceedings* No 27, 1975, 275 pp. Field observations at 14 rock tunnelling projects have enabled empirical correlations to be drawn between the rock quality parameters, average discontinuity spacing, RQD and seismic velocity ratio, and the rock bolt-shotcrete supports used in loosening ground conditions.

261. ROBERTS, G.T. and ANDRIC, M. Geological factors in the location of the power station and associated works, Gordon River Power Development, Stage 1, South-West Tasmania. *Proceedings, Second Australia-New Zealand Conference on Geomechanics*, Brisbane, 1975, pages 213-217. The Gordon power station is located 200m underground in foliated quartzite and schist. The investigations leading to the choice of this site over several alternatives are described. See items 262 for geotechnical details of the chosen site.

262. LACK, L.J., BOWLING, A.J. and KNOOP, B.P. Rock mechanics studies and instrumentation for the Gordon Underground Power Station. *Proceedings Second Australia-New Zealand Conference on Geomechanics*, Brisbane, 1975, pages 274-280. The machine hall is 22m wide, 30m high and 95m long with its long axis oriented approximately east-west. Tests used to classify the rock masses and to determine the in situ stresses, deformation modulus and creep properties of the rock are described. Investigations concentrated on the schist as early tests showed that the other rock type present, foliated quartzite, was sound and unlikely to present problems. Instruments being used to monitor construction are also described.

263. KOHLER, H. Excavation work by means of tunnel miners in the conglomerate for the underground parking halls in the Mönschsberg (Salzburg, Austria). (In German). *Rock Mechanics*, Supplement 4, 1975, pages 85-97.

264. RABCEWICZ, L.V. Tunnel under Alps uses new cost-saving lining method. *Civil Engineering*, ASCE, Vol 45, No 10, Oct 1975, pages 66-68. Describes the application of NATM to the construction of the Tauern highway tunnel through cohesionless rock talus and intensely folded and weakened phyllites in the central Austrian alps. Support was by rock bolting and shotcrete reinforced with steel ribs and mesh. See also item 295.

265. LEMPERIERE, F. and VIGNY, J.P. Civil engineering work for the Cabora Bassa project. *Water Power and Dam Construction*, Vol 27, No 10, Oct 1975, pages 362-3. This project on the Zambezi River in Mozambique is one of the largest hydro schemes in the world. The power station chamber is 27m wide, 57m high and 220m long. There are two downstream surge chambers 21m wide, 72m high with lengths of 72 and 76m. No details of geotechnical aspects of the project are given.

266. YOSHIDA, M. Okutataragi pumped storage power station. *Water Power and Dam Construction*, Vol 27, No 11, Nov 1975, pages 399-406. The power station cavity is located 200m underground and is 24.9m wide, 49.2m high and 133.4m long. Rock types encountered during excavation were quartz-porphyry, diabase and rhyolite. Excavation was by the bench-cut method in stages of 2-3m lifts. After each bench cut rock bolts varying in length from 5-15m were installed, one every $3m^2$. Concrete lining was subsequently applied.

267. JAEGER, C. Assessing problems of underground structures. *Water Power and Dam Construction*, Vol 27, No 12, Dec 1975, pages 443-450 (Part 1); Vol 28, No 1, Jan 1976, pages 29-36 (Part 2).

268. NIGAM, P.S., JAIN, O.P. and KANCHI, M.B. Three-dimensional analysis of hydroelectric power station. *Journal of the Power Division*, ASCE, Vol 102, No PO1, Jan 1976, pages 35-52. Gives the results of a photoelastic model study of the elastic stress distribution around an underground excavation of complex three-dimensional shape.

269. ANON. Europe's largest pumped-storage scheme will set new records. *Water Power and Dam Construction*, Vol 28, No 3, Mar 1976, pages 30-34. General description of the Dinorwic scheme in North Wales. See items 290, 291 and for geotechnical details.

270. SATO, Y. et al. Construction of Shin-Kanmon tunnel on Sanyo Shinkansen. *Proceedings, 1976 Rapid Excavation and Tunneling Conference*, AIME, New York, 1976, pages 335-354. The Shin-Kanmon undersea tunnel, completed in mid-1974, forms part of the rail link between Tokyo and Hakata. The most difficult portion of the 19 km long tunnel was the

880m undersea section, part of which was excavated in a major fault zone. Full details of the multi-stage excavation, chemical grouting and support techniques used in this and other less troublesome sections of the tunnel are given.

271. MURPHY, D.K., LEVAY, J. and RANCOURT, M. The LG-2 underground powerhouse. *Proceedings, 1976 Rapid Excavation and Tunneling Conference*, AIME, New York, 1976, pages 515-533. When completed as planned in 1982, the LG-2 project which is one element of the La Grande Complex in the James Bay area of north-western Quebec, will be the largest underground power station in the world. The machine hall is 26.4m wide and 493m long, and is divided into two sections each housing eight 333 MW units, by a central erection bay and control room. Geological conditions in the granitic rock mass are generally good and as expected from site investigations. Rock bolting serves as temporary and permanent support except in faulted zones in the tailrace tunnels where steel sets and shotcreting have been used. See also items 320 and 328.

272. IMRIE, A.S. and CAMPBELL, D.D. Engineering geology of the Mica underground powerplant. *Proceedings 1976 Rapid Excavation and Tunneling Conference*, AIME, New York, 1976, pages 534-569. The Mica underground powerplant is located in the right abutment of the Mica dam on the Columbia River, British Columbia, Canada. The machine hall is 24.4m wide, 237m long and 44.2m high, and is located in quartzite and mica gneisses. The behaviour of the rock was excellent and problems were minimal, the only local failures occurring where faults or fracture planes intersected the excavations to isolate unsupported slabs or wedges of rock. Generally, roof support was provided by 6-7m rockbolts on a 1.5m x 1.5m pattern. Full details of the geology, rock mechanics studies and extensometer measurements are given.

273. HYND, J.G.S., CLELLAND, I.S. and GILLETT, P.A. The construction of large underground excavations in Arizona copper mines. *Proceedings 1976 Rapid Excavation and Tunneling Conference*, AIME, New York, 1976, pages 570-590. Gives construction details of two complex permanent underground mining structures, a rotary dump-ore pocket - ore pass complex in a monzonite porphyry at the Magma Copper Company's San Manuel mine, and the main station level in a two station - loading pocket - ore pass complex in a volcanic agglomerate. The first job involved heavily water bearing ground and heavy ground pressures which required extensive grouting and steel/concrete support. The ground conditions on the second job were good, and extensive use was made of shotcrete.

274. JACOBY, C.H. Creation and stability of large sized openings in salt. *Proceedings, 1976 Rapid Excavation and Tunneling Conference*, AIME, New York, 1976, pages 591-608. Describes the development of room and pillar salt mining techniques in the Gulf Coast region, U.S.A. Experience with 46m rooms and 46m square pillars suggests that 60m square rooms are possible.

275. FURSTNER, J.M.M. Engineering geology of the Ramu 1 hydro-electric project in Papua New Guinea. *Bulletin, International Association of Engineering Geology*, No 13, June, 1976, pages 71-77. The Ramu 1 scheme is located in the Eastern Highlands of Papua New Guinea in gently dipping Oligocene marble with small dolerite intrusions with an overlying interbedded shale-siltstone-greywacke sequence. Excavation problems were caused by the solution and collapse of cave structures in the marble, by block and wedge failures along bedding planes and dolerite dykes in the shale-siltstone - greywacke, and by densely fractured or crushed rock in the vicinity of the contact between the two rock units.

276. BROCH, E. and RYGH, J.A. Permanent underground openings in Norway - design approach and some examples. *Underground Space*, Vol 1, No 2, July-Aug 1976, pages 87-100. Outlines the extensive use of the underground for permanent installations in Norway, and gives the procedure followed in the design of the excavations - choice of location, orientation of the opening, selection of opening shale and dimensioning. Four recent examples of underground excavations are given - a sports centre, a swimming pool, a telecommunication centre, and a drinking water reservoir. Some cost data are given, but geotechnical data are limited. See also item 332.

277. KUHNHENN, K. and SPAUN, G. Damage in a powerhouse cavern and a penstock caused by rock slides. (In German). *Rock Mechanics*, Supplement 5, 1976, pages 245-262. Excessive movements and cracking of the concrete lining in the Centrale Belviso power plant cavern could not be stabilised by grouting and prestressed anchors. The cavern in

schist and quartzite is 93m long, 15m wide and open to the valley at its ends and was constructed in the period 1942-45. The movement has now been stabilised by anchoring the crane beam and heavily injecting the perimeter of the excavation.

278. PALMER, W.T., BAILEY, S.G. and FULLER, P.G. Experience with pre-placed supports in timber and cut and fill stopes. *Symposium on Influence of Excavation and Ground Support on Underground Mining Efficiency and Costs*, Australian Mineral Industries Research Association, Melbourne, 1976, pages 45-71. Describes the evolution of preplaced support techniques at the C.S.A. mine, Cobar, and the A.M. & S mines, Broken Hill. Grouted cable dowels 15-20m long are placed in the roof ahead of stoping which then proceeds through successive lifts. Cables are made up of 7mm dia. high tensile steel wire or 16mm dia. Dywidag threadbar. This technique has reduced support costs and eliminated many difficulties in large stopes (10m wide by 100m long with 4-5m lifts).

279. JOHANSSON, S. and LAHTINEN, R. Oil storage in rock caverns in Finland. *Tunnelling '76*. IMM, London, 1976, pages 41-58. The general features of the planning, geological investigations and construction of oil storages in Finland are given in some detail, together with three brief case histories - a light fuel oil storage with a volume of 300,000m^3 in porphyritic granodiorite at Nokia, a heavy fuel oil storage with a volume of 280,000m^3 in granite-gneiss at Rauma, and a crude oil storage with a volume of 1,000,000m^3 in migmatite at Porvoo.

280. BARTON, N. Recent experiences with the Q-system of tunnel support design. *Proceedings, Symposium on Exploration for Rock Engineering*, Johannesburg, 1976, Vol 1, pages 107-115. Sets out details of the application of the Q-system to the design of support for a 19m span underground power station and a 9m diameter tunnel.

281. DUNHAM, R.K. A rock mechanics investigation into room and pillar mining at Impala Platinum Mines. *Proceedings, Symposium on Exploration for Rock Engineering*, Johannesburg, 1976, Vol 1, pages 263-270. Investigates possible causes of hanging wall collapse over room and pillar workings with 25m rooms and 5m square pillars. Rock types are pyroxenites and feldspathic pyroxenites. A joint survey, stress measurements, pillar deformation, and convergence measurements were carried out. A suggested pillar design method is presented.

282. BIENIAWSKI, Z.T. Elandsberg Pumped Storage Scheme. *Proceedings, Symposium on Exploration for Rock Engineering*, Johannesburg, 1976, Vol 1, pages 273-289. Describes the rock engineering investigations for this project in Cape Province, South Africa. Construction is expected to commence in 1980. The underground caverns are in greywacke with minor amounts of phyllite, quartzite and quartz veins. The main cavern is expected to be 47.5m high, 117m long and up to 22m wide. The rock conditions for the power house are at best "fair" (Class III on the CSIR classification). A trial test enlargement with a span of 22m is being constructed. Considerable groundwater problems are anticipated.

283. CHUNNETT, E.R.P. Ruacana hydro-power scheme - rock engineering studies. *Proceedings, Symposium on Exploration for Rock Engineering*, Johannesburg, 1976, Vol 1, pages 313-324. Description of geology and layout of the 320 MW Ruacana scheme on the Cunene River at the South West Africa-Angola border. The principal rock type is apparently randomly fractured porphyroblastic gneiss. Very few engineering, construction or support details given.

284. BOWCOCK, J.B., BOYD, J.M., HOEK, E. and SHARP, J.C. Drakensberg pumped storage scheme - rock engineering aspects. *Proceedings, Symposium on Exploration for Rock Engineering*, Johannesburg, 1976, Vol 2, pages 121-139. The Drakensberg scheme is a multi-purpose project being undertaken in Natal, South Africa. It involves three major underground excavations for pumping and generating plant and associated equipment. The machine hall is 16.3m wide, 193m long and 45m high with 150m of cover; the rocks are horizontally bedded sandstones, mudstones and siltstones. This paper describes the geological and geotechnical investigations including in-situ testing and evaluation of rock reinforcement and pneumatically applied concrete. The testing of an enlargement to the full cross-sectional dimensions of the future machine hall and a full scale penstock test chamber are described. See also items 298 and 350.

285. MACKELLAR, D.C.R. Geomechanics aspects of the Steenbras pumped storage scheme. *Proceedings, Symposium on Exploration for Rock Engineering*, Johannesburg, 1976, Vol 2, pages 141-147. Describes the geological and geotechnical factors involved in making the choice between three alternative underground schemes. The CSIR Geomechanics Classification was used in making the assessment.

286. BOWCOCK, J.B. Vanderkloof hydro-electric power station. *Proceedings, Symposium on Exploration for Rock Engineering*, Johannesburg, 1976, Vol 2, pages 149-158. The underground machine hall with maximum width and depth of 24m and 48.6m is excavated in dolerite with very low cover. Initially 8m by 8m headings were driven along each springing to permit examination of the rock. Primary roof support was by fully resin bonded rock bolts on 2m centres; wall support was by grouted rock bolts. Pre-splitting techniques were generally used and concrete lining followed close behind the excavation.

287. BARTON, J., LIEN, R. and LUNDE, J. Estimation of support requirements for underground excavations. *Design Methods in Rock Mechanics - Proceedings, 16th Symposium on Rock Mechanics*, ASCE, 1977, pages 163-177. An analysis of some 200 case records has revealed a useful correlation between the amount and type of permanent support and the rock mass quality index Q. Methods of estimating permanent roof and wall support and temporary support requirements in underground excavations are given. Support measures considered include various combinations of shotcrete, bolting, and cast concrete arches.

288. HAIMSON, B.C. Design of underground powerhouses and the importance of pre-excavation stress measurements. *Design Methods in Rock Mechanics - Proceedings, 16th Symposium on Rock Mechanics*, ASCE, 1977, pages 197-204. The newly developed hydrofracturing technique has been used to measure in-situ stresses at two pumped storage underground powerhouse sites - the Helms project in the Sierra Nevada Mountains, California, and the Bad Creek project in South Carolina.

289. HARDY, M.P. and AGAPITO, J.F.T. Pillar design in underground oil shale mines. *Design Methods in Rock Mechanics - Proceedings, 16th Symposium on Rock Mechanics*, ASCE, 1977, pages 257-266. The Colony pilot oil shale mine in Colorado provided valuable pillar strength values that should form the basis of any subsequent large room and pillar mine designs. The average strength of nominal 17.7m square pillars, 18.3m high was 21 MPa.

290. WHITTLE, R.A. Geotechnical aspects of tunnel construction for the Dinorwic Power Station. *Ground Engineering*, Vol 10, No 1, 1977, pages 15-20. See items 269, 291 and 297 for fuller details of this pumped storage project being constructed in slate in North Wales.

291. ANDERSON, J.G.C., ARTHUR, J. and POWELL, D.B. The engineering geology of the Dinorwic underground complex and its approach tunnels. *Proceedings, Conference on Rock Engineering*, Newcastle-upon-Tyne, April 1977, pages 491-510. Describes the general geology of the site, the various sets of discontinuities present in the vicinity of the underground works, in situ stress measurements and gives a layout of the main underground complex and some details of support systems recommended for tunnels. See also items 269, 290, 297 and 344.

292. HOBST, L. and ZAJIC, J. *Anchoring in Rock*, Elsevier, Amsterdam, 1977, 390 pages. This book contains an abundance of details about the practice of rock anchoring. One section of 20 pages is devoted to the support of underground excavations. Anchoring in the Lipno (Czechoslovakia), Vianden (Luxemburg) and Lutz (Austria) underground power stations described.

293. BRADY, B.H.G. An analysis of rock behaviour in an experimental stoping block at the Mount Isa Mine, Queensland, Australia. *International Journal of Rock Mechanics and Mining Sciences*, Vol 14, No 2, Mar 1977, pages 59-66. Describes an experimental stoping programme carried out in a steeply dipping tabular silver-lead-zinc orebody conformable with shale host rock. The two trial stopes 10m wide and with a down dip span of 124m were advanced along strike 30m and 38m respectively. They were separated by a pillar containing a bored raise, the rock on the periphery of which progressively failed as stoping advanced. A stress analysis of this pillar permitted a failure criterion for the rock mass to be back-calculated for use in mining design.

294. WITTKE, W. A new design concept for underground openings in jointed rock. *Publications of the Institute for Foundation Engineering, Soil Mechanics, Rock Mechanics and Waterways Construction, RWTH Aachen*, Vol 1, 1976, pages 46-117. Discusses the author's approach to cavern design based on finite element analyses of jointed rock and field measurements, and its application to the Wehr underground power station cavern (see items 253 and 304) and the Bremm trial cavern (see items 304 and 308).

295. HERBECK, J. Construction of the Tauern summit tunnels. *Underground Space*, Vol 1, No 3, 1977, pages 201-226 (Part 1); Vol 1, No 4, 1977, pages 373-391 (Part 2). Gives very complete details of the construction of the 6.4 km tunnel in the summit section of the Tauern highway in the Austrian Alps. The excavated cross-section area is 93m^2. The rocks are weakened phyllites but loose slope deposits (talus) encountered for 350m from the northern portal gave extremely difficult tunnelling conditions. Details of methods of tunnelling through this cohesionless ground, support, monitoring, construction equipment and site organisation are given. See also item 264.

296. LANE, K.S. Instrumented tunnel tests: a key to progress and cost saving. *Underground Space*, Vol 1, No 3, 1977, pages 247-259. This review of case histories shows how the use of test sections can result in major cost savings in underground construction by validating design concepts particularly in the costly area of tunnel support. The documented experience from a wide variety of locations and projects suggests that test sections should be more widely used.

297. DOUGLAS, T.H., RICHARDS, L.R. and O'NEILL, D. Site investigation for main underground complex - Dinorwic pumped storage scheme. *Field Measurements in Rock Mechanics, Proceedings of the International Symposium*, K. Kovari (ed.), A.A. Balkema, 1977, Vol 2, pages 551-567. Gives full details of stress measurements and determination of elastic and strength properties of the rock at the site of this large pumped storage scheme being constructed in slate in N. Wales. See also items 269, 290, 291 and 344.

298. SHARP, J.C., RICHARDS, L.R. and BYRNE, R.J. Instrumentation considerations for large underground trial openings in civil engineering. *Field Measurements in Rock Mechanics, Proceedings of the International Symposium*, Zurich, K. Kovari (ed.), A.A. Balkema, 1977, Vol 2, pages 587-609. Sets out principles involved and techniques available for monitoring the performance of trial excavations. Gives an excellent case history study of the trial enlargement made in horizontally bedded sandstones and siltstones for the Drakensberg Pumped Storage Scheme in South Africa (see item 284). The results of this trial opening were used to evaluate reinforcement design.

299. CORDING, E.J., MAHAR, J.W. and BRIERLEY, G.S. Observations for shallow chambers in rock. *Field Measurements in Rock Mechanics, Proceedings of the International Symposium*, Zurich, K. Kovari (ed.), A.A. Balkema, 1977, Vol 2, pages 485-508. Reports observations made during construction of stations in rock for the Washington, D.C. Metro. The stations are commonly excavated at depths of less than 30m below ground surface. The schistose gneiss in many of the stations is unweathered and contains four or five sets of joints that are planar, continuous, and often slickensided. See also items 152, 215, 243 and 319.

300. MARTINETTI, S. Experience in field measurements for underground power stations in Italy. *Field Measurements in Rock Mechanics, Proceedings of the International Symposium*, Zurich, K. Kovari (ed.), A.A. Balkema, 1977, Vol 2, pages 509-534. Gives major geotechnical features including rock quality, stress analysis, support, and deformation measurements for the Roncovalgrande, San Fiorano, Taloro and Timpagrande underground power stations in Italy.

301. DESCOEUDRES, F. and EGGER, P. Monitoring system for large underground openings - experiences from the Grimsel-Oberaar scheme. *Field Measurements in Rock Mechanics, Proceedings of the International Symposium*, Zurich, K. Kovari (ed.), A.A. Balkema, 1977, Vol 2, pages 535-549. Describes the monitoring system and results obtained in the first stage of construction of this pumped storage project on Lake Grimsel. Excavation is mainly in granodiorite with local aplitic zones; some granites and gneisses are also encountered. The machine cavern is 29m wide, 19m high, and 140m long with 100-300m cover. Rock quality is generally good except for some unfavourable joint orientations and a gneiss zone. No major geotechnical problems are envisaged.

Results of measurements verify design assumptions. High horizontal stresses justify the adoption of a span greater than the height.

302. WISSER, E. Design and observation of the underground power station Langenegg. *Field Measurements in Rock Mechanics, Proceedings of the International Symposium,* Zurich, K. Kovari (ed.), A.A. Balkema, 1977, Vol 2, pages 569-575. The underground excavations consist of a machine cavern 18m wide, 33m high and 27m long, a transformer hall 13m wide, 12m high and 26m long, a 50m long access tunnel and a 120m long tailwater gallery. The rocks are an alternating sequence of marl, conglomerate and sandstone; the power station was designed to fit into a 40m thick band of sandstone dipping at 60° because the marl is very weak. Support in both roof and walls was by rock bolts, steel mesh and shotcrete. Horizontal in-situ stress found to be twice the vertical stress, not 0.3 times as originally assumed in design.

303. LETSCH, U. Seelisberg tunnel: Huttegg ventilation chamber. *Field Measurements in Rock Mechanics, Proceedings of the International Symposium,* Zurich, K. Kovari (ed.), A.A. Balkema, 1977, Vol 2, pages 577-586. The 9.25km long Seelisberg tunnel forms part of the Swiss national motorway N2. The Huttegg underground ventilation chamber located in Valangin marl consists of two parallel tubes 18m dia. and 52m long connected by a 14m wide by 16m high tube. A trial chamber 12m wide, 8m high and 21m long was excavated and monitored. Support for the completed chamber was by reinforced shotcrete (15cm thick but up to 100cm in the difficult central section), Perfo rock bolts 3.8m long, and pre-stressed rock anchors 15m long. Overburden pressure approximately reaches the unconfined compressive strength of the marl. See also item 314.

304. WITTKE, W. New design concept for underground openings in rock. In *Finite Elements in Geomechanics,* (G. Gudehus, ed.), John Wiley, London, 1977, pages 413-478. A design concept for underground openings based on a three-dimensional finite element method is described. The approach involves determination of stress-strain relationships for the jointed rock mass and construction monitoring to check the results of design calculations. The method is applied to three projects - the Wehr underground power house (see items 253 and 294), the 6.5m dia. Nurnberg water tunnel, and the 24m wide, 30m long and 9m high trial excavation for an underground power station at Bremm, West Germany (see items 294 and 308).

305. LIEN, R. and LOSET, F. A review of Norwegian rock caverns storing oil products or gas under high pressure or low temperature. *Storage in Excavated Rock Caverns,* M. Bergman (ed.), Pergamon Press, Oxford, 1977, Vol 1, pages 199-201. Twenty three Norwegian rock caverns are briefly described and details tabulated. Fourteen are constructed in Precambrian gneiss, seven in rocks of Cambro-Silurian age and two in Permian syenite. Most of the caverns are constructed in good quality rock and few stability or operating problems have occurred. The caverns as a rule are unlined, but rock-bolting and shotcrete are frequently used for support and grouting to reduce leakage.

306. BELLO, A.A. Simplified method for stability analysis of underground openings. *Storage in Excavated Rock Caverns,* M. Bergman (ed.), Pergamon Press, Oxford, 1977, Vol 2, pages 289-294. Gives brief details, in useful tabular form, of three large mining excavations in limestone in Mexico. Two are supported by pillars and the third, 120m long by 55m wide, is reported as being unsupported.

307. NELSON, C.R. The Minnesota - RANN underground test room. *Storage in Excavated Rock Caverns,* M. Bergman (ed.), Pergamon Press, Oxford, 1977, Vol 2, pages 337-342. An underground flat-roofed test room 15m wide, 30m long and 25m below surface was excavated in friable sandstone below a bedded limestone. Instrumentation including piezometers, stressmeters and extensometers were installed before excavation. Methods of strengthening the friable sandstone with sprayed grout and a rational design method for flat roofs in bedded formations were developed.

308. WITTKE, W., PIERAU, B. and SCHETELIG, K. Planning of a compressed-air pumped-storage scheme at Vainden, Luxembourg. *Storage in Excavated Rock Caverns,* M. Bergman (ed.), Pergamon Press, Oxford, 1977, Vol 2, pages 367-376. In addition to describing the geotechnical investigations for this new compressed-air pumped-storage scheme in Devonian clay slate, details are given of the measurements made at the Bremm test cavern in similar rock in West Germany. See also items 294 and 304.

309. JACOBSSON, U. Storage for liquified gases in unlined refrigerated rock caverns. *Storage in Excavated Rock Caverns*, M. Bergman (ed.), Pergamon Press, Oxford, 1977, Vol 2, pages 449-458. Underground storage chambers for propylene and LPG excavated in good quality gneissic granite under low rock cover, have operated satisfactorily for several years at temperatures down to -40°C. A 15m wide, 45m long and 20m high cavern in the same rock intended for the storage of ethylene at temperatures at or below -100°C failed on cooling. Following very detailed rock mechanics investigations, the cavern was repaired using a polyethylene glycol-water solution sprayed lining and external injection grouting of the cracks using an ethylene glycol water mixture. The cavern has since operated satisfactorily for the storage of propylene at -40°C.

310. ANTTIKOSKI, U.V. and SARASTE, A.E. The effect of horizontal stress in the rock on the construction of the Salmisaari oil caverns. *Storage in Excavated Rock Caverns*, M. Bergman (ed.), Pergamon Press, Oxford, 1977, Vol 3, pages 593-598. Three oil storage caverns 14m wide, 28m high and up to 260m long were excavated in mixed granite, amphibolite and gneiss near the centre of Helsinki, Finland. High horizontal in-situ stresses caused rockbursts and tensile failure of rock pillars during construction. Systematic roof bolting or concrete grouting reduced rock falls from the cavern roofs.

311. BIENIAWSKI, Z.T. Design investigations for rock caverns in South Africa. *Storage in Excavated Rock Caverns*, M. Bergman (ed.), Pergamon Press, Oxford, 1977, Vol 3, pages 657-662. Gives brief but useful details of ten large underground excacations in South Africa, six mining chambers some of which are the deepest in the world, and four underground power stations.

312. HANSAGI, I. and HEDBERG, B. Large rock caverns at LKAB's mines. *Storage in Excavated Rock Caverns*, M. Bergman (ed.), Pergamon Press, Oxford, 1977, Vol 3, pages 697-704. Reprinted in *Underground Space*, Vol 4, No 2, 1979, pages 103-110. Describes the construction of large underground excavations for mining equipment storage, service areas, crushing stations, canteens, ore passes, crusher bins and storage bunkers at Kiruna and Malmberget.

313. MARGISON, A.D. Canadian underground NORAD economically achieved. *Underground Space*, Vol 2, No 1, 1977, pages 9-17. A major underground installation for the North American Air Defence system was constructed near North Bay, Ontario, Canada, in granitic gneiss. Few geotechnical details are given in this paper. The main installation caverns, which have a width of approximately 12m, were generally supported by 3.4m long rockbolts on 1.5m centres in the roof, and 3.0m long bolts on 3.0m centres in the walls, with heavy wire mesh being added throughout.

314. BURI, F., HERRENKNECHT, M. and GRINDAT, W. Seelisberg Middle Section construction. *Tunnels and Tunnelling*, Vol 9, No 5, 1977, pages 40-44. The middle or Huttegg section of the Seelisberg tunnel of the Swiss National Motorway N2 is excavated in marly rocks with high overburden pressures and contains an underground ventilation complex composed on two caverns each 52m long and 18m in diameter. The rock exposed by excavation was immediately treated with shotcrete and anchored with 4m long Perfo rock bolts on 1.1m centres. This was followed by 16-18m long anchors on 4.5m centres stressed to 80 tonnes. See also item 303.

315. AGAPITO, J.F.T. and PAGE, J.B. A case study of long-term stability in the Colony Oil Shale Mine, Piceance Creek Basin, Colorado. *Monograph on Rock Mechanics Applications in Mining*, W.S. Brown, S.J. Green and W.A. Hustrulid (eds.), AIME, New York, 1977, pages 138-143. Stress determinations and convergence measurements were conducted to assess the long term stability of an experimental room-and-pillar oil shale mine. Mining was conducted in a 20m thick seam at depths of 180-260m. Pillars are 17.7m square and rooms 16.8m wide. A well defined system of joints and bedding planes exists in the flat lying oil shale beds. The pillars contain two joint sets approximately at right angles to each other. Only one joint set appears in the roof. See also items 289 and 325.

316. HEGAN, B.D. Engineering geological aspects of Rangipo underground powerhouse. *Papers presented to the Symposium on Tunnelling in New Zealand, Proceedings of Technical Groups*, NZIE, Vol 3, No 3 (G), 1977, pages 6.23 - 6.32. The Rangipo powerhouse chambers under construction in the Kaimanawa Mountains are sited in indurated greywacke

sandstone and argillite of poor rock mass quality. The dominant rock defects are N-S striking, steeply westward-dipping crush zones. The machine hall axis orientation of 272° was selected to minimize the effect of these zones on roof stability. See also items 317 and 333.

317. MILLAR, P.J. The design of Rangipo underground powerhouse. *Papers presented to the Symposium on Tunnelling in New Zealand, Proceedings of Technical Groups*, NZIE, Vol 3, No 3(G), 1977, pages 6.46-6.59. Analyses of the caverns were carried out using elastic and elasto-plastic finite element programs. A plastic zone up to 8m from the machine hall boundary was predicted. Empirical methods of rock bolt support design were considered and found to be not applicable to the poor quality rock mass. In addition to rock bolting installed ahead of excavation, a thin concrete lining is used throughout. See also items 316 and 333.

318. DREW, G.E. Applications of prestressing techniques to hydroelectric projects. *Energy Resources and Excavation Technology, Proceedings, 18th U.S. Symposium on Rock Mechanics*, Colorado School of Mines, Golden, Colorado, 1977, pages 3A3-1 to 3A3-7. A short but useful description is given of the prestressed rock anchors used to support the main cavern of the Hongrin underground pumped storage scheme in Switzerland. The cavern is approximately 30m wide, 27m high and 137m long, and has a semi-circular arch roof. Shotcrete and some 650 anchors, 12 to 14m long with working loads of 1110 to 1380kN, were used for support. See also item 161.

319. MAHAR, J.W. Effect of geology and construction on behaviour of a large, shallow underground opening in rock. Ph.D. Thesis, University of Illinois, Urbana-Champaign, 1977, 333 pages. Reports a study carried out during construction of the Dupont Circle Station of the Washington DC Metro, to relate measured rock movements to geological and construction conditions.

320. HAMEL, L. and NIXON, D. Field control replaces design conservation at world's largest underground powerhouse. *Civil Engineering, ASCE*, Vol 48, No 2, Feb 1978, pages 42-44. Gives brief description of the excavation and support of the LG-2 powerhouse cavern on the La Grande River in Quebec, Canada. For fuller details see items 271 and 328.

321. LANDER, J.H. JOHNSON, F.G., CRICHTON, J.R. and BALDWIN, M.W. Foyers pumped storage project: planning and design. *Proceedings of The Institution of Civil Engineers*, Part 1, Vol 64, Feb 1978, pages 103-117. This scheme incorporates the first shaft-type power station in the U.K. and is located on the eastern bank of Loch Ness, Scotland. In addition to the two 19.35m diameter fully lined machine shafts, the project involves a 18m diameter surge chamber and low and high pressure tunnels. A trial shaft was sunk on the centreline of one of the machine shafts through hard gritty sandstone that was badly broken with bands of black, limey, closely jointed shale. Considerable inflows of water were encountered indicating shaft sinking and tunnel driving difficulties. Construction of the project is described in item 322.

322. LAND, D.D. and HITCHINGS, D.C. Foyers pumped storage project: construction. *Proceedings of The Institution of Civil Engineers*, Part 1, Vol 64, Feb 1978, pages 119-136. Gives construction details of the tunnels and shafts described in item 321.

323. ADAMS, J.W. Lessons learned at Eisenhower Memorial Tunnel. *Tunnels and Tunnelling*, Vol 10, No 4, 1978, pages 20-23. The lessons learned from the first bore of the twin-bore four-lane Eisenhower Tunnel 97km west of Denver, Colorado, aided in the construction of the second bore beginning in 1975. The excavation is 14.5m wide, 13.5m high and 2.7km long. The rock is 75% granite and 20+% gneiss and schist with migmatite inclusions and augite diorite dykes. A shear zone of "squeezing rock" produced exceptionally difficult conditions. In this section full perimeter drifts filled with concrete were required for support. Elsewhere, combinations of steel supports, concrete linings and resin grouted rebar reinforcement were used.

324. WILLOUGHBY, D.F. and HOVLAND, H.J. Finite element analysis of stages of excavation of Helms underground powerhouse. *Preprint - Proceedings, 19th U.S. Symposium on Rock Mechanics*, Stateline, Nevada, 1978, Vol 1, pages 159-164. Results of a plane strain linear elastic finite element analysis of the Helms project are presented. The main section of the machine hall will be 25m wide, 38m high, 98m long and is located 300m

underground in good quality granite. See also item 288.

325. HARDY, M.P., AGAPITO, J.F.T. and PAGE, J. Roof design considerations in underground oil shale mining. *Preprint - Proceedings, 19th U.S. Symposium on Rock Mechanics*, Stateline, Nevada, 1978, Vol 1, pages 370-377. The design of roof spans for the Colony pilot oil shale mine is considered on the basis of stress determinations, observed rock conditions and numerical analysis. Room spans of up to 27m were studied, but such large spans involved greater bed separation and required more artificial support than the conventional bolting used in the standard 16.8m roof spans used in the room and pillar mining operation. See also items 289 and 315.

326. DODDS, R.K., ERIKSSON, K. and KUESEL, T.R. Mined rock caverns for underground nuclear plants in California. *Preprint - Proceedings, 19th U.S. Symposium on Rock Mechanics*, Stateline, Nevada, 1978, Vol 1, pages 444-451. Describes the conceptual design of a 1300MW nuclear powerplant to be constructed in a granite foothill location. The plant involves four major underground caverns, the largest in cross-section being 30.8m wide and 64.6m high. A review of stabilization methods is given.

327. KORBIN, G.E. and BREKKE, T.L. Field study of tunnel prereinforcement. *Journal of the Geotechnical Engineering Division, ASCE*, Vol 104, No GT8, Aug 1978, pages 1091-1108. A program of field instrumentation was used at two tunnel sites to examine the mechanisms by which prereinforcement and, in particular, spiling reinforcement work. Results are presented and analysed for the South Bore of the Eisenhower Tunnel in Colorado. The 2700m long tunnel is excavated in extremely varied ground ranging from massive granite to wide zones of squeezing fault gouge. The maximum overburden depth is 442m and the opening is approximately 14m wide and 12 to 14m high. Support varied widely with ground conditions. See also item 323.

328. HAMEL, L. and NIXON, D. Excavation of world's largest underground powerhouse. *Journal of the Construction Division, ASCE*, Vol 104, No CO3, 1978, pages 333-351. Excavation of the LG-2 powerhouse cavern in northwestern Quebec, Canada,is described. The machine hall is 26.5m wide, 47.3m high and 483.4m long. A nearby surge chamber is 22m wide, 45m high and 451m long. The rock is exceptionally good quality granitic gneiss and rock support is entirely by rock-bolts with shotcrete applied locally. In the arches, 6.1m long grouted rock bolts on 2m centres are tensioned to 200kN. See also items 271 and 32.

329. CROOKSTON, R.B. Mining oil shale. *Underground Space*, Vol 2, No 4, 1978, pages 229-241. The history of test mining of oil shale deposits in the western United States is given. Basic room and pillar designs have been used with rooms up to 20m wide being commonly used. Data obtained from these experimental excavations will be of great value in full-scale oil shale mine design. See also items 289, 315 and 325.

330. CRAIG, R.N. and MUIR WOOD, A.M. Tunnel lining practice in the UK. *Transport and Road Research Laboratory Supplementary Report* 335, 1978. Gives full details of tunnel lining practice in the U.K. with case history details of all major tunnels constructed in the U.K., including a number of historical interest. The diameters of the tunnels discussed rarely exceed 10m.

331. PLICHON, J.N., LE MAY, Y. and COMES, G. Centrales hydroelectriques souterraines - realisations récentes d'Electricité de France. *Tunnels et Ouvrages Souterrains*, Issue 30, Nov-Dec, 1978, pages 272-289. Describes several recent underground hydroelectric power plants constructed by Electricité de France. Plants described include Brommat II, Echaillon, Montézic, Orelle, Revin, Saussez II and Sisteron.

332. RYGH, J.A. Rock installations in Norway - sport halls, telecommunication centres, swimming pools in rock. *Tunnelling under Difficult Conditions*, I. Kitamura (ed.), Pergamon Press, Oxford, 1979, pages 31-44. Brief details are given of an underground sports centre excavated in granites and gneisses at Odda, the Gjovik public bath and swimming pool, and the Skien telecommunication centre excavated in limestone interbedded with shale. These excavations are 15-25m wide and are reinforced with rockbolts, shotcrete and some in situ concrete. See also item 276.

333. KENNEDY, H.J.F. New Zealand's Rangipo project. *Water Power and Dam Construction*, Vol 31, No 3, 1979, pages 45-46. New Zealand's second underground powerhouse has a main cavern 20m wide, 60m long and 38m in maximum height excavated in greywacke and argillite. Rock support consists of grouted rock bolts, 6m to 9m long, with expansion shell anchors and generally on 1m centres. The project is scheduled for completion in 1982. See also items 316 and 317.

334. DOLEZALOVA, M. The influence of construction work sequence on the stability of underground openings. *Proceedings, Third International Conference on Numerical Methods in Geomechanics*, W. Wittke (ed.), A.A. Blakema, Rotterdam, 1979, Vol 2, pages 561-570. Discusses the design of the Mantaro-Restitution underground hydro-electric scheme in the western Andes, Peru, with emphasis on the influence of construction sequence on the stress distribution and provision of support around the cavities. The main cavern is 22.0m wide, 34.5m high and 104.5m long. It is separated from the 13m wide, 18m high and 88m long transformer hall by a 26m wide rock pillar. Both cavities are protected by concrete roof arches, 0.9-1.5m thick in the machine hall and 0.6-1.0m thick in the transformer hall, and are located under 226m of cover in variable quality granite containing kaolinized and mylonitized zones and steeply dipping joints. The in-situ stresses considerably exceed the overburden stress. See also item 348.

335. EINSTEIN, H.H. Tunnelling in swelling rock. *Underground Space*, Vol 4, No 1, 1979, pages 51-61. Although the excavations described in this case history review are not large, the paper provides an excellent review of European tunnelling experience in swelling rocks, generally marls and anhydrites.

336. LEGGET, R.F. The world's first underground water power station. *Underground Space*, Vol 4, No 2, 1979, pages 91-94. Gives brief details of the Snoqualmie Falls Power Station No 1 built near Seattle in 1899 and still operating. The power house chamber is 12m wide, 60m long and 9m high and remains unlined. See also item 13.

337. PROVOST, A.G. and GRISWOLD, G.G. Excavation and support of the Norad expansion project. *Tunnelling '79*, The Institution of Mining and Metallurgy, London, 1979, pages 379-390. The Norad expansion project provides an expanded power generating facility for the Norad installation located within Cheyenne Mountain, Colorado (see items 125, 130, 138, 217, 218). The project includes power plant, cooling tower and exhaust valve chambers, approximately 500m of interconnecting tunnels and adits and 400m of small diameter mechanically mined shafts. The power plant chamber is approximately 20m wide, 16m high and 52m long. The host rock is a coarse-grained biotite granite with a fine- to medium-grained granite intrusive creating shear zones of highly fractured rock throughout the excavation area. This created a major problem at the intersection of an adit and the power plant chamber. Support was by grouted rockbolts including Perfo bolts. See also item 342.

338. FREIRE, F.C.V. and SOUZA, R.J.B. Lining, support and instrumentation of the cavern for the Paulo Afonso IV power station, Brazil. *Tunnelling '79*, The Institution of Mining and Metallurgy, London, 1979, pages 182-192. The 24m wide, 54m high and 210m long main cavern is located in a good quality crystalline Precambrian rock mass containing migmatite, granite, biotite gneiss, amphibolite and biotite schist. Cavern roof support was by systematic rock-bolting using 9m long, 32mm diameter bolts on a 1.5m square grid, tensioned to 22.5m tons each and cement grouted. A 4cm thick shotcrete layer was then applied followed by a 10cm square wire mesh and a second shotcrete layer. Wall support was by rock-bolts and 18m long tendons with only untensioned 32mm grouted dowels being used near the bottom of the excavation.

339. ROSENSTROM, S., SPARRMAN, G. and AHLGREN, B. Two underground hydroelectric power plants in East Africa. *Proceedings, 1979 Rapid Excavation and Tunnelling Conference*, A.C. Maevis and W.A. Hustrulid (eds), AIME, New York, 1979, Vol 2, pages 1354-1371. Gives considerable detail of the Kafue Gorge power station, Zambia (see also item 211) and the Kidatu power station, Great Ruaha project, Tanzania.

340. LEE, C.F., OBERTH, R.C. and TAYLOR, E.M. Some geotechnical and planning considerations in underground siting of nuclear power plant. *Proceedings, 1979 Rapid Excavation and Tunnelling Conference*, A.C. Maevis and W.A. Hustrulid (eds), AIME, New York, 1979, Vol 2, pages 1386-1408. Describes the conceptual design of an underground nuclear

power plant for Ontario, Canada. Brief details of existing underground nuclear plants in France, Norway, Sweden and Switzerland are given.

341. KIPP, T.R., CORDING, E.J., MERRITT, A.H. and KENNEDY, R.P. Feasibility evaluation for the excavation of large hemispherical cavities at the Nevada Test Site. *Proceedings, 1979 Rapid Excavation and Tunnelling Conference*, A.C. Maevis and W.A. Hustrulid (eds), AIME, New York, 1979, Vol 2, pages 1446-1465. Preliminary design of hemispherical chambers from 24.4 to 91.4m in diameter is discussed. The rocks are bedded tuffs. Two cavity support systems were considered - internally installed rockbolts for spans up to 54.9m, and annular tendon galleries located above the larger diameter chambers and from which primary support tendons are introduced prior to general excavation. Precedent practice for the support of large caverns in bedded rocks is reviewed.

342. PROVOST, A.G. and GRISWOLD, G.G. Design and construction of NORAD expansion project "excavation phase". *Proceedings, 1979 Rapid Excavation and Tunnelling Conference*, A.C. Maevis and W.A. Hustrulid (eds), AIME, New York, 1979, pages 1485-1504. Gives many of the same details as those given in item 337.

343. KUESEL, T.R. and KING, E.H. MARTA's Peachtree Center Station. *Proceedings, 1979 Rapid Excavation and Tunnelling Conference*, A.C. Maevis and W.A. Hustrulid (eds), AIME, New York, 1979, Vol 2, pages 1521-1544. Gives full details of the construction of this subway station in Atlanta. The main cavern is 18m wide, 14m high and 200m long. It is located about 25m underground in biotite gneiss and hornblende in which the foliation strikes diagonally across the station axis and dips at about 15°. The rock is strong and of good quality and contains four dominant joint sets. The arch roof is concrete lined and reinforced by 25mm dia. untensioned dowels, 6.1m and 7.6m long, on 1.5m centres. Dowels are also used in the walls, major portions of which are unlined.

344. DOUGLAS, T.H., RICHARDS, L.R. and ARTHUR, L.J. Dinorwic power station - rock support of the underground caverns. *Proceedings, Fourth Congress, International Society for Rock Mechanics*, Montreux, 1979, Vol 1, pages 361-369. The design approach used for the nomination of the rock support for a complex of nine underground caverns in slate is described. Geological and site investigations, stability analyses and the results obtained from a fully instrumented trial enlargement are discussed. Various aspects of this important project are also discussed in items 269, 290, 291 and 297.

345. LAUFFER, H. and PIRCHER, W. Sinking and supporting of an 80m deep power station shaft of 30m diameter. (In German). *Proceedings, Fourth Congress, International Society for Rock Mechanics*, Montreux, 1979, Vol 1, pages 467-474. The upper part of the shaft lining for the Kuhtai shaft power station near Innsbruck, Austria, was constructed in open cut and backfilled. Subsequent shaft sinking in slaty gneiss and amphibolite was carried out along a helical surface of 3m pitch, advancing by sectors. The permanent lining followed the excavation closely as a helical concrete strip 3m high and a minimum of 60cm in thickness, rendering any temporary support unnecessary.

346. WALDECK, H.G. The design and support of large underground chambers at depth in gold mines of the Gold Fields Group of South Africa. *Proceedings, Fourth Congress, International Society for Rock Mechanics*, Montreux, 1979, Vol 1, pages 565-571. The methods used to design, excavate and support large excavations at depths of up to 2000m are described. These excavations are located in good quality brittle quartzite but the in-situ stresses are very high compared with those under which most large underground chambers are excavated. Emphasis is placed on hoist chambers, the largest of which has a volume of 14000m^3. Roof support is generally by long cable anchors with intermediate rock bolts or grouted wire rope tendons about 3m long.

347. GEIS, H-P. Cave-in and subsidence at the Rodsand Mine, Western Norway. *Proceedings, Fourth Congress, International Society for Rock Mechanics*, Montreux, 1979, Vol 1, pages 645-648. The Rodsand mine is working an iron-titanium-vanadium deposit which occurs in irregular lenses 2 to 70m thick. Until the beginning of 1974 mining was by shrinkage stoping without fill. The empty stopes, the largest of which was 50m wide, 250m high and 200m long, were stable for many years. However, following robbing of the pillars around raises and the main levels, many of these open stopes collapsed. This paper describes the progressive failure mechanism involved and shows how it may be arrested by filling the empty stopes.

348. DOLEZALOVA, M. and DROZD, K. Site investigations and FEM calculations for two underground caverns in Peru. *Proceedings, Fourth Congress, International Society for Rock Mechanics*, Montreux, 1979, Vol 2, pages 105-112. Describes the geotechnical and design investigations of the Machu Picchu and Mantaro-Restitution (Mantaro III) underground hydroelectric power plants in the Andes, Peru. The Mantaro caverns are also discussed in item 334. The Machu Picchu cavern, in fine-grained granodiorite, will be 18 to 22m wide, 28m high and 46m long, and will be located only about 25m from an existing machine hall cavern.

349. MIZUKOSHI, T. and MIMAKI, Y. The behaviour of bedrock by the large cavern opening and comparison with the analysis. *Proceedings, Fourth Congress, International Society for Rock Mechanics*, Montreux, 1979, Vol 2, pages 439-446. The behaviour of the rock around two adjacent caverns is discussed in terms of field observations and analytical results. The main power station cavern of the Shintakasegawa pumped storage scheme in Japan is 27m wide, 54.5m high and 165m long, and is constructed in good quality granite. A transformer hall 20m wide, 35.3m high and 109m long is located 41.5m away from the main cavern. Particular attention is paid to the formation of relaxed zones around the caverns and to the performance of the rock anchors used for wall support. See also item 252.

350. SHARP, J.C., PINE, R.J., MOY, D. and BYRNE, R.J. The use of a trial enlargement for the underground cavern design of the Drakensberg Pumped Storage Scheme. *Proceedings, Fourth Congress, International Society for Rock Mechanics*, Montreux, 1979, Vol 2, pages 617-626. A 16.3m wide trial excavation was made at about mid-height of the proposed machine hall as part of the exploratory studies for the scheme. The rocks in the vicinity of the power station complex are horizontally bedded sandstones, siltstones and mudstones. The trial excavation was carried out in several stages with both tensioned and untensioned rock reinforcement being installed at each stage. The beneficial effect of tensioned reinforcement in bedded strata was demonstrated. See also items 284, 298 and 311.

351. WILLET, D.C. The economic use of underground space. *Tunnels and Tunnelling*, Vol 11, No 8, 1979, pages 23-29. Gives useful tabulations of some major underground hydroelectric plants and existing underground nuclear plants, and outlines other uses of underground space.

352. YOSHIDA, M. The Okuyoshino pumped-storage station. *Water Power and Dam Contruction*, Vol 31, No 10, 1979, pages 41-48. The powerhouse excavation is approximately 180m underground and is 20.5m wide, 41.6m high and 1578m long. The rocks are sandstones and shales, sometimes interbedded and dipping at 40°. Ten fault zones, generally dipping in the same direction as the bedding, intersect the powerhouse cavern site. Maximum fault width is 1.5m. The cavern has a concrete arch roof, and wall support is by 5m long stressed bars and 10-20m long stressed cables.

353. MYRSET, O. Underground hydroelectric power stations in Norway. *Subsurface Space - Rockstore 80*, M. Bergman (ed.), Pergamon Press, Oxford, 1980, Vol 2, pages 691-699. Almost all of Norway's hydroelectric plants are located underground. The layout, design and excavation of these underground power plants and of associated lined and unlined pressure shafts are described. Power stations referred to include Sima and Tonstad.

435

PROJECT NAME, LOCATION AND TYPE	APPROX. COMPLETION DATE	APPROX. DEPTH BELOW SURFACE (metres)	MAXIMUM EXCAVATION DIMENSIONS Width x Height x Length (metres)	ROCK TYPES AND CONDITIONS	SUPPORT DETAILS	REFERENCE NOS.*
AUSTRALIA						
Kareeya power station, Queensland.	1958	40	13.5 x 13.5 x 84.4	Hard, massive rhyolite.	Concrete barrel arch roof 30-38cm thick; end walls fully concrete lined, side walls concrete lined over half their area.	94
Kiewa No. 1 power station, Victoria	1960	60	13.7 x 21.4 x 74.4	Generally sound granodiorite but containing a major set of joints dipping at 15°. A major vertical fault-dyke complex some 15m wide influenced final location.	Concrete barrel arch roof; walls concrete lined to a minimum thickness of 30cm. 25mm dia. grouted rock bolts on a 2.4 x 3.0m pattern.	104
Poatina power station, Tasmania	1965	150	13.7 x 25.9 x 91.4	Horizontally bedded mudstone. Horizontal stresses approximately twice vertical. In situ mudstone compressive strength approx. 35MPa.	Trapezoidal roof with 3.7 and 4.3m rock bolts on 1m centres. Stress relief slots cut in roof haunches. Side walls supported by 3.7m bolts on 1m pattern. Roof and walls meshed and sprayed with 10cm gunite.	118, 149, 203.
Tumut 1 power station, Snowy Mountains, N.S.W.	1958	335	18 x 32 x 93	Fair to good quality granite and granitic gneiss. Horizontal stress approximately equal to vertical. High angle fault intersects one wall. Smooth, slickensided and graphite coated joints spaced at 0.15 to 0.6 m.	Ungrouted 4.6m long, 25mm dia. bolts on 1.2m centres in roof. Roof arch concreted and supplementary steel sets placed at request of contractor. Walls supported by 3.7m long, 25mm dia. bolts on 1.5m centres.	27,50,53, 58,61,95, 99,103,120, 147,203.
Tumut 2 power station, Snowy Mountains, N.S.W.	1962	230	18.3 x 33.5 x 97.5	Good quality granite and granitic gneiss with some porphyry dykes and minor faulting. Horizontal in situ stress 1.2 times vertical.	Support by 4.3m long grouted bolts on 1.2m centres. Concrete ribs 3m wide and 0.6m thick also used for permanent roof support.	120,124, 203
AUSTRIA						
Langenegg power station	1976	60	18 x 33 x 27	Alternating sequence of marl, conglomerate and sandstone. Power station excavation in 40m thick sandstone band dipping at 60°. Horizontal stress twice vertical.	Roof and wall support by rock bolts, steel mesh and shotcrete.	301

* Note: The reference numbers quoted here refer to items in the Bibliography presented in this Appendix and not to references cited in the main text.

PROJECT NAME, LOCATION AND TYPE	APPROX. COMPLETION DATE	APPROX. DEPTH BELOW SURFACE (metres)	MAXIMUM EXCAVATION DIMENSIONS Width x Height x Length (metres)	ROCK TYPES AND CONDITIONS	SUPPORT DETAILS	REFERENCE NOS.
Tauern highway tunnel	1975	Up to 800	11.8 x 10.75 x 6400	Weakened phyllites of variable quality. Loose talus deposits for 350m from portal. Very heavy squeezing ground encountered.	Support varies with rock quality classification. All categories use shotcrete and rock bolts. For the poorer ground conditions steel sets and forepoling also used.	225, 226, 264, 295
BRAZIL						
Nilo Pecanha (originally Forcacava) power station.	1954	100	18.9 x 25.9 x 115	Gneiss of fair quality traversed by faults partially or fully kaolinized. Joint set strikes parallel to excavation axis and dips at 80°.	Concrete barrel arch roof. Roof and walls supported by 25mm dia. mechanically anchored rockbolts; walls concrete lined.	25, 31, 33, 39, 43, 53, 64
Paulo Afonso power station	1952	50	15 x 31.3 x 60.2	Migmatite formed by intrusion of pegmatite and aplite into a biotite schist.	Generally no support required.	91
Paulo Afonso IV power station	1978	55	24 x 54 x 210	Good quality migmatite containing granite, biotite gneiss, amphibolite and biotite schist.	Roof support by 9m long, 32mm dia. rock bolts on 1-5m grid tensioned to 225 kN and grouted. A 4cm thick shotcrete layer followed by 10cm square wire mesh and a further shotcrete layer. Wall support by rock bolts and 18m long tendons.	338
CANADA						
Bersimis I power station, northern Quebec	1956	150	19.8 x 24.4 x 172	Granite and paragneiss.	Concrete roof arch, suspended ceiling and bare walls. Some spot bolting of roof.	50, 53, 59, 65, 87, 240
Chute-des-Passes power station, northern Quebec	1959	165	19.8 x 32.0 x 144	Generally sound granite, gneiss and paragneiss. After drilling, proposed cavern was rotated by 90° to obtain more favourable conditions.	Concrete arch and aluminium rock bolts for roof support.	87, 222, 240

PROJECT NAME, LOCATION AND TYPE	APPROX. COMPLETION DATE	APPROX. DEPTH BELOW SURFACE (metres)	MAXIMUM EXCAVATION DIMENSIONS Width x Height x Length (metres)	ROCK TYPES AND CONDITIONS	SUPPORT DETAILS	REFERENCE NOS.
Churchill Falls Power station, Labrador	1970	300	24.7 x 36.6 x 296	Hard gneissic assemblage. Horizontal stresses 1.7 times vertical.	Roof support by 4.6, 6.1 and 7.6m long, 28mm dia. expansion shell type rock bolts on 1.5m centres with chain link mesh. Walls similarly bolted but mesh not used.	148, 165, 180, 201
Kimano power station	1955	300	25 x 42.4 x 347.5	Granite and granodiorite.	Concrete barrel arch roof with rockbolts.	14, 15, 20, 21, 34, 35, 38, 45, 50, 53, 222
La Grande LG-2 power station, northwestern Quebec.	1979	100	26.5 x 47.3 x 483.4	Very good quality granitic gneiss with bands of diorite, gneiss, migmatite and pegmatite dykes and sills. Three joint sets present with sub-horizontal joints containing gouge the most significant.	Roof supported by 6.1m long grouted rock bolts tensioned to 200 kN. Wall support by bolting with shotcrete applied locally. Bolts on 1.5 or 2.1m square grids.	271, 320, 328
Mica dam power station, British Columbia.	1976	220	24.4 x 44.2 x 237	Good quality quartzitic gneiss with compressive strength of approx. 150 MPa. Vertical and horizontal in situ stresses 6.9 and 10.3 MPa. Some local slab or wedge failures encountered.	Roof supported by 6-7m long 25mm dia. grouted rock bolts on 1.5m centres and shotcrete cover. 6m bolts on 3m grid in walls.	272
Portage Mountain power station, Peace River, British Columbia	1965	150	20.4 x 43.9 x 271.3	Interbedded sandstone, shale and Coal Measures rocks dipping at 150. Horizontal in situ stress twice vertical. Roof in shale, walls in sandstone.	Thin concrete barrel arch roof. Grouted expansion shell bolts, 4.25 to 6.1m long on 1.5m centres for wall support.	142, 146
CHILE						
El Toro power station.	-	-	24.4 x 38.4 x 102	Granodiorite with orthogonal jointing.	15 to 17m long tendons on 6.1m pattern tensioned to 1.8MN and 4m long rock bolts on 2.4m pattern tensioned to 180 kN in roof. 15.2m long tendons on 6.1m pattern in walls	203
COLUMBIA						
Alto Anchicaya power station		180	18 x 30 x 60	Massive and competent quartz diorite. At end of construction loose rock caused by surface stress relief held up by wire mesh.	Roof arch and walls reinforced by 4.9m long rock bolts with wire mesh added in roof.	212

438

PROJECT NAME, LOCATION AND TYPE	APPROX. COMPLETION DATE	APPROX. DEPTH BELOW SURFACE (metres)	MAXIMUM EXCAVATION DIMENSIONS Width x Height x Length (metres)	ROCK TYPES AND CONDITIONS	SUPPORT DETAILS	REFERENCE NOS.
CZECHOSLOVAKIA						
Lipno power station.	1957	140	22 x 35 x 55	Heavily jointed and faulted granite of poor rock mass quality. Steeply dipping joints caused construction difficulties.	Thick concrete roof arch. Grouted 4 to 9m long, 36mm dia. rock bolts spaced on average one per $4.2m^2$.	153, 292
FINLAND						
Pirttikoski power station.	1959	40	16 x 43 x 500 (surge chamber) 22.5 x 32 x ? (machine hall) Tailrace tunnel cross-section 16x23.	Granite with joints diagonally across machine hall roof.	Rock bolted throughout with 22mm dia. bolts. Three concrete arches cast against rock in fissured areas. Mesh with 10cm. thick shotcrete in roof.	109, 116
Salmisaari oil storage caverns, Helsinki.	1973	15	14 x 28 x 175 to 260 (three parallel caverns)	Mixed granite, amphibolite and gneiss. High horizontal stresses caused rockbursts and failure of pillars during construction. Significant sets of inclined and horizontal joints.	Systematic rock bolting and concrete grouting.	310
Porvoo crude oil storage.	1975	30	12 to 18 x 32 x 500 to 600 (four parallel caverns)	Migmatite, principally gneiss and granite of variable grain size and moderate degree of schistosity. Relatively little jointing.	Weak zones locally supported by bolting and shotcrete.	279
FRANCE						
La Bathie (Rose-land) power station.	1960	–	25 x 32 x 130	Good quality fresh gneiss.	Reinforced concrete barrel arch roof 40cm thick. Concrete screen walls on sides. No rock wall supports other than crane beam supports.	
Randens (Isere-Arc) power station.	1954		16.8 x 31.4 x 109	Tough mica schist dipping at 20-60°, striking 45-85° to machine hall axis.	Concrete barrel arch roof; walls partly gunited.	30, 50, 53, 222.

PROJECT NAME, LOCATION AND TYPE	APPROX. COMPLETION DATE	APPROX. DEPTH BELOW SURFACE (metres)	MAXIMUM EXCAVATION DIMENSIONS Width x Height x Length (metres)	ROCK TYPES AND CONDITIONS	SUPPORT DETAILS	REFERENCE NOS.
Revin power station	1975	70	17 x 41 x 114	Phyllites and schists of good quality.	Roof lined with concrete 70cm. thick; walls rockbolted and gunited. Wall stability aided by the 120 crane beam anchors of 100 tonne capacity on 2m centres.	331
GERMANY (WEST)						
Bremm trial cavern for power station.	1970	200	24 x 9 x 30	Clay slate with intercalations of quartzitic sandstone. Cleavage approximately parallel to bedding strikes at 60° to the tunnel axis and dips at 50°.	20cm thick reinforced shotcrete and 6 to 8m long anchors on 1.3m spacing pretensioned to 150 kN.	194, 304, 308
Sackingen power station	1966	400	23.2 x 30 x 162	Granite and gneiss faulted over 25% of machine hall length. Excavation of elliptical shape for more favourable stress distribution.	3m long double cone anchor and perfo bolts used for roof support with 3 to 15cm shotcrete. 60cm thick concrete rib used in one faulted zone.	141
Waldeck II power station.	1973	350	33.5 x 50 x 105	Interbedded shale and greywacke, faulted and jointed. Bedding strikes perpendicular to cavern axis and dips at 32°. Faults form wedges in cavern roof.	Support by 18-24cm thick shotcrete reinforced with wire mesh and extensive system of 996 prestressed rock anchors up to 23m long and 3800 4 or 6m long rockbolts. Cavern walls curved in horseshoe shape.	207, 228, 233, 235, 244, 248.
Wehr power station.	1973	350	19.5 x 33 x 219	Gneiss with 5 sets of joints of varying intensity. One fault and 4 master joints transect the cavern Have considerable thicknesses of baryte, calcite and clay fillings. One master joint and the upstream wall isolate an unstable 30m high rock wedge.	Generally support by 15cm shotcrete reinforced by one layer of wire mesh with 22-24mm dia. Perfo bolts, 4m long spaced at 1.5-2.5m centres. In the wedge area extensive grouting and 82 prestressed anchors, 13 to 30m long with 1.7 MN capacity each, used.	253, 294, 304
INDIA						
Chibro power station, Yamuna Hydel scheme.	1972	230	18.2 x 32.5 x 113	Thinly bedded limestones and slates Joints, shear zones and bedding planes isolate potentially unstable blocks. Bedding dips at 45 to 50°.	Roof supported by steel arches. Walls supported by 350 prestressed anchors of average length 23.5m, capacity 600 kN, spaced at 2 to 5m; 7.5cm reinforced shotcrete used where necessary.	187, 188, 189

PROJECT NAME, LOCATION AND TYPE	APPROX. COMPLETION DATE	APPROX. DEPTH BELOW SURFACE	MAXIMUM EXCAVATION DIMENSIONS Width x Height x Length (metres)	ROCK TYPES AND CONDITIONS	SUPPORT DETAILS	REFERENCE NOS.
IRELAND						
Turlough Hills power station	1971	100	23 x 32 x 82	Coarse grained granite containing thin veins of pegmatite, aplite and quartz. Some zones of intensely jointed and partly decomposed rock.	Systematic rock bolting using 3, 4 and 5m bolts on 1.75 or 2m square grids and up to 25cm of gunite reinforced with steel fabric.	205, 237
ITALY						
Lago Delio (Roncovalgrande) power station	1970	130	21 x 60.5 x 195.5	Gneiss with sub-vertical foliation at right angles to cavern axis. Several joint sets in rock mass.	Concrete arch roof. Walls heavily reinforced with 3-5m long tensioned rock bolts, 5-25m long prestressed cables of up to 800 kN capacity and reinforced shotcrete.	173, 181, 197, 203, 221, 300
San Fiorano power station	1972	210	19.0 x 64.7 x 96.7	Phyllite with sub-vertical closely spaced schistosity planes forming major discontinuities.	Concrete arch roof. Cables up to 33m long tensioned up to 800 kN and 5m long, 50 kN perfo bolts used for wall reinforcement.	181, 197, 221, 300
San Massenza power station	1953	-	29 x 28 x 198	Good quality limestone. Construction difficulties caused by water inflow.	Concrete arch roof; walls not reinforced but concrete lined.	12, 32, 50, 53, 222
JAPAN						
Okutataragi power station	1973	200	24.9 x 49.2 x 133.4	Quartz porphyry, diabase and rhyolite.	Concrete arch roof. Walls supported by 5 to 15m long rock bolts, one per 3m^2 of wall area, tensioned from 80 to 350 kN.	266
Okuyoshino power station	1978	180	20.5 x 41.6 x 157.8	Sandstones and shales, sometimes interbedded, dipping at 40°. Ten fault zones, with maximum width 1.5m, intersect cavern.	Concrete arch roof. Wall support by 5m long prestressed bars and 10-20m long prestressed cable anchors.	352
Shintakasegawa power station	1978	250	27 x 54.5 x 163 (machine hall) 20.0 x 35.3 x 109 (transformer hall)	Good quality granite with major faults nearby. Average joint spacing 20cm. Horizontal in situ stress 1.8 times vertical. Long axis of cavities rotated towards horizontal major principal stress direction.	Machine hall roof reinforced with 5m long 25mm dia. and 2m long, 22mm dia. rock bolts. Walls reinforced by bolts and 15 and 20m long anchors tensioned to 1200 kN. 16 to 24cm thickness of mesh reinforced shotcrete on roof and upper walls. Lower walls concreted.	252, 349

PROJECT NAME, LOCATION AND TYPE	APPROX. COMPLETION DATE	APPROX. DEPTH BELOW SURFACE (metres)	MAXIMUM EXCAVATION DIMENSIONS Width x Height x Length (metres)	ROCK TYPES AND CONDITIONS	SUPPORT DETAILS	REFERENCE NOS.
LUXEMBURG Vianden power station	1970	100	17 x 30 x 330	Devonian schists with bedding planes containing mylonite dipping at 57°.	Concrete arch roof. Anchored reinforced concrete vertical beams support walls.	206, 292
MALAYA Sultan Idriss II (formerly Woh) power station, Batang Padang scheme, Cameron Highlands	1968	300	20.7 x 27.4 x 97.5	Good quality jointed granite. Exfoliation of walls of excavation on exposure. In situ stress field approximately hydrostatic.	Mainly 3m with some 4.6m and 2.9m long, 25mm dia., expansion shell rock bolts on 1.5m grid with 75mm square wire mesh and gunite throughout.	170
Sultan Yussuf (formerly Jor) power station, Cameron Highlands scheme	1962	260	15 x 24.5 x 80	Good quality granite, extensively jointed. Horizontal in situ stresses 1.8 and 2.6 times the vertical stress of 7 MPa.	Concrete roof arch. Walls unlined. Upper 5m of side walls below haunches rock bolted.	119, 128
MOZAMBIQUE Cabora Bassa power station	1975	160	27 x 57 x 220 (machine hall) 21 x 72 x 76 (two surge chambers)	Excellent quality granitic gneiss with slight schistosity.	No reinforcement other than concrete lining.	
NEW ZEALAND Manapouri power station, south-west South Island.	1969	150	18 x 34 x 111	Gneiss and pegmatite of good quality. In situ stress field approximately hydrostatic.	Roof and walls supported by 25mm dia., 4.6m long grouted rock bolts on 1.8m centres with wire mesh reinforced gunite added throughout.	204
Rangipo power station	1982	-	21 x 38 x 61 (machine hall) 14 x 30 x 51 (surge chamber)	Closely jointed, indurated grey-wacke sandstone and argillite of poor rock-mass quality. Dominant defects are steeply westward dipping crush zones. In-situ stress field approximately hydrostatic.	Grouted rock bolts 6-9m long with expansion shell anchors, generally on 1m centres throughout. Concrete roof arch.	316, 317, 333

PROJECT NAME LOCATION AND TYPE	APPROX. COMPLETION DATE	APPROX. DEPTH BELOW SURFACE (metres)	MAXIMUM EXCAVATION DIMENSIONS Width x Height x Length (metres)	ROCK TYPES AND CONDITIONS	SUPPORT DETAILS	REFERENCE NOS.
NORWAY						
Aura power station	1953	250	Two parallel machine halls, 18x17x123 (North) and 17x15.3x95 (South)	Gneiss with three joint sets. Foliation joints, rough, 25-100mm spacing; second set smooth, continuous, 50-150mm spacing; third set rough, less well developed. One non swelling clay-filled zone.	Multi-planar concrete roof with cast pillars between rock and roof installed in North hall before excavation completed. Rock bolts used for initial roof support in South hall; concrete pillars omitted.	73
Rendalen power station	1969		16 x ? x 43	Sparagmite (feldspar-rich sandstone), thinly bedded, very blocky and seamy. Joints, seams and flat-lying bedding planes carry montmorillonite with low activity.	Multiple drift excavation using 3m long perfo bolts on 1.5m spacing, wire mesh and shotcrete. Final concrete lining 40cm thick.	151
Skjomen power station	1974		13.4 x ? x 95	Granitic gneiss; in situ principal stresses 17, 2.3 and -1.6 MPa. Rock-burst problems in tunnels and main excavations.	Rock bolts on 1.0m centres in roof and 1.5m centres in walls plus 12-15cm reinforced shotcrete.	224
Tafjord IV power station		150	13 x ? x 47	Banded and folded gneiss. Rough schistosity planes, 5-10cm spacing. Two other less well developed joint sets with spacings from 25cm to 1m. Some slabbing due to high stresses.	25cm thick concrete arch roof with ribs cast between arch and rock. Localized support by bolting and mesh.	
PAKISTAN						
Tarbela dam diversion tunnels	1972	Up to 270	Four tunnels up to 19m dia. and 660-770m long.	Poor quality, closely jointed gabbro core flanked by alternating layers of limestone and chloritic schists with some carbonaceous and graphitic schists and phyllites.	About 10cm. shotcrete applied immediately on excavation. Wide flanged steel ribs then wedged against shotcrete. Up to 9m long, 36mm dia. perfo anchors also used in transition sections.	220, 227, 241, 255
PAPUA-NEW GUINEA						
Ramu I power station	1975	200	15 x 24 x 51	Gently dipping marble with small dolerite intrusions and overlying interbedded shale-siltstone-greywacke sequence. Difficulties caused by solution features in marble, block and wedge failures and fractured zone in contact area.	Roof pattern bolted with 4.3m long rock bolts on 1.2m centres. Roof and parts of end walls lined with wire mesh and gunite.	275

PROJECT NAME, LOCATION AND TYPE	APPROX. COMPLETION DATE	APPROX. DEPTH BELOW SURFACE (metres)	MAXIMUM EXCAVATION DIMENSIONS Width × Height × Length (metres)	ROCK TYPES AND CONDITIONS	SUPPORT DETAILS	REFERENCE NOS.
PERU						
Mantaro III power station		230	22.0 × 34.5 × 104.5 (machine hall) 13.0 × 18.0 × 88.0 (transformer hall)	Variable quality granite containing kaolinized and mylonitized zones and steeply dipping joints. Loosened zones in roof and in pillar between excavations.	Concrete roof arches, 0.9-1.5m thick in machine hall, 0.6-1.0m thick in transformer hall. Anchors provide an average support pressure of 160 kPa in roof and walls.	334, 348
SOUTH AFRICA						
Drakensberg power station	1981	130	16.3 × 45 × 193	Horizontally bedded mudstones, sandstones and siltstones. Rock mass of fair quality. Horizontal in situ stress three times vertical.	Grouted and tensioned rock bolts 32mm dia., 2 to 5m long on 1m grid with shotcrete and wire mesh.	284, 298, 311, 350
Durban Roodepoort Deep gold mine hoist chamber.	1948	1500	16.5 × 8.9 × 19.0	Hard, unjointed quartzite of very good quality with tight but pronounced bedding planes.	Concrete roof arch 0.5m thick reinforced with steel arches at 1.2m centres and resting on reinforced concrete beams 0.9 × 0.75m located on 0.5m ledges on either side. Sidewalls covered with bricks back-filled with concrete.	2, 311
Elandsberg power station	1987 (pilot excavations 1977)	150	22 × 47 × 197	Greywacke with minor amounts of phyllite. Rock mass of good quality. Approximately hydrostatic in situ stress field.	Resin grouted 6m long rock bolts on 2m centres in roof with 15cm thick shotcrete and wire mesh.	282, 311
Kloof gold mine hoist chamber	1967	1800	16.8 × 16.8 × 60	Massive lava of excellent rock mass quality. Compressive strength > 300 MPa.	Local spot bolting of roof and walls by 1.8m long, 16mm dia. bolts. Permanent support by 8.5m long tendons on 3m centres tensioned to 180 kN and grouted. Chamber meshed and gunited.	311
President Steyn gold mine hoist chamber	1975	2090	16 × 13 × 27	Quartzite and conglomerate	10m long prestressed tendons at 4m centres plus 3m long bolts at 2m centres with wire mesh reinforced shotcrete 5-7cm thick.	311
Raucana power station, South West Africa	1977	100	16 × 36 × 142	Porphyroblastic gneiss intruded by amphibolite. Rock mass quality good. Horizontal in situ stress approximately twice vertical.	In roof, 1MN capacity anchors, 20m long, on 3.3m centres with intermediate 5m long bolts. Wire mesh and shotcrete placed subsequently.	283, 311

PROJECT NAME, LOCATION AND TYPE	APPROX. COMPLETION DATE	APPROX. DEPTH BELOW SURFACE (metres)	MAXIMUM EXCAVATION DIMENSIONS Width x Height x Length (metres)	ROCK TYPES AND CONDITIONS	SUPPORT DETAILS	REFERENCE NOS.
Vaal Reefs gold mine hoist chamber	1966	1700	10 x 12 x 13.5	Very good quality quartzite.	Reinforced concrete roof; occasional rock bolts as temporary support. Mass concrete in side walls.	137
Vanderkloof power station	1976	30	25 x 49 x 100	Very good quality dolerite of high strength.	Concrete arch vault reinforced with 3m long resin bonded bolts at 2m centres. Grouted 6m long bolts at 2.5m centres in walls.	286, 311
Western Deep Levels gold mine hoist chamber	1974	2750	16.6 x 12 x 32	Hard quartzite under very high stress; rock burst hazard.	Steel ropes 32mm dia., cement grouted, 7m long at 1.5m centres with wire mesh and 7cm shotcrete.	311
SWEDEN						
Arstodal, Stockholm, wine and liquor storage rooms		20	20 x 24.5 x 300 (two parallel caverns, separated by 16m)	Massive gneiss with few joints (RQD =100); major normal faulting in the vicinity.	No support other than 50 spot bolts.	260
Baltic coast naval storage chambers		30	15 x 30 x up to 500	Massive granite and gneiss; normal faulting in the region.	No structural support required for static loading.	260
Harspranget power station	1951	70	18.6 x 31 x 104	Sound granite	Roof reinforced by 25mm dia. grouted bars, 3 to 4.6m long on a 1m pattern covered by reinforced gunite.	11, 19, 50, 53
Seitevare power station		170	13 x 11 x 40	Massive granite with a few widely spaced vertical joints. Minor inflows of water through joints caused operational problems.	No support required; gunite roof coating.	260
Stornorrfors power station	1958	200	18.5 x 29 x 124 (machine hall) 14.4 to 16.0 wide 26.5 high (tail-race tunnel)	Strong, massive gneiss with a few very rough joints.	Roof supported by cement mortar anchored rock bolts, reinforcing wire mesh and at least 3cm gunite. Generally not reinforced. Weak zones reinforced with concrete; rockbolts used in a loosened granitic zone.	55, 92, 114 80, 89, 114

PROJECT NAME, LOCATION AND TYPE	APPROX. COMPLETION DATE	APPROX. DEPTH BELOW SURFACE (metres)	MAXIMUM EXCAVATION DIMENSIONS Width x Height x Length (metres)	ROCK TYPES AND CONDITIONS	SUPPORT DETAILS	REFERENCE NOS.
SWITZERLAND						
Cavergno power station	1955		28 x 22 x 103	Mica schist with foliation dipping at 42°. Machine hall axis perpendicular to foliation strike.	Fully concrete lined.	40, 50, 53
Grimsel II East power station	1978	100	29 x 19 x 140	Mainly granodiorite of good quality except for some unfavourable joint orientations and a gneissic zone. High in-situ horizontal stresses.	Machine hall fully concrete lined.	301
Hongrin power station	1970	55-150	30 x 27 x 137	Limestone and limestone-schist containing several sets of vertical fractures and a considerable amount of clay.	Profile excavated by series of small longitudinal galleries. Support by 650 pre-stressed anchors 11 to 13m long and shotcrete installed soon after excavation.	161, 318
Huttegg ventilation chamber, Seelisberg highway tunnel	1977		Two parallel tubes 18m dia. x 52m long connected by a tube 14m wide x 16m high.	Marl with very high overburden pressure, reaching the uniaxial compressive strength of the rock.	Reinforced shotcrete applied immediately with 3.8m long Perfo bolts on 1.1m centres followed by final anchoring with 16-18m long anchors on 4.5m centres stressed to 80 tonnes.	303, 314
Innertkirchen power station	1942	> 40	19.5 x 26.8 x 100	Very good quality gneiss.	No rock reinforcement required but 50cm thick concrete roof arch used.	1, 32, 40, 49, 50, 53, 222
UNITED KINGDOM						
Cruachan power station, Scotland.	1965	320	23.5 x 38 x 91.5	Excavation in diorite near boundary with phyllite. Diorite well jointed with some kaolinization; 6m wide "crush" band runs transversely to machine hall axis. Vertical and horizontal in situ stresses both approx. three times overburden stress.	Concrete arch roof. Other support by 4.6m long, 25mm dia. expansion shell rockbolts on 2.3m centres. Bolts up to 9.2m long on 1.5m spacing used in faulted and closely jointed zones. Prestressed anchors, 11m long on 3.5m centres used in conjunction with concrete beam on north wall.	136

PROJECT NAME, LOCATION AND TYPE	APPROX. COMPLETION DATE	APPROX. DEPTH BELOW SURFACE (metres)	MAXIMUM EXCAVATION DIMENSIONS Width x Height x Length (metres)	ROCK TYPES AND CONDITIONS	SUPPORT DETAILS	REFERENCE NOS.
Dinorwic power station, North Wales	1980	300	24.5 x 52.2 x 180.3 (machine hall) 24.5 x 18.9 x 155 (transformer hall)	Good quality slate with numerous dolerite dykes and some faults. Five principal discontinuity sets identified. Most major discontinuities have low shear strength. Wedge or block failures the main geotechnical problem.	Primary reinforcement by 1200 kN capacity anchors, generally 12m long on 6m centres. Secondary reinforcement by 3.7m long rock bolts on 2m centres tensioned to 10 tonnes. Tertiary support by 50mm minimum thickness of mesh-reinforced shotcrete.	269, 290, 291, 297, 344
UNITED STATES OF AMERICA						
Bear Swamp power station, Massachusetts	1973		24 x 46 x 69	Good quality chlorite mica-schist with well developed schistosity. Major principal stress parallel to cavern axis.	Roof supported by 10cm shotcrete and 6.1m long, 25mm dia., hollow groutable bolts on 1.5m centres. Same pattern in upper 5m of walls; 6m long bolts on 2.4m centres perpendicular to foliation in downstream wall.	238, 242
Boundary Dam power station, Washington	1965	200	22.5 x 58 x 146	Good to excellent quality limestone.	Roof support by 4.6m long grouted bolts on 1.8m centres with wire mesh and some gunite. Walls bolted only where required. Additional 9.2m long bolts used in roof where wedges appeared to form.	132, 139, 158, 203
Edward Hyatt power station, Oroville Dam, California.	1966	90	21 x 43 x 170	Generally massive amphibolite with three prominent joint sets. Shears and schistose zones strike 600-800 across the chamber axis. They dip steeply, are from 2 to 7m apart, are 2 to 15cm wide and contain crushed rock, schist and clay gouge.	Fully grouted 25mm dia., 6.1m long, expansion shell rock bolts on 1.2m centres in roof and 1.8m centres in walls. Additional bolts angled across slabby rock where shears intersect roof. Surface covered with chain link mesh and 10cm thickness of gunited applied.	127, 150, 191, 199, 203, 259
Haas power station, California	1957	150	17 x 30 x 53	Massive granite with widely spaced fractures.	Fully grouted perforated-sleeve rock anchors of 25mm dia. reinforcing bar, 3, 3.8 and 4.6m long spaced on 1m centres in arch roof. 10cm square wire mesh attached to bars and 10cm thickness of gunite applied.	85

PROJECT NAME, LOCATION AND TYPE	APPROX. COMPLETED DATE	APPROX. DEPTH BELOW SURFACE (metres)	MAXIMUM EXCAVATION DIMENSIONS Width x Height x Length (metres)	ROCK TYPES AND CONDITIONS	SUPPORT DETAILS	REFERENCE NOS.
Morrow Point power station, Colorado	1963	120	17.4 x 42 x 63	Micaceous quartzite and mica schist with some pegmatite intrusions. During construction, a 30m wide wedge isolated by two intersecting shear zones moved 5cm into excavation.	Fully grouted 25mm dia. expansion shell rock bolts on 1.2m centres, 6.1m long in roof and 3.7m long in walls. Special reinforcement to stabilize side-wall wedge: 9 off 60 tonne grouted rebars, 18-30m long; 25 off 110 tonne tendons; 27 off 35mm dia. bolts, 8-24m long, tensioned to 27 tonnes.	134, 140, 167, 190, 203
Nevada Test Site, Test Cavities I and II, Ranier Mesa, Nevada	1965	400	24 x 43 x 37 (end hemispherical; crown dome-shaped)	Alternating layers of low strength tuff, predominantly massive and thick-bedded, dipping at 8-15°. Occasional thin beds (7-45cm thick) of soft, friable white tuff. In Cavity II, high angle joints more prevalent. Water content of tuff =21%.	In roof, 10m long, 28mm dia. bolts on 0.9m centres. In wall, 7.3m long bolts on 1.8m centres. Bolts tensioned with back 2.4m grouted. Gunite applied to prevent drying and cracking. Some chain link fabric supported from 1.8m long Perfo-bolts used for primary support. Additional support by 14.6m long bolts on 0.9m centres in large areas of Cavity II where unstable wedges formed.	203, 341
North American Air Defence Command (NORAD) Cheyne Mountain Complex (NCMC), Colorado Springs, Colorado	1964	400	Three parallel main chambers 14 x 18 x 180 plus several other major excavations	Coarse-grained biotite granite intruded by fine to medium grained granite and thin basalt dykes. Granite is closely jointed and mostly unaltered. Two steeply dipping joint sets predominate.	Fully grouted 25mm dia. slot and wedge rock bolts, 3m long on 1.2m centres in roof; 2.4, 3.0 and 3.7m long on 1.5m centres in walls. Chain link fabric 5cm x 5cm over rock surface. Extra bolts up to 9m long on 0.6 to 1.8m centres used to stabilize wedges at intersections.	125, 130, 138, 203
NCMC expansion, Colorado Springs, Colorado	1971	400	Power plant chamber 20 x 16 x 52 plus several other excavations	As above. A 10m wide shear zone created by an intrusion, crosses the power plant chamber.	32mm dia. 4.9m long recessed rock anchors installed over 1100 arc of roof on 1.4m centres, inclined at 60° to direction of advance. 25mm dia. 7.3m long Perfo bolts on 1.4m centres installed in roof and walls. Chain link fabric over roof and 35% wall area below spring-line.	217, 218

PROJECT NAME, LOCATION AND TYPE	APPROX. COMPLETED DATE	APPROX. DEPTH BELOW SURFACE (metres)	MAXIMUM EXCAVATION DIMENSIONS Width x Height x Length (metres)	ROCK TYPES AND CONDITIONS	SUPPORT DETAILS	REFERENCE NOS.
Northfield Mountain power station, Massachusetts.	1970	170	21 x 47 x 100	Interbedded layers of gneiss, quartzite and schist. Two major joint sets steeply dipping and widely spread.	In roof, 25mm dia. fully grouted expansion shell rock bolts on 1.5m centres, 10.7m long in central part of arch, 7.6m long in lower arch. In walls, same pattern except top row 6.1m long; remainder 4.9m. 7.5cm square welded wire mesh with 10cm thickness of gunite on roof.	192, 198
Racoon Mountain power station, Tennessee.	1973	200	22 x 50 x 150	Good quality, horizontally bedded limestone with two major near vertical joint sets. Machine hall axis parallel to major principal in-situ stress which is 2 to 3 times the vertical stress.	All excavations supported by Williams groutable bolts, with wire mesh and shotcrete as required.	208
Rosslyn Station, Washington D.C. Metro	1973	16-21	25 x 17 x 220	Gneiss, highly fractured and weathered at top, improving in quality with depth; 5 prominent joint sets identified; shear zones and clay filled joints common. Joint pattern caused overbreak during blasting.	Roof pre-reinforced with "drilled in" 6.1m long spiles. At least 15cm shotcrete and steel ribs on 1.2m centres provide roof support. Walls reinforced by horizontal and angled 5m long bolts on 1.5 and 2.4m centres.	243
ZAMBIA Kafue Gorge power station	1972	500	15 x 32 x 130 (machine hall) 17 x 21 x 125 (transformer hall)	Good quality granite, granitic gneiss and gneiss. Cavern axes aligned approx. parallel to major principal stress. Horizontal in situ stress approx. 1.5 times vertical.	25mm dia., 4.5m long rock bolts on 2m centres and mesh in roof; 6m long bolts on 2m centres in walls. Crane beams and their anchors provide additional wall support.	211
ZIMBABWE (formerly Rhodesia) Kariba I power station.			23 x 40 x 143 (machine hall) 17 x 18 x 163 (transformer hall) Three 19m dia. x 52m high surge chambers.	Generally sound biotite gneiss with granitic pegmatite dykes. A faulted zone affected a 30m length at west end of north wall of machine hall. Fractured and altered gneiss with slickensides in upper sections of surge chambers.	Main excavations upstream of dam; rock cover extensively grouted. Rock bolting used during excavation. Excavations fully concrete lined except for downstream machine hall wall. Voids between concrete walls and rock and joints in rock grouted.	71, 97, 108

Appendix 2: Isometric drawing charts

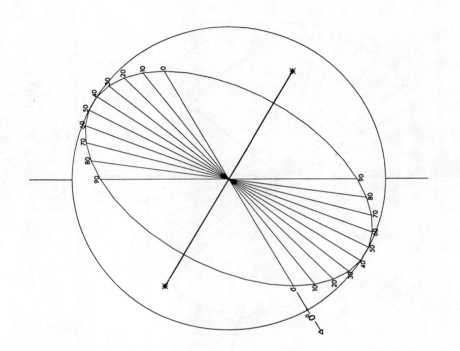

STRIKE AND DIP LINES OF PLANES OF DIP DIRECTION 0.

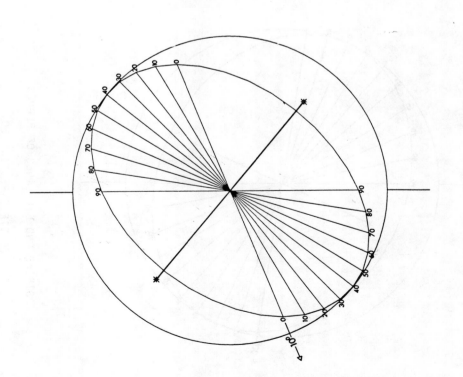

STRIKE AND DIP LINES OF PLANES OF DIP DIRECTION 10.

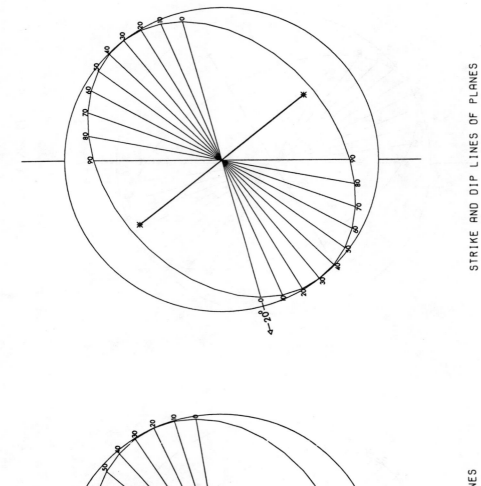

STRIKE AND DIP LINES OF PLANES
OF DIP DIRECTION 20.

STRIKE AND DIP LINES OF PLANES
OF DIP DIRECTION 30.

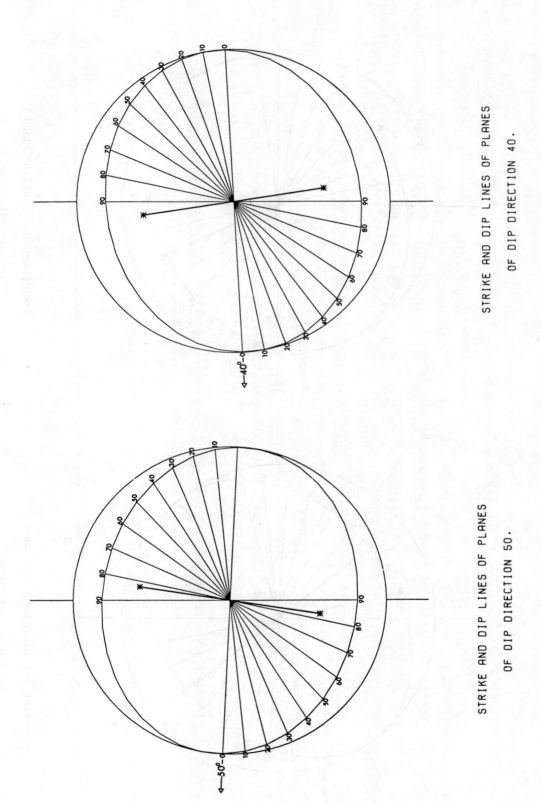

STRIKE AND DIP LINES OF PLANES
OF DIP DIRECTION 40.

STRIKE AND DIP LINES OF PLANES
OF DIP DIRECTION 50.

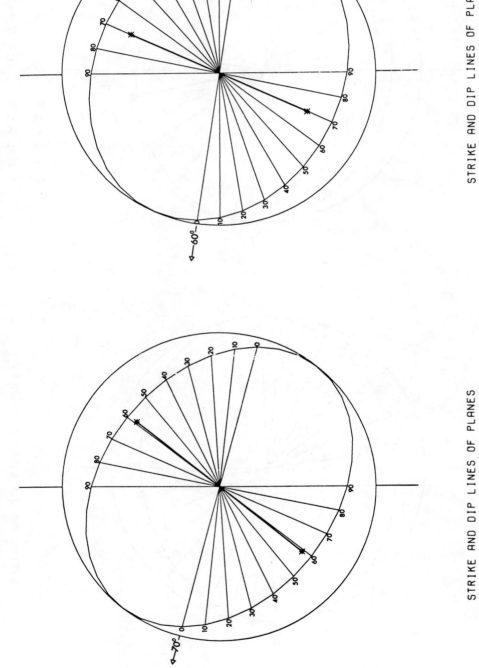

STRIKE AND DIP LINES OF PLANES
OF DIP DIRECTION 60.

STRIKE AND DIP LINES OF PLANES
OF DIP DIRECTION 70.

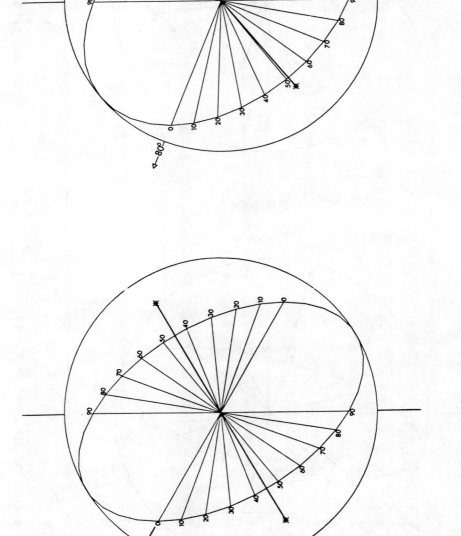

STRIKE AND DIP LINES OF PLANES
OF DIP DIRECTION 80.

STRIKE AND DIP LINES OF PLANES
OF DIP DIRECTION 90.

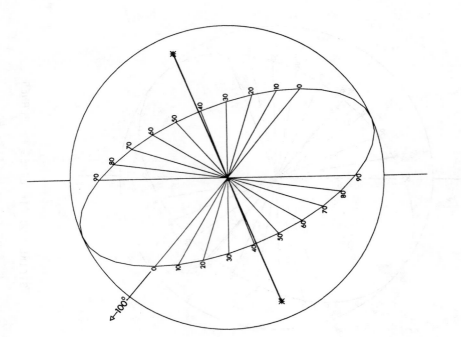

STRIKE AND DIP LINES OF PLANES
OF DIP DIRECTION 100.

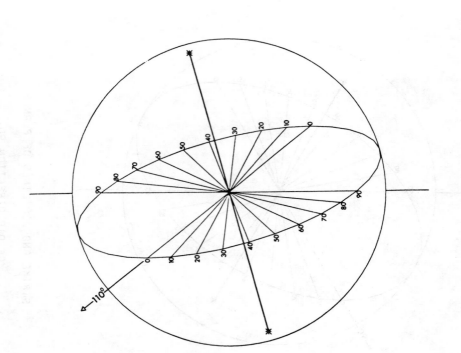

STRIKE AND DIP LINES OF PLANES
OF DIP DIRECTION 110.

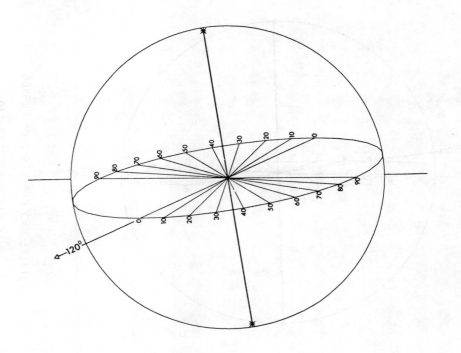

STRIKE AND DIP LINES OF PLANES
OF DIP DIRECTION 120.

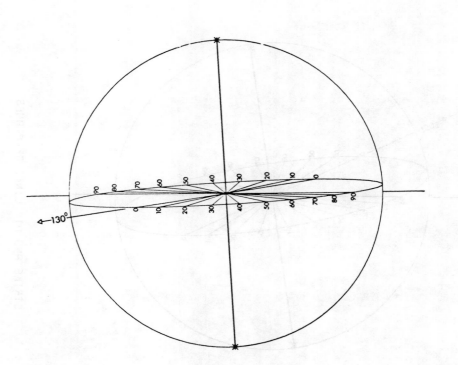

STRIKE AND DIP LINES OF PLANES
OF DIP DIRECTION 130.

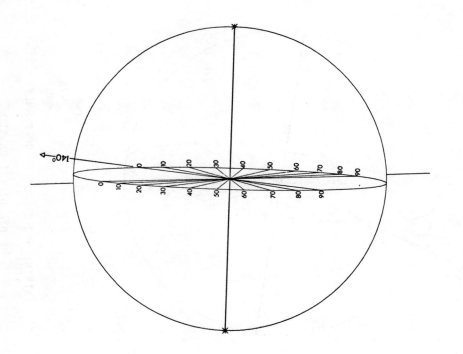

STRIKE AND DIP LINES OF PLANES OF DIP DIRECTION 140.

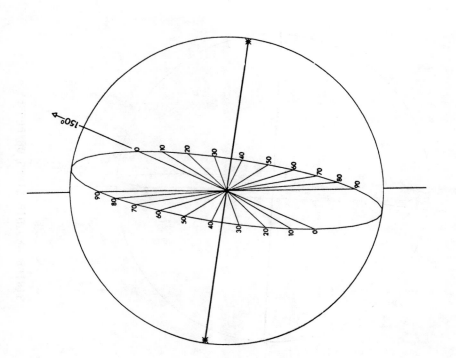

STRIKE AND DIP LINES OF PLANES OF DIP DIRECTION 150.

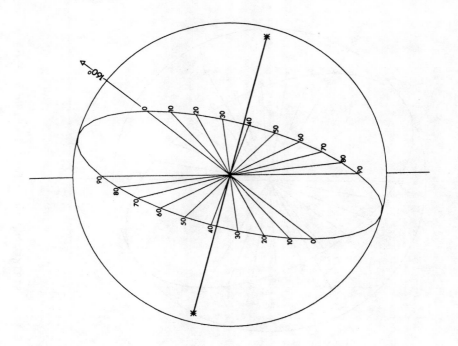

STRIKE AND DIP LINES OF PLANES
OF DIP DIRECTION 160.

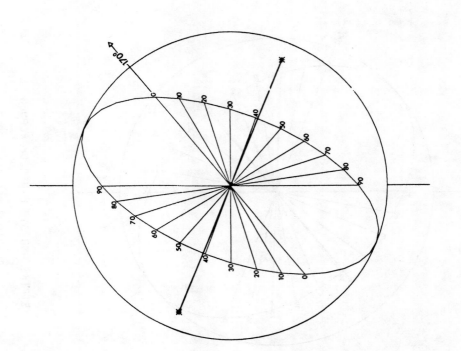

STRIKE AND DIP LINES OF PLANES
OF DIP DIRECTION 170.

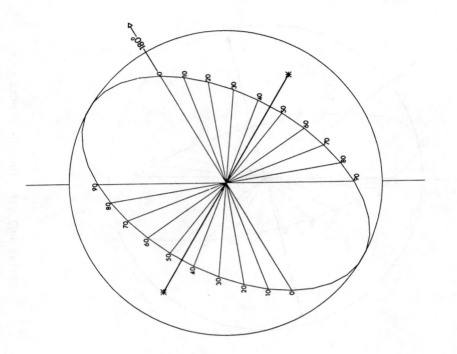

STRIKE AND DIP LINES OF PLANES OF DIP DIRECTION 180.

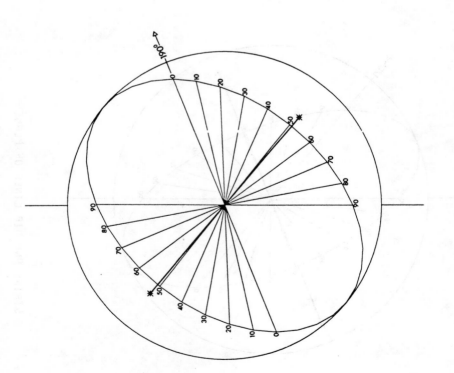

STRIKE AND DIP LINES OF PLANES OF DIP DIRECTION 190.

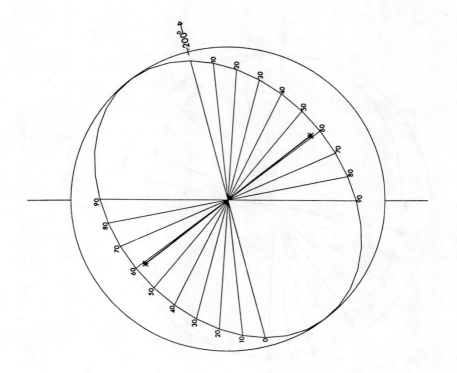

STRIKE AND DIP LINES OF PLANES
OF DIP DIRECTION 200.

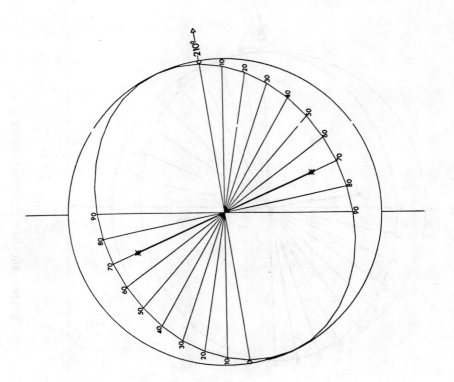

STRIKE AND DIP LINES OF PLANES
OF DIP DIRECTION 210.

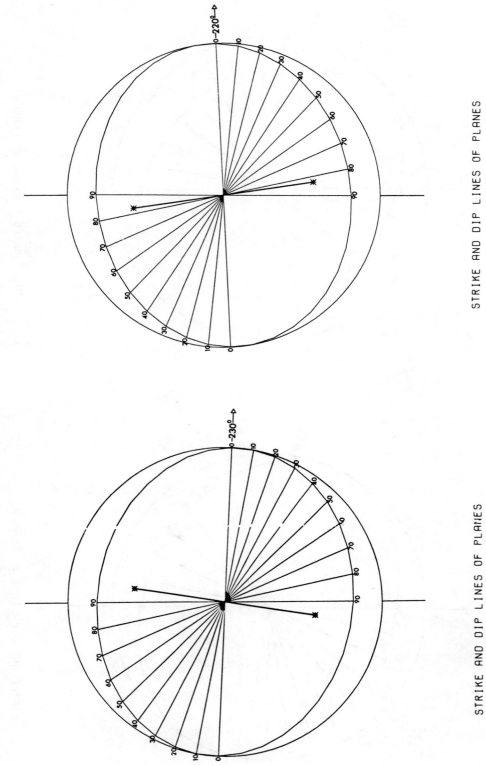

STRIKE AND DIP LINES OF PLANES
OF DIP DIRECTION 220.

STRIKE AND DIP LINES OF PLANES
OF DIP DIRECTION 230.

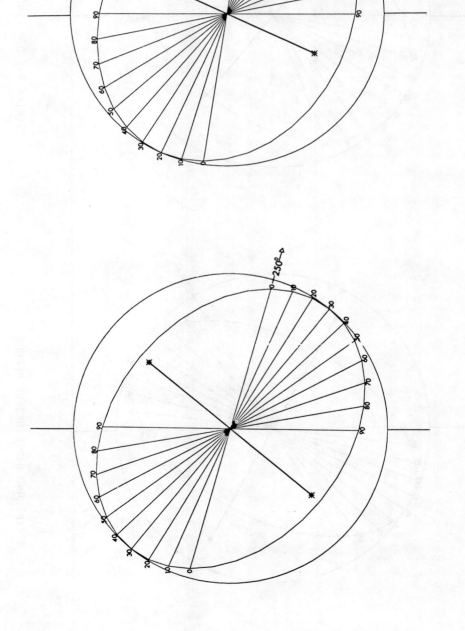

STRIKE AND DIP LINES OF PLANES OF DIP DIRECTION 240.

STRIKE AND DIP LINES OF PLANES OF DIP DIRECTION 250.

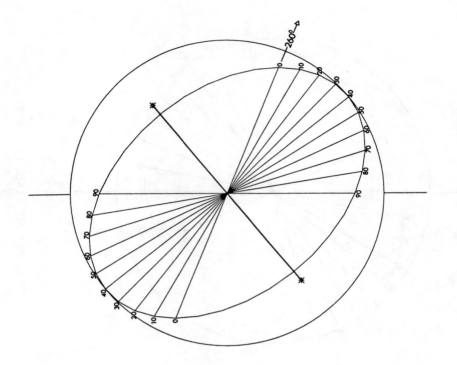

STRIKE AND DIP LINES OF PLANES
OF DIP DIRECTION 260.

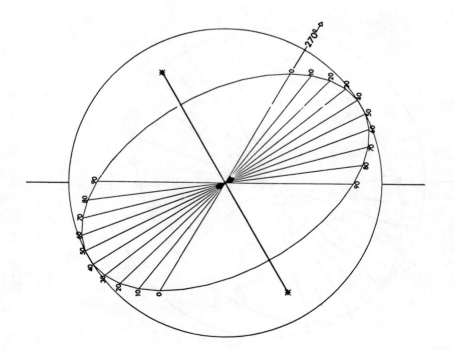

STRIKE AND DIP LINES OF PLANES
OF DIP DIRECTION 270.

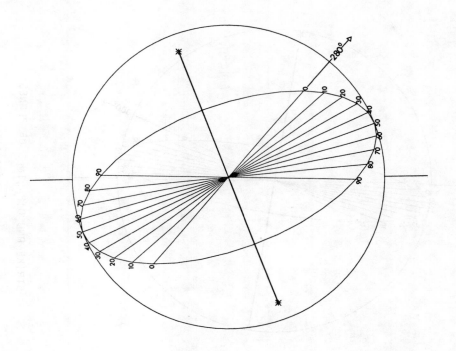

STRIKE AND DIP LINES OF PLANES
OF DIP DIRECTION 280.

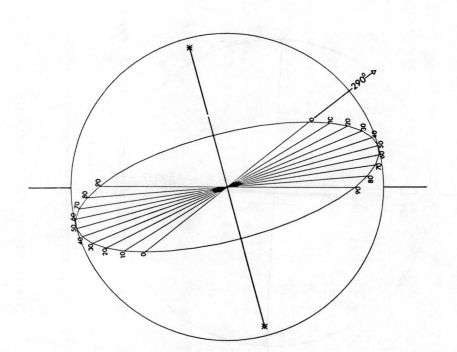

STRIKE AND DIP LINES OF PLANES
OF DIP DIRECTION 290.

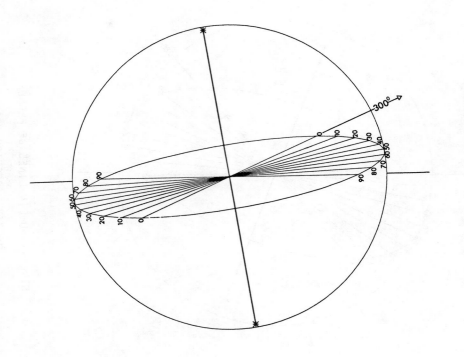

STRIKE AND DIP LINES OF PLANES
OF DIP DIRECTION 300.

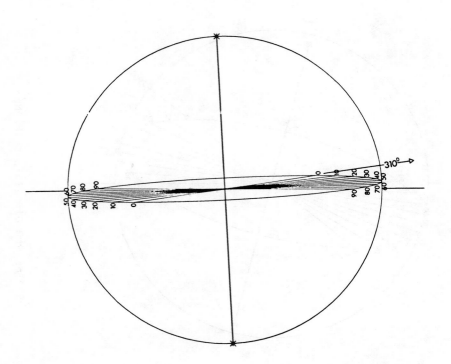

STRIKE AND DIP LINES OF PLANES
OF DIP DIRECTION 310.

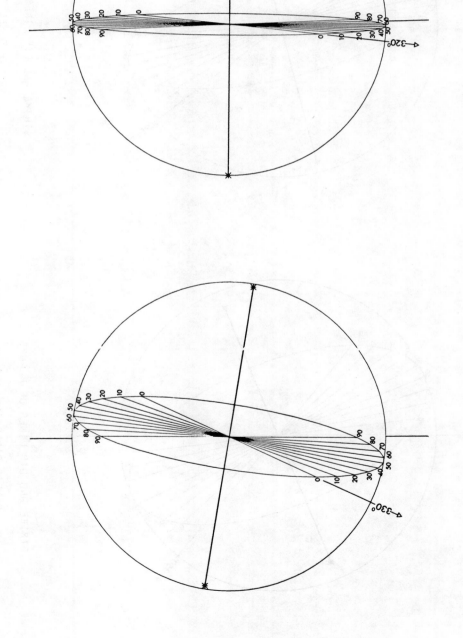

STRIKE AND DIP LINES OF PLANES OF DIP DIRECTION 320.

STRIKE AND DIP LINES OF PLANES OF DIP DIRECTION 330.

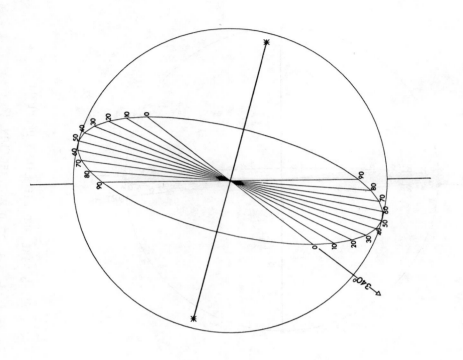

STRIKE AND DIP LINES OF PLANES
OF DIP-DIRECTION 340.

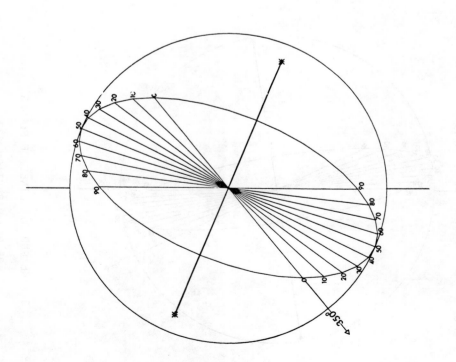

STRIKE AND DIP LINES OF PLANES
OF DIP DIRECTION 350.

Appendix 3: Stresses around single openings

The stress distributions presented on the following pages were prepared at Imperial College by Dr Elsayed Ahmed Eissa under the direction of Dr J.W.Bray. The two-dimensional boundary element method of stress anaylsis presented in Appendix 4 was used to derive these stress distributions.

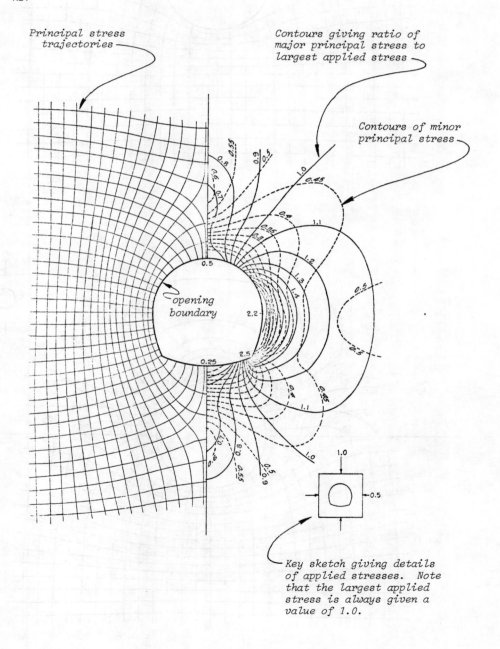

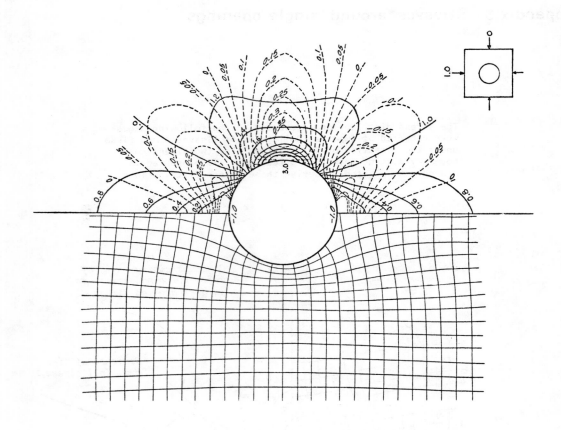
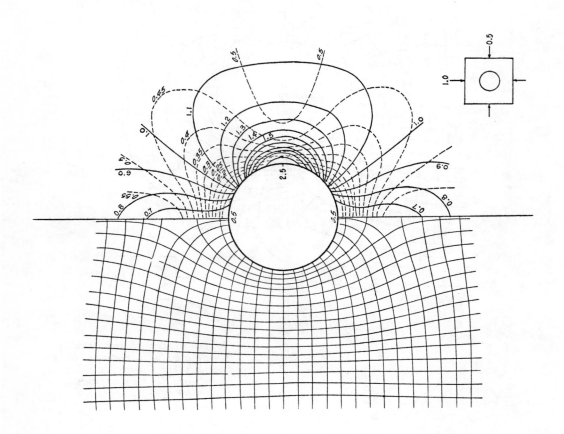

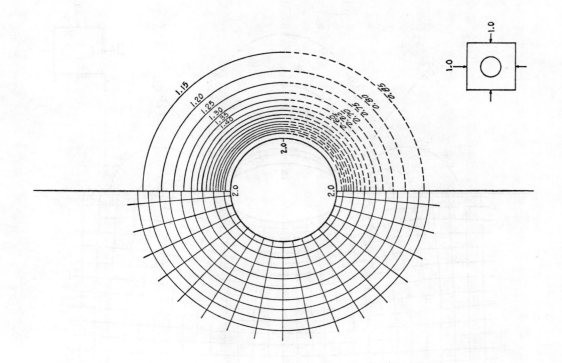
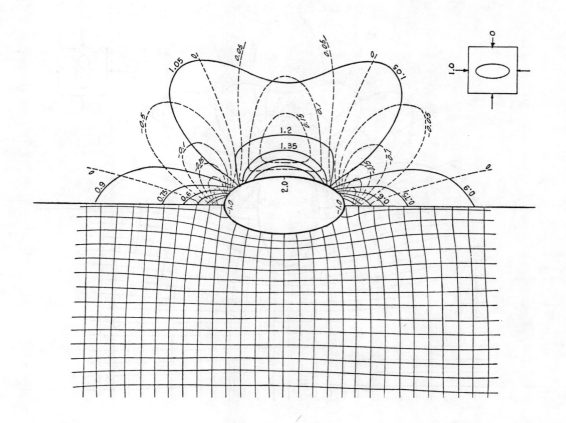

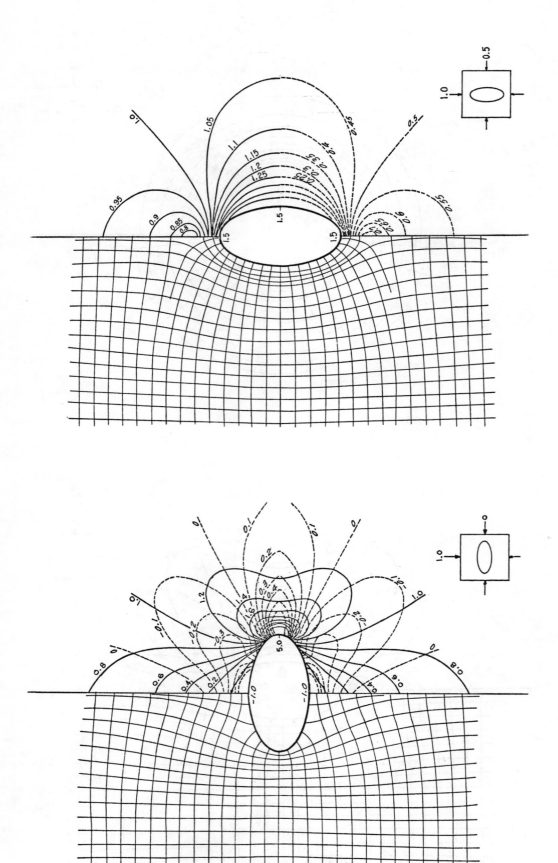

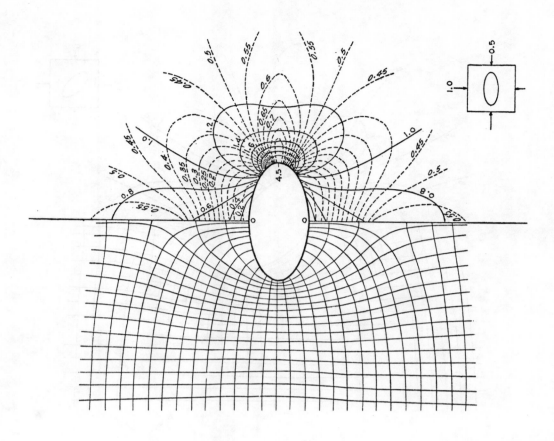
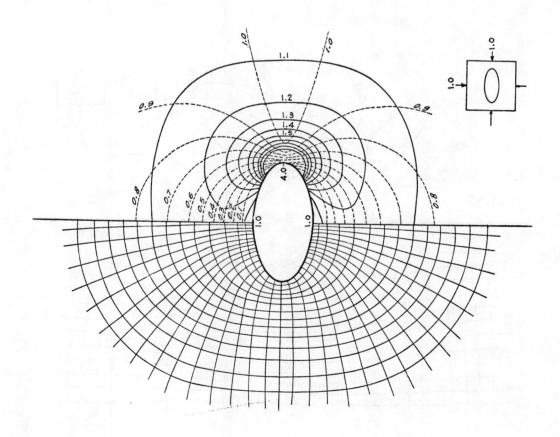

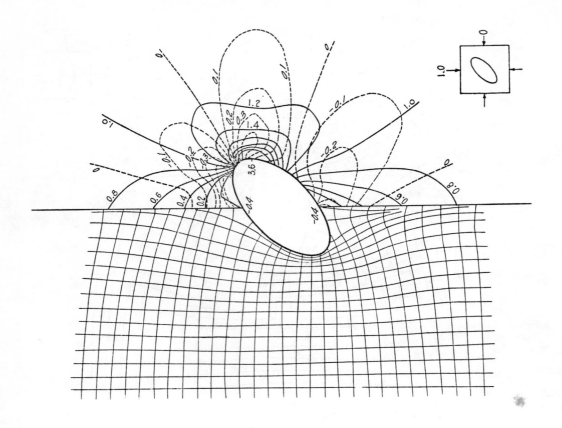
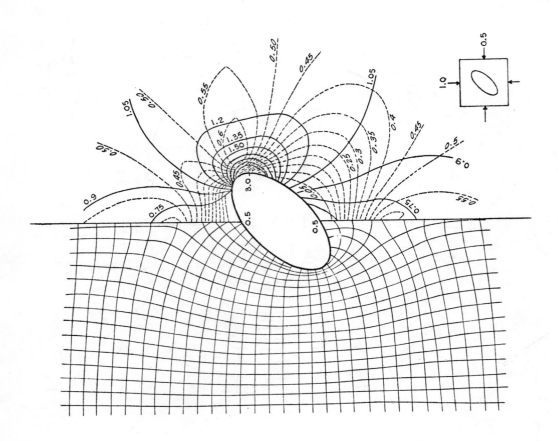

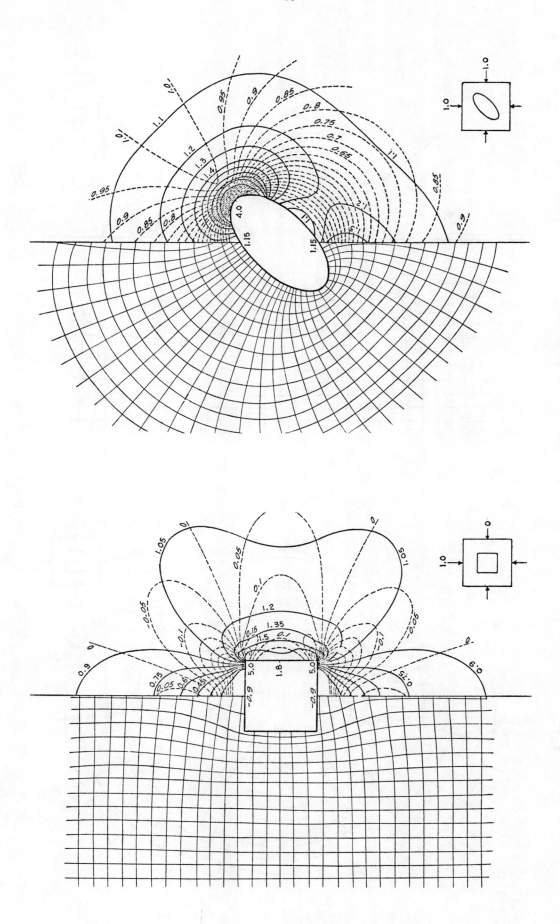

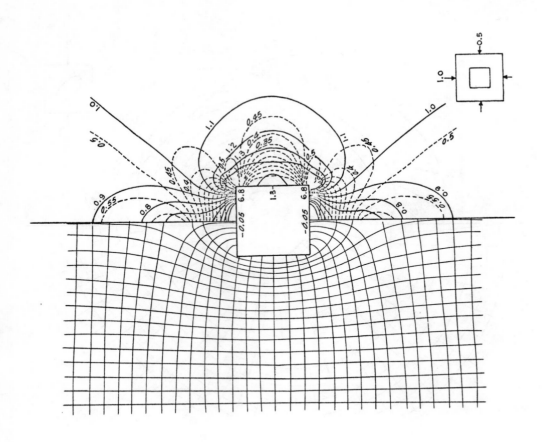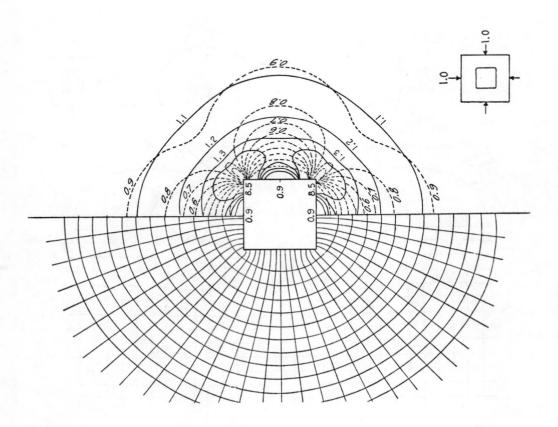

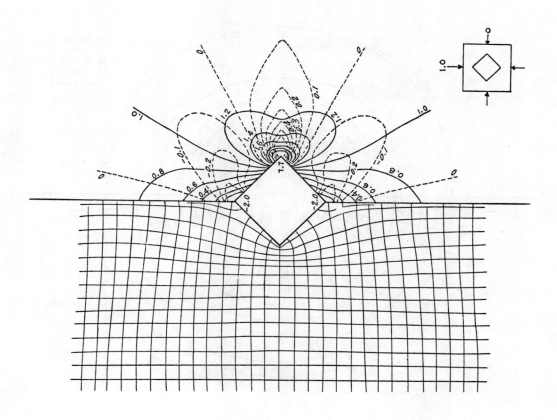
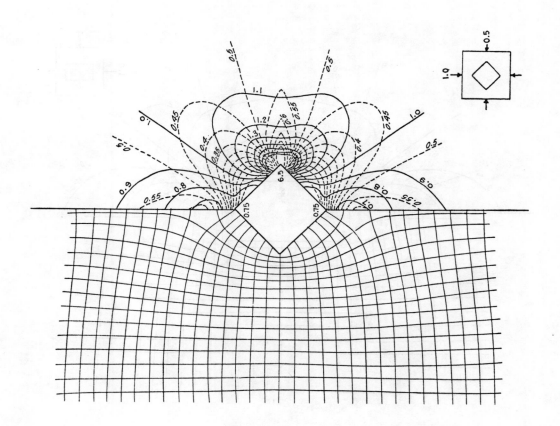

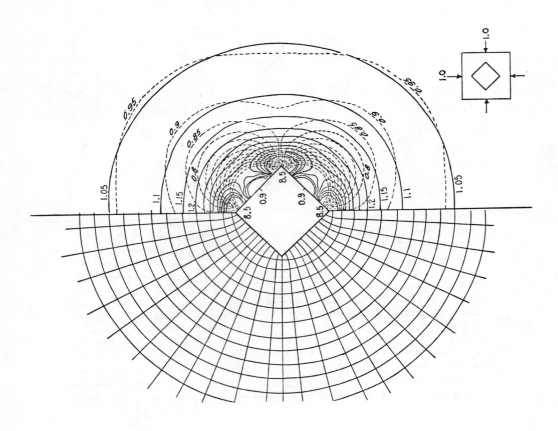
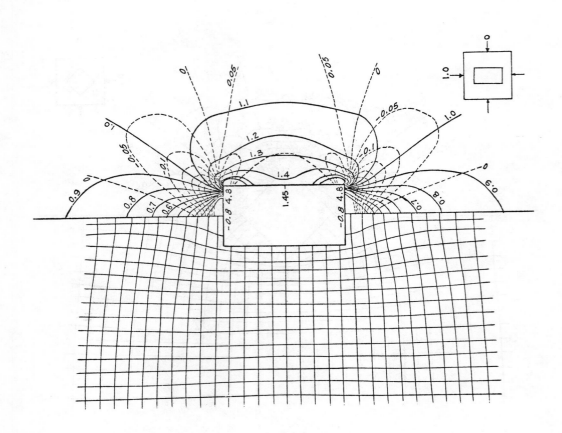

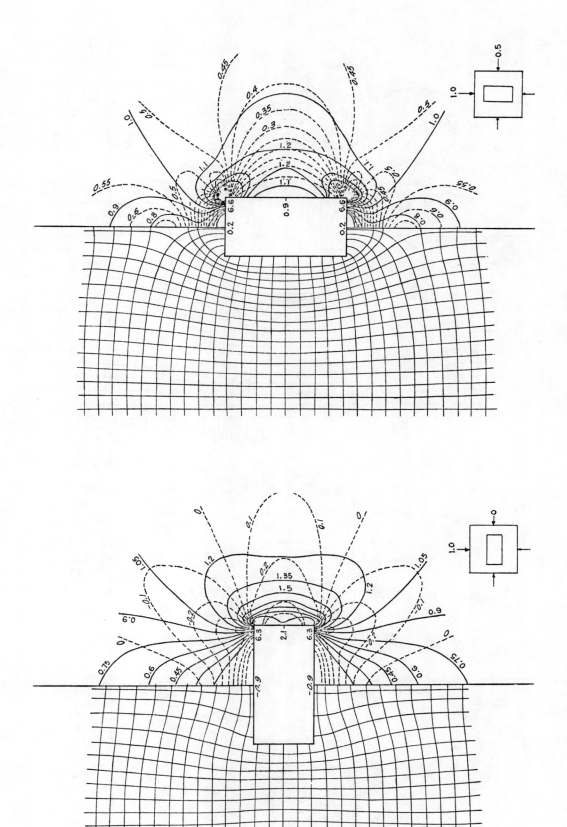

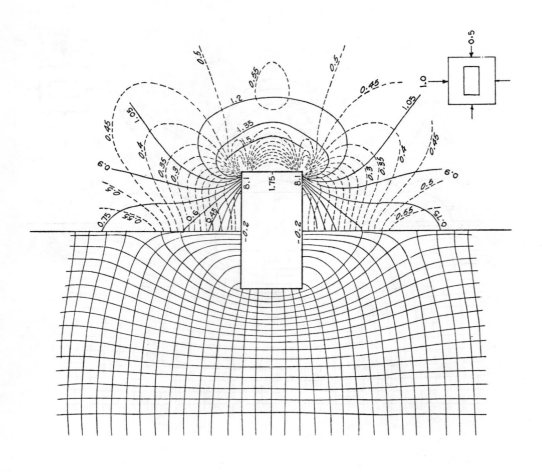
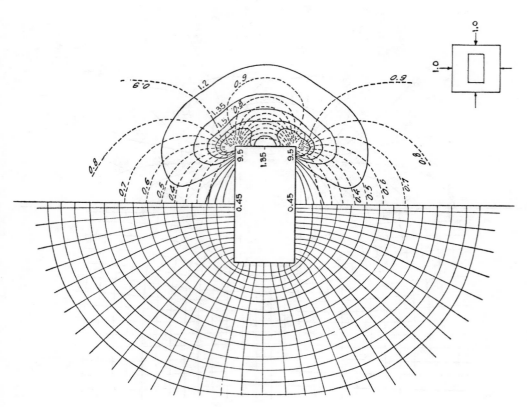

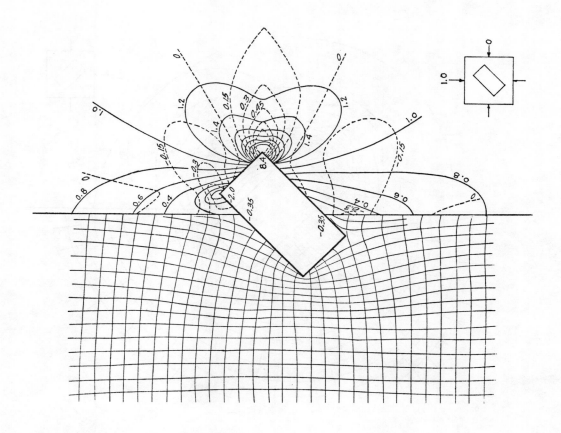
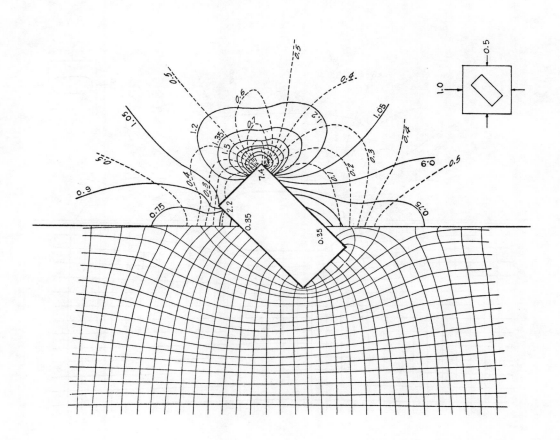

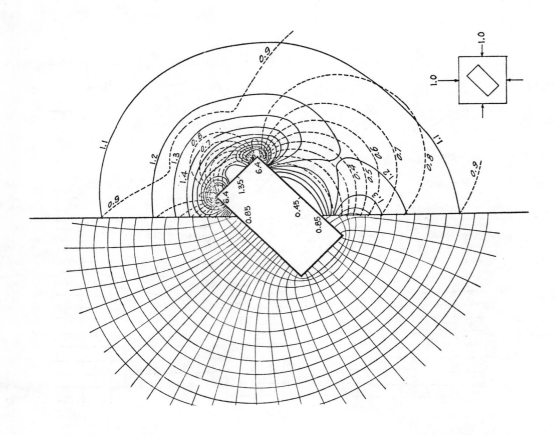
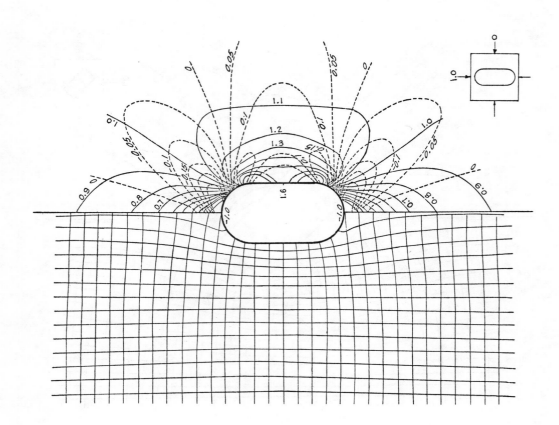

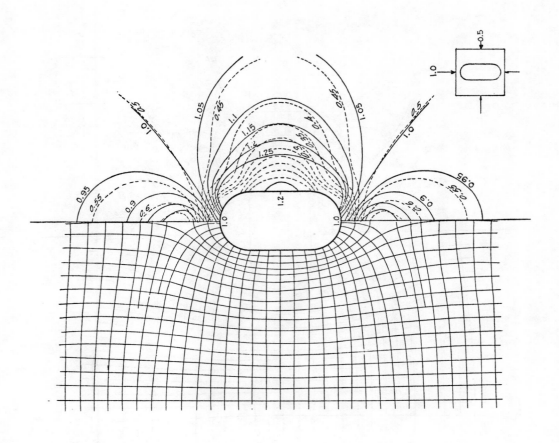
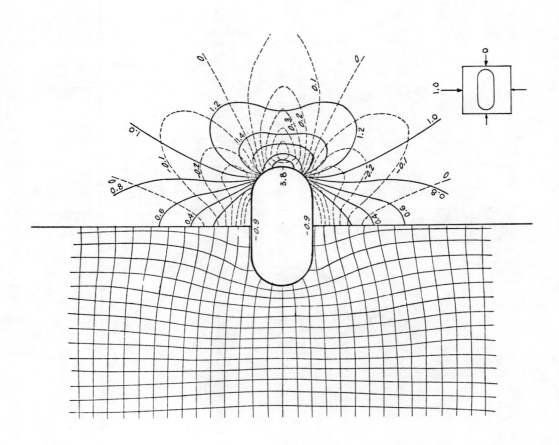

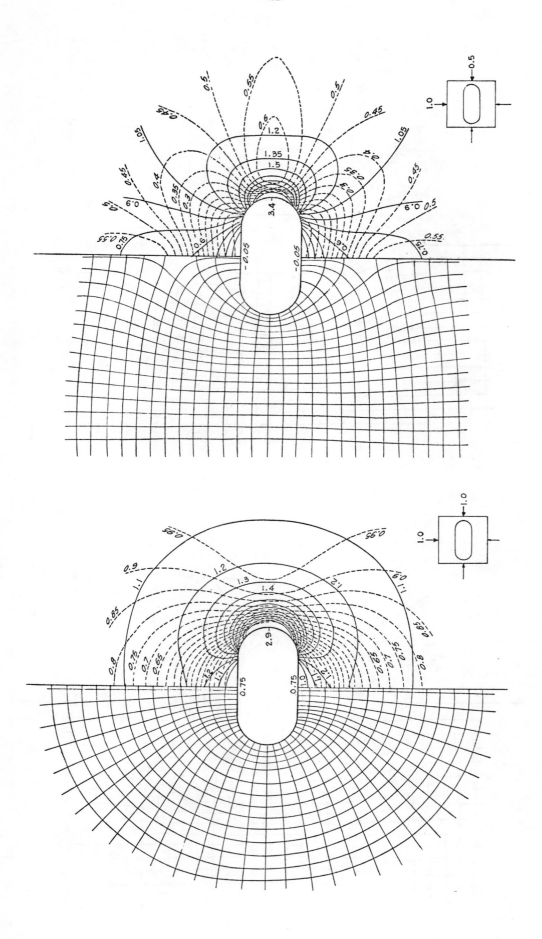

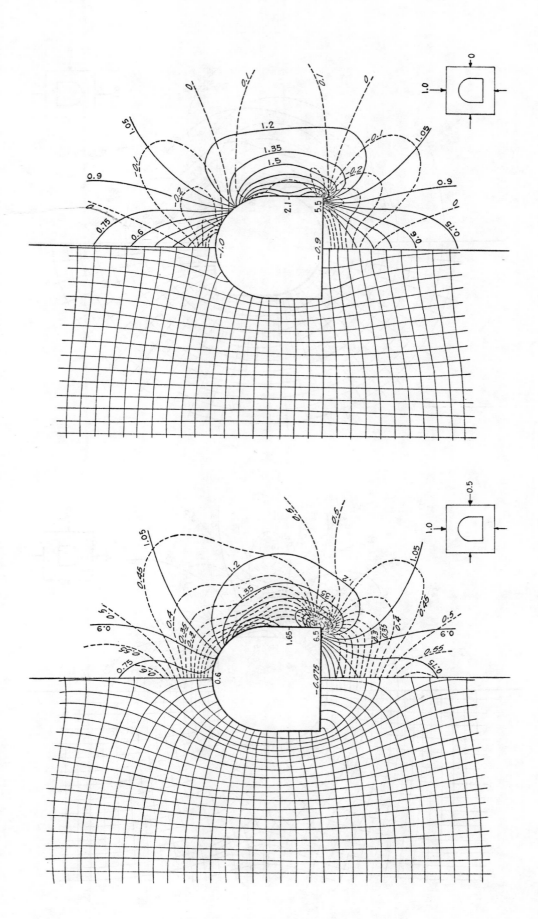

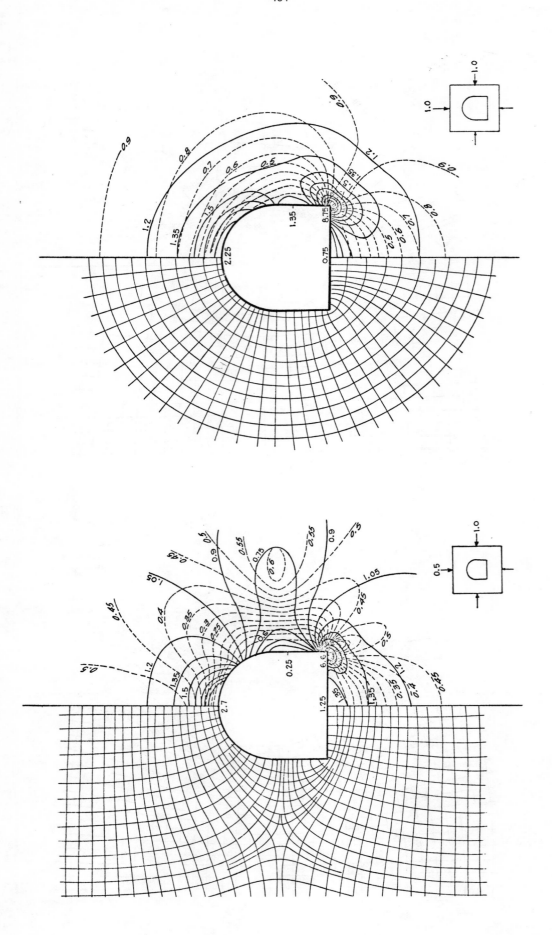

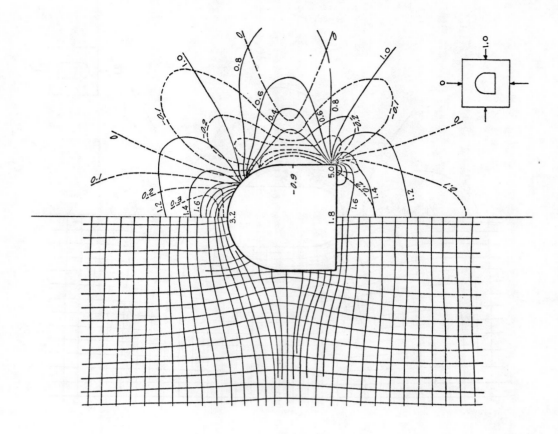
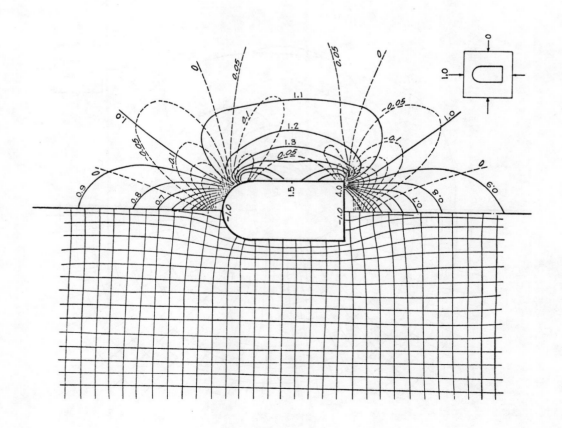

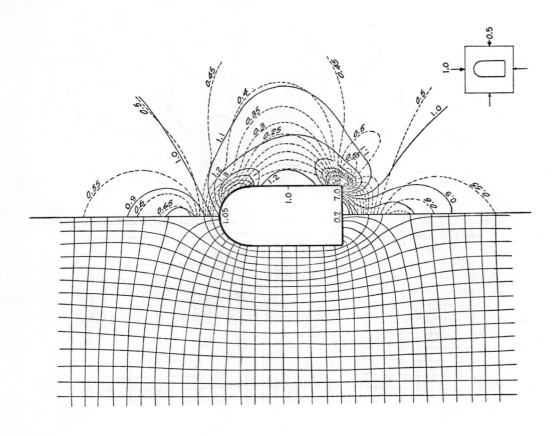

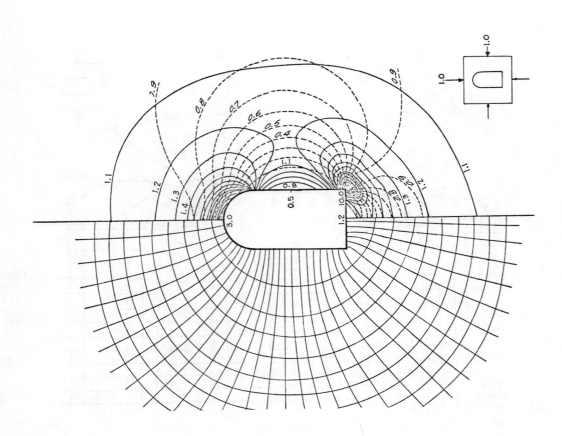

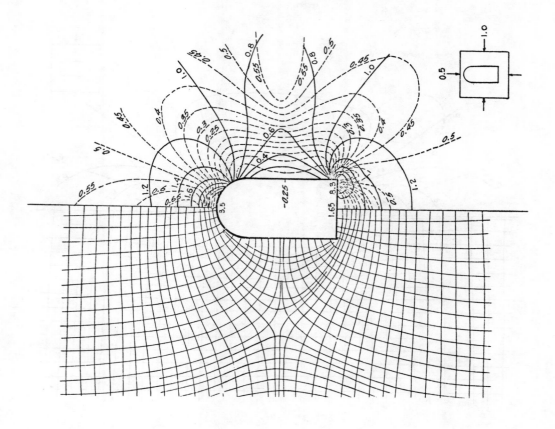
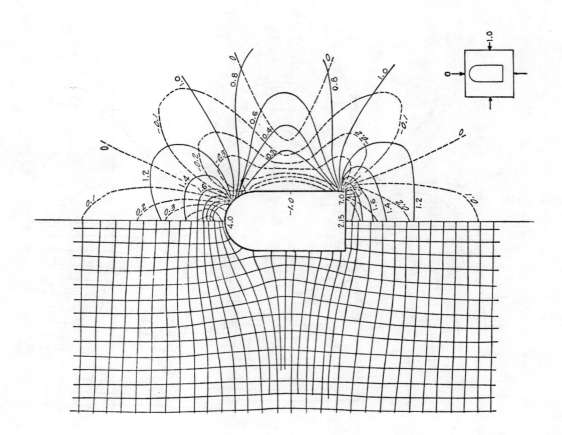

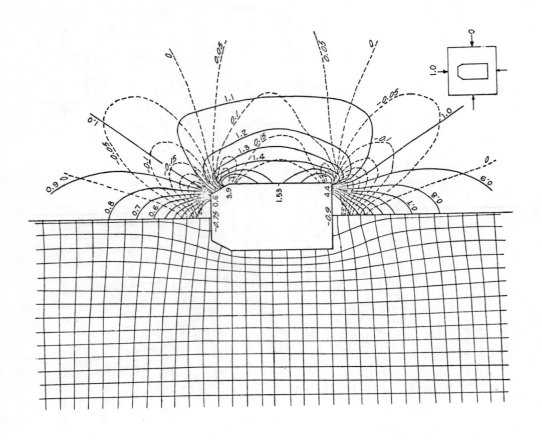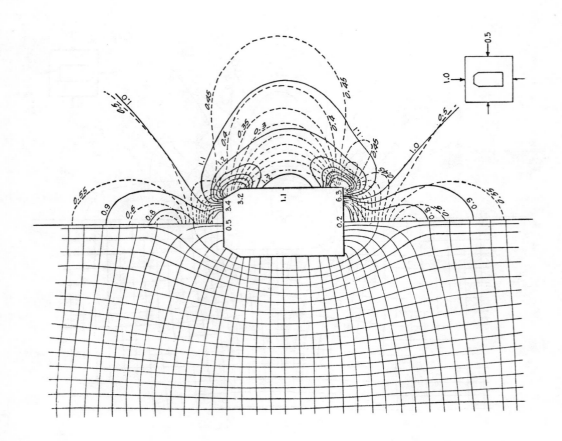

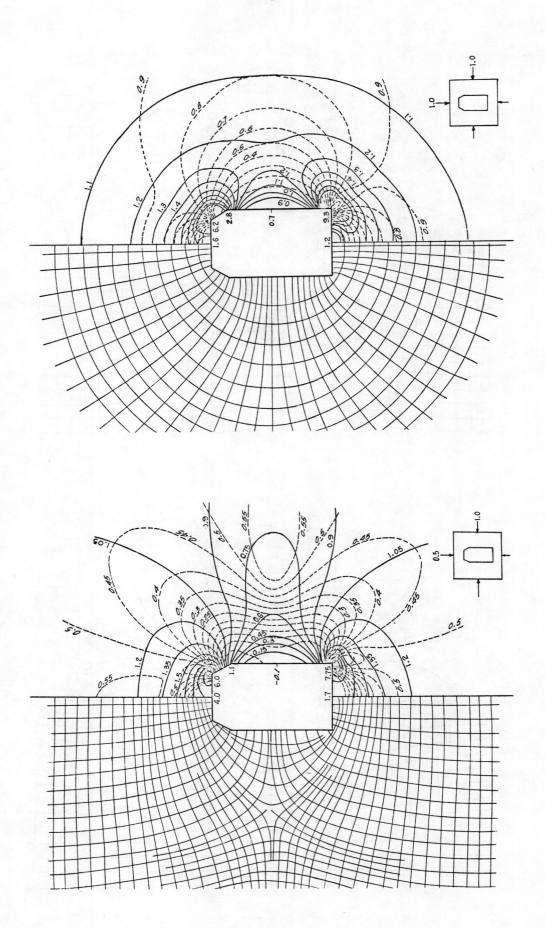

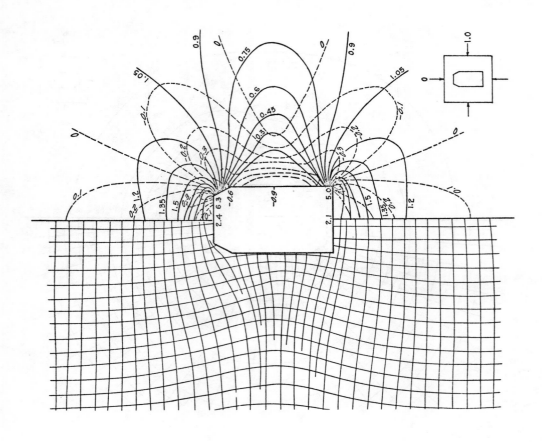
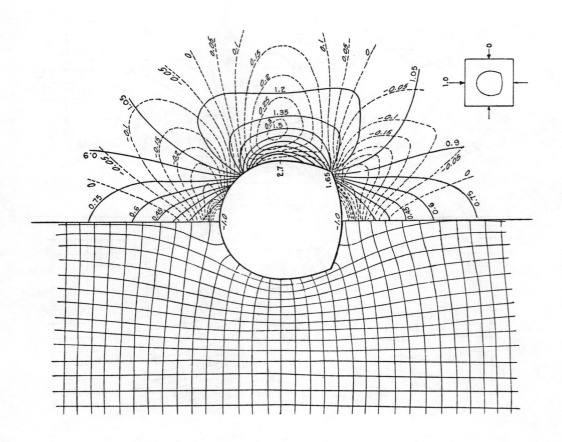

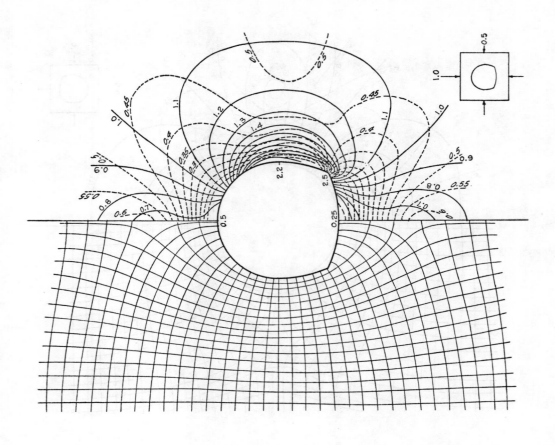
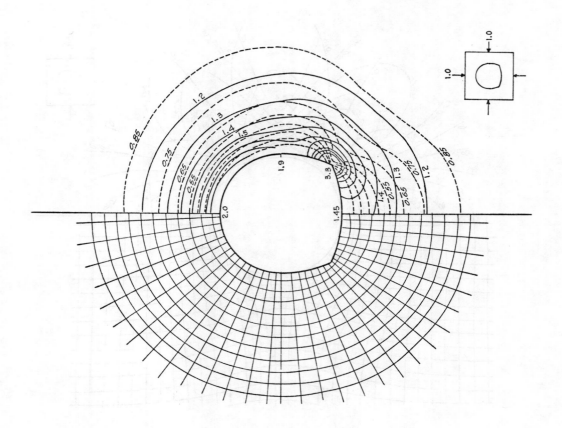

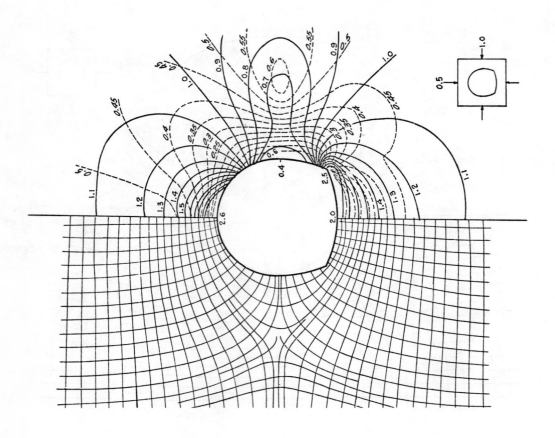
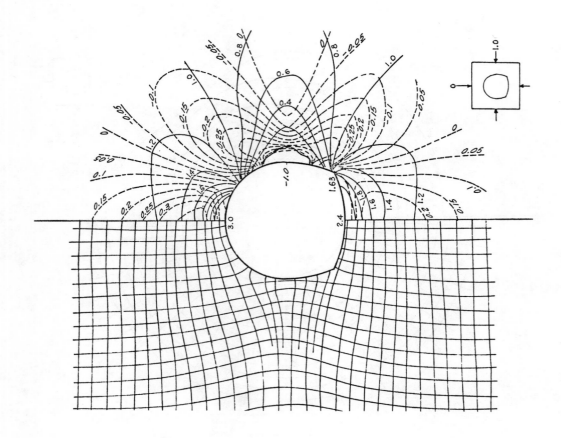

Appendix 4: Two-dimensional boundary element stress analysis

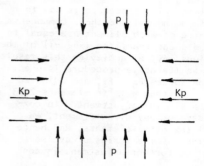

The problem to be solved

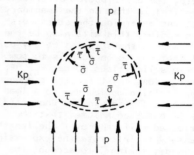

Tractions on potential boundary before excavation of hole in an infinite plate

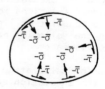

Negative tractions representing effects of excavation

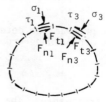

Fictitious forces and stresses on elements of imaginary surface on infinite plate

Introduction

The two-dimensional boundary element program presented in this appendix was prepared by Drs J.W.Bray, G.Hocking and E.S.A.Eissa at Imperial College, London and by Dr R.D.Hammett of Golder Associates, Vancouver. It can be used to calculate stresses and displacements for the following conditions :

a. the material is homogeneous, isotropic and linearly elastic;

b. the conditions are those of plane strain as defined on page 91;

c. the medium is infinite, or closed by a finite external boundary of arbitrary shape;

d. the medium may contain a number of holes of arbitrary shape;

e. the loading may consist of any combination of uniform field stresses or of uniformly distributed loads on the boundaries. Gravitational loading is simulated by increasing the field stresses with depth.

In most of the literature on boundary element or boundary integral methods, the principles of the method, which in reality are quite simple, are often obscured by the formal mathematical presentation. Bray has attempted to overcome this problem with the following simplified explanation of his indirect formulation of the uniform field stress case.

The problem is to determine the stresses around a long excavation of specified cross-section, given the field stresses p (vertical) and Kp (horizontal) as shown in the margin sketch. Prior to excavation, the rock that is to be removed provides support for the surrounding rock. This support may be represented in terms of normal and tangential tractions ($\bar{\sigma}$, $\bar{\tau}$) on the potential boundary of the excavation. The magnitude of these tractions will vary from point to point depending upon the orientation of the various parts of the potential boundary. When the hole is excavated these tractions are reduced to zero, and thus the excavation can be regarded as being equivalent to applying a system of negative tractions to the hole. The resultant state of stress can then be considered as the superposition to two stress systems - a) the original uniform stress state and b) the stresses induced by negative surface tractions ($-\bar{\sigma}$, $-\bar{\tau}$). We will now seek to determine the distribution of induced stresses corresponding to the negative surface tractions.

The "real" situation shown in the third margin sketch is now compared with an imaginary situation, shown in the lower margin sketch, where we have another infinite plate, only in this case the plate is unperforated ; there is no hole. Instead, we imagine a line to be inscribed on the face of the plate corresponding to the boundary of the hole in the first plate. The line is divided into a series of elemental lengths and the elements are numbered consecutively, 1, 2, 3, etc. We now imagine that each of the elements is subjected to an external force, whose line of action lies in the plane of the plate. The force is resolved into components F_n and F_t, normal and tangential to the element, and these are taken to be uniformly distributed over the length of the element. The forces are referred to as fictitious forces, since they in no way correspond to the forces applied to the boundary of the real plate. The procedure is now to adjust these forces so that the normal and shear components of stress ($\bar{\sigma}$, $\bar{\tau}$) at the centre of each element are equal to the corresponding normal and shear tractions ($-\bar{\sigma}$, $-\bar{\tau}$) of the real plate. There are various ways of achieving this result, but in the present program an iterative procedure is used. Starting with element 1, the forces F_{n1}, F_{t1} are adjusted so that $\sigma_1 = -\bar{\sigma}_1$ and $\tau_1 = -\bar{\tau}_1$. We then pass to elements 2, 3, etc. in turn and carry out a similar adjustment. In correcting the values of σ and τ for any given element, we disturb the stresses on all the other elements, and hence the procedure must be continued for a series of cycles around the "boundary" until no further adjustment is considered necessary.

Once this process is complete, the distribution of tractions on the real boundary is identical to that on the imaginary boundary. Since these tractions determine the stress distribution in the surrounding medium, this will also be the same for the two cases. To compute the stresses at any point in the imaginary plate, all that has to be done is to sum the contributions of the various fictitious forces F_{n1}, F_{t1}, F_{n2}, F_{t2} etc. Standard expressions exist for the components of stress at any point in an infinite medium due to a load applied at some other point (see Timoshenko and Goodier[73] page 113, for example). As it has been assumed that the fictitious forces are distributed uniformly over the length of each element, it is necessary to integrate the expressions for the influence of point loads, taking into account the orientation of stress components.

Once the stresses due to the negative surface tractions have been determined, they may be added to those of the original stress field to give the required stresses following excavation. Elastic displacements around the excavation can be calculated by making use of standard solutions for displacements in an infinite medium due to point or line loads. The listing of the program given on the following pages includes the calculation of elastic displacements.

As a consequence of the method used for calculating displacements, there may be a tendency for the excavation and its surrounds to "float" numerically in space causing small rigid body movements of the problem field to occur. This can happen for unsymmetrical excavation shapes in which there is an imbalance of elements on opposing sides of the excavation boundary. In these cases, all calculated relative displacements will be accurate, but the absolute values of displacements at particular points may be in error. This problem does not arise for symmetrical excavation shapes.

A two-dimensional boundary element program for determining elastic stresses and displacements around underground excavations.

Segments

In representing a boundary of arbitrary shape, the boundary is divided into a number of segments (NSEG). These segments may be of three types: a) straight lines, b) circular arcs, and c) elliptical arcs. In the following specifications, reference is made to the end points of the segment, described as the initial and final points. In deciding which is which,

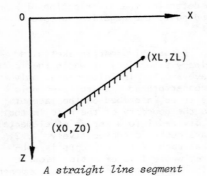

A straight line segment

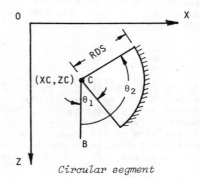
Circular segment

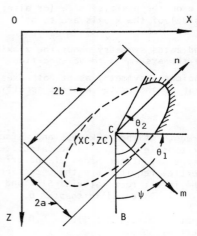
Elliptical segment

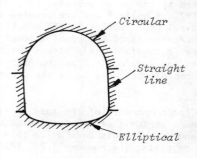

the rule is that when the boundary is traced from the initial to the final point and one faces the direction of travel, the solid material lies on the right hand side.

Straight line segments

    X0, Z0   = co-ordinates of the initial point

    XL, ZL   = co-ordinates of the final point.

Circular segments

    XC, ZC   = co-ordinates of centre of circle

    RDS   = radius of circle

    THET1   = polar angle of initial point

    THET2   = polar angle of final point

Line CB is drawn from the centre C in the direction of the +Z axis.
The polar angles are measured in a counter-clockwise direction from CB.

Elliptical segments

    XC, ZC   = co-ordinates of centre C

    SEMIAX   = length of one semi-axis (a)

    RATIO   = b/a, where b is the length of the other semi-axis

    PSI   = polar angle of axis a

    THET1   = polar angle of initial point

    THET2   = polar angle of final point

These polar angles are measured as described above.

These segments may be combined to form a boundary of virtually any arbitrary shape as illustrated in the lower margin sketch.

    NSEG   = the total number of segments used in defining all the boundary surfaces of the problem.

Elements

Each segment is divided into a number (NELR) of elements. In the case of straight line and circular segments, the elements are all of equal length. To understand the position with elliptical segments, note is made of the fact that an ellipse may be represented in parametric form by the equations :

$$m = a.\cos \chi$$
$$n = b.\sin \chi$$

where m and n represent co-ordinates measured parallel to the a and b axes of the ellipse. Each element corresponds to the same increment of the angle χ. It follows that the length of the element varies, being small where the radius of curvature is small.

It may be necessary to divide a single segment into a number of separate segments, a) to achieve a variation in length along the original segment, and b) to allow for the application of loads to specified portions of the segment which would not be possible with a uniform division of the original segment. In connection with the first reason, it should be pointed out that a relatively high density of elements is desirable in regions of high stress gradient. The elements are numbered progressively from I = 1 to I = IMAX, over all segments.

Symmetry

In order to minimise the storage requirements for doubly subscripted variables, advantage should be taken of any symmetry exhibited by the system under investigation. This can be done only if the symmetry refers to both the geometry of all boundaries and the disposition of all loads. The presence or absence of symmetry is indicated in the program by the values accorded to the codes KXS, KZS. The following rules apply :

   a) KXS = 0, KZS = 0 indicates that there is no symmetry.

   b) KXS = 1, KZS = 0 indicates symmetry about the x-axis, in which case only the boundaries or parts of the boundaries which lie on the positive side (or alternatively negative side) of the x-axis are to be specified.

   c) KXS = 0, KZS = 1 indicates symmetry about the z-axis and again only half the system is to be specified.

   d) KXS = 1, KZS = 1 indicates symmetry about both axes, and only one quadrant of the system must be specified.

Material properties

The program allows the strength of the rock mass and a factor of safety against shear or tensile failure to be calculated at the centre of each element and at the nodes of the grid lines. The material properties required are the unconfined compressive strength of the intact rock material SIGC and the constants m (RM) and s (S) discussed in chapter 6.

If a failure analysis is not required, put SIGC = 0.0 or use a blank material properties card.

Loading

FPX and FPZ denote the principal field stresses, parallel to the x- and z- axes respectively. The program solves two classes of problem defined by the value assigned to ICODE. For ICODE = 1, FPX and FPZ are uniform and the problem is of the infinite medium type. For ICODE = 2, FPX and FPZ vary with depth and the problem is of the gravitational stress field type. In this case, a constant field stress ratio (FPX/FPZ) must be defined and a value assigned to FSR. For ICODE = 2, the origin of coordinates must be at the ground surface.

BPX, (BPZ) denote the x (z) components of the load applied to a given boundary element per unit of area projected onto the x (z) plane. The range of elements over which this load is applied is specified by giving the numbers (LP1, LP2) of the first and last elements of the loaded section.

Cycles

Part of the analytical procedure involves the inversion of a matrix. This is carried out by a simple iteration, and the number of cycles of iteration has to be decided. For most problems, 20 cycles (NCYC = 20) is sufficient, but a greater number may be required where the boundaries are approximately parallel and close together (eg. a thin cylinder).

Grid lines

Internal stresses are determined at the nodes of grid lines consisting of two intersecting families of curves. One set of lines is drawn normal to the boundaries, starting at the centre of every boundary element. The second set is drawn parallel to the boundaries, with a spacing which is constant between any two adjacent lines. DELN is the distance of the first line from the boundary. The outermost grid line which corresponds to the largest value of N, is denoted by NSL.

The experienced user of the program may wish to vary the spacing of the lines drawn parallel to the boundary. For example, the spacing can be increased in an arithmetic progression by making the spacing of the Nth line from the boundary equal to $\frac{1}{2}(N)*(N + 1)*DELN$.

DELN is chosen to provide a sufficient density of lines close to the boundary. Experience in using this program suggests that the value of DELN should be no less than the element length. NSL is chosen with a value which is just large enough to ensure that the area of interest is covered.

Data cards

These are assembled in the following order :

 a) Title

 b) Data card

 c) Material properties card

 d) Field stress card

 e) Segment cards

 f) Boundary load card (if no boundary loads, card can be omitted)

 g) Blank card.

It is desirable that the segment cards relating to any given boundary element be grouped together and placed in the order in which the boundary is traced.

Information on the data cards is arranged as shown on the following page, using the following definitions :

TITLE	= Any title up to a maximum of 80 charatters
ICODE	= Code for problem type (= 1 for infinite medium = 2 for gravitational medium) When ICODE = 1, use card F. When ICODE = 2, use card G.
NSEG	= Total number of segments specified = number of segment cards
SIGC	= Uniaxial compressive strength of intact rock material
RM	= Strength parameter, m
S	= Strength parameter, s

GAMMA	=	Unit weight of rock mass	
FSR	=	Field stress ratio, k = FPX/FPZ	
KXS, KZS	=	Codes of symmetry	
NCYC	=	Number of cycles of iteration	
NSL	=	Number of grid lines parallel to the excavation boundary	
DELN	=	Normal distance between first grid line and excavation boundary	
RNU, E	=	Poisson's ratio and Young's modulus of the rock mass	
FPX, FPZ	=	Principal field stresses	
X0, Z0	=	Co-ordinates of the initial point on a straight line segment	
XL, ZL	=	Co-ordinates of the final point on a straight line segment	
NELR	=	Number of elements required per segment	
XC, ZC	=	Co-ordinates of centre of circle, ellipse	
THET1, THET2	=	Polar angles of initial and final points on circular or elliptical segments	
RDS	=	Radius of circle	
SEMIAX	=	Length of one semi-axis (a) of ellipse	
RATIO	=	b/a = ratio of semi-axes	
PSI	=	Polar angle of semi-axis (a)	
LP1, LP2	=	Initial and final element numbers to which load is applied	
BPX, BPZ	=	Components of load per unit of projected area in x and z directions. Note that BPX and BPZ are positive into the material.	

Card TYPE		1	6	11	16	21	26	31	41	51	61	71
T	TITLE											
D	DATA CARD	ICODE	NSEG	KXS	KZS	NCYC	NSL	DELN	E	RNU		
P	Material properties	SIGC		RM		S						
F	Field stresses Infin. medium	FPX		FPZ								
G	Field stresses Gravitation.	GAMMA		FSR								
S1	Straight Line Seg.	NELR		X0		Z0		XL	ZL			
S2	Circular Segment	NELR		XC		ZC		THET1	THET2	RDS		
S3	Elliptical Segment	NELR		XC		ZC		THET1	THET2	SEMIAX (a)	RATIO (b/a)	PSI (of a)
BL	Boundary Load	LP1		LP2		BPX		BPZ				

All integer specifications are I5 except NELR (I10)

Output

The following information is printed out :

a) Input data

b) Boundary stress distribution, tabulated as follows:

```
    I    CX    CZ    SIG1    SIG3    ALPHA    UX    UZ    F.O.S.    BETA    FAILURE
```

where I = element number

CX,CZ = co-ordinates of centre of element

SIG1, SIG3 = principal stresses at centre of element

ALPHA = angle that σ_1 makes with the normal to the boundary

UX = displacement in the x- direction

UZ = displacement in the z- direction

F.O.S. = factor of safety

BETA = angle made by the failure plane with the σ_1 direction

FAILURE = failure mode (shear or tensile).

If F.O.S., BETA and FAILURE are not required, put SIGC = 0.0 or use a blank material properties card.

Note that examination of SIG1, SIG3 and ALPHA values in b) enables one to determine whether a sufficient number of cycles of iteration have been used. Thus, for an element which carries only normal loading (or no load at all), the value of σ_1 or σ_3 should be equal to the intensity of loading (or zero), and ALPHA should be zero or $90°$. In other cases, it is necessary to determine whether the values of σ_1, σ_3 and ALPHA produce the required intensities of normal and shear loading.

c) Internal stresses, tabulated under the same headings as in b), but now I, CX, CZ and ALPHA have different meanings :

I = normal grid line number

CX, CZ = co-ordinates of an interior point

ALPHA = angle that σ_1 makes with the z- axis.

Example

A cavern with a circular roof is located at an equivalent depth of 300m below surface. The properties of the rock mass are defined by :

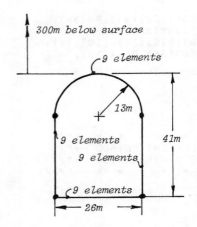

Uniaxial compressive strength of intact rock	σ_c = 150 MPa
Material constants	m = 2.5
	s = 0.004
Unit weight of rock mass	γ = 0.027 MN/m^3
Young' modulus of rock mass	E = 7000 MPa
Poisson's ratio of rock mass	ν = 0.25
Horizontal/vertical in situ stress ratio	k = 2.0

Note that there are 9 input cards, the last being a blank

card. In this case the symmetry that this problem possesses is not taken advantage of and the complete excavation boundary is represented by 36 elements. If the compiler used does not set input data to zero if a blank card is read, then it will be necessary to place zeros on this card.

Card type

T	UNDERGROUND CAVERN - CIRCULAR ROOF - GRAVITATIONAL STRESS FIELD										
D	2	4	0	0	20	3	2.0	7000.0	0.25		
P	150.0		2.5		0.004						
G	0.027		2.0								
S1	9		0.0		341.0		26.0	341.0			
S1	9		26.0		341.0		26.0	313.0			
S2	9		13.0		313.0		90.0	270.0	13.0		
S1	9		0.0		313.0		0.0	341.0			
	BLANK CARD										

The results of this analysis are presented after the program listing given below.

Boundary element program listing

```
C         INPUT = TAPE1, OUTPUT = TAPE7
C
C         ****************************************************************
C         BOUNDARY ELEMENT PROGRAM FOR TWO-DIMENSIONAL STRESS DISPLACEMENT
C         AND FAILURE ANALYSIS IN HOMOGENEOUS ISOTROPIC MEDIA UNDER UNIFORM
C         OR GRAVITATIONAL LOADING.
C         ****************************************************************
C
          DIMENSION CX(36),CZ(36),EX1(36),EX2(36),EZ1(36),EZ2(36),PN(36),
         1 PNM(36),QM(36),QN(36),BMM(36,36),BMN(36,36),BNM(36,36),
         2 BNN(36,36),DM(36,36),DN(36,36),SB(36),CB(36),SIG1(36),SIG3(36),
         3 ALPHA(36),SINB(36),COSB(36),UX(36),UZ(36),FOS(36),BETA(36),
         4 FPX(36),FPZ(36),TITLE(20)
C
C         READING AND PRINTING OF INPUT DATA
C
          WRITE(7,12)
    12    FORMAT("1",///
         110X,"*  *  *  *  *  *  *  *  *  *  *  *  *  *  *  *  *"/
         110X,"*    2-D STRESS ANALYSIS BY THE BOUNDARY ELEMENT METHOD    *"/
         110X,"*      PLANE STRAIN CONDITIONS,  FICTITIOUS STRIP LOADS    *"/
         110X,"*  *  *  *  *  *  *  *  *  *  *  *  *  *  *  *  *"///)
          READ(1,10) (TITLE(I),I=1,20)
    10    FORMAT(20A4)
          WRITE(7,9) (TITLE(I),I=1,20)
     9    FORMAT(10X,20A4)
          READ(1,11) ICODE,NSEG,KXS,KZS,NCYC,NSL,DELN,E,RNU
    11    FORMAT(6(2X,I3),3F10.3)
          WRITE(7,317) ICODE
```

```
  317 FORMAT(1H ///,7X,23HCODE OF PROBLEM TYPE = ,I3)
      IF(ICODE.GE.1.AND.ICODE.LE.2) GO TO 319
      WRITE(7,318)
  318 FORMAT(1H ///,7X,29HCODE OF PROBLEM TYPE IS WRONG)
      GO TO 503
  319 CONTINUE
      READ(1,35) SIGC,RM,S
   35 FORMAT(3F10.5)
      IF(SIGC.EQ.0.0) GO TO 299
      WRITE(7,36) SIGC,RM,S
   36 FORMAT(1H ///,7X,41HROCK PROPERTIES.      SIGC, RM, S = ,
     13F10.5)
  299 WRITE (7,13) NSEG,KXS,KZS,NCYC,NSL,DELN,E,RNU
   13 FORMAT(1H ///,7X,28HNSEG, KXS, KZS, NCYC, NSL = ,5I6 ///,
     1 7X,23HDELN, E MODULUS, RNU = ,3F10.3 /)
      IF(ICODE.EQ.2) GO TO 322
  320 WRITE(7,321)
  321 FORMAT(1H ///,7X,48HCODE OF THE ANALYSIS IN INFINITE ISOTROPIC MED
     1IA)
      READ(1,21) FPXX,FPZZ
   21 FORMAT(2F10.3)
      WRITE(7,22) FPXX,FPZZ
   22 FORMAT(1H ///,7X,11HFPX, FPZ = ,2F10.3 ///)
      GO TO 336
  322 WRITE(7,323)
  323 FORMAT(1H ///,7X,55HCODE OF THE ANALYSIS IN HOMOGENOUS GRAVITATION
     1AL MEDIA  )
      READ(1,324) GAMMA,FSR
  324 FORMAT(2F10.3)
      WRITE(7,325) GAMMA,FSR
  325 FORMAT(1H ///,7X,25HGAMMA, HORZ. STR RATIO = ,2F10.3)
  336 CONTINUE
      NN = 0
      PI = ATAN(1.0) * 4.0
      TA = 2.0*(1.0 - RNU)
      TJ = 1.0/(2.0*PI*TA)
      TU=3.0-4.0*RNU
      G=E/(2.0*(1.0+RNU))
      TV=0.5*TJ/G
C
C     INTERPRETATION OF SYMMETRY CODE
C
      KAS = 0
      IF(KZS.EQ.-1) KAS=1
      KXT = 2*KXS + 1
      KZT = 2*(KZS + KAS) + 1
C
C     DIVISISON OF BOUNDARY INTO SEGMENTS
C
      I = 0
      NSEGG = 0
      WRITE(7,16)
  16  FORMAT(///
     *  "     ELEMENTS CENT X     CENT Z     THET1      THET2",
     *  "     RADIUS      RATIO      PSI"/
     *  "                 (FIRST-X) (FIRST-Z)  (LAST-X)  (LAST-Z)"/)
C
  700 IF(NSEGG.EQ.NSEG) GO TO 50
      NSEGG = NSEGG + 1
      NELG = 0
      RELG = NELG
      READ(1,17) NELR,XO,ZO,XL,ZL,RDS,RATIO,PSI
   17 FORMAT(7X,I3,7F10.0)
      RELR = NELR
      IF(RDS.EQ.0.0) GO TO 800
C
C     DIVISION OF ELLIPTICAL OR CIRCULAR SEGMENTS INTO ELEMENTS
C
      IF(RATIO.EQ.0.0) RATIO = 1.0
      WRITE(7,18) NELR,XO,ZO,XL,ZL,RDS,RATIO,PSI
```

```
   18 FORMAT(1H ,6X,I3,7F10.3)
      SINPSI = SIN(PSI*PI/180.)
      COSPSI = COS(PSI*PI/180.)
      GD = .1E-10
      GA = RATIO*COS((XL-PSI)*PI/180.)
      IF(ABS(GA).LT.GD) GA=GD
      GB=RATIO*COS((ZL-PSI)*PI/180.)
      IF(ABS(GB).LT.GD) GB=GD
      CHI1=ATAN2(SIN((XL-PSI)*PI/180.),GA)
      CHI2 = ATAN2(SIN((ZL-PSI)*PI/180.),GB)
      DCHI = (CHI2 - CHI1)/RELR
      IF(ABS(DCHI).LT.GD) GO TO 606
      GC = DCHI/ABS(DCHI)
      GO TO 605
  606 GC=-1.0
  605 DCHI = DCHI + ((ZL-XL)/ABS(ZL-XL)-GC)*PI/RELR
  600 I = I + 1
      CHI = CHI1 + RELG*DCHI
      EX1(I) = RDS*(COS(CHI)*SINPSI+SIN(CHI)*COSPSI*RATIO)+XO

      EZ1(I) = RDS*(COS(CHI)*COSPSI - SIN(CHI)*SINPSI*RATIO) + ZO
      CHI = CHI + DCHI
      EX2(I) = RDS*(COS(CHI)*SINPSI + SIN(CHI)*COSPSI*RATIO) + XO
      EZ2(I) = RDS*(COS(CHI)*COSPSI - SIN(CHI)*SINPSI*RATIO) + ZO
      CX(I) = 0.5*(EX1(I) + EX2(I))
      CZ(I) = 0.5*(EZ1(I) + EZ2(I))
      DX = EX2(I) - EX1(I)
      DZ = EZ2(I) - EZ1(I)
      IF(ABS(DX).LT.(.1E-13)*RDS) DX = 0.0
      IF(ABS(DZ).LT.(.1E-13)*RDS) DZ = 0.0
      DS = SQRT(DX*DX + DZ*DZ)
      SINB(I) = -DZ/DS
      COSB(I) = DX/DS
      NELG = NELG + 1
      RELG = NELG
      IF(NELG.LT.NELR) GO TO 600
      GO TO 700
C
C     DIVISION OF STRAIGHT LINE SEGMENTS INTO ELEMENTS
C
  800 CONTINUE
      WRITE(7,15) NELR,XO,ZO,XL,ZL
   15 FORMAT(1H ,6X,I3,4F10.3)
      DX = (XL-XO)/RELR
      DZ = (ZL-ZO)/RELR
      DS = SQRT(DX*DX+DZ*DZ)
  900 I = I + 1
      SINB(I) = -DZ/DS
      COSB(I) = DX/DS
      EX1(I) = XO + RELG*DX
      EZ1(I) = ZO + RELG*DZ
      CX(I) = EX1(I) + 0.5*DX
      CZ(I) = EZ1(I) + 0.5*DZ
      EX2(I) = EX1(I) + DX
      EZ2(I) = EZ1(I) + DZ
      NELG = NELG + 1
      RELG = NELG
      IF(NELG.LT.NELR) GO TO 900
      GO TO 700
   50 MAXI = I
      MAXJ = I
C
C     DETERMINATION OF BOUNDARY TRACTIONS EQUIVALENT TO FIELD STRESSES
C
      DO 100 I = 1,MAXI
      IF(ICODE.EQ.1) GO TO 328
      FPZ(I) = GAMMA*CZ(I)
      FPX(I) = FSR*FPZ(I)
      GO TO 329
  328 FPZ(I) = FPZZ
      FPX(I) = FPXX
  329 PN(I) = 2.0*(FPZ(I)*(COSB(I))**2 + FPX(I)*(SINB(I))**2)
      PNM(I) = 2.0*(FPX(I) -FPZ(I))*SINB(I)*COSB(I)
```

```
              QM(I) = PNM(I)
              QN(I) = PN(I)
      100 CONTINUE
C
C         ADDITION OF BOUNDARY TRACTIONS DUE TO BOUNDARY LOADS
C
      106 READ(1,19) LP1,LP2,BPX,BPZ
       19 FORMAT(5X,I5,5X,I5,2F10.0)
          IF(LP1.EQ.0) GO TO 105
          WRITE(7,20) LP1,LP2,BPX,BPZ
   20     FORMAT(///6X,"LP1 =",I6/6X,"LP2 =",I6/6X,
        1               "BPX =",F10.3/6X,"BPZ =",F10.3///)
          DO 107 I = LP1,LP2
          PN(I) = PN(I)-2.0*((BPX-BPZ)*(SINB(I))**2+BPZ)
          PNM(I) = PNM(I) - 2.*(BPX-BPZ)*SINB(I)*COSB(I)
      107 CONTINUE
          GO TO 106
C
C         DETERMINATION OF COEFFICIENTS IN EXPRESSIONS FOR STRESSES
C         INDUCED BY FICTITIOUS LOADS
C
      105 DO 101 I = 1,MAXI
          CXI = CX(I)
          CZI = CZ(I)
          IF(NN.GT.0) GO TO 104
          COS2BI = 2.0 * (COSB(I))**2 - 1.0
          SIN2BI = 2.0 * SINB(I) * COSB(I)
      104 DO 118 J = 1,MAXJ
          TK = 0.0
          TL = 0.0
          TM = 0.0
          TN = 0.0
          TO = 0.0
          TP = 0.0
          DO 102  KXU = 1,KXT,2
          KX = 2- KXU
          RX = KX
          DO 102 KZU=1,KZT,2
          KZ = (2-KZU)*(1-KAS)+KAS*KX
          RZ = KZ
          COSBJ = RX*COSB(J)
          SINBJ = RZ*SINB(J)
          EX1J = RZ*EX1(J)
          EX2J = RZ*EX2(J)
          EZ1J = RX*EZ1(J)
          EZ2J = RX*EZ2(J)
          LL = KX+KZ-2+10*(I-J)+1000*NN
          IF(LL.EQ.0) GO TO 135
          RN = (CZI - EZ1J)*COSBJ + (CXI - EX1J)*SINBJ
          RM1 = (CXI - EX1J)*COSBJ - (CZI - EZ1J)*SINBJ
          RM2 = (CXI -EX2J)*COSBJ - (CZI - EZ2J)*SINBJ
          RSQ1 = RM1*RM1 + RN*RN
          RSQ2 = RM2*RM2 + RN*RN
          TB = 2.0*ATAN2((RM1 - RM2)*RN,(RN*RN + RM1*RM2))
          TC = 2.0 *RN * (RM1/RSQ1 - RM2/RSQ2)
          TD = (RN**2 - RM1**2)/RSQ1 - (RN**2 - RM2**2)/RSQ2
          TE = ALOG(RSQ1/RSQ2)
          COS2F = 2.0*(COS2BI*(COSBJ**2 - 0.5) + SIN2BI*SINBJ*COSBJ)
          SIN2F = 2.0*(SIN2BI*(COSBJ**2 - 0.5) - COS2BI*SINBJ*COSBJ)
          GO TO 137
      135 TB = 2.0*PI
          TC = 0.0
          TD = 0.0
          TE = 0.0
          COS2F = 1.0
          SIN2F = 0.0
      137 CONTINUE
```

```
              TK = TK + TB*RX*RZ
              TL = TL + TE
              TM = TM + (TD + TA*TE)*COS2F + (TC - TA*TB)*SIN2F
              TN = TN + (((1.0-TA)*TB-TC)*COS2F + (TD+(1.0-TA)*TE)*SIN2F)*RX*RZ
              TO = TO+(TD+TA*TE)*SIN2F-(TC-TA*TB)*COS2F
              TP = TP + (((1.0-TA)*TB-TC)*SIN2F - (TD+(1.0-TA)*TE)*COS2F)*RX*RZ
  102     CONTINUE
          BMM(I,J) = (TL + TM) * TJ
          BMN(I,J) = (TK + TN) * TJ
          BNM(I,J) = (TL - TM) *TJ
          BNN(I,J) = (TK -TN) * TJ
          DM(I,J) = TO *TJ
          DN(I,J) = TP *TJ
  118     CONTINUE
  101   CONTINUE
        IF(NN.GT.0) GO TO 404
        M = 0
C
C       DETERMINATION OF FICTITIOUS LOADS
C
  400   DO 401 I = 1,MAXI
        QMI = PNM(I)
        QNI = PN(I)
        DO 402 J = 1,MAXJ
        IF(I.EQ.J) GO TO 402
        QMI = QMI - DM(I,J)*QM(J) - DN(I,J)*QN(J)
        QNI = QNI - BNM(I,J)*QM(J) -BNN(I,J)*QN(J)
  402   CONTINUE
        DENOM = BNN(I,I)*DM(I,I) - DN(I,I)*BNM(I,I)
        QM(I) = (QMI*BNN(I,I) - QNI*DN(I,I))/DENOM
        QN(I) = (QNI*DM(I,I) - QMI*BNM(I,I))/DENOM
  401   CONTINUE
        M = M + 1
        IF(M.LT.NCYC) GO TO 400
C
C       DETERMINATION OF STRESS COMPONENTS, PRINCIPAL STRESSES
C       AND DIRECTIONS
C
  404   DO 500  I =1, MAXI
        IF(ICODE.EQ.1) GO TO 403
        FPZ(I) = GAMMA*CZ(I)
        FPX(I) = FSR*FPZ(I)
  403   SMI = 2.0*FPX(I)
        SNI = 2.0*FPZ(I)
        SNMI = 0.0
        IF(NN.GT.0) GO TO 405
        SMI = 2.0*((FPX(I) - FPZ(I))*(COSB(I))**2 + FPZ(I))
        SNI = 2.0*((FPX(I) - FPZ(I))*(SINB(I))**2 + FPZ(I))
        SNMI = 2.0*(FPX(I) - FPZ(I))*SINB(I)*COSB(I)
  405   DO 501 J = 1,MAXJ
        SNI = SNI - BNM(I,J)*QM(J) - BNN(I,J)*QN(J)
        SMI = SMI-BMM(I,J)*QM(J)-BMN(I,J)*QN(J)
        SNMI = SNMI - DM(I,J)*QM(J) - DN(I,J) *QN(J)
  501   CONTINUE
        SDI = 0.5*(SMI-SNI)
        TAUMAX = 0.5*SQRT(SDI**2 + SNMI**2)
        SIG1(I) = 0.25*(SMI+SNI) + TAUMAX
        SIG3(I) = 0.25*(SMI+SNI) - TAUMAX
        TR = 2.0*TAUMAX + SNMI - SDI
        IF(TR.EQ.0.0) TR = 0.00001
        ALPHA (I) = (180./PI)*ATAN(1.0 + 2.*SDI/TR)
  500   CONTINUE
C
C       DETERMINATION OF INDUCED DISPLACEMENTS
C
        DO 601 I=1,MAXI
        CXI=CX(I)
        CZI=CZ(I)
        UXX=0.0
        UZZ=0.0
        DO 602 J=1,MAXJ
        DO 603 KXU=1,KXT,2
```

```
      KX=2-KXU
      RX = KX
      DO 603 KZU=1,KZT,2
      KZ=(2-KZU)*(1-KAS)+KAS*KX
      RZ = KZ
      COSBJ = RX*COSB(J)
      SINBJ = RZ*SINB(J)
      EX1J = RZ*EX1(J)
      EX2J = RZ*EX2(J)
      EZ1J = RX*EZ1(J)
      EZ2J = RX*EZ2(J)
      RN = (CZI-EZ1J)*COSBJ+(CXI-EX1J)*SINBJ
      LL=KX+KZ-2+10*(I-J)+1000*NN
      IF(LL.EQ.0) RN = 0.0001
      RM1 = (CXI - EX1J)*COSBJ - (CZI - EZ1J)*SINBJ
      RM2 = (CXI - EX2J)*COSBJ - (CZI - EZ2J)*SINBJ
      RSQ1 = SQRT(RM1*RM1 + RN*RN)
      RSQ2 = SQRT(RM2*RM2 + RN*RN)
      RNO = -EZ1J*COSBJ - EX1J*SINBJ
      IF(ABS(RNO).LT.0.0001) RNO = 0.0001
      RM10 = EX1J*COSBJ - EZ1J*SINBJ
      RM20 = EX2J*COSBJ - EZ2J*SINBJ
      RSQ10 = SQRT(RM10*RM10 + RNO*RNO)
      RSQ20 = SQRT(RM20*RM20 + RNO*RNO)
      IF(RSQ10.LT.0.001) RSQ10 = 0.001
      IF(RSQ20.LT.0.001) RSQ20 = 0.001
      IF(ABS(RNO).LT.0.0001.AND.RSQ20.LT.0.001) RSQ20 = 0.001
      TS = -RN*ALOG(RSQ2/RSQ1)
      TR = -RN*ATAN2((RM2-RM1)*RN, (RN*RN+RM1*RM2))
      TT = TU *(-RM2*ALOG(RSQ2)
    1   +RM1*ALOG(RSQ1)+RM2-RM1+TR)
      UM = TV*(TS*QN(J)*RX*RZ - (TT+TR)*QM(J))
      UN = TV*(TS*QM(J) - (TT-RM2+RM1-TR)*RX*RZ*QN(J))
      UXX = UXX-UM*COSBJ-UN*SINBJ
      UZZ = UZZ + UM*SINBJ - UN*COSBJ
  603 CONTINUE
  602 CONTINUE
      UX(I) = UXX
      UZ(I) = UZZ
  601 CONTINUE
      WRITE(7,119) NN
  119 FORMAT("1",5X,"NN =",I3)
      IF(SIGC.EQ.0.0) GO TO 420
      IF(NN.GT.0) GO TO 611
      WRITE(7,612)
  612 FORMAT(1H ///,4X,82HSTRESSES AND DISPLACEMENTS , AND FAILURE CRITE
     1RION AT CENTERS OF BOUNDARY ELEMENTS)
      GO TO 613
  611 WRITE(7,614)
  614 FORMAT(1H ///,4X,68HSTRESSES AND DISPLACEMENTS, AND FAILURE CRITER
     1ION AT INTERIOR POINTS)
  613 WRITE(7,25)
   25 FORMAT(1H ///,3X,2H I,6X,2HCX,9X,2HCZ,6X,4HSIG1,6X,4HSIG3,5X,
     1 5HALPHA,6X,3H UX,7X,3H UZ,7X,5HF.O.S,6X,4HBETA)
C
C     FAILURE CRITERION AND FACTOR OF SAFETY DETERMINATIONS
C
      DO 620 I=1,MAXI
      IF(ABS(SIG1(I)).LT.0.0001) SIG1(I) = 0.0001
      SIGT = 0.5*SIGC*(RM-SQRT(RM*RM+4.0*S))
      IF(SIG3(I).LE.SIGT) GO TO 622
C
C     CHECK FOR SHEAR FAILURE
C
      TSC = SQRT(RM*SIGC*SIG3(I)+S*SIGC*SIGC)
      FOS(I) = (SIG3(I) + TSC)/SIG1(I)
      IF(FOS(I).GT.1.0) GO TO 621
      TAUM = 0.5*(SIG1(I) - SIG3(I))
      TCM = 0.25*SIGC*RM
      TCM1 = SQRT(TAUM*TAUM + TAUM*TCM)
      TCM2 = 0.5*TCM
      BETA(I) = 0.5*(180.0/PI)*ATAN(TCM1/TCM2)
```

```
          WRITE(7,616) (I,CX(I),CZ(I),SIG1(I),SIG3(I),ALPHA(I),
     1  UX(I),UZ(I),FOS(I),BETA(I))
  616 FORMAT(1H ,I4,2F10.3,2G10.3,F10.3,2G10.3,2F10.3,4X,
     1       13HSHEAR FAILURE)
      GO TO 620
C
C     CHECK FOR TENSILE FAILURE
C
  622 TCT = RM - SQRT(RM*RM + 4.0*S)
      FOS(I) = SIGC*TCT/(2.0*SIG3(I))
      IF(FOS(I).GT.1.0) GO TO 621
      BETA(I) = 0.0
      WRITE(7,617) (I,CX(I),CZ(I),SIG1(I),SIG3(I),ALPHA(I),
     1  UX(I),UZ(I),FOS(I),BETA(I))
  617 FORMAT(1H ,I4,2F10.3,2G10.3,F10.3,2G10.3,2F10.3,4X,
     1       15HTENSILE FAILURE)
      GO TO 620
  621 WRITE(7,625)(I,CX(I),CZ(I),SIG1(I),SIG3(I),ALPHA(I),UX(I),UZ(I),
     1  FOS(I))
  625 FORMAT(1H ,I4,2F10.3,2G10.3,F10.3,2G10.3,F10.3,7X,3HN.A,6X,
     1       11HNO  FAILURE)
  620 CONTINUE
      GO TO 450
  420 IF(NN.GT.0) GO TO 422
      WRITE(7,421)
  421 FORMAT(1H ///,4X,58HSTRESSES AND DISPLACEMENTS AT CENTERS OF BOUND
     1ARY ELEMENTS)
      GO TO 425
  422 WRITE(7,423)
  423 FORMAT(1H ///,4X,45HSTRESSES AND DISPLACEMENTS AT INTERIOR POINTS)
  425 WRITE(7,430)
  430 FORMAT(1H ///,3X,2H I,6X,2HCX,9X,2HCZ,6X,4HSIG1,6X,4HSIG3,5X,
     1  5HALPHA,6X,3H UX,7X,3H UZ)
      WRITE(7,440)(I,CX(I),CZ(I),SIG1(I),SIG3(I),ALPHA(I),UX(I),
     1        UZ(I),I=1,MAXI)
  440 FORMAT(1H ,I4,2F10.3,2G10.3,F10.3,2G10.3)
  450 CONTINUE
      IF(NN.EQ.NSL) GO TO 503
C
C     GENERATION OF INTERIOR POINTS, STRESSES AND DISPLACEMENTS
C     DETERMINED BY LOOPING TO 105
C
      NN = NN + 1
      IF(NN.GT.1) GO TO 504
      COS2BI = 1.0
      SIN2BI = 0.0
      MAXI = MAXJ
      DO 502 I = 1,MAXI
      K = I
      CX(I) = CX(K)
      CZ(I) = CZ(K)
      SB(I) = SINB(K)
      CB(I) = COSB(K)
  502 CONTINUE
  504 TQ = DELN
      DO 505 I = 1,MAXI
      CX(I) = CX(I) + SB(I)*TQ
      CZ(I) = CZ(I) + CB(I)*TQ
  505 CONTINUE
      GO TO 105
  503 CONTINUE
      STOP
      END
```

UNDERGROUND CAVERN-CIRCULAR ROOF- GRAVITATIONAL STRESS FIELD

CODE OF PROBLEM TYPE =   2
ROCK PROPERTIES.   SIGC, RM, S = ,   150.00000   2.50000   .00400

NSEG, KXS, KZS, NCYC, NSL =   4   0   0   20   3
DELN, E MODULUS, RNU =   2.000   7000.000   .250
CODE OF THE ANALYSIS IN HOMOGENOUS GRAVITATIONAL MEDIA
GAMMA, HORZ. STR RATIO =   .027   2.000

ELEMENTS	CENT X (FIRST-X)	CENT Z (FIRST-Z)	THET1 (LAST-X)	THET2 (LAST-Z)	RADIUS	RATIO	PSI
9	0.000	341.000	26.000	341.000			
9	26.000	341.000	26.000	313.000			
9	13.000	313.000	90.000	270.000	13.000	1.000	0.000
9	0.000	313.000	0.000	341.000			

NN = 0

STRESSES AND DISPLACEMENTS, AND FAILURE CRITERION AT CENTERS OF BOUNDARY ELEMENTS

I	CX	CZ	SIG1	SIG3	ALPHA	UX	UZ	F.O.S	BETA	
1	1.444	341.000	57.4	-.948E-05	-90.000	.203E-01	-.836E-02	.165	25.835	SHEAR FAILURE
2	4.333	341.000	33.3	.121E-04	-90.000	.124E-01	-.169E-01	.285	21.218	SHEAR FAILURE
3	7.222	341.000	28.5	.161E-04	-90.000	.759E-02	-.213E-01	.333	19.965	SHEAR FAILURE
4	10.111	341.000	26.7	.186E-04	-90.000	.363E-02	-.235E-01	.355	19.453	SHEAR FAILURE
5	13.000	341.000	26.2	.218E-04	-90.000	.295E-06	-.243E-01	.362	19.306	SHEAR FAILURE
6	15.889	341.000	26.7	.267E-04	-90.000	-.363E-02	-.235E-01	.355	19.453	SHEAR FAILURE
7	18.778	341.000	28.5	.353E-04	-90.000	-.759E-02	-.213E-01	.333	19.965	SHEAR FAILURE
8	21.667	341.000	33.3	.525E-04	-90.000	-.124E-01	-.169E-01	.285	21.218	SHEAR FAILURE
9	24.556	341.000	57.4	.896E-04	-90.000	-.203E-01	-.836E-02	.165	25.835	SHEAR FAILURE
10	26.000	339.444	27.2	.380E-04	-90.000	-.518E-01	.926E-02	.348	19.604	SHEAR FAILURE
11	26.000	336.333	4.86	-.357E-04	-90.000	-.715E-01	.108E-01	1.954	N.A	NO FAILURE
12	26.000	333.222	.972	-.344E-04	-90.000	-.831E-01	.970E-02	9.762	N.A	NO FAILURE
13	26.000	330.111	.100E-03	.498	-.001	-.906E-01	.801E-02	.481	0.000	TENSILE FAILURE
14	26.000	327.000	.100E-03	-1.08	-.001	-.951E-01	.611E-02	.222	0.000	TENSILE FAILURE
15	26.000	323.889	.100E-03	-1.15	-.001	-.972E-01	.418E-02	.208	0.000	TENSILE FAILURE
16	26.000	320.778	.100E-03	-.794	-.001	-.971E-01	.237E-02	.302	0.000	TENSILE FAILURE
17	26.000	317.667	-.766E-01	-.268E-04	-89.988	-.950E-01	.844E-03	123.826	N.A	NO FAILURE
18	26.000	314.556	1.49	-.244E-04	-89.999	-.907E-01	-.153E-03	6.367	N.A	NO FAILURE
19	25.608	310.777	6.72	-.257E-04	-90.000	-.823E-01	-.727E-03	1.412	17.343	SHEAR FAILURE
20	24.087	306.599	20.3	-.262E-04	90.000	-.677E-01	-.174E-03	.468	21.832	SHEAR FAILURE
21	21.229	303.193	35.8	-.211E-04	90.000	-.482E-01	-.117E-02	.265	24.299	SHEAR FAILURE
22	17.379	300.970	48.0	-.120E-04	90.000	-.252E-01	-.247E-02	.198	25.082	SHEAR FAILURE
23	13.000	300.197	52.6	-.154E-05	90.000	-.211E-06	-.300E-02	.180	25.082	SHEAR FAILURE
24	8.621	300.970	48.0	.695E-05	90.000	.252E-01	-.247E-02	.198	24.299	SHEAR FAILURE
25	4.771	303.193	35.8	.108E-04	90.000	.482E-01	-.117E-02	.265	21.832	SHEAR FAILURE
26	1.913	306.599	20.3	.867E-05	90.000	.677E-01	-.174E-03	.468	17.343	SHEAR FAILURE
27	.392	310.777	6.72	-.255E-05	-90.000	.823E-01	-.727E-03	1.412	N.A	NO FAILURE
28	0.000	314.556	1.49	-.988E-06	-89.999	.907E-01	-.153E-03	6.367	N.A	NO FAILURE
29	0.000	317.667	-.766E-01	-.103E-05	-89.990	.950E-01	.844E-03	123.803	N.A	NO FAILURE
30	0.000	320.778	.100E-03	-.794	-.001	.971E-01	.237E-02	.302	0.000	TENSILE FAILURE
31	0.000	323.889	.100E-03	-1.15	-.001	.972E-01	.418E-02	.208	0.000	TENSILE FAILURE
32	0.000	327.000	.100E-03	-1.08	-.001	.951E-01	.611E-02	.222	0.000	TENSILE FAILURE
33	0.000	330.111	.100E-03	-.498	-.001	.906E-01	.801E-02	.481	0.000	TENSILE FAILURE
34	0.000	333.222	.972	-.520E-05	-89.999	.831E-01	.970E-02	9.763	N.A	NO FAILURE
35	0.000	336.333	4.86	-.755E-05	-90.000	.715E-01	.108E-01	1.954	N.A	NO FAILURE
36	0.000	339.444	27.2	0.	90.000	.518E-01	.926E-02	.348	19.604	SHEAR FAILURE

NN = 1

STRESSES AND DISPLACEMENTS, AND FAILURE CRITERION AT INTERIOR POINTS

I	CX	CZ	SIG1	SIG3	ALPHA	UX	UZ	F.O.S	BETA	
1	1.444	343.000	41.1	7.92	68.600	.224E-01	-.552E-02	1.538	N.A	NO FAILURE
2	4.333	343.000	33.5	1.56	82.670	.152E-01	-.135E-01	.822	20.894	SHEAR FAILURE
3	7.222	343.000	29.2	.755	86.802	-.933E-02	-.181E-01	.688	19.947	SHEAR FAILURE
4	10.111	343.000	27.4	.555	88.688	.447E-02	-.205E-01	.651	19.481	SHEAR FAILURE
5	13.000	343.000	26.9	.508	-90.000	-.285E-06	-.212E-01	.642	19.342	SHEAR FAILURE
6	15.889	343.000	27.4	.555	-88.688	-.446E-02	-.205E-01	.651	19.481	SHEAR FAILURE
7	18.778	343.000	29.2	.755	-86.802	-.933E-02	-.181E-01	.688	19.947	SHEAR FAILURE
8	21.667	343.000	33.5	1.56	-82.670	-.152E-01	-.135E-01	.822	20.894	SHEAR FAILURE
9	24.556	343.000	41.1	7.92	-68.600	-.224E-01	-.552E-02	1.538	N.A	NO FAILURE
10	28.000	339.444	29.7	1.94	-41.873	-.479E-01	-.134E-02	1.028	N.A	NO FAILURE
11	28.000	336.333	11.5	.284	-21.864	-.671E-01	-.255E-02	1.248	N.A	NO FAILURE
12	28.000	333.222	5.35	.222	-16.311	-.790E-01	-.413E-02	2.501	N.A	NO FAILURE
13	28.000	330.111	2.88	.222	-12.534	-.866E-01	-.438E-02	4.645	N.A	NO FAILURE
14	28.000	327.000	1.81	.232	-5.271	-.912E-01	-.403E-02	7.497	N.A	NO FAILURE
15	28.000	323.889	1.55	.191	5.036	-.912E-01	-.346E-02	8.334	N.A	NO FAILURE
16	28.000	320.778	1.94	.900E-01	15.721	-.933E-01	-.290E-02	5.772	N.A	NO FAILURE
17	28.000	317.667	2.96	-.764E-02	19.035	-.911E-01	-.257E-02	3.147	N.A	NO FAILURE
18	28.000	314.556	4.85	-.247	19.439	-.867E-01	-.273E-02	.971	0.000	TENSILE FAILURE
19	27.578	310.430	9.72	-.343	27.564	-.772E-01	-.362E-02	.700	0.000	TENSILE FAILURE
20	25.819	305.599	19.9	.790	42.555	-.600E-01	-.412E-02	1.029	N.A	NO FAILURE
21	22.515	301.661	31.0	2.74	57.693	-.398E-01	-.239E-02	1.165	N.A	NO FAILURE
22	18.063	299.090	39.7	4.28	73.666	-.195E-01	-.157E-03	1.144	N.A	NO FAILURE
23	13.000	298.197	43.0	4.86	90.000	-.232E-06	-.136E-01	1.131	N.A	NO FAILURE
24	7.937	299.090	39.7	4.28	-73.666	.195E-01	-.157E-03	1.144	N.A	NO FAILURE
25	3.485	301.661	31.0	2.74	-57.693	.398E-01	-.239E-02	1.165	N.A	NO FAILURE
26	.181	305.599	19.9	.790	-42.555	.600E-01	-.412E-02	1.029	N.A	NO FAILURE
27	-1.578	310.430	9.72	-.343	-27.564	.772E-01	-.362E-02	.700	0.000	TENSILE FAILURE
28	-2.000	314.556	4.85	-.247	-19.439	.867E-01	-.273E-02	.971	0.000	TENSILE FAILURE
29	-2.000	317.667	2.96	-.760E-02	-19.035	.911E-01	-.257E-02	3.147	N.A	NO FAILURE
30	-2.000	320.778	1.94	.900E-01	-15.720	.933E-01	-.290E-02	5.772	N.A	NO FAILURE
31	-2.000	323.889	1.55	.191	-7.035	.933E-01	-.346E-02	8.334	N.A	NO FAILURE
32	-2.000	327.000	1.81	.232	5.272	.912E-01	-.403E-02	7.498	N.A	NO FAILURE
33	-2.000	330.111	2.88	.222	12.534	.866E-01	-.438E-02	4.645	N.A	NO FAILURE
34	-2.000	333.222	5.35	.222	16.312	.790E-01	-.413E-02	2.501	N.A	NO FAILURE
35	-2.000	336.333	11.5	.284	21.864	.671E-01	-.255E-02	1.248	N.A	NO FAILURE
36	-2.000	339.444	29.7	1.94	41.873	.479E-01	-.134E-02	1.028	N.A	NO FAILURE

NN = 2

STRESSES AND DISPLACEMENTS, AND FAILURE CRITERION AT INTERIOR POINTS

I	CX	CZ	SIG1	SIG3	ALPHA	UX	UZ	F.O.S	BETA		
1	1.444	345.000	32.8	9.21	68.040	.202E-01	-.521E-02	2.098	N.A	NO	FAILURE
2	4.333	345.000	32.3	3.60	77.609	.153E-01	-.108E-01	1.285	N.A	NO	FAILURE
3	7.222	345.000	30.0	1.78	83.748	.993E-02	-.150E-01	.977	19.886	SHEAR	FAILURE
4	10.111	345.000	28.6	1.22	87.333	.484E-02	-.174E-01	.861	19.636	SHEAR	FAILURE
5	13.000	345.000	28.1	1.08	-90.000	.280E-06	-.181E-01	.830	19.543	SHEAR	FAILURE
6	15.889	345.000	28.6	1.22	-87.333	-.484E-02	-.174E-01	.861	19.636	SHEAR	FAILURE
7	18.778	345.000	30.0	1.78	-83.748	-.993E-02	-.150E-01	.977	19.886	SHEAR	FAILURE
8	21.667	345.000	32.3	3.60	-77.609	-.153E-01	-.108E-01	1.285	N.A	NO	FAILURE
9	24.556	345.000	32.8	9.21	-68.039	-.202E-01	-.521E-02	2.098	N.A	NO	FAILURE
10	30.000	339.444	26.5	2.18	-51.004	-.468E-01	-.508E-02	1.218	N.A	NO	FAILURE
11	30.000	336.333	16.2	.468	-33.197	-.630E-01	-.265E-02	1.034	N.A	NO	FAILURE
12	30.000	333.222	9.71	.382	-23.337	-.746E-01	-.239E-03	1.612	N.A	NO	FAILURE
13	30.000	330.111	6.41	.464	-15.950	-.823E-01	.126E-02	2.608	N.A	NO	FAILURE
14	30.000	327.000	4.79	.524	-7.604	-.870E-01	.211E-02	3.646	N.A	NO	FAILURE
15	30.000	323.889	4.22	.504	2.507	-.892E-01	.266E-02	4.077	N.A	NO	FAILURE
16	30.000	320.778	4.49	.387	12.110	-.892E-01	.313E-02	3.503	N.A	NO	FAILURE
17	30.000	317.667	5.53	.209	19.324	-.870E-01	.375E-02	2.387	N.A	NO	FAILURE
18	30.000	314.556	7.51	-.636E-01	25.136	-.826E-01	.472E-02	1.074	13.939	SHEAR	FAILURE
19	29.547	310.082	12.1	-.222	35.670	-.725E-01	.636E-02	.200	N.A	NO	FAILURE
20	27.551	304.599	20.1	1.17	51.406	-.541E-01	.661E-02	1.204	N.A	NO	FAILURE
21	23.800	300.129	27.9	4.13	65.252	-.339E-01	.311E-02	1.601	N.A	NO	FAILURE
22	18.747	297.211	33.5	6.81	77.908	-.158E-01	-.163E-01	1.738	N.A	NO	FAILURE
23	13.000	296.197	35.6	7.88	90.000	-.248E-06	-.380E-01	1.774	N.A	NO	FAILURE
24	7.253	297.211	33.5	6.81	-77.908	.158E-01	-.163E-01	1.738	N.A	NO	FAILURE
25	2.200	300.129	27.9	4.13	-65.252	.339E-01	-.311E-02	1.601	N.A	NO	FAILURE
26	-1.551	304.599	20.1	1.17	-51.406	.541E-01	.661E-02	1.204	N.A	NO	FAILURE
27	-3.547	310.082	12.1	-.221	-35.670	.725E-01	.636E-02	.200	13.939	SHEAR	FAILURE
28	-4.000	314.556	7.51	-.636E-01	-25.136	.826E-01	.472E-02	1.074	N.A	NO	FAILURE
29	-4.000	317.667	5.53	.209	-19.324	.870E-01	.375E-02	2.387	N.A	NO	FAILURE
30	-4.000	320.778	4.49	.387	-12.109	.892E-01	.313E-02	3.503	N.A	NO	FAILURE
31	-4.000	323.889	4.22	.504	-2.506	.892E-01	.266E-02	4.077	N.A	NO	FAILURE
32	-4.000	327.000	4.79	.524	7.604	.870E-01	.211E-02	3.646	N.A	NO	FAILURE
33	-4.000	330.111	6.41	.464	15.950	.823E-01	.126E-02	2.608	N.A	NO	FAILURE
34	-4.000	333.222	9.71	.382	23.338	.746E-01	-.239E-03	1.612	N.A	NO	FAILURE
35	-4.000	336.333	16.2	.468	33.198	.630E-01	-.265E-02	1.034	N.A	NO	FAILURE
36	-4.000	339.444	26.5	2.18	51.004	.468E-01	-.508E-02	1.218	N.A	NO	FAILURE

NN = 3

STRESSES AND DISPLACEMENTS, AND FAILURE CRITERION AT INTERIOR POINTS

I	CX	CZ	SIG1	SIG3	ALPHA	UX	UZ	F.O.S	BETA	
1	1.444	347.000	28.9	9.30	70.561	.179E-01	-.511E-02	2.390	N.A	NO FAILURE
2	4.333	347.000	30.1	5.32	76.153	.142E-01	-.888E-02	1.696	N.A	NO FAILURE
3	7.222	347.000	29.5	3.18	82.011	.963E-02	-.123E-01	1.323	N.A	NO FAILURE
4	10.111	347.000	28.7	2.30	86.398	.481E-02	-.145E-01	1.155	N.A	NO FAILURE
5	13.000	347.000	28.4	2.06	-90.000	-.280E-06	-.152E-01	1.106	N.A	NO FAILURE
6	15.889	347.000	28.7	2.30	-86.398	-.481E-02	-.145E-01	1.155	N.A	NO FAILURE
7	18.778	347.000	29.5	3.18	-82.011	-.963E-02	-.123E-01	1.323	N.A	NO FAILURE
8	21.667	347.000	30.1	5.32	-76.153	-.142E-01	-.888E-02	1.696	N.A	NO FAILURE
9	24.556	347.000	28.9	9.30	-70.561	-.179E-01	-.511E-02	2.390	N.A	NO FAILURE
10	32.000	339.444	24.1	2.66	-54.407	-.463E-01	-.710E-02	1.480	N.A	NO FAILURE
11	32.000	336.333	17.8	1.21	-39.823	-.597E-01	-.560E-02	1.382	N.A	NO FAILURE
12	32.000	333.222	12.4	.960	-28.908	-.705E-01	-.326E-02	1.791	N.A	NO FAILURE
13	32.000	330.111	9.00	1.01	-19.795	-.781E-01	-.117E-02	2.515	N.A	NO FAILURE
14	32.000	327.000	7.13	1.07	-10.218	-.828E-01	.493E-03	3.263	N.A	NO FAILURE
15	32.000	323.889	6.37	1.06	.574	-.851E-01	.187E-02	3.629	N.A	NO FAILURE
16	32.000	320.778	6.51	.923	11.198	-.851E-01	.314E-02	3.348	N.A	NO FAILURE
17	32.000	317.667	7.47	.707	20.152	-.830E-01	.445E-02	2.616	N.A	NO FAILURE
18	32.000	314.556	9.26	.463	27.912	-.787E-01	.591E-02	1.802	N.A	NO FAILURE
19	31.517	309.735	13.4	.329	40.061	-.684E-01	.801E-02	1.116	N.A	NO FAILURE
20	29.283	303.599	20.1	1.62	56.583	-.495E-01	.805E-02	1.394	N.A	NO FAILURE
21	25.086	298.596	25.9	4.71	70.504	-.297E-01	.353E-02	1.843	N.A	NO FAILURE
22	19.431	295.331	29.6	7.91	81.417	-.132E-01	.244E-02	2.134	N.A	NO FAILURE
23	13.000	294.197	30.8	9.29	90.000	.259E-06	.515E-02	2.240	N.A	NO FAILURE
24	6.569	295.331	29.6	7.91	-81.417	.132E-01	.244E-02	2.134	N.A	NO FAILURE
25	.914	298.596	25.9	4.71	-70.504	.297E-01	.353E-02	1.843	N.A	NO FAILURE
26	-3.283	303.599	20.1	1.62	-56.583	.495E-01	.805E-02	1.394	N.A	NO FAILURE
27	-5.517	309.735	13.4	.329	-40.061	.684E-01	.801E-02	1.116	N.A	NO FAILURE
28	-6.000	314.556	9.26	.463	-27.912	.787E-01	.591E-02	1.802	N.A	NO FAILURE
29	-6.000	317.667	7.47	.707	-20.152	.830E-01	.445E-02	2.617	N.A	NO FAILURE
30	-6.000	320.778	6.51	.923	-11.198	.851E-01	.314E-02	3.348	N.A	NO FAILURE
31	-6.000	323.889	6.37	1.06	-.574	.851E-01	.187E-02	3.629	N.A	NO FAILURE
32	-6.000	327.000	7.13	1.07	10.218	.828E-01	.493E-03	3.263	N.A	NO FAILURE
33	-6.000	330.111	9.00	1.01	19.795	.781E-01	-.117E-02	2.515	N.A	NO FAILURE
34	-6.000	333.222	12.4	.960	28.909	.705E-01	-.326E-02	1.791	N.A	NO FAILURE
35	-6.000	336.333	17.8	1.21	39.823	.597E-01	-.560E-02	1.382	N.A	NO FAILURE
36	-6.000	339.444	24.1	2.66	54.407	.463E-01	-.710E-02	1.480	N.A	NO FAILURE

Appendix 5 : Determination of material constants

Part 1 - Intact rock

The empirical criterion given by equation 43 :

$$\sigma_1 = \sigma_3 + \sqrt{m\sigma_c \cdot \sigma_3 + s\sigma_c^2}$$

may be rewritten as:

$$y = m\sigma_c \cdot x + s\sigma_c^2 \qquad (A.1)$$

where $y = (\sigma_1 - \sigma_3)^2$ and $x = \sigma_3$.

For intact rock, $s = 1$ and the uniaxial compressive strength σ_c and the material constant m are given by :

$$\sigma_c^2 = \frac{\Sigma y_i}{n} - \left[\frac{\Sigma x_i y_i - \frac{\Sigma x_i \Sigma y_i}{n}}{\Sigma x_i^2 - \frac{(\Sigma x_i)^2}{n}}\right] \frac{\Sigma x_i}{n} \qquad (A.2)$$

$$m = \frac{1}{\sigma_c}\left[\frac{\Sigma x_i y_i - \frac{\Sigma x_i \Sigma y_i}{n}}{\Sigma x_i^2 - \frac{(\Sigma x_i)^2}{n}}\right] \qquad (A.3)$$

where x_i and y_i are successive data pairs and n is the total number of such data pairs.

The coefficient of determination r^2 is given by :

$$r^2 = \frac{\left[\Sigma x_i y_i - \frac{\Sigma x_i \Sigma y_i}{n}\right]^2}{\left[\Sigma x_i^2 - \frac{(\Sigma x_i)^2}{n}\right]\left[\Sigma y_i^2 - \frac{(\Sigma y_i)^2}{n}\right]} \qquad (A.4)$$

The closer the value of r^2 is to 1.00, the better the fit of the empirical equation to the triaxial test data.

Part 2 - Broken or heavily jointed rock

For broken or heavily jointed rock, the strength of the intact rock pieces, σ_c, is determined from an analysis such as that presented above or it is estimated from a test such as the Point Load Index (see page 52). The value of m for the broken or heavily jointed rock is found from equation A.3 and the value of the constant s is given by :

$$s = \frac{1}{\sigma_c^2}\left[\frac{\Sigma y_i}{n} - m\sigma_c \frac{\Sigma x_i}{n}\right] \qquad (A.5)$$

The coefficient of determination r^2 is found from equation A.4.

When the value of the constant s is very close to zero, equation A.5 will sometimes give a negative value. In such a case, put $s = 0$ and calculate m as follows :

$$m = \frac{\Sigma y_i}{\sigma_c \Sigma x_i} \qquad (A.6)$$

Note that equation A.4 cannot be used to calculate the coefficient of determination r^2 when s is negative.

Part 3 - Mohr envelope

Balmer[135] expressed the equation for a system of Mohr circles (see page 94) as :

$$\left(\sigma - \tfrac{1}{2}(\sigma_1 + \sigma_3) \right)^2 + \tau^2 = \tfrac{1}{4}(\sigma_1 - \sigma_3)^2 \qquad (A.7)$$

If the partial derivative of σ_1 with respect to σ_3 is determined and solved for σ, the following equation is obtained :

$$\sigma = \sigma_3 + \frac{\sigma_1 - \sigma_3}{\frac{\partial \sigma_1}{\partial \sigma_3} + 1} \qquad (A.8)$$

Substituting equation A.8 into equation A.7 and solving for τ gives :

$$\tau = \frac{\sigma_1 - \sigma_3}{\frac{\partial \sigma_1}{\partial \sigma_3} + 1} \cdot \sqrt{\frac{\partial \sigma_1}{\partial \sigma_3}} \qquad (A.9)$$

For the relationship given by equation 43, the partial derivative of σ_1 with respect to σ_3 is :

$$\frac{\partial \sigma_1}{\partial \sigma_3} = 1 + \frac{m \sigma_c}{2(\sigma_1 - \sigma_3)} \qquad (A.10)$$

Substitution of equation A.10 into equations A.8 and A.9 gives :

$$\sigma = \sigma_3 + \frac{\tau_m^2}{\tau_m + m\sigma_c/8} \qquad (A.11)$$

$$\tau = (\sigma - \sigma_3)\sqrt{1 + m\sigma_c/4\tau_m} \qquad (A.12)$$

where $\tau_m = \tfrac{1}{2}(\sigma_1 - \sigma_3)$.

By substituting successive pairs of σ_1 and σ_3 values into equations A.11 and A.12, a complete Mohr envelope can be generated. While this process is convenient for some applications, it is inconvenient for calculations such as those involved in slope stability assessment where the shear strength is required for a specified normal stress level. For these applications, a more useful expression for the Mohr envelope is :

$$\tau/\sigma_c = A\left(\sigma/\sigma_c - \sigma_t/\sigma_c \right)^B \qquad (A.13)$$

where A and B are empirical constants which are determined by means of the following analysis :

Rewrite equation A.13 as follows :

$$y = ax + b \qquad (A.14)$$

where
$\quad y = \log \tau/\sigma_c$
$\quad x = \log(\sigma/\sigma_c - \sigma_t/\sigma_c)$
$\quad a = B$
$\quad b = \log A$
$\quad \sigma_t/\sigma_c = \tfrac{1}{2}(m - \sqrt{m^2 + 4s}\ )$

The values of the constants A and B are found by linear regression analysis of equation A.14 for a range of values of σ and τ calculated from equations A.11 and A.12.

$$B = \frac{\Sigma x_i y_i - \frac{\Sigma x_i \Sigma y_i}{n}}{\Sigma x_i^2 - \frac{(\Sigma x_i)^2}{n}} \qquad (A.15)$$

$$\log A = \frac{\Sigma y_i}{n} - B \cdot \frac{\Sigma x_i}{n} \qquad (A.16)$$

The instantaneous friction angle ϕ_i and the instantaneous cohesion c_i for a given value of σ are defined by:

$$\operatorname{Tan} \phi_i = AB(\sigma/\sigma_c - \sigma_t/\sigma_c)^{B-1} \qquad (A.17)$$

$$c_i = \tau - \sigma \operatorname{Tan} \phi_i \qquad (A.18)$$

Calculation sequence

The analysis presented in this appendix can be carried out with the aid of a programmable calculator. The following calculation sequence can be used:

Intact rock

1. Enter triaxial data in the form $x = \sigma_3$, $y = (\sigma_1 - \sigma_3)^2$,
2. Calculate and accumulate:
     Σx_i, Σx_i^2, Σy_i, Σy_i^2 and $\Sigma x_i y_i$,
3. Calculate σ_c from equation A.2,
4. Calculate m from equation A.3,
5. Calculate r^2 from equation A.4,
6. Note that s = 1 for intact rock.

Broken or heavily jointed rock

1. Enter value of σ_c for intact rock,
2. Enter triaxial data in the form $x = \sigma_3$, $y = (\sigma_1 - \sigma_3)^2$,
3. Calculate and accumulate:
     Σx_i, Σx_i^2, Σy_i, Σy_i^2 and $\Sigma x_i y_i$,
4. Calculate m from equation A.3,
5. Calculate s from equation A.5,
6. Calculate r^2 from equation A.4,
7. When s < 0 in step 5, put s = 0 and calculate m from equation A.6,
8. Note that equation A.4 is not valid when s < 0.

Generation of Mohr envelope

1. Enter values of σ_c, m and s,
2. Calculate σ_1 for a chosen value of σ_3 from equation 43,
3. Calculate $\tau_m = \frac{1}{2}(\sigma_1 - \sigma_3)$,
4. Calculate σ from equation A.11,
5. Calculate τ from equation A.12,
6. Repeat for new values of σ_3 until the complete curve has been generated.

Calculation of constants A *and* B

1. Input values of σ_c, m and s; calculate $\sigma_t = \frac{1}{2}\sigma_c(m - \sqrt{m^2 + 4s})$,
2. Calculate values of σ and τ from equations A.11 and A.12 for values of σ_3 equal to :

   σ_{3m}, $\sigma_{3m}/2$, $\sigma_{3m}/4$, $\sigma_{3m}/8$, $\sigma_{3m}/16$, $\sigma_{3m}/32$, $\sigma_{3m}/64$, $\sigma_{3m}/128$, $\sigma_{3m}/256$,
   $\sigma_t/4$, $\sigma_t/2$, $3\sigma_t/4$ and σ_t,

   where σ_{3m} is the maximum value of σ_3 in the problem under consideration.
3. Enter each of the calculated values in the form:

   $$x = \log(\sigma/\sigma_c - \sigma_t/\sigma_c) \text{ and } y = \log \tau/\sigma_c,$$

4. Calculate and accumulate :

   $$\Sigma x_i, \quad \Sigma x_i^2, \quad \Sigma y_i, \quad \Sigma y_i^2 \text{ and } \Sigma x_i y_i,$$

5. Calculate the constants A and B from equations A.15 and A.16.
6. Calculate the values of ϕ_i and c_i from equations A.17 and A.18 if required.

Appendix 6 : Underground wedge analysis

Introduction

The analysis presented in this appendix was prepared by Dr J.W. Bray of Imperial College in London. Some modifications were carried out by Dr R.D. Hammett of Golder Associates in Vancouver, Canada.

The analysis can be used for wedges defined by three intersecting discontinuities in either the sidewalls or the roof of an underground excavation. For a defined wedge, the analysis determines whether : (a) the wedge floats or falls out, (b) the wedge is inherently stable (the resultant force tending to push it into the rock mass), (c) the wedge tends to slide out on two planes simultaneously, or (d) the wedge tends to slide out on one plane. In the latter two cases, a factor of safety is calculated for given values of the cohesive strength c and the angle of friction ϕ. Rotational instability is not considered.

Note that wedges can always be formed at the wall of an underground excavation, provided that none of the planes 1, 2, 3 and 4 (see diagram on page 518) are parallel, and the discontinuities can be taken to be ubiquitous. This is different from the case of slopes where the existence of wedges has to be established.

The calculation procedure has been specially designed so that the results obtained are independent of whether the normals to the four wedge faces ($\vec{a}$, $\vec{b}$, $\vec{d}$ and $\vec{f}$ in the diagram on page 518) are directed into or out of the wedge, and independent of the order in which the planes 1, 2 and 3 are arranged around the wedge.

Two solutions are given for wedge sizes defined by

a) the largest wedge that can be released in a tunnel of span or sidewall height W_b, and

b) a wedge defined by the length W_t of the trace formed by the intersection of discontinuity plane 3 and the free face 4.

Notation

1, 2, 3	-	discontinuities in the rock mass
4	-	free face (roof or sidewall)
W_b	-	span of free face between boundary edges
W_t	-	length of trace of plane 3 on free face
ψ, α	-	dip and dip direction of plane
ψ_5, α_5	-	plunge and trend of boundary edges of free face
ψ_6, α_6	-	plunge and trend of bolt force T. Note that plunge ψ_6 is defined as *negative* when bolt acts *upwards*.
p_x, p_y, p_z q_x, q_y, q_z t_x, t_y, t_z	-	coordinates of vertices of wedge on free face using coordinate system : y axis pointing north, z axis pointing upwards, origin at internal vertex
A_1, A_2, A_3	-	areas of faces
W	-	weight of wedge

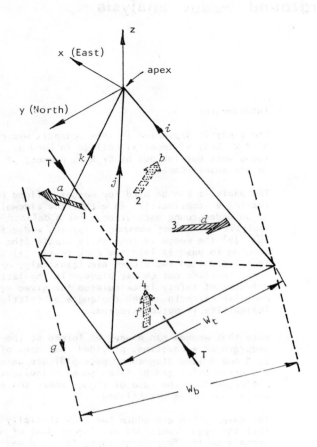

T	-	bolt force
γ_r	-	unit weight of rock
u	-	average water pressure on discontinuities
N, S	-	normal and shear reactions on plane carried by rock contacts
c, ϕ	-	cohesion and angle of friction
F_s	-	factor of safety

Calculation sequence

For wedge defined by span or sidewall height W_b

1. $a_x, a_y, a_z = \sin\psi_1 \sin\alpha_1, \quad \sin\psi_1 \cos\alpha_1, \quad \cos\psi_1$

2. $b_x, b_y, b_z = \sin\psi_2 \sin\alpha_2, \quad \sin\psi_2 \cos\alpha_2, \quad \cos\psi_2$

3. $d_x, d_y, d_z = \sin\psi_3 \sin\alpha_3, \quad \sin\psi_3 \cos\alpha_3, \quad \cos\psi_3$

4. *For overhanging faces (note for vertical faces assume a slight overhang in order to make the dip direction determinate)*

   $f_x, f_y, f_z = \sin\psi_4 \sin\alpha_4, \quad \sin\psi_4 \cos\alpha_4, \quad \cos\psi_4$

   For faces with no overhang

   $f_x, f_y, f_z = -\sin\psi_4 \sin\alpha_4, \; -\sin\psi_4 \cos\alpha_4, \; -\cos\psi_4$

5. $g_x, g_y, g_z = \cos\psi_5 \sin\alpha_5, \quad \cos\psi_5 \cos\alpha_5, \quad -\sin\psi_5$

6. $c_x, c_y, c_z = \cos\psi_6 \sin\alpha_6, \quad \cos\psi_6 \cos\alpha_6, \quad -\sin\psi_6$

7. $h_x, h_y, h_z = (f_y g_z - f_z g_y), (f_z g_x - f_x g_z), (f_x g_y - f_y g_x)$

8. $i_x, i_y, i_z = (d_y b_z - d_z b_y), (d_z b_x - d_x b_z), (d_x b_y - d_y b_x)$

9. $j_x, j_y, j_z = (a_y d_z - a_z d_y), (a_z d_x - a_x d_z), (a_x d_y - a_y d_x)$

10. $k_x, k_y, k_z = (b_y a_z - b_z a_y), (b_z a_x - b_x a_z), (b_x a_y - b_y a_x)$

11. $l = -(a_x i_x + a_y i_y + a_z i_z)$

12. $l_1 = f_x i_x + f_y i_y + f_z i_z$

13. $l_2 = f_x j_x + f_y j_y + f_z j_z$

14. $l_3 = f_x k_x + f_y k_y + f_z k_z$

15. $m_1 = (h_x i_x + h_y i_y + h_z i_z)/l_1$

16. $m_2 = (h_x j_x + h_y j_y + h_z j_z)/l_2$

17. $m_3 = (h_x k_x + h_y k_y + h_z k_z)/l_3$

18. $e_1 = W_b/|m_1 - m_2|, \quad e_2 = W_b/|m_2 - m_3|, \quad e_3 = W_b/|m_3 - m_1|$

19. e = least of e_1, e_2, e_3

20. $k_1 = -e/l_1, \quad k_2 = -e/l_2, \quad k_3 = -e/l_3$

21. $A_1 = l \cdot k_2 k_3/2, \quad A_2 = k_1 A_1/k_2, \quad A_3 = k_1 A_1/k_3$

22. $W = -\gamma_r l A_1 k_1/3$

23. Areas of faces = $|A_1|, |A_2|, |A_3|$

24. Weight of wedge = $|W|$

25. $U_1 = u_1 A_1, \quad U_2 = u_2 A_2, \quad U_3 = u_3 A_3$

26. $r_1 = d_x b_x + d_y b_y + d_z b_z, \quad t_1 = 1 - r_1^2$

27. $r_2 = a_x d_x + a_y d_y + a_z d_z, \quad t_2 = 1 - r_2^2$

28. $r_3 = b_x a_x + b_y a_y + b_z a_z, \quad t_3 = 1 - r_3^2$

29. $T_x, T_y, T_z = T \cdot c_x, T \cdot c_y, T \cdot c_z$

30. $R_x = -U_1 a_x - U_2 b_x - U_3 d_x - T_x \cdot W/|W|$

31. $R_y = -U_1 a_y - U_2 b_y - U_3 d_y - T_y \cdot W/|W|$

32. $R_z = -U_1 a_z - U_2 b_z - U_3 d_z + W - T_z \cdot W/|W|$

33. $g_1 = R_x a_x + R_y a_y + R_z a_z, \quad h_1 = R_x i_x + R_y i_y + R_z i_z$

34. $g_2 = R_x b_x + R_y b_y + R_z b_z, \quad h_2 = R_x j_x + R_y j_y + R_z j_z$

35. $g_3 = R_x d_x + R_y d_y + R_z d_z, \quad h_3 = R_x k_x + R_y k_y + R_z k_z$

36. $N_1 = (g_1 - r_3 g_2) A_1 / t_3 |A_1|, \quad N_2 = (g_2 - r_3 g_1) A_2 / t_3 |A_2|$

37. $S = h_3 A_1 A_2 / |A_1 A_2| \sqrt{t_3}$

38. If $N_1 > 0$, $N_2 > 0$ and $S < 0$, the active forces drive the wedge into the rock mass and there is no tendency to fall or slide. The factor of safety has no meaning and the wedge is stable. Terminate computation.

39. If $N_1 > 0$, $N_2 > 0$ and $S > 0$, contact occurs on planes 1 and 2 with separation on plane 3.

    Factor of safety $F_s = (N_1 \tan\phi_1 + c_1|A_1| + N_2 \tan\phi_2 + c_2|A_2|)/S$

    Terminate computation.

40. $N_2 = (g_2 - r_1 g_3)A_2/t_1|A_2|$, $N_3 = (g_3 - r_1 g_2)A_3/t_1|A_3|$

41. $S = h_1 A_2 A_3/|A_2 A_3|\sqrt{t_1}$

42. If $N_2 > 0$, $N_3 > 0$ and $S < 0$, same conditions apply as for step 38, wedge is stable. Terminate computation.

43. If $N_2 > 0$, $N_3 > 0$ and $S > 0$, contact occurs on planes 2 and 3 with separation on plane 1.

    Factor of Safety $F_s = (N_2 \tan\phi_2 + c_2|A_2| + N_3 \tan\phi_3 + c_3|A_3|)/S$

    Terminate computation.

44. $N_3 = (g_3 - r_2 g_1)A_3/t_2|A_3|$, $N_1 = (g_1 - r_2 g_3)A_1/t_2|A_1|$

45. $S = h_2 A_3 A_1/|A_3 A_1|\sqrt{t_2}$

46. If $N_3 > 0$, $N_1 > 0$ and $S < 0$, same conditions apply as for step 38, wedge is stable. Terminate computation.

47. If $N_3 > 0$, $N_1 > 0$ and $S > 0$, contact occurs on planes 3 and 1 with separation on plane 2.

    Factor of Safety $F_s = (N_3 \tan\phi_3 + c_3|A_3| + N_1 \tan\phi_1 + c_1|A_1|)/S$

    Terminate computation.

48. $N_1 = g_1 A_1/|A_1|$

49. If $N_1 < 0$, separation on plane 1, continue with step 53

50. $S_x = R_x - g_1 a_x$, $S_y = R_y - g_1 a_y$, $S_z = R_z - g_1 a_z$

51. $S = (S_x^2 + S_y^2 + S_z^2)^{\frac{1}{2}}$

52. Contact on plane 1, separation on planes 2 and 3,

    Factor of Safety $F_s = (N_1 \tan\phi_1 + c_1|A_1|)/S$

    Terminate computation.

53. $N_2 = g_2 A_2/|A_2|$

54. If $N_2 < 0$, separation on plane 2, continue with step 58

55. $S_x = R_x - g_2 b_x$, $S_y = R_y - g_2 b_y$, $S_z = R_z - g_2 b_z$

56. $S = (S_x^2 + S_y^2 + S_z^2)^{\frac{1}{2}}$

57. Contact on plane 2, separation on planes 3 and 1,
    Factor of Safety $F_s = (N_2 \tan\phi_2 + c_2|A_2|)/S$
    Terminate computation.

58. $N_3 = g_3 A_3/|A_3|$

59. If $N_3 < 0$, separation on plane 3, wedge is unstable since it falls out, floats out or is forced out by water pressure. Terminate computation.

60. $S_x = R_x - g_3 d_x$, $S_y = R_y - g_3 d_y$, $S_z = R_z - g_3 d_z$

61. $S = (S_x^2 + S_y^2 + S_z^2)^{\frac{1}{2}}$

62. Contact on plane 3, separation on planes 1 and 2,
    Factor of Safety $F_s = (N_3 \tan\phi_3 + c_3|A_3|)/S$

For wedge defined by length W_t of trace of plane 3 on free face

Calculation is identical to that described above except for the following steps:

Step 5 - not required

Step 7 - h'_x , h'_y , $h'_z = (f_y d_z - f_z d_y)$, $(f_z d_x - f_x d_z)$, $(f_x d_y - f_y d_x)$

$s = (h'^2_x + h'^2_y + h'^2_z)^{\frac{1}{2}}$

h_x , h_y , $h_z = h'_x/s$, h'_y/s , h'_z/s

Step 17 - not required

Step 18 - not required

Step 19 - $e = W_t/|m_1 - m_2|$

Examples

Problem 1:

ψ_1	α_1	ψ_2	α_2	ψ_3	α_3	over-hang	ψ_4	α_4	ψ_5	α_5	ψ_6	α_6
45	240	45	120	45	0	no	90	180	45	90	0	0

c_1	ϕ_1	c_2	ϕ_2	c_3	ϕ_3	γ_r	u	W_b	T
0	30	0	30	0	30	0.027	0	10	0

Output:

$A_1 = 20.41241$, $A_2 = 20.41241$, $A_3 = 40.82483$
$W = 1.59099$
$N_1 = 0.9000$, $N_2 = 0.9000$, $S = 0.71151$
Contact on planes 1 and 2, separation on plane 3
Factor of Safety $F_s = 1.46059$

Problem 2 :

Identical to problem 1 except that overhang exists

Areas and wedge weight identical to problem 1.

N_3 = 1.125 ,   S = 1.125

Contact on plane 3 , separation on planes 1 and 2

Factor of Safety F_s = 0.57735

The above two cases represent opposite vertical faces of an excavation and show the marked instability of the wedge on one side as compared with the other.

Problem 3 :

ψ_1	α_1	ψ_2	α_2	ψ_3	α_3	over-hang	ψ_4	α_4	ψ_5	α_5	ψ_6	α_6
50	290	60	345	35	140	yes	0	180	0	353	0	0

c_1	ϕ_1	c_2	ϕ_2	c_3	ϕ_3	γ_r	u	W_b	T
0	30	0	30	0	30	1	0	4	0

Output:

A_1 = 0.89273 A_2 = 0.93425 A_3 = 2.31098

W = 0.54361

N_1 = -0.34943 N_2 = -0.27181 N_3 = -0.44530

All normal reactions are negative and hence wedge falls out.

Problem 4 :

Identical to problem 3 but with no overhang.

Areas and weight of wedge identical to problem 3.

N_1 = 0.31271 N_2 = 0.05231   S = -0.41476

Wedge is stable under conditions defined in step 38.

The above two cases represent wedges in the roof (problem 3) and floor (problem 4) of an excavation.

Appendix 7: Conversion factors

	Imperial	Metric	SI
Length	1 mile	1.609 km	1.609 km
	1 ft	0.3048 m	0.3048 m
	1 in	2.54 cm	25.40 mm
Area	1 mile^2	2.590 km^2	2.590 km^2
	1 acre	0.4047 hectare	4046.9 m^2
	1 ft^2	0.0929 m^2	0.0929 m^2
	1 in^2	6.452 cm^2	6.452 cm^2
Volume	1 yd^3	0.7646 m^3	0.7646 m^3
	1 ft^3	0.0283 m^3	0.0283 m^3
	1 ft^3	28.32 litres	0.0283 m^3
	1 UK gal.	4.546 litres	4546 cm^3
	1 US gal.	3.785 litres	3785 cm^3
	1 in^3	16.387 cm^3	16.387 cm^3
Mass	1 ton	1.016 tonne	1.016 Mg
	1 lb	0.4536 kg	0.4536 kg
	1 oz	28.352 g	28.352 g
Density	1 lb/ft^3	16.019 kg/m^3	16.019 kg/m^3
Unit weight	1 lbf/ft^3	16.019 kgf/m^3	0.1571 kN/m^3
Force	1 ton f	1.016 tonne f	9.964 kN
	1 lb f	0.4536 kg f	4.448 N
Pressure or stress	1 ton f/in^2	157.47 kg f/cm^2	15.44 MPa
	1 ton f/ft^2	10.936 tonne f/m^2	107.3 kPa
	1 lb f/in^2	0.0703 kg f/cm^2	6.895 kPa
	1 lb f/ft^2	4.882 kg f/m^2	0.04788 kPa
	1 standard atmosphere	1.033 kg f/m^2	101.325 kPa
	14.495 lb f/in^2	1.019 kg f/cm^2	1 bar
	1 ft water	0.0305 kg f/cm^2	2.989 kPa
	1 in mercury	0.0345 kg f/cm^2	3.386 kPa
Permeability	1 ft/year	0.9659 × 10^{-6} cm/s	0.9659 × 10^{-8} m/s
Rate of flow	1 ft^3/s	0.02832 m^3/s	0.02832 m^3/s
Moment	1 lbf ft	0.1383 kgf m	1.3558 Nm
Energy	1 ft lbf	1.3558 J	1.3558 J
Frequency	1 c/s	1 c/s	1 Hz

SI unit prefixes

Prefix	tera	giga	mega	kilo	milli	micro	nano	pico
Symbol	T	G	M	k	m	μ	n	p
Multiplier	10^{12}	10^9	10^6	10^3	10^{-3}	10^{-6}	10^{-9}	10^{-12}

SI symbols and definitions

N = Newton = kg m/s^2
Pa = Pascal = N/m^2
J = Joule = m.N

Index

Active span, definition of 18
Adits, exploratory 57-58
Amphibolite, triaxial test data on 141, 143
Anchoring of rockbolts 345-346
Andesite, Panguna, see Panguna, Bougainville, Papua New Guinea, andesite from, strength of
Anisotropic rock strength 157-159
Atlas Copco-Craelius core orientation system 50

Barton-Lien-Lunde index, see NGI tunnelling quality index
Bibliography on large underground excavations 12, 397-434
Bieniawski rock mass classification system, see CSIR rock mass classification system
Blocked steel sets, available support of 254
Blocks, unstable, support of 246-248
Borehole extensometers, use of, in measurement of rock mass displacements 391-392
Borehole methods of stress measurement 384-389
Boundary element method of stress analysis 110, 122, 493-511
Brekke-Howard discontinuity infilling classification system 20-23
Brittle-ductile transition 133, 139-140
Brittle failure, definition of 133
Buckling of slabs parallel to excavation boundaries 234-235

Cables, grouted, support provided by 256-258
Cable support of wedges or blocks free to fall or slide 246-248
Case histories, bibliography and tabulation of 397-448
Chainlink mesh, use of, as support 351
Chert, triaxial test data on 141, 143
Christensen-Hugel core orientation method 50
Circular excavations, stresses around 103-112
Circular tunnel, rock surrounding, fracture propagation in 211-215
Coal pillar design 206-207
Cobar mine, Australia 317
Clar compass 42
Components of stress 87-89
Computer analysis of structurally controlled instability 191, 194
Computer processing of structural data 79
Computer program for calculating elastic stresses and displacements by the boundary element method 494-511
Concrete lining, available support of 253-254
Convergence measurements 390
Conversion factors for units 523
Core logging 55-56
Core orientation 48-52
Core photography 55
Core storage 55, 57
Counting nets 66, 75, 78
CSIR rock mass classification system 18, 22-27, 34, 35, 171, 173, 303, 319, 380
  use of, for estimating support 296-298

Damage caused by blasting 370-371, 378-380
Deere's rock quality designation (RQD) 18 ff
Deformability of rock masses 173-177
Design data, collection of, instrumentation for 384-389
Design process, flow chart of 10
Diamond drilling, sub-surface exploration by 45 ff
Dinorwic Pumped Storage Scheme, Wales 191-193-6
Dolerite, triaxial test data on 141,144
Dolomite, triaxial test data on 141,144
Drilling, see Diamond drilling, sub-surface exploration by
Ductile behaviour, definition of 133

Effective stress, Terzaghi principle of 153
Elasticity, definition of 87
Elliptical excavation, stresses on boundary of 110-112
Equal angle projections, see Stereographic projections
Equal area projections 61-62
Excavation shape, influence of, on induced stresses 221-231
Explosive rock breaking, basic mechanics of 367-368
Extension test on rock samples 135
Extensometers, borehole 391-392

Failure criterion for rock, empirical 137 ff
Failure, definition of 150
Failure mechanisms of rock around excavations 183 ff
Fan cut 368
Fault, influence of, on excavation stability 232-234
Fibre reinforced shotcrete 363-364
Finite element method of stress analysis 108-110
Flat jacks, use of, in direct stress measurement 384
Forepoling 312-313
Four Fathom Limestone, Weardale Valley, UK 300, 303
Four Fathom Mudstone, Weardale Valley, UK 300, 303, 306, 311
Free face, creation of 368-370
'Friction set' rock anchors 295, 335
Frozen stress photoelasticity 122

Gabbro, triaxial test data on 141, 145
Geological data, collection of 38 ff
Geological data, graphical presentation of 61 ff
Geomechanics classification, see CSIR rock mass classification system
Geophysical exploration 43
Gneiss, triaxial test data on 141, 145
Granite, progressive fracture of 166-168
Granite, triaxial test data on 141, 146
'Granulated' marble, strength of 168-169

Gravitational stresses 125
Great Limestone, Weardale Valley, UK 300, 303
Groundwater pressure, measurement of 389
Grouted rockbolts and cables, support provided by 256-258
Grouting of rockbolts 345, 349-350
'Gunite', see Shotcrete, use of, as support

'Harmonic holes' 223
Heavily jointed rock masses, strength of 166-171
Heim's rule 95
Hollow inclusion stress cell 386-389
Horizontally bedded rock, excavations in 235-236
Hydraulic fracturing method of stress measurement 384

Inclination, influence of, on pillar stresses 125
Inclination of failure surfaces 151-152
Inclination of orebody, influence of, on pillar stability 210-211
Indiana limestone, triaxial tests on 139
In situ states of stress 93-101
In situ stress, influence of, on structurally controlled instability 199-200
In situ stress, measurement of 384-389
Instability, structurally controlled, computer analysis of 191-194
Instability, types of 11-12, 183 ff
Instrumentation 382 ff
Integral sampling method 51
Isometric drawings of structural planes 79-84, 449-466

Kemano powerhouse, Canada 223
Kielder experimental tunnel, UK 298-312
Kirsch equations 103
Kiruna iron ore mine, Sweden 378

Lambert projections, see Equal area projections
Lauffer rock mass classification 18, 27
Limestone, triaxial test data on 141, 146
Local mine-stiffness 237-239

Mapping, geological 40-43
Marble, 'granulated', strength of 168-169
Marble, triaxial test data on 142, 147
Martinsburg slate, triaxial tests on 158
Mesh, see Wire mesh, use of, as support
Models, plastic demonstration 84-85
Mohr's circle diagram 93-94
Monitoring of underground excavations 389-394
Mount Isa mine, Queensland, Australia 197, 309, 317
Mudstone, triaxial test data on 142, 147
Multiple discontinuities, rock with, failure of 163-164
Multiple excavations, stresses around 112-122

Natrass Gill Sandstone, Weardale Valley, UK 300, 303
New Broken Hill Consolidated mine, Australia 317-319
NGI tunnelling quality index 27-35, 171, 303, 319, 380
  use of, for estimating support 289-295
Norite, triaxial test data on 142, 148

Optical surveying, use of, in monitoring rock mass displacements 390
Orebody inclination, influence of, on pillar stability 210-211
Orientation of core 48-52
Orientation, optimum, of excavations in jointed rock 194-197

Panguna, Bougainville, Papua New Guinea, andesite from, strength of 170-171, 177
Parallel hole cuts 369-370
'Perfobolts' 282, 295, 335, 345
Photoelastic stress analysis 107, 122
Photogeology 39
Pillar failure 200-211
Pillar strength, influence of shape and size on 207-210
Pillars, stresses in 112 ff
Pillar stiffness 236-239
Plane strain, definition of 91-92
Plane stress, definition of 90-91
Plough cut 369
Point load index 52
Pore fluid, influence of, on strength of rock 153-154
Pre-reinforcement of rock masses 312-319
Presplitting 372-373, 376
Principal stress, definition of 90
Projects, underground excavation, tabulation of 434-448

Quartzdiorite, triaxial test data on 142, 148
Quartzite, triaxial test data on 142, 149

Regional geology, study of 38-39
Rockbolt installation 342-350
Rockbolt length, empirical rules for 321
Rockbolt spacing, empirical rules for 321
Rockbolt support of wedges or blocks free to fall or slide 246-248
Rockbolt systems 332-341
Rockbolting programme, organisation of 329-332
Rockbolts, grouted, support provided by 256-258
Rockbolts, ungrouted, available support of 254-256
Rock mass classifications 14 ff
Rock mass classifications, use of, for estimating support 286-298
Rock-support interaction analysis 248-286
RQD, see Deere's rock quality designation

Sandstone, triaxial test data on 142, 149
Scaling after blasting 342
Schmidt projections, see Equal area projections
Shafts, exploratory 57-58
Shape, excavation, influence of, on stresses in surrounding rock 221 ff
Shape, optimum, of excavations in jointed rock 194-197
Shotcrete, engineering properties of 360
Shotcrete, mix design for 355-359
Shotcrete, placement of 360-363
Shotcrete, use of, as support 353-364
Shotcrete lining, available support of 253-254
Sidewall failure analysis 187-191
Sidewall failure in circular tunnels 215
Sidewall failure in square tunnels 217-221
Single excavations, stress distributions around 101-112, 467-492
Size of excavation, influence of, on structurally controlled failure 197-199
Slabs parallel to excavation boundaries, buckling of 234-235
Smooth blasting 372-374
Snowy Mountains Engineering Corporation, Australia, tests by, on Panguna andesite 170-171
Snowy Mountains hydroelectric scheme, Australia 8, 321
Spiling 312-313
'Split set' rock anchors 295, 335
Square tunnels, sidewall failure in 217-221
Stand-up time, definition of 18
Stereographic projection 61 ff
Stereographic nets 65-66, 75, 78
Stiffness of pillars 236-239
Stini rock mass classification 18
Stone Mountain granite, triaxial tests on 151
Streamline analogy for principal stress trajectories 101-103
Stress, components of 87-89
Stress, transformation of 89-90, 92-94
Stress, two-dimensional state of 90-93
Stress distributions around multiple excavations 112-122
Stress distributions around single excavations, 101-112, 467-492.
Stress shadows 124-125
Structurally controlled failure 183-200
Support design 244 ff

Tensioning of rockbolts 347-349
Terzaghi joint sampling correction 79
Terzaghi principle of effective stress 153
Terzaghi-Richart theory of in-situ stress 93-95
Terzaghi rock load classification 14-17, 18, 20, 35, 246.
Three-dimensional pillar stress problems 122
Transformation of stress 89-90, 92-94
Trial excavations 194-195, 393-394
Triaxial compression-tensile tests on rock samples 134-135

Triaxial test data for intact rock specimens 140-149
Tributary theory 112-114
True dip and apparent dip, relationship between 74
Two-dimensional elastic stress distributions 101 ff, 467-492
Two-dimensional state of stress 90-93
Types of underground excavation 8-9

Uniaxial tensile tests on rock samples 134
Units, conversion factors for 523

V-cut 369

Wedges, analysis of,
  by computer 191,194
  by stereographic projection 185-192
  by vector method 517-522
Wedges, unstable, support of 246-248
Weldmesh, use of, as support 352
Westerley granite, triaxial tests on 151, 155, 166
Wire mesh, use of, as support 351-352
Wombeyan marble, strength of 168-169
Wulff projections, see Stereographic projections